PREFACE

This textbook is a complete course book for the examination in Intermediate 2 Mathematics (Units 1, 2, and 3 or 4 [Application of Maths]), a National Qualification of the Scottish Qualifications Authority. While it has been written as a course book in its own right, it will also serve as a rich source of additional exercises for those using any other Intermediate 2 textbooks.

Teachers using this text have complete freedom regarding the introduction of each topic. Proofs of introductory theory have been omitted and only reminders of relevant theory have been included. Naturally, teachers will want to introduce each lesson in their own way, providing their own notes or recommending other revision guides containing the relevant theory.

The order of topics follows the syllabus listing provided by SQA. This order was never intended to define a teaching syllabus, so there is no need for teachers to stick slavishly to it. Teaching two topics at a time has its merits.

Each exercise is preceded by a brief reminder of the theory, where necessary, and appropriate worked examples to guide the pupil through the exercise. The worked examples include guidance on good exam technique, especially adequate communication, which may be necessary for full marks in an exam question.

The exercises have been carefully graded with some examples coded 'B' to indicate a greater level of difficulty. Such examples should interest those pupils aspiring to a B or A grade pass. Some other examples have been coded 'H' to indicate relevance to the Higher Mathematics course. Some pupils using this textbook will aspire to Higher Mathematics after passing the Intermediate 2 exam. Each exercise hopefully contains enough material for pupils of all levels of ability on this course, and teachers should select the relevant examples for each pupil or class, as there is probably insufficient time to work through every example given.

Each topic concludes with a 45 minute test, so that progress may be monitored.

Calculators should be used sensibly. Some calculations are impossible without them, but the first paper in the exam is a non-calculator paper, so pupils should make a habit of dispensing with their calculators where possible. It is much quicker, for example, to do 4 - (-1) mentally.

The author wishes success to all who use this book, and hopes that it will be of great assistance.

Another work by the same author relevant to Intermediate 2 Mathematics is

Practice Papers for Intermediate 2 Mathematics
(ISBN 0 340 81208 7)
(6 typical exam papers with answers)

CONTENTS

Unit 1

Percentages

Volume

Coordinate Geometry

Algebra

Circle Properties

Unit 2

Trigonometry

Simultaneous Equations

Graphs, Charts and Tables

Statistics

Unit 3

Further Algebra

Quadratic Functions

Further Trigonometry

CONTENTS

PERCENTAGES

1 Equivalence of Percentages and Fractions

Reminder

Here are some common percentages expressed as vulgar fractions (i.e. common fractions) and decimal fractions:

$50\% = \frac{1}{2} = 0{\cdot}5$ $\quad 10\% = \frac{1}{10} = 0{\cdot}1$ $\quad 33\frac{1}{3}\% = \frac{1}{3} = 0{\cdot}33333\ldots\ (= 0{\cdot}\dot{3})$

$25\% = \frac{1}{4} = 0{\cdot}25$ $\quad 20\% = \frac{1}{5} = 0{\cdot}2$ $\quad 66\frac{2}{3}\% = \frac{2}{3} = 0{\cdot}66666\ldots\ (= 0{\cdot}\dot{6})$.

Example Express a) 0.4 as a percentage and as a vulgar fraction
b) 32% as a decimal fraction and as a vulgar fraction
c) $\frac{17}{25}$ as a percentage and as a decimal fraction.

Solution

a) $0{\cdot}4 = 0{\cdot}4 \times 100\% = 40\%$

$0{\cdot}4 = \frac{4}{10} = \frac{2}{5}$

b) $32\% = \frac{32}{100} = 0.32$

$32\% = \frac{32}{100} = \frac{8}{25}$

c) $\frac{17}{25} = \frac{17 \times 4}{25 \times 4} = \frac{68}{100} = 68\%$

or

$\frac{17}{25} = \frac{17}{25} \times \frac{100}{1}\% = 17 \times 4\% = 68\%$

$\frac{17}{25} = \frac{17 \times 4}{25 \times 4} = \frac{68}{100} = 0{\cdot}68.$

Exercise 1 no calculator to be used for this exercise

1 Express these percentages as decimal fractions:
a) 27% b) 89% c) 6% d) 35·5%.

2 Express these vulgar fractions as percentages:
a) $\frac{3}{10}$ b) $\frac{2}{5}$ c) $\frac{3}{4}$ d) $\frac{3}{8}$.

3 Express these decimal fractions as percentages:

a) 0·28 b) 0·6 c) 0·08 d) 0·265.

4 Express these percentages as vulgar fractions in their lowest terms:

a) 37% b) 36% c) 84% d) $2\frac{1}{2}$%.

5 Copy and complete this table of equivalent fractions and percentages:

percentage	vulgar fraction	decimal fraction
(e.g. 50%	$\frac{1}{2}$	0·5)
22%		
$37\frac{1}{2}$%		
	$\frac{17}{100}$	
	$\frac{9}{10}$	
		0·34
		0·005

Reminder

Example Express $\frac{4}{11}$ as a percentage correct to 3 decimal places.

Solution $\frac{4}{11} = \frac{4}{11} \times \frac{100}{1}\% = \frac{400}{11}\%$ $\quad 11\overline{)400{\cdot}0000}$ = 36·3636 hence 36·364%

6B Express these fractions as percentages correct to 3 decimal places:

a) $\frac{2}{3}$ b) $\frac{4}{9}$ c) $\frac{5}{6}$ d) $\frac{3}{7}$.

2 Calculating a Percentage of a Quantity (without a calculator)

Reminder

Example Calculate 30% of £25·40.

Solution (Method 1) $30\% = \frac{3}{10}$

so 30% of £25·40 = $\frac{3}{10}$ × £25·40 = 3 × £2·54 = £7·62

(Method 2) (the method of practice) 100% = £25·40
10% = £2·54
30% = £7·62

(Method 3) (with a calculator) comes in Exercise 6

Exercise 2 no calculator to be used for this exercise

1 Calculate a) 30% of £120 b) 40% of £65 c) 80% of 75 km d) 25% of 70 kg.

2 Bobby earns £436 per week. He is given a 10% pay rise. How much more money is Bobby earning every week?

3 Sandra earns £6·27 per hour as a carer in a nursing home. She secures an 8% pay rise. How much more is she now earning per hour?

4 David earns $2\frac{1}{2}$% commission on the cars he sells. Last month he sold £127 000 worth of cars. Calculate his commission for that month.

5 The annual insurance premium on Sally's antique brooch is 0·75% of the value of the brooch. Calculate the premium when the brooch is valued at £1500.

6 A trawler skipper has been forced to reduce his catch by 22%. His catch used to be 1400 tonnes. How much less is it now?

7 Some nails are sold in $\frac{1}{2}$ kg packs. The mass of the nails in a pack is never short by more than $\frac{1}{2}$%. Calculate (in grams) the greatest possible shortage of nails.

8 The McCallums are retired and have a £20 000 investment which pays 1·5% every quarter. Calculate their quarterly income from this source.

9 Use the method of practice to calculate the V.A.T. ($17\frac{1}{2}$%) on a garage bill of £326.

10 Use the method of practice to calculate the V.A.T. ($17\frac{1}{2}$%) on the hotel's invoice of £11 420 for the wedding of Julia and Michael.

> **Reminder**
>
> **Example** Calculate the simple interest on £500 for 3 years at 7% per annum.
>
> *Solution* (Method 1) 7% of £500 = £35
> hence for 3 years, the interest = 3 × £35 = £105
>
> (Method 2) use the formula $I = \frac{PTR}{100}$ where I denotes the **Interest**, P the **Principal**, £500 in this case, T the length of **Time** in years, 3, and R the **Rate**, 7.
>
> $$I = \frac{500 \times 3 \times 7}{100} = £105$$
>
> Note : If required, A is the **Amount** now in the bank (£605).

11 Calculate the simple interest due on
- a) £400 at 5% invested for 2 years
- b) £750 at 3% invested for 3 years
- c) £1200 at 7·5% invested for 2 years
- d) £950 at 4·5% invested for 4 years.

12 Calculate the simple interest due on
- a) £200 at 6% invested for 6 months
- b) £360 at 5% invested for 4 months
- c) £2400 at 7% invested for 3 months
- d) £1800 at 2·5% invested for 18 months.

3 Expressing one Quantity as a Percentage of Another (without a calculator)

> **Reminder**
>
> **Example** Out of 560 flights scheduled to land at Glenbank airfield, 336 had to be diverted.
> What percentage of flights was diverted?
>
> *Solution* $\frac{336}{560} \times \frac{100}{1}\% = \frac{33600}{560}\% = \frac{3360}{56}\% = \frac{420}{7}\% = 60\%$

Exercise 3

no calculator to be used for this exercise

1 What percentage of
- a) 30 is 6
- b) 80 is 32
- c) 96 is 72
- d) 40 is 16?

2 There are 160 people living in McNaught Avenue in Forthbank. Only 72 of them were born in Forthbank. What percentage were born locally?

3 A 250 g packet of Kornphlaiks contains 7 g of protein and 8 g of iron. What are the percentages of protein and iron?

4 There were 5600 books in 'The Worm' second hand bookshop and 3080 of them were paperbacks. What percentage of books in the shop were paperbacks?

5 What percentage of

a) 240 is 204 b) 112 is 14 c) 1080 is 90
d) 240 is 64 e) 425 is 170 f) 650 is 104?

Working backwards

Reminder

Example When Tommy's Mum left him in the house for a few minutes to run an errand, Tommy ate 5 chocolate biscuits. This was 20% of the chocolate biscuits in the house. How many chocolate biscuits were in the house before Tommy's Mum went out?

Solution (Method 1)

$$20\% = 5 \text{ biscuits}$$
$$\Rightarrow 1\% = \frac{5}{20} \text{ biscuits}$$
$$\Rightarrow 100\% = \frac{5}{20} \times \frac{100}{1} = 5 \times 5 = 25 \text{ biscuits}$$

(Method 2)

$$20\% = 5 \text{ biscuits}$$
$$\Rightarrow 4\% = 1 \text{ biscuit}$$
$$\Rightarrow 100\% = 25 \text{ biscuits}$$

Note that since the numbers involved are easy, the second line (in both methods) can be omitted, and the last line can be obtained by multiplying the first line by 5.

Note also that these solutions depend simply on direct proportion, rather than any great knowledge of percentages.

6 Yesterday in Dorothy's Maths class 3 pupils were absent. This was 15% of the class. How many are in the whole class?

7 A hotel received its delivery of eggs from the local grocer and three of them were broken. The chef complained that 5% of his eggs were broken. How many had he ordered?

8 In a pack of wood screws I found 6 with no slot in the head. This was 3% of the screws. How many screws were in the pack?

9 I found 4 black ones in my packet of jelly babies. This was $12\frac{1}{2}\%$ of them. How many jelly babies were in the pack?

10B Calculate the rate of simple interest when £1250 earns £125 in 2 years.

11B Calculate the rate of simple interest when £2500 earns £125 in 5 years.

12B For how long will £2400 have to be invested to earn £192 simple interest at 4% p.a.?

13B For how long will £3000 have to be invested to earn £270 simple interest at 3% p.a.?

14B Patrick's 5% pay rise brought his daily pay up to £84. What had he been earning previously?

4 Percentage Increase or Decrease (without a calculator)

Reminder

Example The cost of bread has risen by 20%. It started off at 75p per loaf. What does it cost now?

Solution (Method 1) 20% of $75p = \frac{1}{5}$ of $75p = 15p$

$\Rightarrow$ new cost $= (75 + 15)p = 90p$

(Method 2) the new price is 120% (100 + 20) of the old price

120% of $75p = \frac{120}{100} \times 75p = \frac{12}{10} \times 75p = \frac{6}{5} \times 75p =$

$6 \times 15p = 90p$

Example Since this time last year, the cost of a DVD player has fallen by 40%. It started off at £90. What does it cost now?

Solution (Method 1) 40% of $£90 = 0{\cdot}4 \times £90 = 4 \times £9 = £36$

$\Rightarrow$ new cost $= £90 - £36 = £54$

(Method 2) the new price is 60% (100 – 40) of the old price

60% of $£90 = 0{\cdot}6 \times £90 = 6 \times £9 = £54$

Exercise 4

no calculator to be used for this exercise

1 Last year there were 4 adult pairs of ospreys nesting on Loch Mamhade. There has been a 75% increase this year. How many adult pairs of ospreys are nesting on the loch this year?

2 The number of cars owned by the residents of Inverotter rose by 2% from 2300 in a year. What was the new number of cars?

3 Last year there were 150 geese living on the park pond. The goose population has increased by 6%. How many geese are living on the pond this year?

4 Dr. Thomson told Mr. Burns that he wanted to see a 20% reduction in his cholesterol reading in the following month. If it was 5·5, what is his target for the next month?

5 One year 350 drivers in Highland Region were charged with drink driving offences. After a publicity campaign, the number the following year was reduced by 8%. How many drivers were charged that year?

6 Morag was encouraged to spend 10% more time on her mathematics homework. If she previously spent 45 minutes each night on it, to what did she increase it?

7 During one school session 250 Fife pupils had German measles. The following session it fell by 4%. How many Fife pupils had German measles in that session?

8 Joanne earns £220 per week. She is given a 5% pay rise. What is her new wage?

9 Because of a fall-off in orders, David is forced to accept a 15% reduction in his hours at a components factory. If he was originally working for 35 hours a week, what are his new weekly hours?

10 St. Fothad's Kirk congregation's annual income rose by 9% from £47 000 last year. What is its income this year?

5 Percentage Profit or Loss (without a calculator)

Reminder

Example A fruiterer bought oranges at 5 for £1 and sold them at 4 for £1. Calculate his percentage profit.

Solution for each orange cost price = 20p selling price = 25p profit = 5p

$$\text{percentage profit} = \frac{\text{profit}}{\text{cost price}} \times 100\% = \frac{5}{20} \times \frac{100}{1}\% = 25\%$$

Example A car salesman gave a customer an allowance of £3000 for his trade-in car. The salesman later sold it for £2700. Calculate the percentage loss.

Solution cost price = £3000 selling price = £2700 loss = £300

$$\text{percentage loss} = \frac{\text{loss}}{\text{cost price}} \times 100\% = \frac{300}{3000} \times 100\% = 10\%$$

Exercise 5 no calculator to be used for this exercise

1 A furniture shop bought in a new leather three-piece suite for £1500 and sold it for £1800. Calculate the percentage profit.

2 A carpet shop buys a 4 metre wide roll of carpet at £6 per (linear) metre and sells it at £9·60 per (linear) metre. Calculate the percentage profit.

3 A joke shop buys Halloween masks at 80p each. They are unpopular and do not sell, so the shopkeeper reduces them to 50p. If they are all sold at that price, calculate the percentage loss.

4 A wine shop buys in Chianti at £10 for 3 bottles and sells it at 4 for £15. Calculate the percentage profit or loss.

5 A greengrocer buys in apples at 10 kg for £12. He sells them at 16p per 100 g. Calculate his percentage profit or loss.

6 A garage sells a second hand car at £5500 having paid £6000 for it. Calculate the percentage loss.

7B A fruiterer buys in 100 kg of plums at £120. He sells half of them at £1·50 per kg but they are ripening quickly so he sells the rest on a special offer of £1 per kg. Calculate his percentage profit or loss.

8B A D.I.Y. shop buys 50 power screwdrivers at £12 each. The first 30 of them are sold at £18 each and the rest are sold as clearance bargains at £10 each. Calculate the percentage profit or loss.

9B A locksmith buys mortice locks at £8 each and wishes to make a 20% profit. At what price should they be sold?

10B A joiner makes a nest of tables for £150 including his time and the materials used. He wants to make a 30% profit. At what price would he have to sell the tables?

6 Percentages using a Calculator

Reminder

Although it is important to be able to deal with simple percentages without a calculator, many percentage calculations are tedious without one. Not all calculators have a % key, but even if you do have one you may choose not to use it. If you do however, you must be very familiar with how it works. The instruction manual should be referred to for this. You should check how it works by doing some easy examples which can be checked mentally.

Example Calculate the V.A.T. (17·5%) on a bill of £35·60.

Solution $17{\cdot}5\%$ of $£35{\cdot}60 = \frac{17{\cdot}5}{100} \times £35{\cdot}60 = 0{\cdot}175 \times £35{\cdot}60 = £6{\cdot}23$

Note. Either the $\frac{17{\cdot}5}{100}$ or the $0{\cdot}175$ could be omitted.

The 0·175 is the preferable method, but be careful with small percentages, e.g. $5\frac{1}{2}\% = 0{\cdot}055$

Exercise 6

1 Calculate
a) 73% of £151
b) 49% of 85 km
c) 16% of 432 kg
d) $23\frac{1}{2}$% of 74 tonnes.

2 What percentage is
a) 723 of 964
b) 392 of 490
c) 663 of 850
d) 92 of 115
e) 437 of 760?

3 Tanya earns £870 per week. Calculate her new wage after a 5·3% increase.

4 A dress costs £125·60 . Calculate the cost in the sale when it is reduced by 35%.

5 A grocer buys bottled water at £7·50 per dozen and sells bottles at 90p each. Calculate the percentage profit.

6 A supermarket has a special offer of fresh orange juice at £1·26 per carton. It was bought in at £18·90 per dozen. Calculate the percentage loss.

7 Calculate:
a) 31·5% of 260
b) 23·6% of 42
c) 32·7% of 370
d) 65·3% of 780.

8 What percentage of:
a) 760 is 437
b) 464 is 377
c) 2320 is 203
d) 2480 is 1271?

9 Calculate the simple interest on
a) £650 invested for 4 years at 2·3%
b) £1500 invested for 3 years at 4·1%
c) £2560 invested for 2 years at 3·6%
d) £4500 invested for 3 years at 4·7%.

10 Calculate the amount in the bank after:
a) £750 is invested for 6 months at 4·1% simple interest
b) £1275 is invested for 9 months at 2·9% simple interest
c) £1820 is invested for 2 months at 5·3% simple interest
d) £2790 is invested for 18 months at 4·9% simple interest.

11 The interest rate in the Fife Linen Bank depends on the principal invested as shown in this table.

Principal	Rate
under £500	3%
£500 – £999	3·5%
£1000 – £2999	3·8%
£3000 and over	4·15%

Calculate the simple interest earned by:
a) £200 in 2 years
b) £750 in 3 years
c) £1400 in 3 years
d) £3000 in 1 year.

12B Calculate the rate of simple interest when:

a) £1250 earns £187·50 in 2 years
b) £521 earns £109·41 in 6 years.

13B In what time will:

a) £475 gain £61·75 at 4% simple interest
b) £456 amount to £499·89 at 3·5% simple interest?

14B Tom had a 6% pay rise which brought his hourly rate up to £6·89. What was his hourly rate before his pay rise?

7 Rounding without a Calculator

Reminder

Many percentages which you calculate, either with or without a calculator, do not work out to be a whole number or a decimal which terminates; e.g. what percentage of 21 is 15?

$$\frac{15}{21} = \frac{15}{21} \times \frac{100}{1}\% = \frac{5 \times 100}{7}\% = \frac{500}{7}\% = 71{\cdot}428\,571\,4 \ldots\ldots\%$$

This is a repeating decimal. You can check that the digits 428571 will be repeated indefinitely.

A calculator might give the answer as 71·428 571 43.

This calculator has rounded the answer. We can indicate the degree of accuracy applied in this rounding by saying that it has been rounded to 8 decimal places or 10 significant figures.

[Notice how the digits are grouped in threes from the decimal in both directions.]

We usually round answers to a far lesser degree of accuracy, but we only do so at the end of our working and not, within reason, in the middle of a calculation.

There are three ways of rounding answers:

a) to the nearest unit
b) to a number of decimal places
c) to a number of significant figures.

Reminder continued

Example Express 47·638 metres correct to:

a) the nearest metre
b) 2 decimal places
c) 3 significant figures.

Solution

a) 48 m to the nearest metre
(in the tenths 6 > 5, so round up)

b) 47·64 m correct to 2 decimal places
(in the thousandths 8 > 5, so round up)

c) 47·6 m correct to 3 significant figures
(in the hundredths 3 < 5, so leave the 6 alone)

Example Investigate rounding 435·6475 to various decimal places and significant figures.

Solution

435·65 correct to 2 decimal places *or* 5 significant figures
435·6 correct to 1 decimal places *or* 4 significant figures

[do not round up the previous answer 435·65; use 435·6475, in the hundredths 4 < 5 so leave the 6]

436 correct to the nearest whole number *or* 3 significant figures
440 correct to the nearest ten *or* 2 significant figures
400 correct to the nearest hundred *or* 1 significant figure

[working backwards, without being told, you cannot tell if 400 has 1, 2 or 3 sig. figs.]

Example Investigate rounding 0·003 706 0 to various decimal places and significant figures.

Solution

Notice that the first three zeros are **not** significant. They are place holders which tell us that the first significant figure, the 3, represents 3 thousandths. The last zero **is** significant. It indicates that this number is correct to the nearest millionth.

0·003 706 correct to 4 significant figures *or* to 6 decimal places
0·003 71 correct to 3 significant figures *or* to 5 decimal places
0·0037 correct to 2 significant figures *or* to 4 decimal places
0·004 correct to 1 significant figure *or* to 3 decimal places

Exercise 7

no calculator to be used for this exercise

1 Round
a) 34·68 to 1 decimal place
b) 147·234 to 2 decimal places
c) 0·0023 to 3 decimal places
d) 27·035 97 to 4 decimal places.

2 Round
a) 23·46 to 3 significant figures
b) 52·8 to 2 significant figures
c) 0·001 59 to 2 significant figures
d) 5147 to 3 significant figures.

3 What percentage of 7 is 4? Express your answer correct to:
a) 2 decimal places
b) 3 significant figures.

4 What percentage of 9 is 5? Express your answer correct to:
a) 2 decimal places
b) 3 significant figures.

5 What percentage is 11 of 32? Express your answer correct to:
a) 1 decimal place
b) 2 significant figures.

6 What percentage is 5 of 6? Express your answer correct to:
a) 2 decimal places
b) 3 significant figures.

7 Calculate 8% of 6·2, expressing your answer correct to:
a) 2 decimal places
b) 1 significant figure.

8 Calculate 7% of 2·7, expressing your answer correct to:
a) 1 decimal place
b) 2 significant figures.

9 Calculate 6% of 13·2, expressing your answer correct to:
a) 1 decimal place
b) 2 significant figures.

10 Calculate 4% of 16·81, expressing your answer correct to:
a) 3 decimal places
b) 2 significant figures.

8 Rounding with a Calculator

Reminder

In the first half of this exercise, you are allowed to do the calculation on your calculator in the normal mode, but I would like you to round your answer in the same way as you did in Exercise 7.

In the second half of the exercise, you should use your calculator to do the calculation including the rounding to the required number of decimal places. Consult your calculator manual or ask your teacher if you need to.

Exercise 8

(In questions **1** to **6**, do the calculation on your calculator in the normal mode and then 'manually' round your answer as required.)

1 Evaluate $11{\cdot}7 \times 2{\cdot}68$ correct to: a) 1 decimal place b) 4 significant figures.

2 Evaluate $189{\cdot}2 \div 76{\cdot}3$ correct to: a) 3 decimal places b) 5 significant figures.

3 Calculate:

a) $23\frac{1}{2}$% of 175 correct to 3 significant figures

b) 37·6% of 243 correct to 4 significant figures

c) 18·7% of 32·9 correct to 2 significant figures

d) 15·3% of 46·8 correct to 4 significant figures.

4 A card shop buys one design of birthday card at £15·50 for 12 and sells them at £1·69 each.
Calculate the percentage profit correct to 3 significant figures.

5 Increase 113 by 24% and express your answer correct to:
a) 1 decimal place b) 3 significant figures.

6 Decrease 243 by 11·7% and express your answer correct to:
a) 2 decimal places b) 3 significant figures.

(In questions **7** to **12**, do the calculation including the rounding on your calculator, with it set to the appropriate number of decimal places.)

7 Set your calculator so that it rounds these numbers to 2 decimal places:
a) 5·267 b) 3·478 c) 15·2769 d) 16·0021.

8 a) Evaluate $3{\cdot}476 \times 12{\cdot}815$ with your calculator set to 2 decimal places.
b) Obtain the correct answer in full by simply changing your calculator to normal mode.
c) Round your answer to 3 decimal places by changing your calculator to another mode.

9 Express each of these vulgar fractions as decimals correct to 3 decimal places:
a) $\frac{7}{11}$ b) $\frac{4}{13}$ c) $\frac{8}{17}$ d) $\frac{5}{19}$.

10 Calculate:

a) 31·2% of 494 correct to 2 decimal places

b) 27·6% of 73 correct to 2 decimal places

c) 32·4% of 47 correct to 2 decimal places

d) 17·8% of 39·6 correct to 3 decimal places.

11 A confectioner buys bars of chocolate at £11·27 for 60 and sells them at 23p each.
Calculate her percentage profit correct to 2 decimal places.

12 Set your calculator to 2 decimal places and calculate:
a) 15% of £63·57 b) 27% of £54·36
c) 18% of £75·73 d) $14\frac{1}{2}$% of £39·41.

9 Rounding and Standard Form

Reminder

To set your calculator to a given number of significant figures, you will probably have to use standard form, and to set your calculator in standard form, you will probably have to specify the number of significant figures,

e.g. $\frac{1}{4}\% = 0{\cdot}25\% = \frac{0{\cdot}25}{100} = 2{\cdot}5 \times 10^{-3}$.

You need to be able to convert back and forward from standard form to the usual form.

Example Teresa scored 17 out of 23 on a spelling test. What was her percentage mark correct to 3 significant figures?

Solution $\frac{17}{23} = \frac{17}{23} \times \frac{100}{1}\% = 7{\cdot}39 \times 10^{1}\%$ (from the calculator, set to 3 significant figures)

$= 73{\cdot}9$ correct to 3 significant figures

Example Write these numbers in full

a) $2{\cdot}5 \times 10^{3}$ b) $6{\cdot}4 \times 10^{-3}$.

Solution a) $2{\cdot}5 \times 10^{3} = 2500$ (the power 3 indicates move the digits 3 places to the left or move the point 3 places to the right)

[Stick to the one rule that you always use.]

b) $6{\cdot}4 \times 10^{-3} = 0{\cdot}0064$ (the power -3 indicates move the digits 3 places to the right or move the point 3 places to the left)

Example Express these numbers in standard form correct to 2 significant figures

a) 3579 b) 0·003 67.

Solution a) $3579 = 3{\cdot}579 \times 10^{3} = 3{\cdot}6 \times 10^{3}$

(move the point or the digits 3 places to find $1 < 3{\cdot}579 < 10$, then round)

b) $0{\cdot}00367 = 3{\cdot}67 \div 10^{3} = 3{\cdot}67 \times 10^{-3} = 3{\cdot}7 \times 10^{-3}$

(move the point or the digits 3 places to find $1 < 3{\cdot}67 < 10$, then round)

Exercise 9

1 Express these numbers in standard form:

a) 235 b) 4367 c) 53·26 d) 0·025
e) 0·007 f) 753 g) 9465 h) 0·0027
i) 0·053 j) 0·685 k) 0·0009 l) 0·83.

2 Write these numbers in full:

a) $2{\cdot}6 \times 10^2$ b) $3{\cdot}2 \times 10^3$ c) $4{\cdot}61 \times 10^2$ d) $3{\cdot}265 \times 10^3$
e) $1{\cdot}1 \times 10^4$ f) $2{\cdot}78 \times 10^2$ g) $5{\cdot}73 \times 10^4$ h) $2{\cdot}4 \times 10^4$
i) $2{\cdot}3 \times 10^{-1}$ j) $3{\cdot}61 \times 10^{-2}$ k) $1{\cdot}4 \times 10^{-3}$ l) $7{\cdot}63 \times 10^{-2}$.

3 Set your calculator to 3 significant figures and evaluate the following, writing each answer *not* in standard form:

a) $456 \div 12{\cdot}3$ b) $123 \div 3{\cdot}6$ c) $3{\cdot}69 \div 42{\cdot}8$ d) $9 \div 11$.

4 A student bought a car for £450 and sold it to a friend for £530. Calculate the percentage profit correct to 3 significant figures.

5 Mary bought a picture for £150 at a car boot sale and auctioned it later for £220. Calculate her percentage profit correct to 3 significant figures.

6 Willie earned £579 per week and obtained a 3% pay rise. Calculate his new pay.

7 Two weeks ago 723 people visited Drumschuchtie Palace. Last week there was an increase of 13%. Calculate how many visitors the palace had last week. (Decide on your own degree of accuracy.)

8 The total mass of a rowing crew before training was 670 kg. After training this had come down to 623 kg. Calculate the percentage reduction in the mass correct to 3 significant figures.

9B Over the last year an insurance company estimates that a particular house has risen in value from £123 000 to £137 000.

a) Calculate the percentage increase correct to 3 significant figures.
b) If the building's insurance premium for last year was £145, what should it be for this year, given that the insurance rate remains the same?

10 Compound Interest

Reminder

With compound interest (added annually) the interest gained in the first year earns interest in the second year, and so on. In dealing with currency, it can be useful to set your calculator to 2 decimal places.

Reminder continued

Example Calculate the total compound interest earned by £4000 at 3% for 3 years.

Solution (Method 1)

Principal for the first year		= £4000
Interest for the first year	= 3% of £4000	= £120
Amount after the first year	= £4000 + £120	= £4120
Interest for the second year	= 3% of £4120	= £123·60
Amount after the second year	= £4120 + £123·60	= £4243·60
Interest for the third year	= 3% of £4243·60*	= £127·31
Amount after the third year	= £4243·60 + £127·31	= £4370·91

Giving a total interest of £4370·91 – £4000 = £370·91.

(Method 2)

3% interest means that 100% grows to 103% in a year.

This means increasing by a factor of $\frac{103}{100}$ or 1·03

So punch 4000 in your calculator followed by 'EXE' or '=' or 'ENTER'

Then 1·03 × Ans gives 4120

Press EXE again and you get 4243·60 (i.e. 1·03 × 4120)

Press EXE again and you get 4370·908 (i.e. 1·03 × 4243·60)

You can even set your calculator to 2 decimal places to round this.

Giving a total interest of £4370·91 – £4000 = £370·91.

* In practice, a bank may only pay interest on complete pounds. The answer has been rounded to the nearest penny, i.e. 2 decimal places.

Exercise 10

1 Calculate the compound interest earned by £200 invested for 2 years at 1·5%.

2 Calculate the compound interest earned by £1200 invested for 2 years at 2%.

3 Calculate the amount after 3 years when £5000 is invested at 3·5% compounded annually.

4 Calculate the amount after 4 years when £3000 is invested at 2·8% compounded annually.

5 Calculate the compound interest earned in 5 years by £1000 invested at 4% per annum.

Reminder

Compound interest can be added more frequently than every year. It can be added quarterly, daily, or even continuously.

Example Calculate the interest earned by investing £6000 for three years at 4·5% p.a. and find how much more interest is earned if the interest is added quarterly.

Solution *annually*:

4·5% means a growth factor of 1·045

$$6000 \xrightarrow{\times 1{\cdot}045} 6270 \xrightarrow{\times 1{\cdot}045} 6552{\cdot}15 \xrightarrow{\times 1{\cdot}045} 6847{\cdot}00 \Rightarrow £847 \text{ interest}$$

quarterly:

the interest due for 3 months is one quarter of the interest due for the year; alternatively, the rate is one quarter of 4·5% or 1·125%, but in the second quarter this interest earns more interest

1·125% means a growth factor of 1·011 25

for the first year:

$$6000 \xrightarrow{\times 1{\cdot}011\,25} 6067{\cdot}50 \xrightarrow{\times 1{\cdot}011\,25} 6135{\cdot}76 \xrightarrow{\times 1{\cdot}011\,25} 6204{\cdot}79 \xrightarrow{\times 1{\cdot}011\,25} 6274{\cdot}59$$

for the second year:

$$6274{\cdot}59 \xrightarrow{\times 1{\cdot}011\,25} 6345{\cdot}18 \xrightarrow{\times 1{\cdot}011\,25} 6416{\cdot}56 \xrightarrow{\times 1{\cdot}011\,25} 6488{\cdot}75 \xrightarrow{\times 1{\cdot}011\,25} 6561{\cdot}75$$

for the third year:

$$6561{\cdot}75 \xrightarrow{\times 1{\cdot}011\,25} 6635{\cdot}57 \xrightarrow{\times 1{\cdot}011\,25} 6710{\cdot}22 \xrightarrow{\times 1{\cdot}011\,25} 6785{\cdot}71 \xrightarrow{\times 1{\cdot}011\,25} 6862{\cdot}05$$

i.e. £862·05 interest, £15·05 more

Again this working can be shortened by using the 'x^y' or '^' button on the calculator.

This may be the first time that you have come across this button. Investigate it. It is very labour saving.

6B Calculate the compound interest added quarterly on £1500 invested for a year at 2% p.a..

7B Calculate the compound interest added quarterly on £2500 invested for 2 years at 3% p.a..

8B Calculate the compound interest added every 6 months on £12 000 invested for 3 years at 8% p.a..

9B Calculate the compound interest added every 6 months on £500 invested for 2 years at 6% p.a..

10H A sum of £700 is invested for three years at 5·2%. Calculate how much more interest is gained when it is added quarterly than when it is added annually.

11 Appreciation and Depreciation

Reminder

The value of a motor car usually reduces as it gets older. This is called **depreciation.**

The value of a house usually increases as time passes. This is called **appreciation.**

Example A car which costs £12 500 today is known to depreciate by 30% in its first year, 20% in its second year and 10% in its third year. What will it be worth after 3 years?

Solution 30% depreciation means a reduction by a factor of 0·7
20% depreciation means a reduction by a factor of 0·8
10% depreciation means a reduction by a factor of 0·9

so $12\,500 \xrightarrow{\times 0\cdot 7} 8750 \xrightarrow{\times 0\cdot 8} 7000 \xrightarrow{\times 0\cdot 9} 6300$
i.e. worth £6300 after 3 years

Example An antique brooch costing £2200 today is expected to appreciate by 5% each year.

What would you expect it to be worth after 5 years?

Solution 5% appreciation means an increase by a factor of 1·05

so $2200 \xrightarrow{\times 1\cdot 05} 2310 \xrightarrow{\times 1\cdot 05} 2425\cdot 50 \xrightarrow{\times 1\cdot 05} 2546\cdot 78 \xrightarrow{\times 1\cdot 05} 2674\cdot 11 \xrightarrow{\times 1\cdot 05} 2807\cdot 82$

an even quicker solution would be $2200 \times (1\cdot 05)^5 = 2807\cdot 82$

i.e. worth £2807·82 after 5 years

Exercise 11

1 Jane has a young plum tree. Its yield is increasing by 10% each year. Last year it yielded 8 kg of plums. What should it yield at the end of this year?

2 The number of cormorants living on Long Craig Rock is falling by 8% every year. At present there are 96 of them. How many will there be after 3 years?

3 The population of Ubongo is 170 000 and is increasing by 9% per annum. Calculate the expected population after 5 years, correct to the nearest hundred.

4 The number of boats fishing out of Pittenstruther harbour has been declining at a rate of 15% every year. There are currently 115 boats. How many will there be 5 years from now?

5 Traffic through the village of Auchtergonie is increasing by 11% each year. Currently an average 823 vehicles pass through the village daily. To what level will the traffic rise 6 years from now?

6 Two years ago the Cupar Stock Exchange stood at 4325. In the next year it fell by 22% and last year it rose again by 17%. At what level does it now stand? Give your answer as a whole number.

7 Sales of 'The Prisoner of Dalkeith' were 2487 in the year 2001. Sales rose by 8% in 2002 and fell back by 4% in 2003. How many were sold in 2003?

8B An antique painting was bought for £10 500. It was expected to increase in value by 9% each year. How many years will it be before it doubles its original value of £10 500?

9B The population of Timbuckhaven is 8000 and falling by 3% every year. How many years will it be till this population is halved?

10H £1000 is invested for 5 years and yields £610·51 compound interest. Find (by trial and improvement) the rate of interest.

12 Revision of Percentages

Exercise 12

1 Calculate 12·7% of 423 g, correct to 4 significant figures.

2 Increase 533 tonnes by 6·2% correct to 4 significant figures.

3 In one Maths class of 28 pupils, 13 were girls. What percentage were boys? (Answer correct to 1 decimal place.)

4 In Aberbuckie High School first year pupils were allowed to choose one of French, German, or Basic Skills. Of the 333 pupils in first year, 62% chose French and 35% chose German. How many pupils chose Basic Skills?

5 After a pay rise of $2\frac{1}{2}$%, Arthur's weekly pay was £988·10. What was his weekly pay before the rise?

6 After the budget, the price of a bottle of wine fell by $2\frac{1}{2}$% to £6·82. What had it been before the budget?

7 A primary class were given two spelling tests. Bill scored 17 out of 23 on the first test and 13 out of 17 on the second test. Was he improving or deteriorating?

8 Of the 320 fourth year pupils who sat an exam, 40% were girls. In the exam, 75% of the girls and $66\frac{2}{3}$% of the boys passed. How many pupils failed?

9 How much money would need to be invested at 4% simple interest to earn £16·80 in 9 months?

10 After spending 25% of his savings, Victor had £24 left. How much savings did he have before?

11B How much money was invested at 3·5% simple interest if the amount after 5 years is £5875?

12B Charlie bought a consignment of bananas, $7\frac{1}{2}$% of which were unsellable. He sold 95% of the remainder, which left him with 37 bananas. How many bananas had he bought?

13B In Glentyre forest, 2% of the trees were blown down in a gale and 6% of those remaining were felled for export. Now, 64 484 trees remain in the forest. How many trees were there prior to the gale?

14B A 10% reduction in the price of sugar means that £25·20 buys 4 kg more of sugar than before. Find the reduced price of the sugar.

15B The (17·5%) V.A.T. inclusive price of a replica Scotland top is £44·99.

a) Calculate the basic price before the addition of V.A.T.

b) If the worst case scenario arises and the Scottish Executive introduces an additional 1% tartan V.A.T., calculate the new price.

13 Test on Percentages

Allow 45 minutes for this test

1 The population of Dunfermline is increasing at 2% per year. Currently it is 55 420. What will it be 3 years from now?

2 The salmon stocks in the River Nethy are dwindling by 7% every year. This year there are estimated to be 1200 salmon in the river. How many will there be 2 years from now?

3 Calum scored 23 out of 29 in his first French vocabulary test and 19 out of 23 in his second. Was his French vocabulary improving or not?

4 Andy's job was to load new concrete slabs on to lorries. He was told that if he broke any more than 4% in any given week, he would be given the sack. Last week he loaded 1527 slabs on to the lorries and broke 52 of them. Is his job secure for another week?

5 A Rover 25 went in for a service. The labour cost £127·50 and the materials £52·36. Calculate the total bill after adding 17·5% V.A.T.

6 In a sale, the seller offered to pay the V.A.T. The goods cost £79·50 including the V.A.T. What was their sale price?

7 £3000 is invested at 4% compound interest. Calculate what the investment is worth after 2 years.

8 Paul has left school and has found a job with a starting salary of £15 000 and with an annual increment of 2%. His father is self-employed, approaching retirement and working less each year. His income, currently £30 000, is reducing by 15% p.a. How long will it be till Paul is earning more than his father?

VOLUME

14 Area

Reminder

The formulae for the areas of these simple shapes should be known thoroughly.

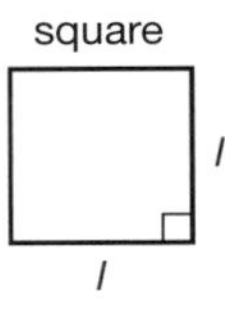

$A = l^2$

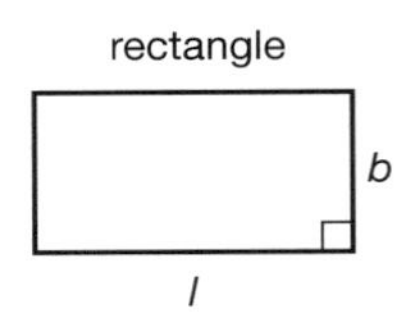

$A = l \times b$

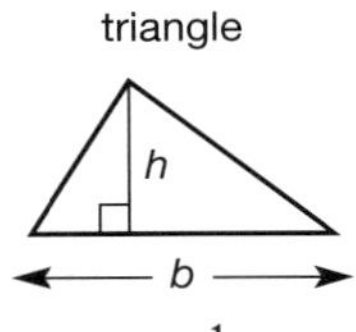

$A = \frac{1}{2}bh$

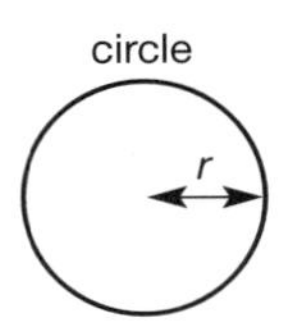

$A = \pi r^2$

Example Calculate the area of this gnomon. (L shape)

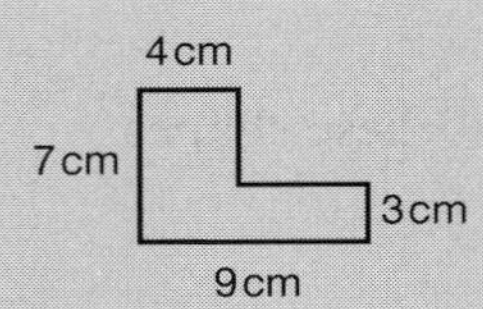

Solution (Method 1)

$(7 \times 4) + (5 \times 3) = 28 + 15 = 43 \text{ cm}^2$

(Method 2)

$(4 \times 4) + (9 \times 3) = 16 + 27 = 43 \text{ cm}^2$

Method 3

$(7 \times 9) - (5 \times 4) = 63 - 20 = 43 \text{ cm}^2$.

Note: Where units are omitted from the diagram given, express lengths in *units*, areas in *units*^{2} and volumes in *units*^{3}.

Exercise 14

1 Calculate the area of each of these shapes:

a) 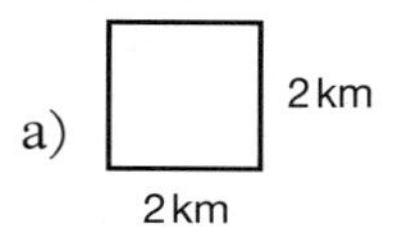

b) 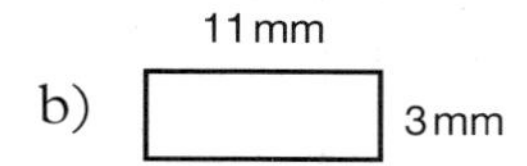

c) 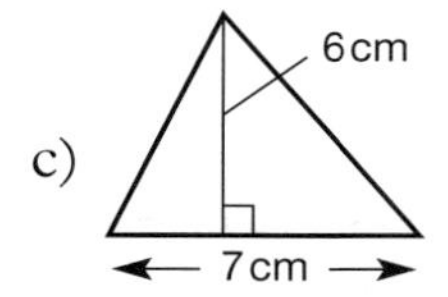

d) 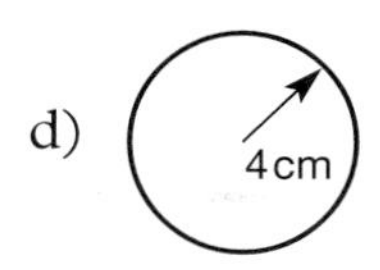

2 Calculate the areas of these composite shapes:

a) 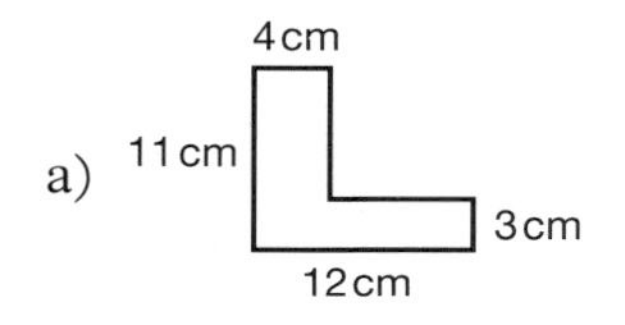

b) 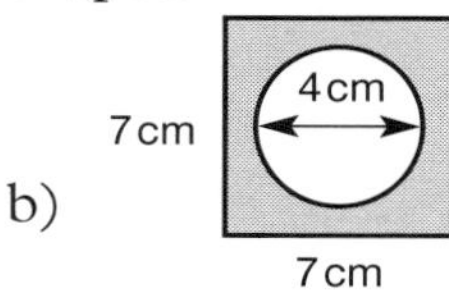

c) 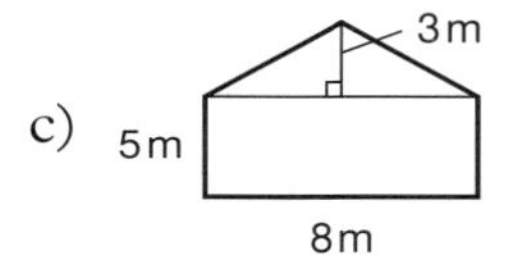

d) 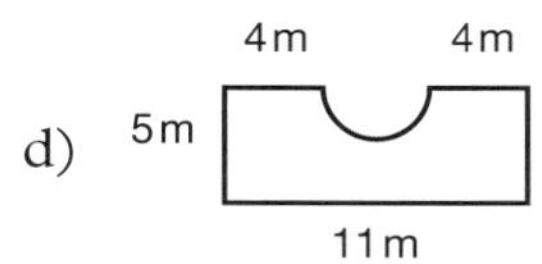

3 Calculate the areas of these shapes:

a) 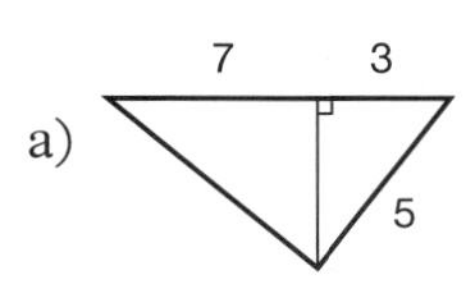

b) 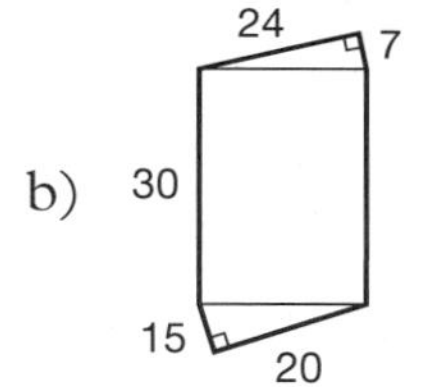

c) 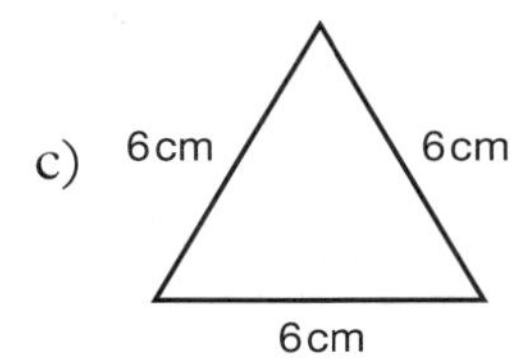

d) 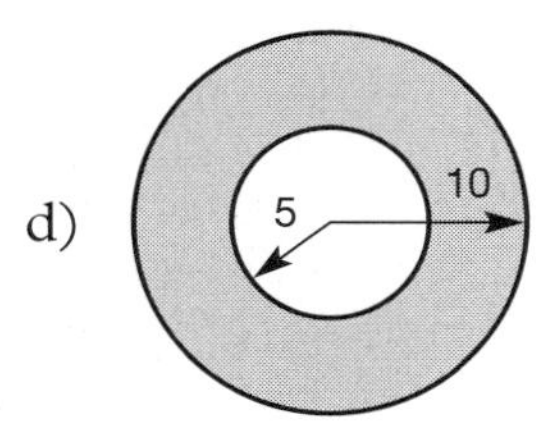

4 Calculate the total surface areas of these solids:

a) 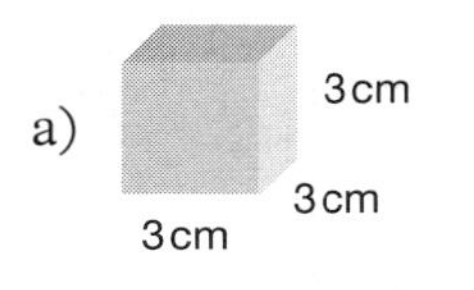

b) 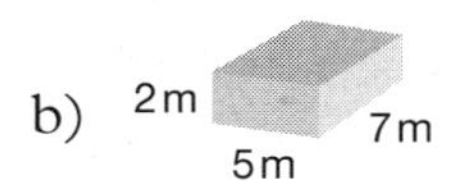

c) 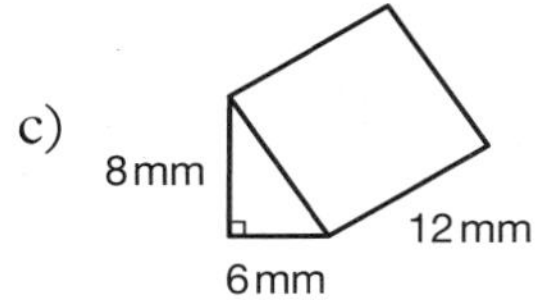

d) 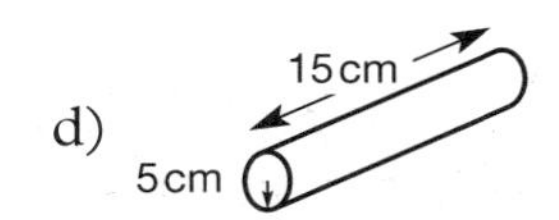

5B Calculate the radius of a circle whose area is:

a) 154 cm^2 b) 531 cm^2 c) 1134 km^2 d) 2642 mm^2.

6B A circle is enlarged so that its area is increased by 20%. If its original radius was 21 cm, calculate the length of the new radius.

15 The Volume of a Prism

Reminder

A prism has congruent parallel end faces, and the edges joining them are parallel to each other and perpendicular to the end faces. For example:

cube 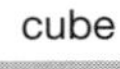

cuboid 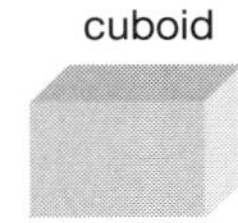

triangular prism 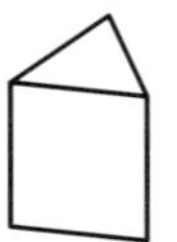

cylinder, hexagonal prism 

The general formula for the volume of a prism, V units^3, is

$$\boldsymbol{V = A \times h}$$

where A units^2 is the area of the base and h units is the height.

Special cases are: a cube $V = l^3$

a cuboid $V = lbh$

a cylinder $V = \pi r^2 h$ (This appears in the Formula List in the exam paper.)

Example Calculate the volume of this triangular prism.

Solution

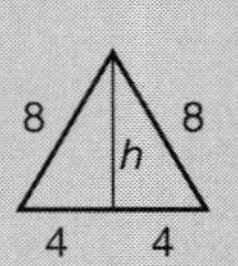

$$h^2 = 8^2 - 4^2$$
$$= 64 - 16$$
$$= 48$$
$$\Rightarrow \quad h = \sqrt{48} = 4\sqrt{3}$$

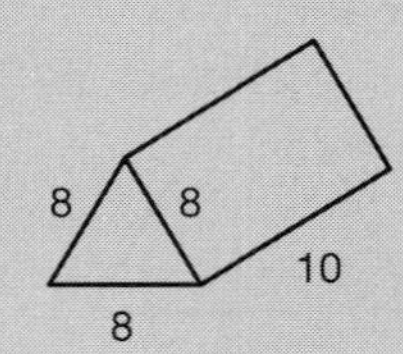

hence area of triangle $= \frac{1}{2}bh = \frac{1}{2} \times 8 \times 4\sqrt{3} = 16\sqrt{3}$,

and $V = A \times h = 16\sqrt{3} \times 10 = 160\sqrt{3} = 277$ units^3.

Exercise 15

1 Calculate the volume of each of these solids, rounding each answer to a whole number:

a) 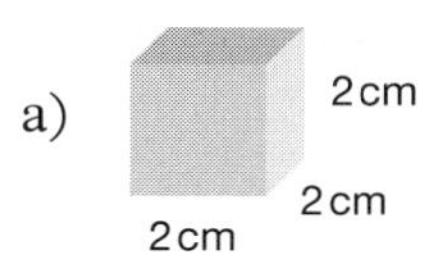

b) 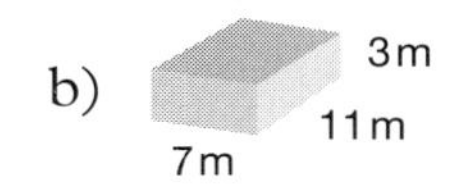

c) 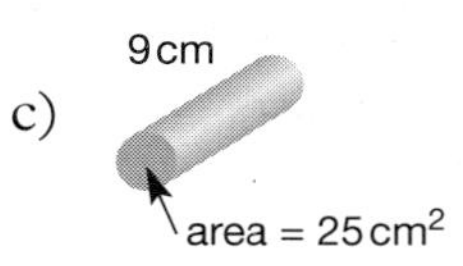

d) 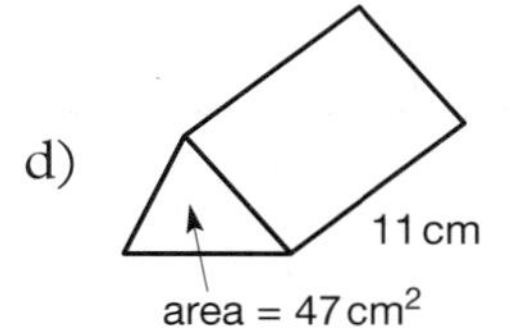

e) 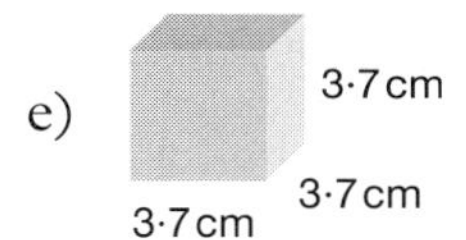

f) 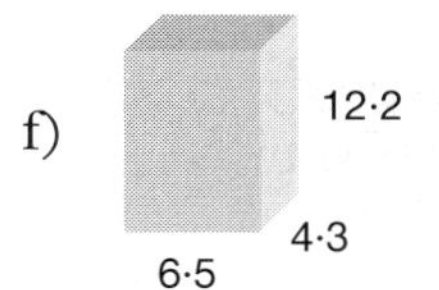

g) 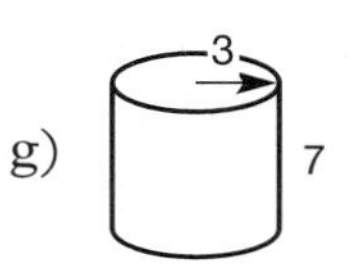

h) 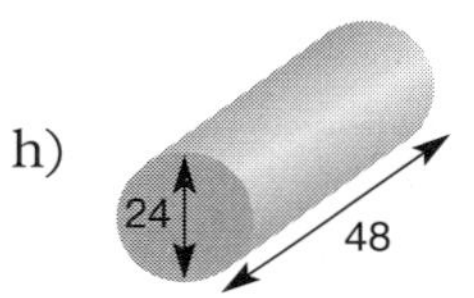

i) 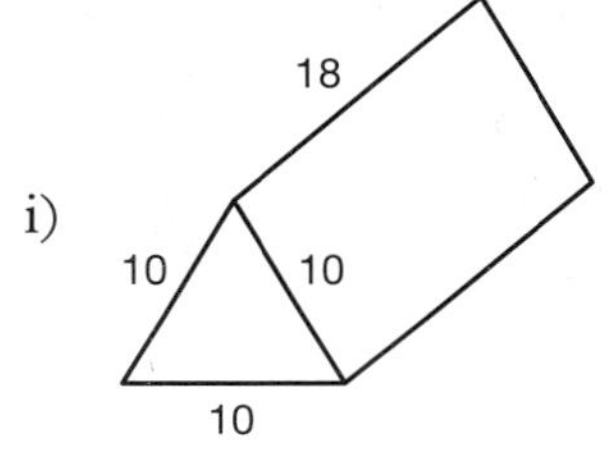

j) 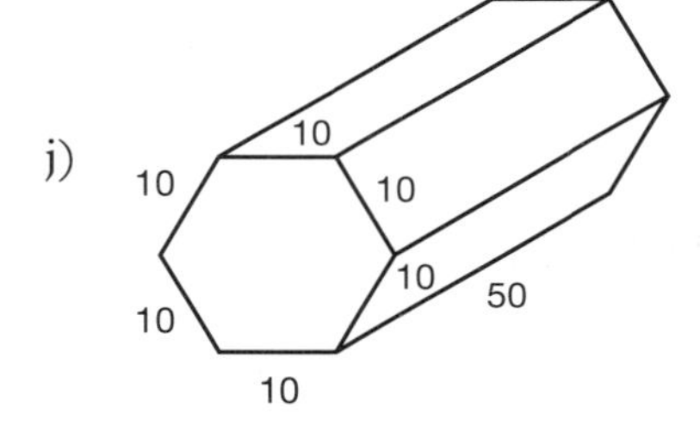

2 Calculate the volume of a cylinder with:

a) base radius 4 cm and height 12 cm
b) base radius 3 cm and height 5 cm
c) base diameter 5·6 cm and height 11 cm
d) base diameter 6·3 cm and height 9·2 cm.

3 Calculate the volume of each right-angled triangle prism:

a) 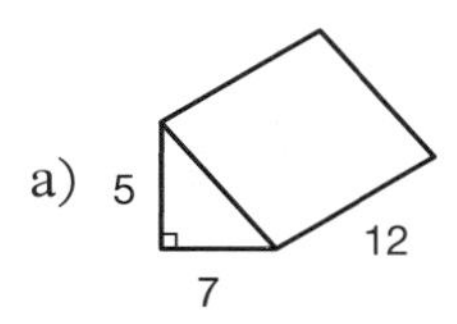

b) 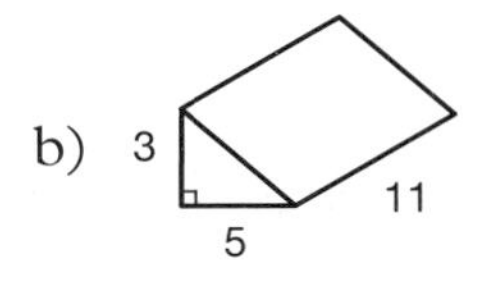

4 Calculate the unknown length in each diagram:

a) 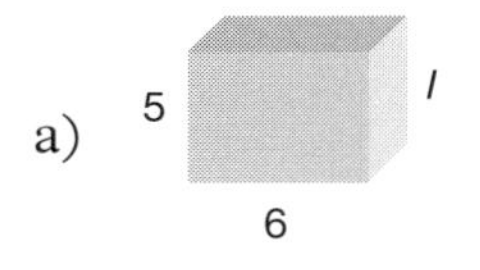

(Volume = 300 units^3)

b) 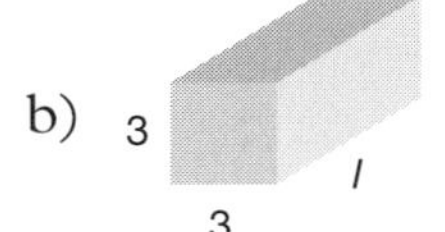

(Volume = 63 units^3)

c) 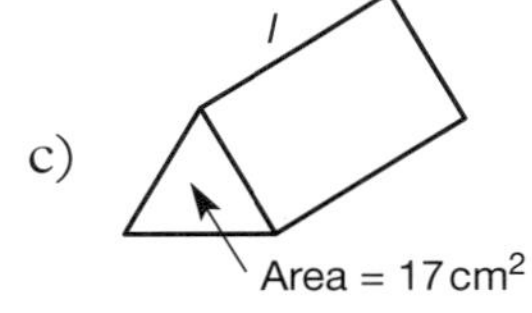

(Volume = 136 cm^3)

d) 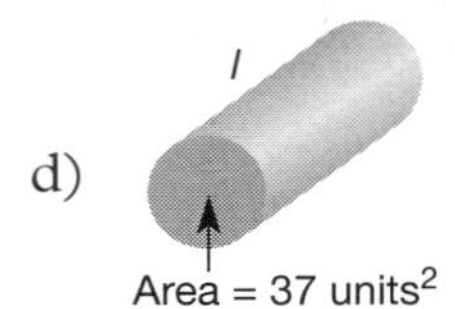

(Volume = 259 units^3)

e) 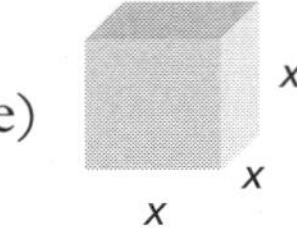

(Volume = 125 units^3)

f) 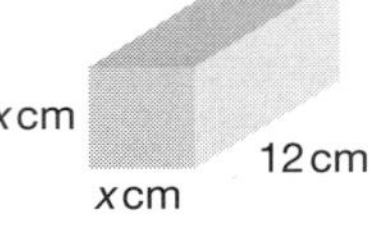

(Volume = 192 cm^3)

5B A cylindrical can of oil is to hold half a litre and to have a diameter of 7 cm. Calculate the height of the can. [Remember that 1000 cm^3 = 1 litre.]

6B A cylindrical carton of spaghetti is to hold half a litre and to have a length of 36 cm. Calculate the radius of the circular end.

16 The Volume of a Sphere

Reminder

Volume = $\frac{4}{3}\pi r^3$

Example Calculate the volume of a sphere of radius 2·6 cm.

Solution $V = \frac{4}{3}\pi r^3 = \frac{4}{3}\pi(2{\cdot}6)^3 = 73{\cdot}6 \text{ cm}^3$

(Make sure that you know how to calculate $(2{\cdot}6)^3$ on your calculator without doing $2{\cdot}6 \times 2{\cdot}6 \times 2{\cdot}6$.)

Exercise 16

1 Calculate correct to one decimal place, the volume of a sphere with radius:

a) 5 cm b) 7 cm c) 21·3 cm d) 56·3 cm.

2 Calculate, correct to one decimal place, the volume of each composite solid:

a) a hemisphere on top of a cylinder

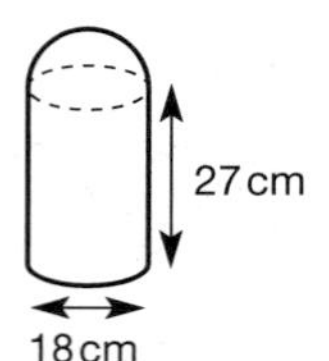

b) a hemisphere on top of a cube

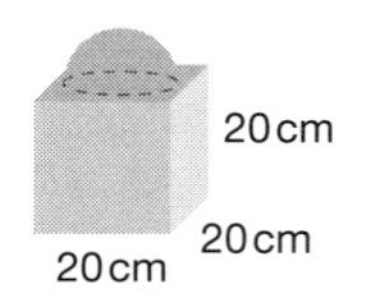

c) a square prism with a hemisphere removed from the top

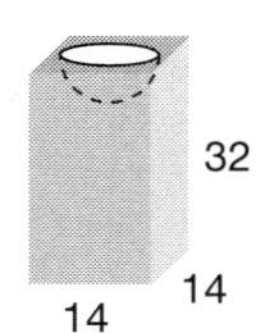

d) a cylinder with a hemisphere removed from one end.

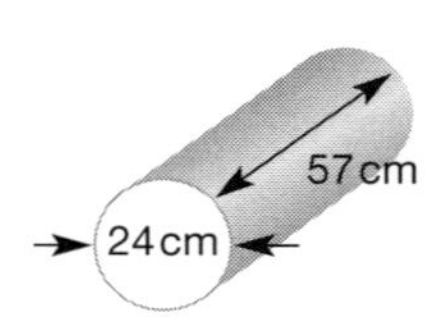

3 A wax sphere of radius 7·5 cm is re-shaped into a square prism. The edges of the square ends are each 7·5 cm. Calculate the length of the prism.

4 A hollow metal sphere has outer and inner radii of 7 cm and 6 cm respectively. Calculate the volume of metal required to make it.

5B A roll of solder is basically the shape of a cylinder of length 5 m and diameter 2 mm. It is melted down and shaped into spheres of radii 1·5 mm. Calculate how many such spheres may be obtained.

6B An ingot of gold in the shape of a cuboid measuring 12 cm × 9 cm × 25 cm is melted down and re-cast as a sphere. Calculate the radius of the sphere, correct to 3 significant figures.

17 The Volume of a Cone

Reminder

Volume = $\frac{1}{3}\pi r^2 h$

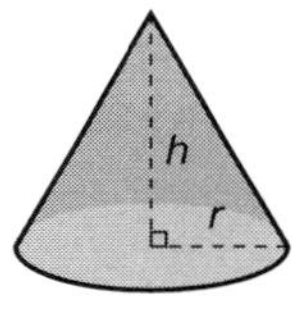

Example Calculate the volume of a cone of height 19 cm and base radius 7·4 cm.

Solution $V = \frac{1}{3}\pi r^2 h = \frac{1}{3}\pi(7\cdot4)^2 \times 19 = 1089\cdot55 \text{ cm}^3 = 1100 \text{ cm}^3$
(correct to 2 significant figs.)

Exercise 17

1 Calculate, correct to one decimal place, the volume of a cone with:

a) height 15 cm and base radius 7 cm
b) height 23 cm and base radius 6·4 cm
c) height 35 cm and base diameter 21 cm
d) height 42 cm and base diameter 12 cm.

2 Calculate, correct to one decimal place, the volume of each composite solid:

a) a cone on top of a cylinder

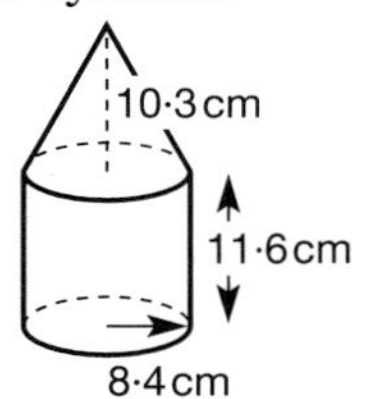

b) a cone on top of a cube

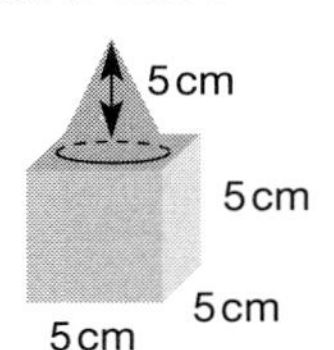

c) a square prism with a conical depression in the top

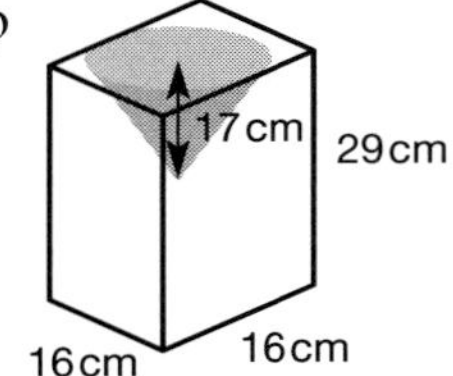

d) a cylinder with a cone removed from one end

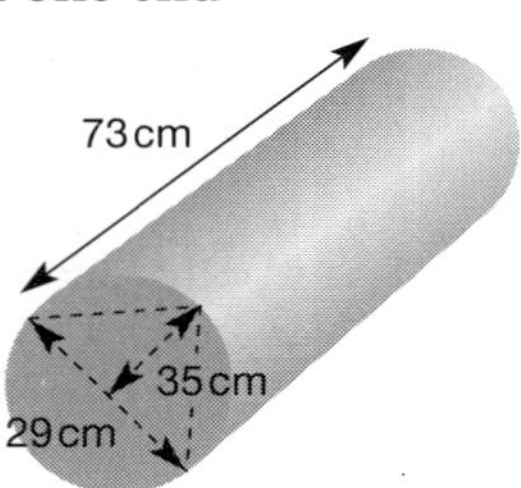

e) a hemisphere on top of a cone.

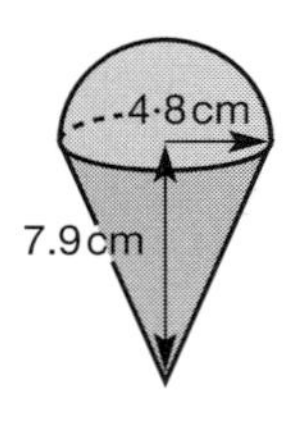

3 A sphere of ice cream is pressed gently into a cone. The ice cream **fills** the cone and has a flat surface level with the top of the cone. The radius of the original sphere of ice cream was 2·5 cm, and the radius of the top of the cone is 2·2 cm. Calculate the height of the cone.

4 A traffic cone without its heavy base is a hollow plastic cone whose external dimensions are diameter 20 cm and height 80 cm. The internal dimensions are diameter 18 cm and height 78 cm. Calculate the volume of plastic required to make this traffic cone.

5 Paper cones are provided beside a cool water supply in a works canteen. They have a depth of 12 cm and a diameter at the top of 6 cm. Jane fills a paper cone with water and pours it into her china mug, which is cylindrical with a diameter of 8 cm. How deep is the water in her mug?

6B Jennifer has a cone of plasticine measuring 7·8 cm across the base and 17 cm high. She rolls it into a sphere. Calculate correct to one significant figure, the radius of the sphere formed.

18 Revision of Volume

Exercise 18

1 Calculate the volume of these shapes, giving each answer correct to a whole number of cm^3:

a) a cube of side 3·8 cm
b) a cuboid measuring 2·7 cm × 3·4 cm × 7·9 cm
c) a cylinder of base radius 4·5 cm and height 7 cm
d) a triangular prism with height 11 cm and area of base 17 cm^2
e) a cone with base radius 3·8 cm and height 17·5 cm
f) a sphere of radius 9·7 cm.

2 Calculate the volume of each composite solid, giving each answer correct to a whole number of cm^3:

a)

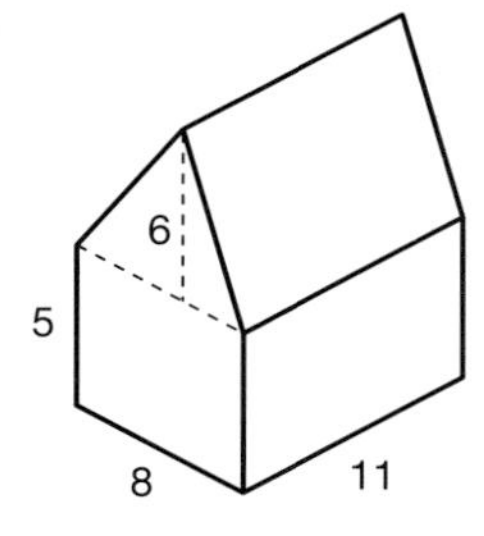

b)

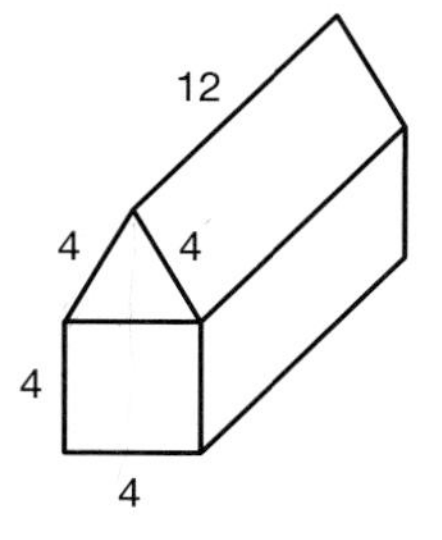

c) (a cylinder with hemispherical ends)

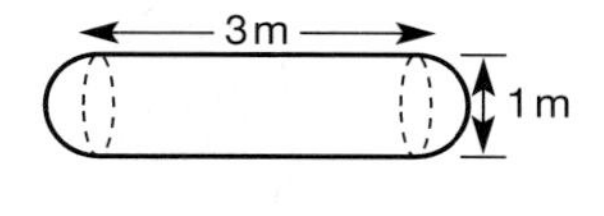

d) a cylinder and a cone

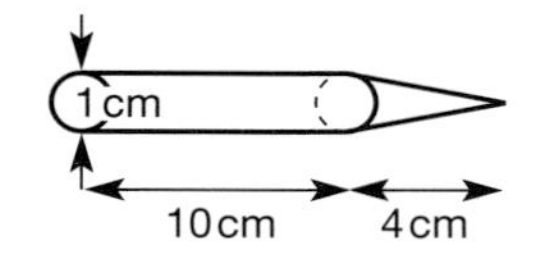

e) a 6 cm cube with a hemisphere on each face.

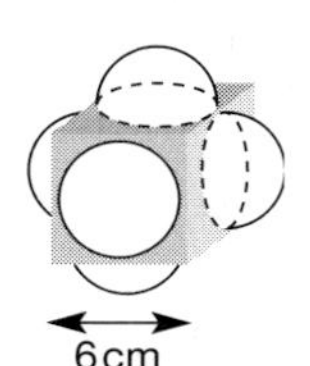

3 Calculate the total volume of this croquet mallet.

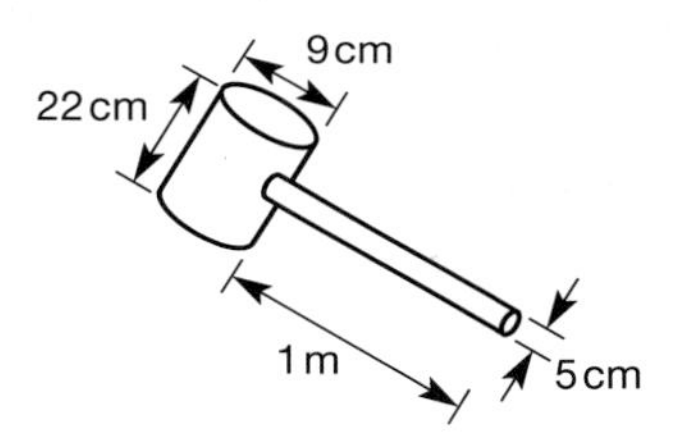

4 This can of condensed soup should be mixed with an equal volume of water, heated and served.

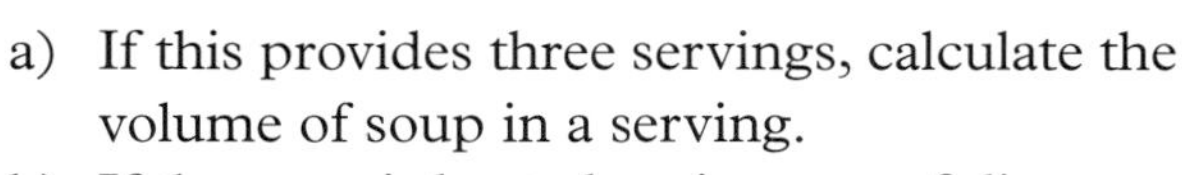

a) If this provides three servings, calculate the volume of soup in a serving.

b) If the soup is heated up in a pot of diameter 17 cm, how deep is the soup in the pot?

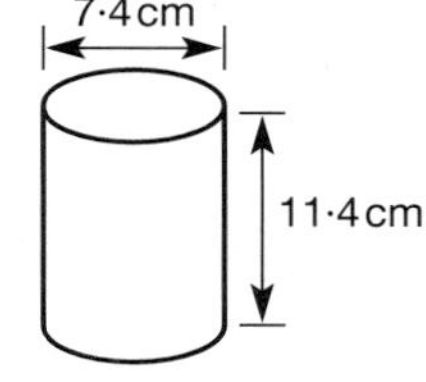

5 a) Calculate the volume of water in this swimming pool when it is filled to the brim.

b) What volume of chlorine in litres, should be added during filling to obtain a 0·01% solution of chlorine in the pool?

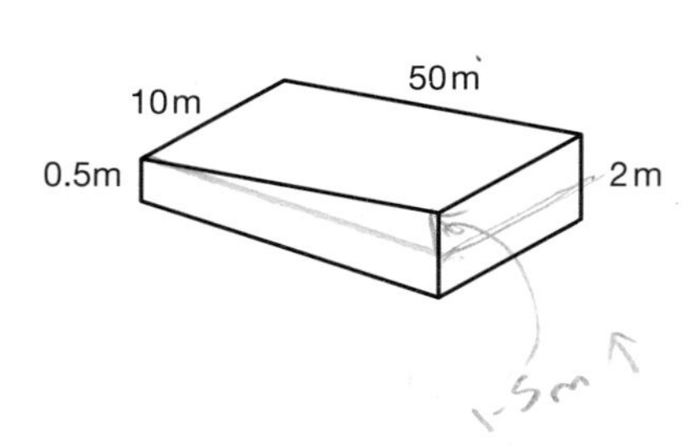

6 Calculate the volume of this coal bunker.

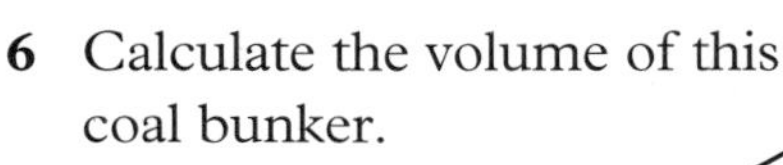

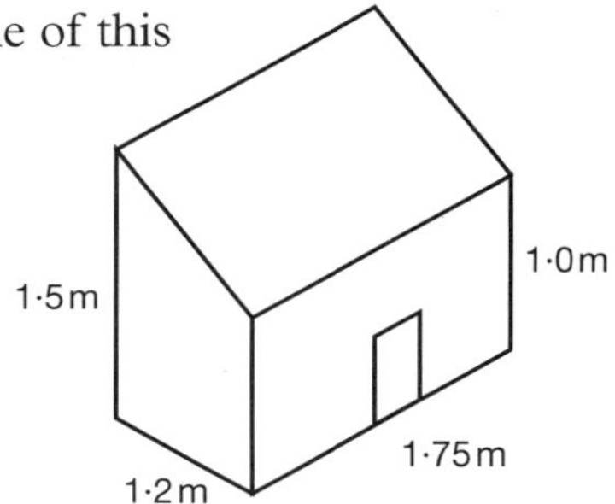

19 Test on Volume

Allow 45 minutes for this test

(Give each answer in this test correct to 2 significant figures)

1 Calculate the volume of:

a) a cube of side 2·7 cm

b) a cuboid measuring 3·4 cm × 2·4 cm × 7·3 cm

c) a triangular prism with the area of the triangle 108 cm^2 and length 40 cm

d) a cylinder with base radius 3·2 cm and height 5·7 cm

e) a sphere of radius 2·5 cm

f) a cone with base radius 5·6 cm and height 17·9 cm.

2 Calculate the volume of each composite solid:

a) a plumb bob consisting of a hemisphere on top of a cone

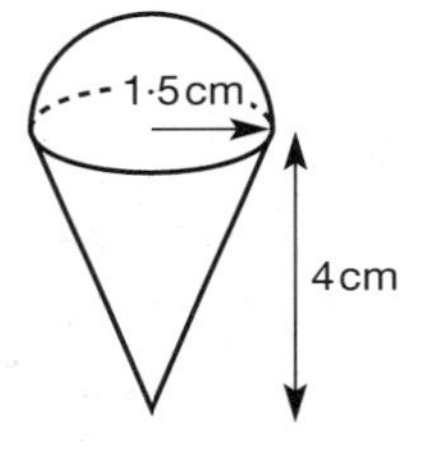

b) an advance factory shed in the shape of a half cylinder on top of a cuboid

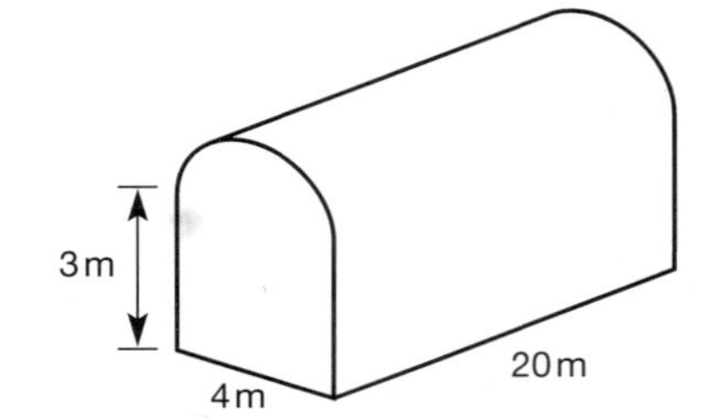

3 The base of a new golf trophy is to be part of a cone with dimensions as shown in the diagram.

Calculate the volume of this trophy base.

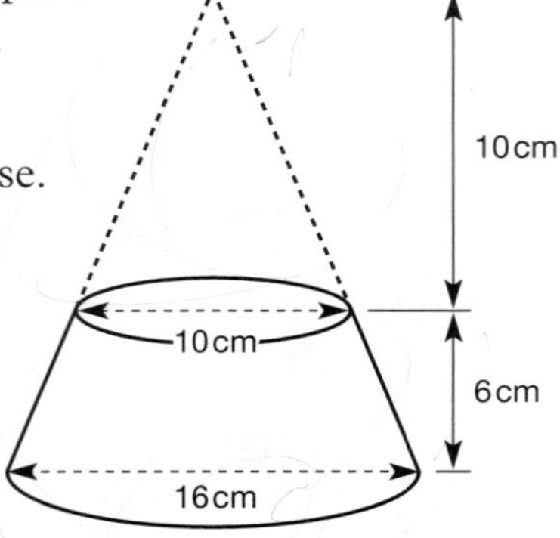

COORDINATE GEOMETRY

20 The Gradient of a Line

Reminder

The gradient of a line is the ratio of the distance up to the distance along.

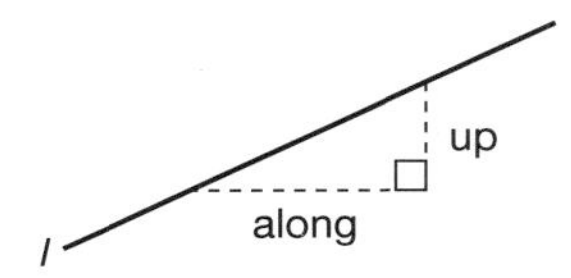

$$\text{the gradient of } l = \frac{\text{distance up}}{\text{distance along}}$$

Example Find the gradient of the line l shown.

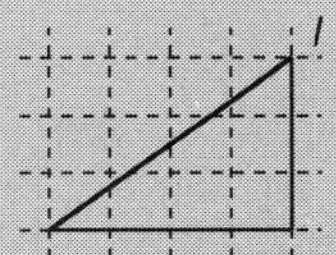

Solution

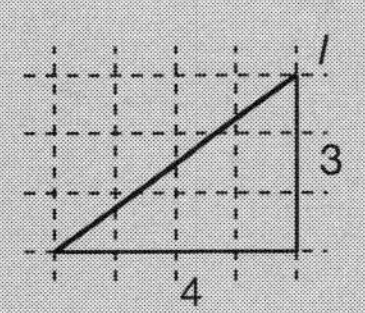

$$\text{gradient} = \frac{\text{distance up}}{\text{distance along}} = \frac{3}{4}$$

Note: A line that slopes up to the right has a positive gradient.
A line that slopes down to the right has a negative gradient.

Example Find the gradient of the line L shown.

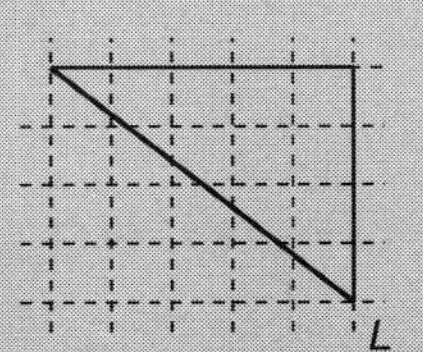

Solution

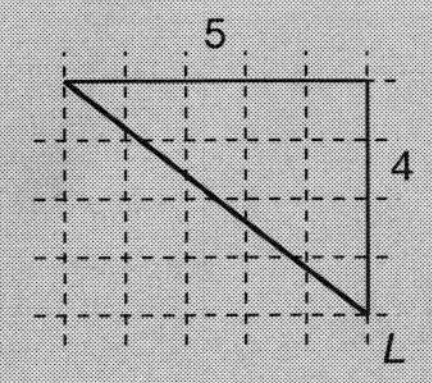

$$\text{gradient} = \frac{\text{distance up}}{\text{distance along}} = \frac{-4}{5} = -\frac{4}{5}$$

Exercise 20

1 Find the gradient of each of these lines:

a) 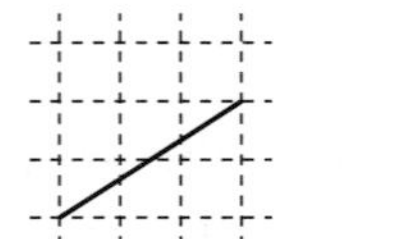

b) 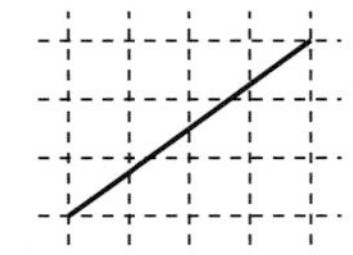

c) 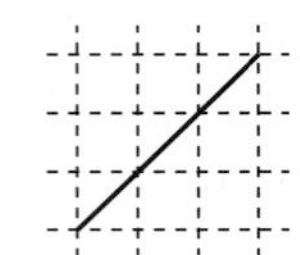

d) 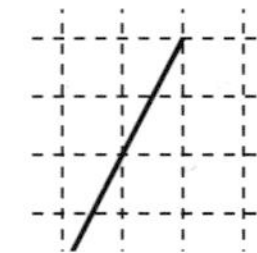

e) 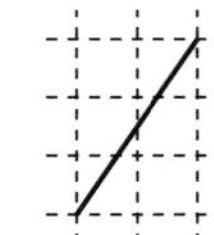

2 Find the gradient of each of these lines:

a) 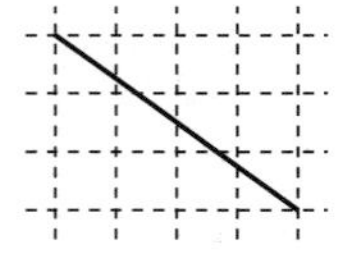

b) 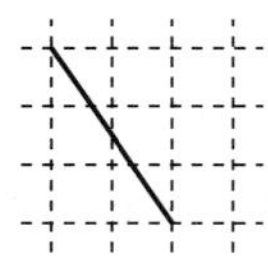

c) 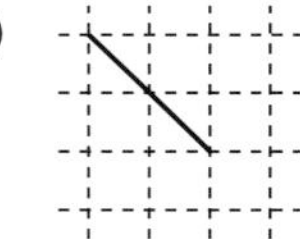

d)

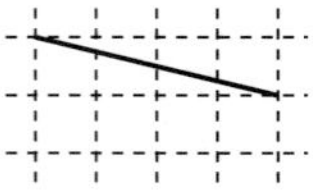

e) 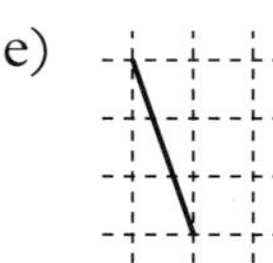

3 Find the gradient of each of these lines:

a) 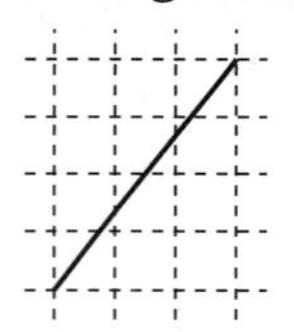

b) 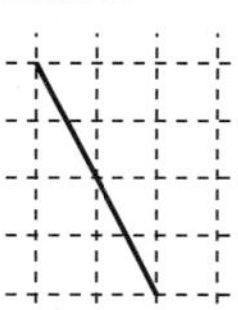

c) 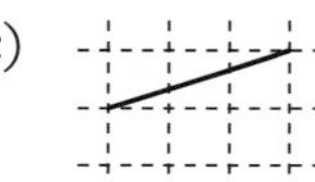

d) 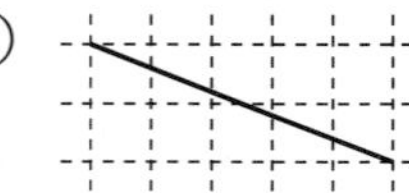

e) 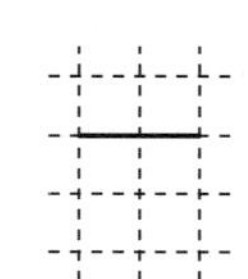

f) 

21 The Gradient Formula

Reminder

The gradient of the line joining A (x_a, y_a) to B (x_b, y_b) is often written m_{AB}

and $m_{AB} = \dfrac{y_a - y_b}{x_a - x_b}$

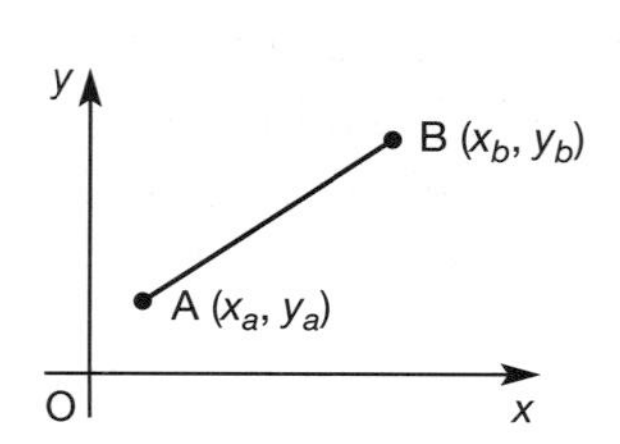

Reminder continued

Note that
- we write the x-coordinate of A as x_a and the y-coordinate of B as y_b
- it is the difference of the y-coordinates which appears on the top line of the formula (putting the x-coordinates on top is a common error – try to avoid it)
- both the top and bottom lines of the formula begin with a coordinate of A (either point could be taken – just be consistent – see the example below)
- great care should be taken with any negative coordinates.

Example Find the gradient of the line joining the points P (3, 5) and Q (4, –1).

Solution

$$m_{PQ} = \frac{y_p - y_q}{x_p - x_q} = \frac{5 - (-1)}{3 - 4} = \frac{6}{-1} = -6$$

Note that we could equally well have written:

$$m_{PQ} = \frac{y_q - y_p}{x_q - x_p} = \frac{-1 - 5}{4 - 3} = \frac{-6}{1} = -6$$

In other words it does not matter whether we think of P or Q as the point 'A', so long as we are consistent.

Exercise 21

1 Find the gradient of the line joining the following pairs of points:

a) A (2, 2) and B (4, 7) b) C (4, 3) and D (7, 9) c) E (1, 2) and F (8, 5)
d) G (2, 3) and H (5, 8) e) K (4, 7) and L (8, 9) f) M (3, 5) and N (7, 11)
g) P (5, 7) and Q (7, 9) h) R (6, 3) and S (9, 5) i) T (7, 4) and U (9, 11).

2 Find the gradient of the line joining the following pairs of points:

a) (5, 6) and (6, 3) b) (2, 3) and (9, 1) c) (5, 9) and (8, 3)
d) (6, 3) and (5, 8) e) (7, 2) and (4, 3) f) (5, 9) and (8, 7)
g) (5, 8) and (3, 10) h) (7, 4) and (9, 1) i) (5, 7) and (8, 3).

3 Find the gradient of the line joining the following pairs of points:

a) (–1, 2) and (3, –1) b) (–2, 5) and (–3, 6) c) (–5, –4) and (–3, –8)
d) (–4, –2) and (3, –1) e) (–3, 5) and (0, –4) f) (2, –7) and (–3, 4)
g) (–8, 6) and (–2, –1) h) (–2, –3) and (–1, –7) i) (–3, –2) and (5, –6).

4 a) Write down the coordinates of A, B and C.
b) Find the gradients of the three sides of triangle ABC.

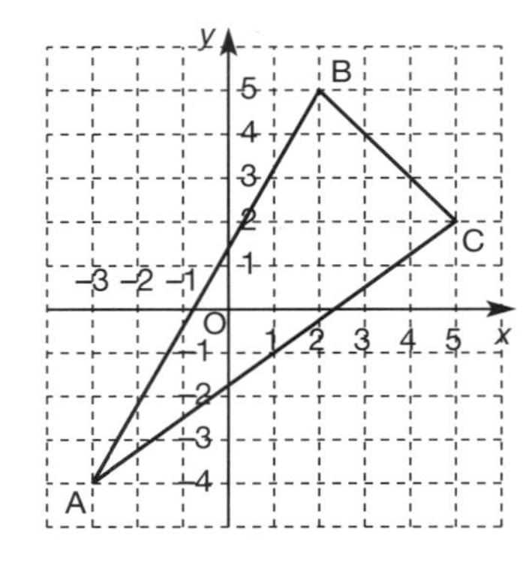

5 a) Write down the coordinates of P, Q, R, and S.
b) Find the gradients of the four sides of the quadrilateral PQRS.
c) Find the gradients of the diagonals of PQRS.

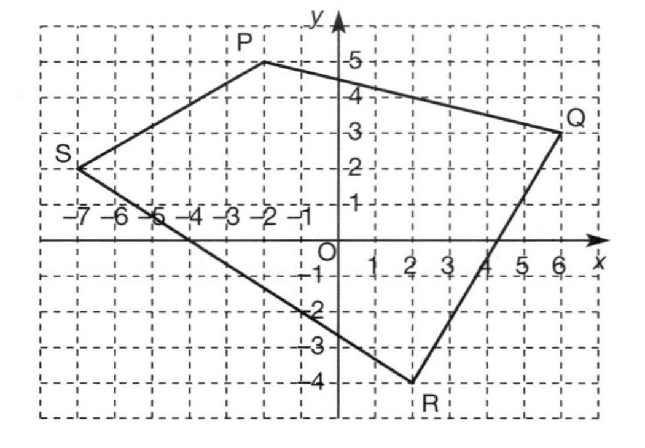

6 The quadrilateral EFGH has vertices E (−2, −1), F (2, 4), G (4, 1), and H (0, −4).
a) Plot EFGH on a coordinate diagram.
b) Show that EFGH is a parallelogram (i.e. opposite sides are parallel).

7 a) Plot the points A (−6, −3), B (−1, 1), C (4, 5), and D (9, 9) on a coordinate diagram.
b) Calculate the gradients of AB, BC, CD, AC, BD, and AD.
c) What do your answers to part (b) tell you about the points A, B, C, and D?

8 a) Plot the points P (−8, −6), Q (2, 1), and R (13, 8).
b) Prove that P, Q and R do not lie in a straight line.

Reminder

Example P is the point (8, 5), Q (11, 3), R (−7, −4) and S (2, t).
Find the value of t for which PQ is parallel to RS.

Solution $m_{PQ} = \frac{3-5}{11-8} = \frac{-2}{3}$ $\quad m_{RS} = \frac{t-(-4)}{2-(-7)} = \frac{t+4}{9}$

parallel lines have equal gradients, so $\frac{-2}{3} = \frac{t+4}{9}$

$\Rightarrow \quad (-2)(9) = 3(t+4) \quad \Rightarrow \quad -18 = 3t + 12 \quad \Rightarrow \quad 3t = -30$

$\Rightarrow \quad t = -10$

9 A is the point (2, 7), B (5, 8), C (11, 7), and D (k, 4).
Find the value of k for which AB is parallel to CD.

10 K is the point (2, –1), L (4, –3), M (6, 5), and N (7, u).
Find the value of u for which KL is parallel to MN.

11 a) (i) Plot the points S (0, 7), T (–3, 0) and V (3, 0).
(ii) Find the gradients of ST, SV and TV.
(iii) What kind of triangle is ΔSTV?
(iv) Name an angle equal to $\angle$STV.
(v) What do you notice about the gradients of ST and SV?
b) (i) Plot the points O (0, 0), K (0, 5), L (9, 0), and M (0, –5).
(ii) Find the gradients of KL, LM and KM.
(iii) What kind of triangle is ΔKLM?
(iv) Name an angle equal to $\angle$OLK.
(v) What do you notice about the gradients of KL and ML?
c) What can be said about the gradients of two (non-parallel) lines which are equally inclined to the x-axis?

12 a) Plot the points A (3, 1) and B (7, 1) and find the gradient of AB.
b) Plot the points P (3, 0) and Q (7, 0) and find the gradient of the x-axis.
c) Plot the points C (2, 3) and D (2, 8) and find the gradient of CD.
d) Plot the points R (0, 3) and S (0, 8) and find the gradient of the y-axis.
[Ask your teacher if you do not understand the answers to this question.]

13 a) Plot the points K (–2, 3), L (0, 3), M (3, 3), and N (5, 3), and observe that they all lie on a line parallel to the x-axis.
b) Write down the gradient of KN.
c) Note that the y-coordinate of each of K, L, M, and N is 3, so the equation of KN is $y = 3$.
Similarly find: (i) the gradient of PQ, and
(ii) the equation of PQ where P is (–2, 4) and Q (3, 4).
d) Write down the equation of the line which is parallel to the x-axis and passes through the point
(i) (7, 5) (ii) (4, –2).
e) Write down the equation of the x-axis.

14 a) Plot the points A (3, –1), B (3, 0), C (3, 3), and D (3, 5), and observe that they all lie on a line parallel to the y-axis.
b) Write down the gradient of AD.
c) Note that the x-coordinate of each of A, B, C, and D is 3, so the equation of AD is $x = 3$.
Similarly find (i) the gradient of KL, and
(ii) the equation of KL where K is (5, –1) and L (5, 5).
d) Write down the equation of the line which is parallel to the y-axis and passes through the point
(i) (4, –3) (ii) (–2, 6).
e) Write down the equation of the y-axis.

15 a) Plot the kite PQRS with vertices P (4, 7), Q (9, 4), R (4, –2), and S (–1, 4).
b) Write down the equations of the diagonals SQ and PR.

16B a) Plot the rhombus WXYZ with vertices W (4, 2), X (1, –2), Y (–2, 2), and Z (1, 6).
b) Find the gradients of OW, OX, OY and OZ.
c) Write down the equations of the diagonals XZ and YW.
d) The line with equation $x = h$ passes through the rhombus WXYZ. Within what range of values must h lie for this to be possible?
e) The line with equation $y = k$ does not intersect the rhombus WXYZ at all. What are the possible values for k?

22 The Straight Line Formula $y = mx$

Reminder

Consider the line with equation $y = mx$.

if $x = 0$, then $y = m \times 0 = 0$ i.e. (0, 0) lies on $y = mx$
if $x = 1$, then $y = m \times 1 = m$ i.e. $(1, m)$ lies on $y = mx$

thus the graph of $y = mx$ looks like this:

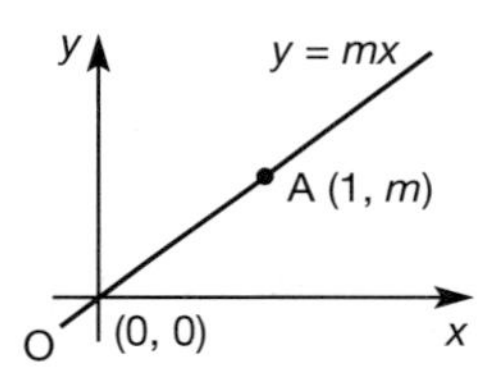

also $m_{OA} = \frac{m - 0}{1 - 0} = \frac{m}{1} = m$

i.e. $y = mx$ passes through the origin with gradient m

Example Find the equation of the line passing through O(0, 0) and A (4, 5).

Solution $m_{OA} = \frac{5 - 0}{4 - 0} = \frac{5}{4} \quad \Rightarrow \quad y = \frac{5}{4}x$

Exercise 22

1 a) Plot the points O (0, 0), A (1, 1), B (1, 2), C (1, 3), D (1, 4), and E (1, 5).
b) Draw the lines OA, OB, OC, OD and OE and find their gradients.
c) Write down the equations of the lines OA, OB, OC, OD, and OE.

2 a) Plot the points O (0, 0), P (1, 1), Q (2, 1), R (3, 1), S (4, 1), and T (5, 1).
b) Draw the lines OP, OQ, OR, OS, and OT and find their gradients.
c) Write down the equations of the lines OP, OQ, OR, OS, and OT.

3 a) Plot the points O (0, 0), F (–5, 1), G (–3, 1), H (–1, 1), K (–1, 3), and L (–1, 5).
b) Draw the lines OF, OG, OH, OK, and OL and find their gradients.
c) Write down the equations of the lines OF, OG, OH, OK, and OL.

4 a) Plot the points O (0, 0), V (−2, 2), W (2, −2), M (2, 2), and N (−2, −2).
b) Draw the lines MON and VOW (i.e. MN and VW) and find their gradients.
c) Write down the equations of the lines MN and VW.

5 Suppose I wish to buy some books at 25p each.
a) Copy and complete this table:

number of books (x)	0	1	2	3	4	5
cost (£y)		0·25				

b) Plot these points on a coordinate diagram with the y-axis going up to 12 (for later use).
c) Draw a straight line through these points (just to emphasise the linear relationship).
d) Find the equation of this line, i.e. the formula for y in terms of x.
e) Complete this table for differently priced books:

number of books (x)	0	1	2	3	4
50p books: cost (£y)		0·50			
£1 books: cost (£y)		1			
£1·50 books: cost (£y)		1·50			
£2 books: cost (£y)		2			
£3 books: cost (£y)		3			
£4 books: cost (£y)		4			
£6 books: cost (£y)		6			

f) Plot the points on the same diagram as you used for part (b) and draw the line graph for each price of book, labelling each line with its equation.
g) State the most important thing that all these lines have in common.
h) Describe the connection between the gradient of a line and its position on the diagram.

6 This diagram shows the line AB, which has equation $y = x$ and gradient 1. It also shows the line CD which has equation $y = -x$ and gradient −1.

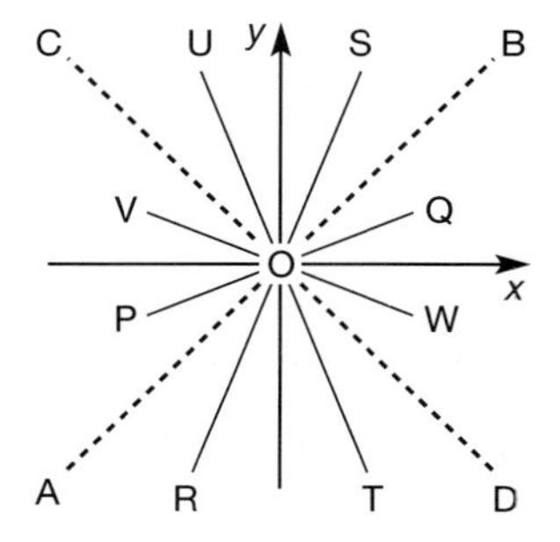

a) PQ is shown lying somewhere between the x-axis and OB. What can you say about the size of the gradient of PQ?
b) RS is shown lying somewhere between OB and the y-axis. What can you say about the size of the gradient of RS?
c) Similarly comment on the gradients of UT and VW.

7 a) Plot the points O (0, 0), A (3, 2), B (2, 5), C (−3, 4), and D (−4, 1).
b) Calculate the gradients of OA, OB, OC, and OD.
c) Write down the equations of OA, OB, OC, and OD.
d) Show that DC and AB are equally inclined to the x-axis.

8 a) Plot the rectangle ABCD with vertices A (5, 4), B (5, –4), C (–5, –4), and D (–5, 4).
b) Write down the equations of the lines AD, AB, CD, and CB.
c) Find the equations of the lines OA and OB.
d) Calculate the area of ABCD.

9 a) Plot the quadrilateral PQRS with vertices P (7, 5), Q (4, –1), R (–7, –5), and S (–4, 1).
b) Calculate the gradients of PQ, QR, RS, and SP.
c) What type of quadrilateral does this prove PQRS to be?
d) Confirm that the diagonals of PQRS bisect each other. (i.e. cut each other in half)
e) Find the equations of the diagonals PR and SQ.

10 a) Plot triangle ABC which has vertices A (4, 4), B (0, –3), and C (–5, 2).
b) Calculate the gradients of AB, BC, and CA.
c) Find the equations of OA, OB, and OC.

11 a) Plot the quadrilateral OKLM with vertices O (0, 0), K (7, 4), L (8, 10), and M (1, 6).
b) Calculate the gradients of OK, OL, OM, KL, LM, and MK.
c) What type of quadrilateral is OKLM?
d) Find the equations of OK, OL, and OM.

12 a) Plot the parallelogram OPQR with vertices O (0, 0), P (5, 10), Q (0, 10), and R (–5, 0).
b) Find the gradient of PR and write down the equations of OP, OQ, and OR.
c) Calculate the area of OPQR.

23 The Straight Line Formula $y = mx + c$

Reminder

Consider the line with equation $y = mx + c$.

if $x = 0$, then $y = m \times 0 + c = c$ i.e. $(0, c)$ lies on $y = mx + c$
if $x = 1$, then $y = m \times 1 + c = m + c$ i.e. $(1, m + c)$ lies on $y = mx + c$

thus the graph of $y = mx + c$ looks like this:

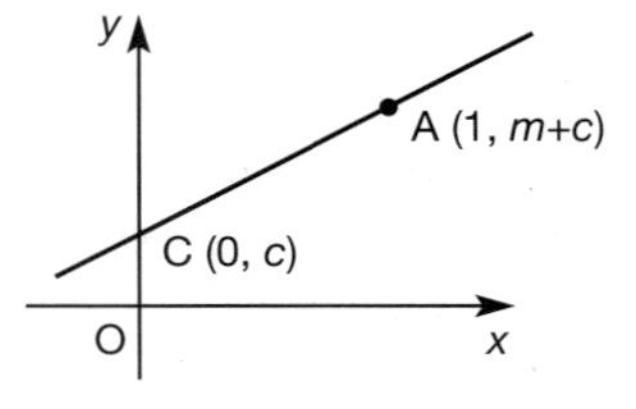

also $m_{AC} = \dfrac{(m + c) - c}{1 - 0} = \dfrac{m}{1} = m$

i.e. $y = mx + c$ passes through $(0, c)$ with gradient m
c is called the intercept (on the y-axis)

Note that if $c = 0$, then the point C is $(0, 0)$ and the equation is $y = mx + 0$ or $y = mx$ i.e. the line passes through the origin with gradient m, hence $y = mx$ (which we studied in the previous exercise) is just the special case of $y = mx + c$ where $c = 0$.

Reminder continued

Example Find the equation of the line passing through the point (0, 7) with gradient 3.

Solution Using the formula $y = mx + c$ with $m = 3$ and $c = 7$ gives $y = 3x + 7$.

Example Write down the gradient of the line with equation $y = 4x - 3$ and the coordinates of the point where it crosses the y-axis.

Solution Compare $y = mx + c$

with $y = 4x - 3 \quad \Rightarrow \quad m = 4$ and $c = -3$

hence the gradient = 4 and the line crosses the y-axis at (0, –3)

Example Find the gradient of the line with equation $3x + 4y = 5$ and where it crosses the y-axis.

Solution Before we can apply the formula we have just learned, the equation has to be re-arranged in the form $y = \ldots\ldots$

Take time to do this carefully – it is a source of many errors.

$$3x + 4y = 5$$
$$\Rightarrow 4y = 5 - 3x \quad \Rightarrow \quad 4y = -3x + 5$$
$$\Rightarrow \quad y = -\frac{3}{4}x + \frac{5}{4} \quad \Rightarrow \quad \text{gradient} = -\frac{3}{4} \text{ and } \left(0, \frac{5}{4}\right)$$

Exercise 23

1 Write down the equation of the line which passes through the point:

a) (0, 2) with gradient 4

b) (0, 1) with gradient 1

c) (0, –1) with gradient 3

d) (0, 3) with gradient –1

e) (0, –2) with gradient $\frac{1}{2}$.

2 Write down the equation of the line which has a gradient of:

a) 2 and has intercept 3

b) –2 and has intercept –1

c) $-\frac{1}{2}$ and has intercept $\frac{1}{2}$

d) $\frac{2}{3}$ and has intercept 4

e) $\frac{1}{4}$ and has intercept $-\frac{1}{3}$.

3 For the equation of each line given, write down the gradient and the intercept:

a) $y = 2x + 3$

b) $y = 3x - 2$

c) $y = -4x + 5$

d) $y = \frac{1}{2}x - 1$

e) $y = -\frac{2}{3}x - \frac{3}{4}$.

4 Find the gradient of each line and where it crosses the y-axis:

a) $2y = 4x + 5$
b) $x = y - 2$
c) $2x + y + 3 = 0$
d) $6x - 3y + 1 = 0$
e) $4x + 5y = 10$.

5 Show that each of the following pairs represents parallel lines:

a) $y = 2x + 1$
$2y = 4x + 5$
b) $y = 5x - 3$
$3y = 15x + 8$
c) $2x + 3y + 5 = 0$
$4x + 6y + 7 = 0$
d) $5x - 2y = 8$
$25x - 10y = 1$.

6 Which of the following lines are parallel to the line with equation $3x - 2y = 5$?

a) $2x - 3y = 5$
b) $6x - 4y = 7$
c) $y = \frac{3}{2}x - \frac{1}{2}$
d) $y = \frac{2}{3}x + \frac{1}{2}$
e) $\frac{1}{3}x - \frac{1}{2}y = \frac{1}{5}$.

7 Find the equation of the line passing through the origin parallel to the given line:

a) $4y = 8x + 7$
b) $x - y = 4$
c) $x = 2y + 1$
d) $2x + 3y = 7$
e) $2x + 5y + 3 = 0$.

8 Find the equation of the line passing through the point (0, 1) parallel to the given line:

a) $2y = 6x - 3$
b) $3x - 4y = 1$
c) $2x + y + 3 = 0$
d) $3x - 2y = 7$
e) $x = 5y + 10$.

Reminder

It is important to be able to tell whether or not a point lies on a line. When a point lies on a line, the coordinates of the point satisfy the equation of the line.

Example Show that the point A (3, –2) lies on the line with equation $3x + 5y + 1 = 0$ but that the point B (–3, 2) does not.

Solution for A, $x = 3, y = -2$,
so $3x + 5y + 1 = 3(3) + 5(-2) + 1 = 9 - 10 + 1 = 0$
$\Rightarrow$ A lies on this line
for B, $x = -3, y = 2$,
so $3x + 5y + 1 = 3(-3) + 5(2) + 1 = -9 + 10 + 1 = 2 \neq 0$
$\Rightarrow$ B does not lie on this line.

9 Does the point (1, 2) lie on the line with equation

a) $2x + 3y = 8$
b) $3x - 2y = 1$
c) $y = 7x - 5$
d) $2y = 3x - 1$
e) $7x - 3y - 1 = 0$?

10 Which of these points lie on the line with equation $3x + 4y = 7$?

P (1, 1) Q (−1, 2) R (3, −2) S (4, −5) T (17, −11)

11 a) On a single diagram on plain paper sketch the lines with equations:

(i) $y = 1$
(ii) $y = x + 1$
(iii) $y = 2x + 1$
(iv) $y = 3x + 1$
(v) $y = -2x + 1$.

b) What do all of these lines have in common?

c) What effect does changing the coefficient of x have on the graph? (The coefficient of x is the number in front of x.)

12 a) On a single diagram on plain paper sketch the lines with equations:

(i) $y = x$
(ii) $y = x + 1$
(iii) $y = x + 2$
(iv) $y = x + 5$
(v) $y = x - 2$.

b) What do all of these lines have in common?

c) What effect does changing the constant term have on the graph? [The constant term is the 'number only' term, no x or y in it.]

13 a) On a single diagram on plain paper sketch the lines with equations:

(i) $y = 2x$
(ii) $y = 2x + 2$
(iii) $y = 2x - 2$.

b) What do all of these lines have in common?

14 a) On a single diagram on plain paper sketch the lines with equations:

(i) $y = -\frac{1}{2}x$
(ii) $y = -\frac{1}{2}x + 3$
(iii) $y = -\frac{1}{2}x - 3$.

b) What do all of these lines have in common?

c) Is the effect of changing the constant term on the graph the same for questions 12, 13, 14?

Reminder

In questions 11 to 14, we were given the equation and had to draw the graph. Now we do the reverse. Given the graph, we find the equation of the line.

Example Find the equation of the line shown.

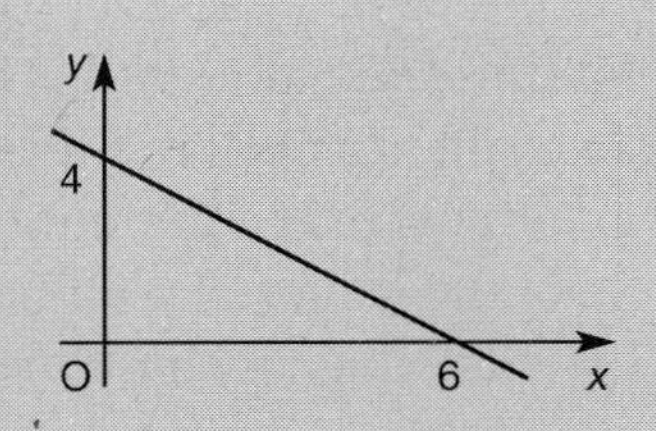

Reminder continued

Solution using the points (0, 4) and (6, 0), $m = \frac{4-0}{0-6} = -\frac{2}{3}$

from (0, 4), $c = 4$

hence using $y = mx + c$, $y = -\frac{2}{3}x + 4$

15 Find the gradient of the line in each of these diagrams:

a)

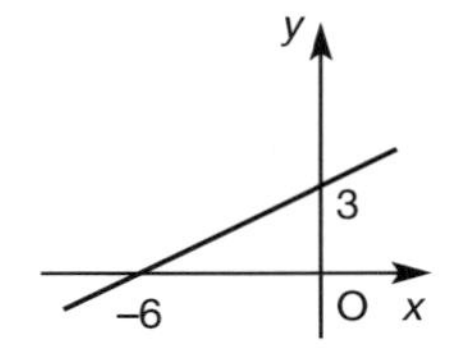

b)

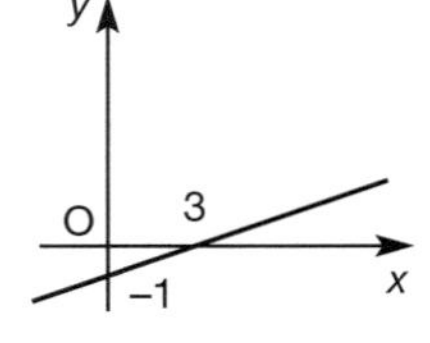

c)

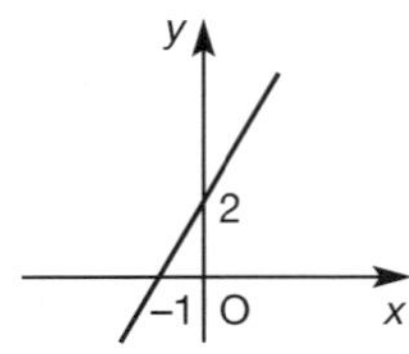

d)

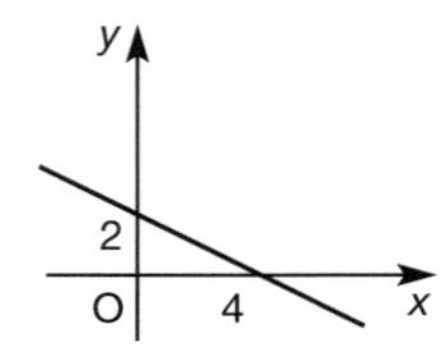

e)

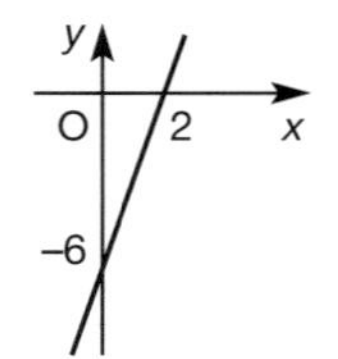

f)

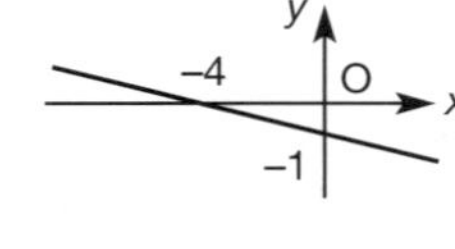

16 I can hire a van from Good Gnus for £30 per day plus £1 for every 5 miles I travel.

a) Complete this table of values for a one day van hire:

distance travelled (x miles)	0	10	20	30	40
total cost (£y)	30	32			

b) Plot these points on a coordinate diagram and draw the line which passes through them.

c) Find the equation of the line you have just drawn.

17 The time to cook a turkey depends on its size. Experience has shown that the appropriate cooking times for certain sizes of turkey are as follows:

size of turkey (x kg)	4	7	11
cooking time (y minutes)	150	240	360

a) Plot these values on a coordinate diagram (possibly on squared paper) and draw the line which passes through the three points.

b) Find the equation of this line.

c) The rule corresponding to these values may be expressed as 'cook your turkey for p minutes plus q minutes per kilogram'.
State the values of p and q.

d) Estimate the cooking time for a 10 kg turkey. Check that you get the same answer from the graph, from the equation, and from the rule.

18 Liz's telephone bill shows a monthly line rental of £10, plus call charges which are 5p per unit.

a) Copy and complete this table of values:

number of units used (x)	0	50	100	150	200
total monthly cost (£y)					

b) Draw a graph to show these values.
c) Find the equation of this line.
d) Use the equation to find the total cost for a month when Liz had used 120 units.

19 Tom's monthly 'phone bill is calculated in a similar fashion to Liz's in question 18.
This graph illustrates the charges used by Tom's telephone provider.

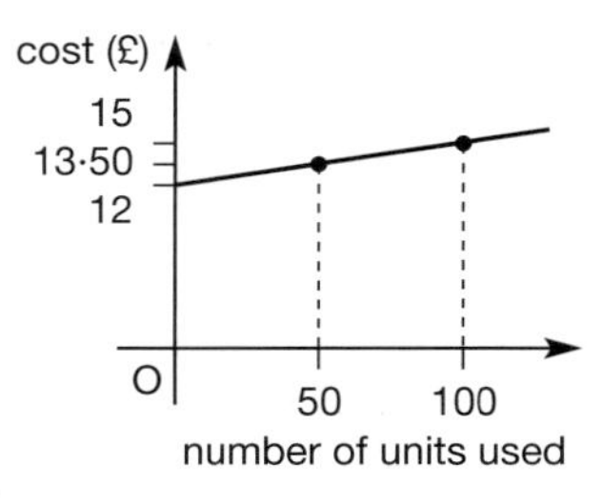

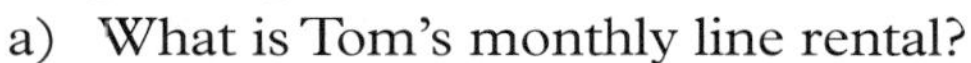
a) What is Tom's monthly line rental?
b) What does Tom pay per 'phone unit?
c) Write down a formula relating the cost (£y per month) to the number (x) of 'phone units used.

20 John took £50 spending money on holiday with him.
The line graph AB shows the relationship between what he spent (£x) and what he had left (£y).

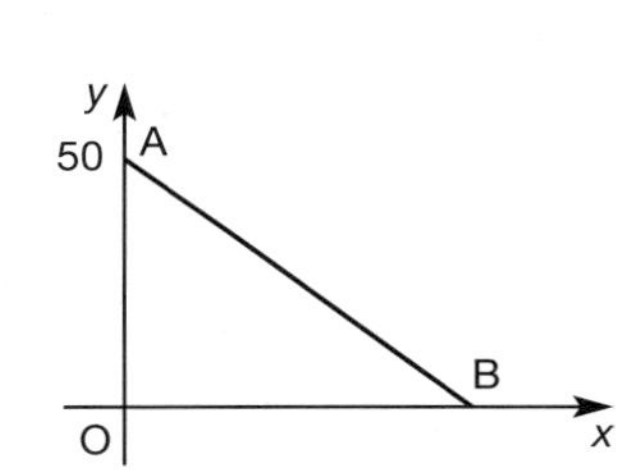

a) Write down the coordinates of B.
b) Find the gradient of the line AB.
c) Find a formula for y in terms of x.

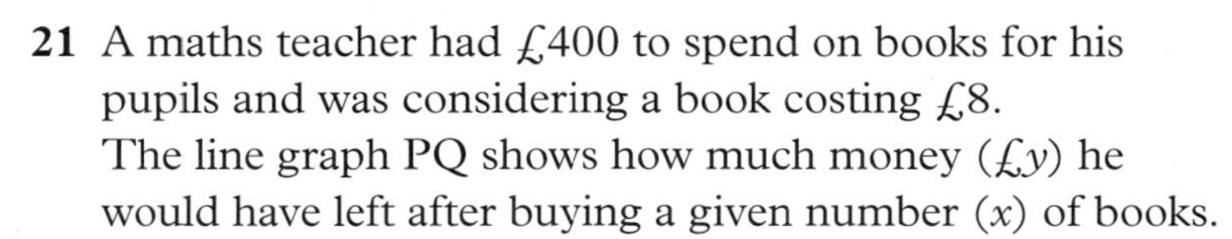
21 A maths teacher had £400 to spend on books for his pupils and was considering a book costing £8.
The line graph PQ shows how much money (£y) he would have left after buying a given number (x) of books.

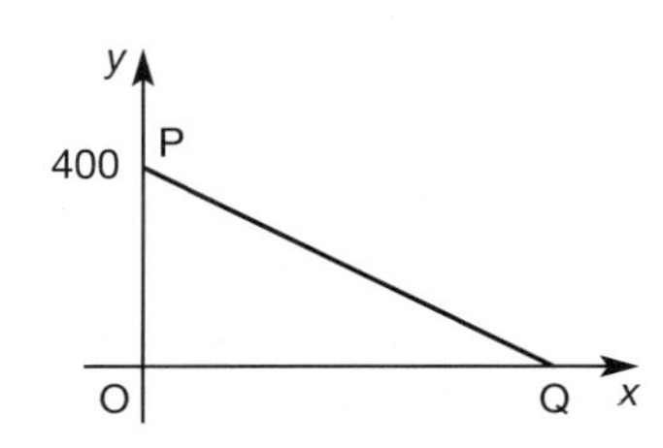

The more books he buys, the less money he has left.

a) Why is a line graph inappropriate for this data?
b) Find the coordinates of Q and hence the gradient of PQ.
c) Find a formula for y in terms of x.

22 a) Plot the points S (0, 7) and T (4, 5).
b) Find the equation of ST.
c) Does the point K (100, –43) lie on the line ST?

23 Plot the triangle whose sides have equations $y = 3x + 1$, $y = x - 1$, and $x + y - 1 = 0$, and hence find the coordinates of the vertices. [Use algebra, not a scale drawing.]

24 Revision of Coordinate Geometry

Exercise 24

1 State the gradient of each of these lines:

a) b) c) d) e)

2 a) Write down the coordinates of the points A, B, C, D, E, F, G, H.

b) Find the gradients of AB, CD, EF, and GH.

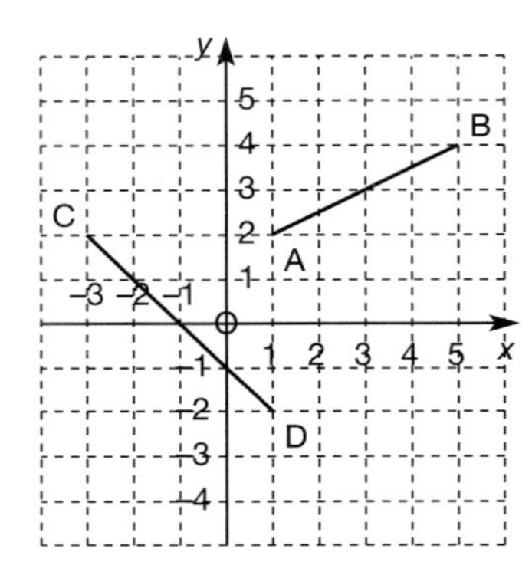

3 Write down the coordinates of P and Q and hence determine whether or not P, O, and Q lie in a straight line.

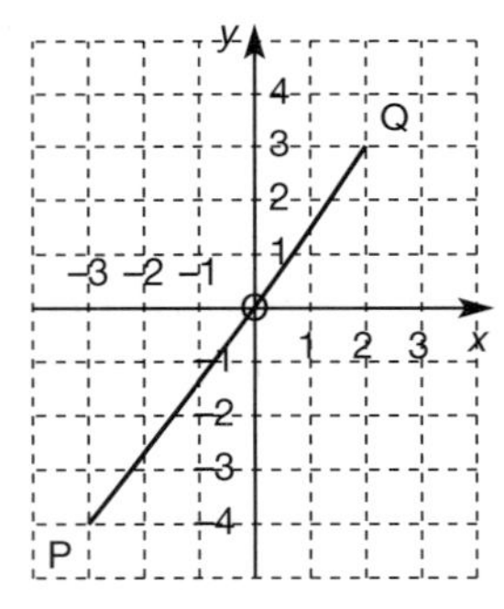

4 A joiner has constructed trusses for the roof of a house using short pieces of wood joined with 'finger comb' joints and flat plate timber connectors.

He lays one of them out on the ground in such a way that the coordinates of A, B and C (as shown in the diagram) are A (–5, 2), B (2, 7) and C (9, 12).

Has the joiner succeeded in making ABC a straight line?

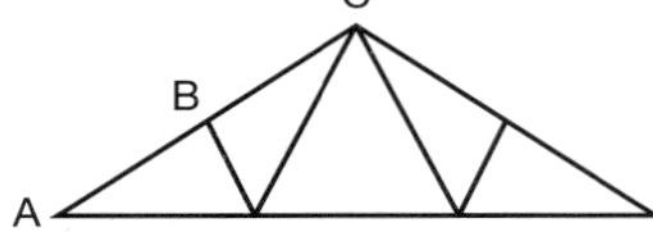

5 a) Find the equation of the line l which passes through the point $(0, -2)$ and is parallel to the line with equation $2x + 5y = 7$.
b) Which of the points A $(5, 2)$, B $(10, -6)$, or C $(-5, 0)$ lie on the line l?

6 a) Plot the points A $(0, 3)$ and B $(2, 7)$.
b) Find the equation of the line AB.
c) Which of these points lie on the line AB: P $(-1, 1)$, Q $(1, 5)$, R $(4, 10)$?

7B P is the point (m, m) and Q $(-n, n)$.
Find the gradient of PQ.

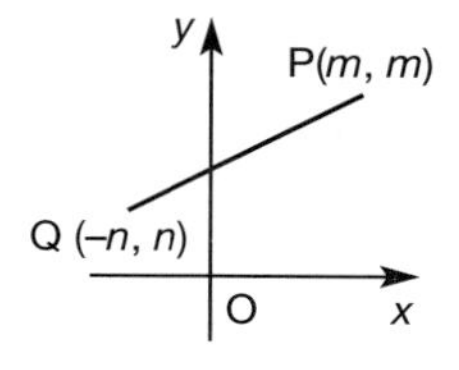

8B R is the point $(m - n, m + n)$ and S is $(-n, n)$.
Find (and simplify) the gradient of RS.

9 A certain school text book costs £6. If I plotted several points to show the cost of 1, 2, 3, ... books, what would the gradient of the line through these points be? [Make up a table of a few values and plot the points if necessary.]

10 Write down the coordinates of A, B, C, and D and hence find the gradients of OA, OB, OC, and OD.

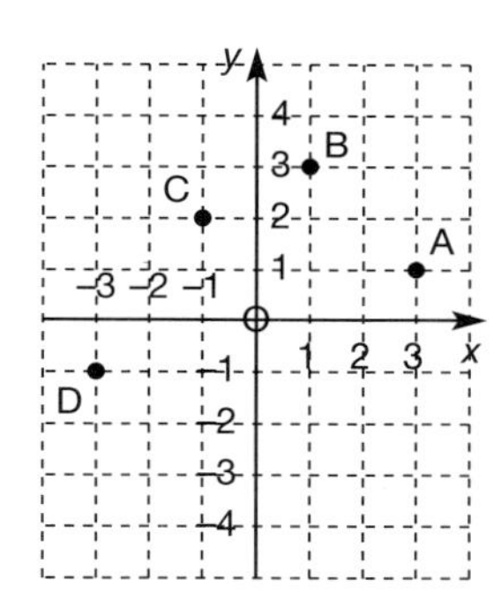

11 The rectangle OABC represents one face of a child's building brick, which is in the shape of a cuboid.
The vertex B is the point $(4, 2)$.
Two other identical bricks are placed on top, as shown.

Find the equation of OD.

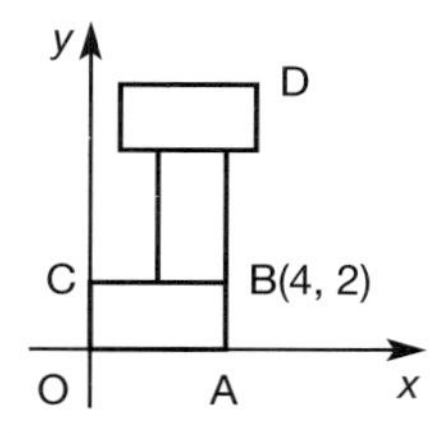

12 Find the coordinates of the point on the line with equation $y = \frac{5}{3}x$ where:

a) $x = 6$ b) $x = -3$ c) $x = 9$.

13 Make two statements about the lines with equations $y = 2x$ and $y = -2x$.

14B P is the point on the line with equation $y = kx$ where $x = 2$.
Q is the point on the line with equation $y = (k + 1)x$ where $x = 3$.
Find the y-coordinates of P and Q and hence the gradient of PQ.

15 Write down the equation of the line with:

a) gradient 2 and intercept 3
b) gradient 1 and intercept –1
c) gradient –2 and intercept 4
d) gradient $\frac{1}{2}$ and intercept –2.

16 Find the gradient and intercept for each line whose equation is given below:

a) $y = 3x - 4$
b) $y = 5 - 2x$
c) $y = \frac{1}{4}x + \frac{1}{3}$
d) $y = 2 - \frac{1}{2}x$
e) $x = 2y - 3$
f) $2x + 3y = 0$
g) $3x - 4y + 12 = 0$
h) $2x + 5y = 7$.

17 Find the equation of the line passing through:

a) (0, 2) and (3, 5)
b) (0, –1) and (2, 3)
c) (0, 7) and (4, 5)
d) (0, –5) and (4, –9)
e) (0, 3) and (–6, 7).

18 Write these equations in the form $ax + by + c = 0$ where a, b and c are whole numbers:

a) $y = 2x + 5$
b) $2y = 4 - 3x$
c) $y = \frac{1}{2}x + 4$
d) $y = 3 - \frac{1}{3}x$.

19 The lines l and m shown in the diagram are equally inclined to the x-axis.

Find the gradient and hence the equation of the line m.

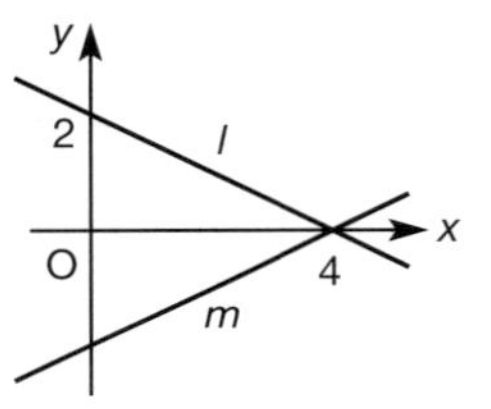

20 The lines p and q shown in the diagram are equally inclined to the y-axis.

Find the equation of the line q.

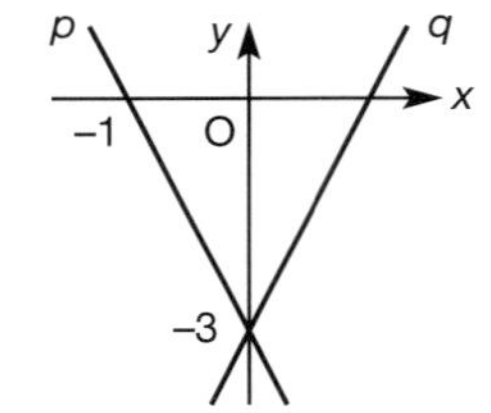

21 ABCD is a rectangle as shown.
C is the point (6, 4) and D (–6, 4).
AC and BD cross at E.

Find a) the coordinates of E
b) the equations of AC and BD.

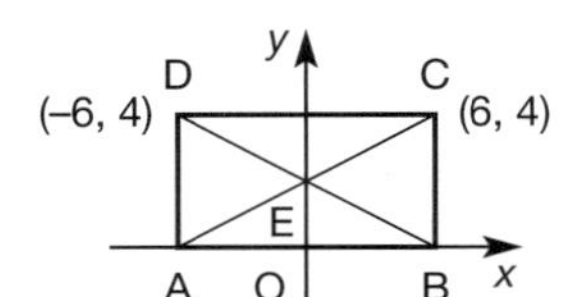

22 ABCD is a parallelogram with

A (–2, –1), B (1, –1) and C (2, 1).

It is part of a parallelogram tiling, as shown.

a) Find the coordinates of D.
b) Find the coordinates of P, Q, R, and S (shown in the diagram).
c) Find the equations of PS and QR.

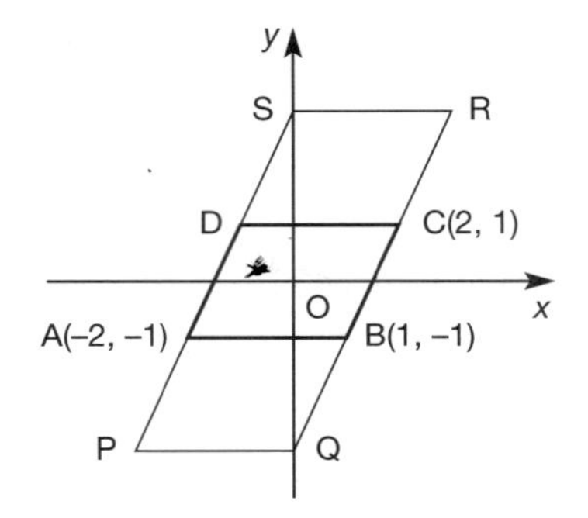

23B The lines with equations $ax + by + c = 0$ and $bx + ay - c = 0$ are parallel.
Show that $a^2 = b^2$.

24B Find in terms of t, the equation of the line passing through the points $(0, t)$ and $(5t, 3t)$.

25B A is the point $(0, -k)$, B $(3k, 0)$ and C $(0, 2k)$.

Find: a) the gradient of AB
b) the equation of the line passing through C parallel to AB
c) the equation of BC.

25 Test on Coordinate Geometry

Allow 45 minutes for this test

1 Find the gradient of each of these lines:

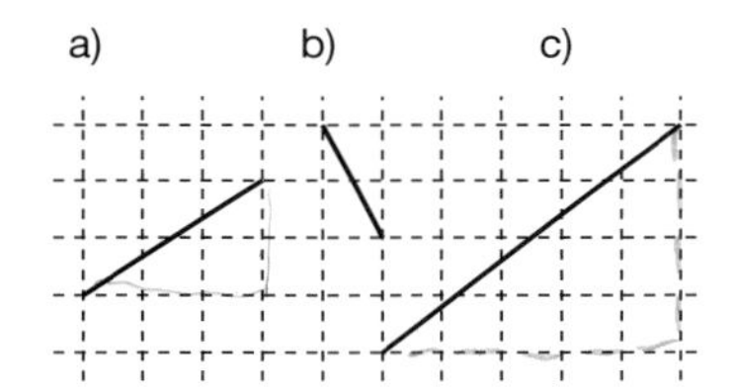

2 a) Write down the coordinates of A, B, C, D, E, and F.
b) Hence find the gradients of AB, CD and EF.

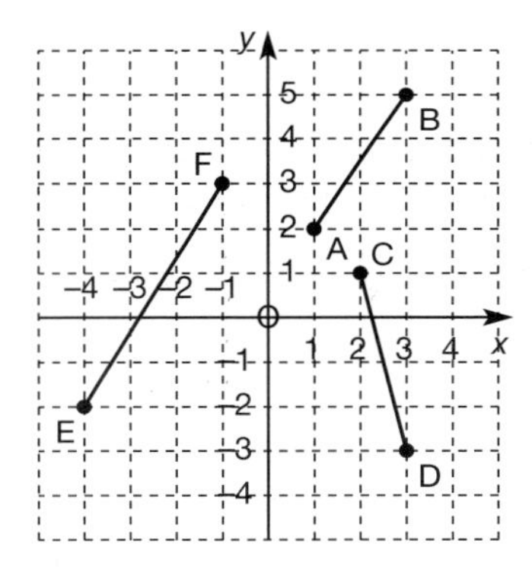

3 Prove that the points K (1, 3), L (6, 5) and M (16, 9) lie in a straight line.

4 Triangle ABC has vertices as shown in the diagram.
a) Write down the equations of AB and BC.
b) Find the equations of OA, OB and OC.

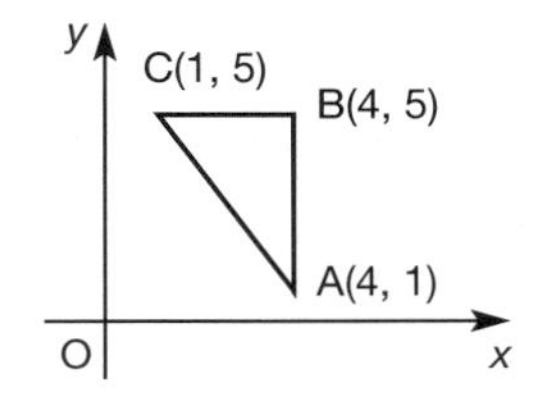

5 ABCD is a parallelogram with
A (−1, 0), B (1, 0) and D (0, 1).
It is part of a tesselation of congruent parallelograms.

Find the coordinates of C, P and Q, as shown in the diagram, and hence the equations of OP and OQ.

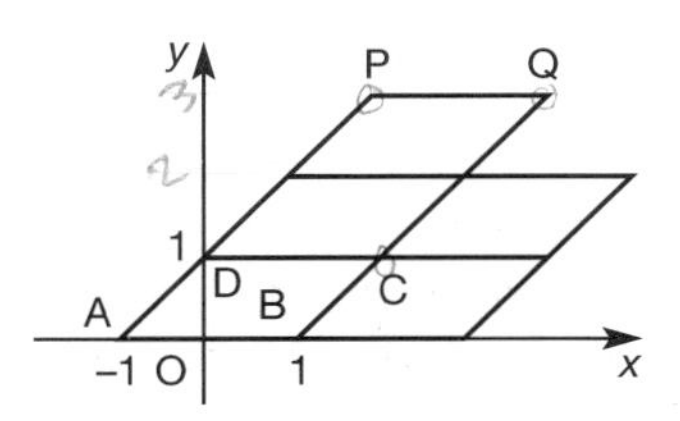

6 Write down the equation of the line with gradient 4 passing through the point $(0, -3)$.

7 State the gradient of the line with equation $3y = 6x + 5$ and the coordinates of the point where it crosses the y-axis.

8 Triangle PQR is isosceles with PQ = PR.
The equation of PR is $y = -3x + 6$.
Find the equation of PQ.

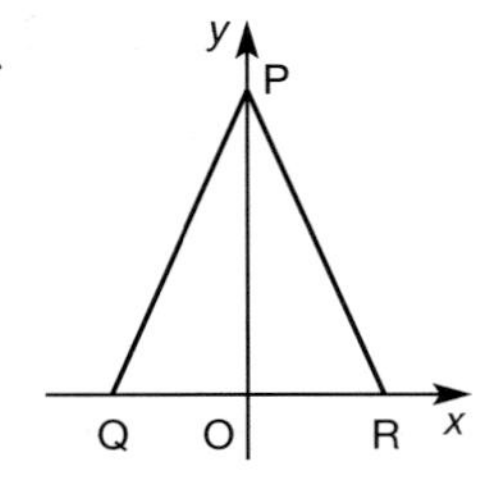

9 EFGH is a parallelogram with vertices E $(-3, -2)$, F $(4, -2)$. and G $(7, 3)$.
Find the coordinates of H and the equation of EH.

ALGEBRA

26 The Distributive Law

Reminder

Suppose that in Lapland, apples cost 11p each and oranges 14p each.
In their Christmas stockings, Santa gave each of three triplets an apple and an orange.
We can work out what this cost Santa in two different ways.

a) each triplet got 1 apple and 1 orange; this cost Santa 25p (11p + 14p) for each triplet; so altogether it cost Santa $3 \times 25\text{p} = 75\text{p}$
b) 3 apples and 3 oranges cost $(3 \times 11\text{p}) + (3 \times 14\text{p})$ i.e. $33\text{p} + 42\text{p} = 75\text{p}$

This shows that $3\,(11 + 14) = (3 \times 11) + (3 \times 14)$.
This is an example of the distributive law.

Now suppose apples cost b pence each and oranges c pence each and that there are a children.
We can still work out what this will cost Santa in the same two ways.

a) each triplet got 1 apple and 1 orange; this cost Santa $(b + c)$ pence for each triplet; so altogether it cost Santa $a \times (b + c)$ pence
b) a apples cost $a \times b$ pence and a oranges cost $a \times c$ pence so altogether $(a \times b) + (a \times c)$ pence

This shows that $a \times (b + c) = (a \times b) + (a \times c)$.
This is known as the distributive law.
The law tells you that when you multiply a binomial (a bracket with two terms in it) by a number, you can remove the bracket and multiply both terms by that number.

Example Expand these brackets

a) $2(x + 3)$ b) $3(x - 2y)$ c) $3x(x + a)$.

Solution

a) $2(x + 3) = 2(x) + 2(3) = 2x + 6$
b) $3(x - 2y) = 3(x) - 3(2y) = 3x - 6y$
c) $3x(x + a) = (3x)(x) + (3x)(a) = 3x^2 + 3ax$

(You are not normally expected to show all the working included here.)

Exercise 26

1 Expand each bracket:

a) $2(x-4)$ b) $3(p+5)$ c) $4(a-3)$ d) $5(t-2)$
e) $4(6-a)$ f) $5(2+t)$ g) $3(4-x)$ h) $5(k+2)$
i) $7(3+z)$ j) $8(m-1)$ k) $6(2-z)$ l) $3(4+p)$
m) $2(3-w)$ n) $3(h+4)$ o) $4(t+3)$ p) $8(2-b)$.

2 Expand each bracket:

a) $3(2-t)$ b) $5(k+1)$ c) $6(3-z)$ d) $7(w+3)$
e) $4(2-3y)$ f) $5(2k+1)$ g) $3(2-3t)$ h) $6(2t+1)$
i) $2(x+3y)$ j) $3(x-2y)$ k) $4(p+2q)$ l) $5(7-3p)$
m) $2(3x+4y)$ n) $3(2x-3y)$ o) $4(2p+3q)$ p) $5(3r-4s)$.

3 Expand each bracket:

a) $x(x+1)$ b) $y(y-1)$ c) $t(t+2)$ d) $p(4-p)$
e) $x(x+2)$ f) $t(2t-1)$ g) $w(4-w)$ h) $z(3-2z)$
i) $a(x+y)$ j) $b(2t-3)$ k) $c(1-4k)$ l) $d(2x+d)$
m) $2x(x+1)$ n) $3y(1-2y)$ o) $4z(2z+3a)$ p) $5w(2x-3w)$.

Reminder

A negative sign outside a bracket changes all the signs inside the bracket.

Example Expand these brackets:

a) $-(p+q)$ b) $-(2p-3q)$
c) $-2(x+y)$ d) $-3(x-2y)$.

Solution

a) $-(p+q) = (-1)(p) + (-1)(q) = -p - q$
b) $-(2p-3q) = (-1)(2p) + (-1)(-3q)$ or $(-1)(2p) - (-1)(3q) = -2p + 3q$
c) $-2(x+y) = (-2)(x) + (-2)(y) = -2x - 2y$
d) $-3(x-2y) = (-3)(x) + (-3)(-2y)$ or $(-3)(x) - (-3)(2y) = -3x + 6y$

4 Expand each bracket:

a) $-(x+y)$ b) $-(p-q)$ c) $-(1-t)$ d) $-(3+r)$
e) $-(a+b)$ f) $-(b-a)$ g) $-(a-b)$ h) $-(-x)$
i) $-(6-r)$ j) $-(c-3)$ k) $-(e+f)$ l) $-(g-2h)$
m) $-(x+2y)$ n) $-(2x-y)$ o) $-(3p+q)$ p) $-(2p-3)$.

5 Expand each bracket:

a) $-2(x+1)$ b) $-3(3-x)$ c) $-2(p-4)$ d) $-5(t+3)$
e) $-4(x+3)$ f) $-3(x-2)$ g) $-4(x-5)$ h) $-5(4-x)$
i) $-3(4a+5b)$ j) $-4(6p-2q)$ k) $-2(2z+3y)$ l) $-3(3m-2n)$
m) $-3a(2a-b)$ n) $-4x(2x-y)$ o) $-5t(2t+4u)$ p) $-v(3u-4v)$.

Reminder

Recall gathering like terms: for example $3x + 4x = 7x$, or $8x - x = 7x$ (because $x = 1x$).
Similarly, $3x + 2y + 4z + x - y - 3z = 3x + x + 2y - y + 4z - 3z = 4x + y + z$.
Sometimes we have to expand brackets prior to collecting like terms.

Example Simplify:

a) $2(x + 3) + 1$

b) $3(4x + 1) - 7x$

c) $1 + 2(x + 3)$

d) $5x - (2x - 1)$.

Solution

a) $2(x + 3) + 1$
$= 2x + 6 + 1$
$= 2x + 7$

b) $3(4x + 1) - 7x$
$= 12x + 3 - 7x$
$= 5x + 3$

c) $1 + 2(x + 3)$
$= 1 + 2x + 6$
$= 2x + 7$

d) $5x - (2x - 1)$
$= 5x - 2x + 1$
$= 3x + 1$

6 Simplify:

a) $2(x - 3) + 1$
b) $3(p - 2) + p$
c) $4(q + 3) - 3q$
d) $5(r + 2) + r$
e) $x(x - 1) + 2x$
f) $y(2y + 3) - y$
g) $z(3 - z) + z^2$
h) $k(2k - 1) + k^2$
i) $1 + 3(x - 2)$
j) $2 + 4(x + 5)$
k) $x + 3(x + 2)$
l) $y + 2(3 - y)$
m) $5 - 3(x - 2)$
n) $7 - 2(t + 1)$
o) $8m - m(2 + m)$
p) $9q - 3(q - 2)$.

7 Expand and simplify:

a) $3(x + 1) - 2(x + 2)$
b) $5(y + 3) - 3(y - 4)$
c) $9(z + 2) - 4(2z + 1)$
d) $11(p - 4) - 5(2p - 8)$
e) $a(a - b) + b(a + b)$
f) $a(a + b) - b(a - b)$
g) $c(d + 4) - d(c - 2)$
h) $f(2g + 1) - g(2f - 1)$
i) $x(x - 2) + 2(x - 1)$
j) $y(2y + 3) - 2(y - 2)$
k) $2z(z - 2) - 3(z + 3)$
l) $2x(x + y) - 3y(x - y)$.

8 Simplify:

a) $3(a + b) + 2(a - b) + (2a + b)$
b) $2(x + y) - 3(x - y) + 4(2x + 3y)$
c) $x(x + 1) - 2(x + 3) - x(x - 2)$
d) $a(b + c) - b(a - c) + c(a + b)$
e) $p(q + 2r) + q(r - p) + r(2p - 3q)$
f) $x(2x + 1) + 2(x + 3) + 3x(x - 1)$
g) $3(p - q) + p(2 - q) + 2q(3 - p)$
h) $r(s + 1) - s(r + 2) + 2(r + s)$.

Reminder

The distributive law still applies when there are more than two terms in the bracket.
Each term in the bracket is multiplied by the multiplier of the bracket,
e.g. $2(x + y - 2) = 2x + 2y - 4$.

Reminder continued

Example Simplify $1 + 3(x + 2y - 1) - 2(3x - y + 4)$.

Solution
$$1 + 3(x + 2y - 1) - 2(3x - y + 4)$$
$$= 1 + 3x + 6y - 3 - 6x + 2y - 8 = 8y - 3x - 10$$

9 Expand each bracket and collect any like terms:

a) $2(2x + 3y + z)$
b) $3(x - 2y + 3z)$
c) $7(2p + 3q - r)$
d) $9(2f + 3g + h)$
e) $3 + 4(a - 2b + 3)$
f) $5 + 6(2p + 3q - 4)$
g) $7 + 5(3x - 4y + 2)$
h) $6 + 7(2 + 3t - 4s)$
i) $7 - 3(x + 2y + 1)$
j) $8 - 2(3x - 4y + 2)$
k) $3 - 4(2a + 3b - 1)$
l) $11 - 9(3r - 4s - 2)$.

10 Simplify:

a) $2(3x + 4y - z + 1) + 3(2x - y + 3z - 2)$
b) $3(2p + 3q - 4r + 2) - 5(p - 2q + 3r - 1)$
c) $5(2a + 3b - 4c + 2) + 3(3a - 2b + c + 3)$
d) $7(2x + 3y - z + 5) - 5(3x - 2y + z - 4)$.

11 This is a plan of a garden.
All the measurements are in metres.

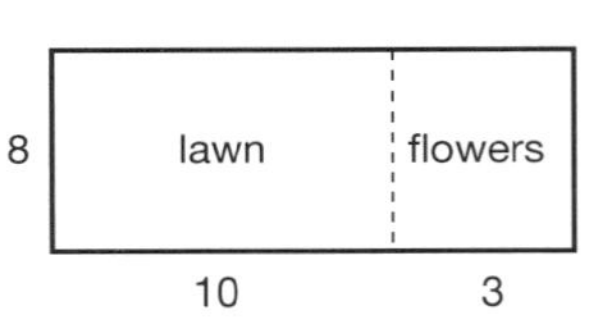

a) Calculate the area of the lawn.
b) Calculate the area of the flower-bed.
c) Calculate the total area of the garden.

d) Calculate the total area of the garden again but by a different method.
e) Write down the appropriate distributive law statement.

12 This is a plan of a garden.
All the measurements are in metres.

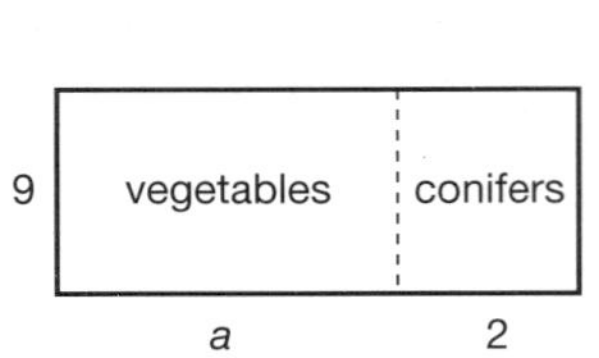

a) Calculate the area of the vegetable plot.
b) Calculate the area for the conifers.
c) Calculate the total area of the garden.
d) Calculate the total area of the garden again but by a different method.
e) Write down the appropriate distributive law statement.

13 This is a diagram of a window with a top opening hopper.
All the measurements are in feet.

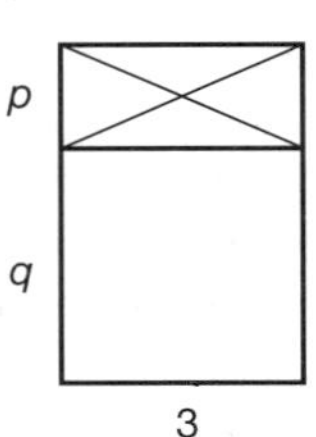

a) Calculate the area of the hopper.
b) Calculate the area of the fixed lower pane.
c) Calculate the total area of the window.
d) Calculate the total area of the window again but by a different method.
e) Write down the appropriate distributive law statement.

14 This is a plan of a caravan.
All the measurements are in feet.
a) Calculate the area of the bedroom.
b) Calculate the area for the kitchen and toilet.
c) Calculate the area of the living room.
d) Calculate the total area of the caravan.
e) Calculate the total area of the caravan again but by a different method.
f) Write down the appropriate distributive law statement.

15 Solve these equations:

a) $3(x-2)=x-4$
b) $2(x+5)=3(x+1)+4$
c) $1+4(x-2)=2(x+4)-5$
d) $7(x+3)+11=2(15+3x)$
e) $5(2x-3)=7(1+3x)$
f) $3(3x+4)=25(x-4)$
g) $5-2(x-3)=11+3(2-x)$
h) $3-4(2x-1)=6(1-x)$.

16 Although it is usual to write $3(x+y)$, $(x+y)3$ sometimes occurs, and means the same thing. Expand these brackets, and simplify where appropriate:

a) $(p+3)r$
b) $(x-1)x$
c) $(q-5)s$
d) $(x+y)x$
e) $(t+2)t$
f) $(t^2+1)t$
g) $(x+5)x$
h) $(y-z)y$
i) $-(u+3)u$
j) $-(v-2)v$
k) $-(2r+1)s$
l) $-(3q-2)q$
m) $2a+(a+1)a$
n) $7b-(b-2)b$
o) $5f+(f-3)f$
p) $8g-(2-g)g$.

27 The Product of Two Binomials

Reminder

A **binomial** is the mathematical name for a bracket with two terms.

In the rule $p(c+d)=pc+pd$, p can be replaced by $(a+b)$ to give

$$(a+b)(c+d)=(a+b)c+(a+b)d$$

The right hand side can be expanded by the distributive law to give

$$(a+b)(c+d)=ac+bc+ad+bd$$

This expansion is used with great regularity in Mathematics. It is essential to remember to obtain these four terms.

$(\mathbf{a}+b)(\mathbf{c}+d)$ $\mathbf{ac}$ is the product of the **first** term in each bracket
$(a+\mathbf{b})(\mathbf{c}+d)$ $\mathbf{bc}$ is the product of the two **inner**most terms
$(\mathbf{a}+b)(c+\mathbf{d})$ $\mathbf{ad}$ is the product of the two **outer**most terms
$(a+\mathbf{b})(c+\mathbf{d})$ $\mathbf{bd}$ is the product of the **last** term in each bracket

This leads to the rule **first, inner, outer, last.**
The acronym **foil** may also help you to remember the rule.

Reminder continued

Example Expand these brackets:

a) $(p + q)(r + s)$ b) $(x - 1)(x + 3)$

c) $(t - 5)^2$ d) $(x + 2)(x^2 + 2x - 3)$.

Solution

a) (first) pr, (inner) qr, (outer) ps, (last) qs hence
$(p + q)(r + s) = pr + qr + ps + qs$

b) to cope with the negative sign, you can think of the first bracket as being
$[x + (-1)]$, so $(x - 1)(x + 3) = x \times x + (-1) \times x + x \times 3 + (-1) \times 3$
i.e. $(x - 1)(x + 3) = x^2 - x + 3x - 3$

The inner and outer products are like terms and can be gathered into a single term, hence $(x - 1)(x + 3) = x^2 + 2x - 3$

c) Remember that x^2 means $x \times x$, so $(t - 5)^2$ means $(t - 5) \times (t - 5)$ also $(-5) \times (-5) = +25$

Thus $(t - 5)^2 = (t - 5)(t - 5) = t \times t - 5 \times t - t \times 5 + (-5) \times (-5)$
$= t^2 - 10t + 25$

d) To multiply this binomial and this trinomial, the 'first, inner, outer, last' rule is not applicable, so we revert to first principles and multiply the second bracket first by x, then by 2, and add them together, as follows:

$$(x + 2)(x^2 + 2x - 3) = x \times (x^2 + 2x - 3) + 2 \times (x^2 + 2x - 3)$$
$$= x^3 + 2x^2 - 3x + 2x^2 + 4x - 6$$
$$= x^3 + 4x^2 + x - 6$$

Exercise 27

1 Expand the brackets:

a) $(x + p)(y + q)$ b) $(x - p)(y + q)$ c) $(x + p)(y - q)$
d) $(x - p)(y - q)$ e) $(a + x)(b + y)$ f) $(a - x)(b + y)$
g) $(a + x)(b - y)$ h) $(a - x)(b - y)$ i) $(x + 1)(a + y)$
j) $(p - 2)(q + r)$ k) $(3 + t)(r - s)$ l) $(k - n)(q - 4)$.

2 Expand the brackets and simplify:

a) $(x + 1)(x + 2)$ b) $(t + 3)(t - 4)$ c) $(y - 2)(y + 3)$
d) $(z - 2)(z - 2)$ e) $(p + 3)(p - 5)$ f) $(q - 2)(q + 4)$
g) $(r + 2)(r + 7)$ h) $(s - 4)(s - 5)$ i) $(x - 3)(x - 4)$
j) $(y + 2)(y + 3)$ k) $(z - 1)(z + 10)$ l) $(w + 3)(w + 7)$.

3 Expand the brackets and simplify:

a) $(2x + 1)(3x + 4)$ b) $(3x - 1)(2x + 3)$ c) $(4y + 3)(5y - 1)$
d) $(2z - 3)(2z - 5)$ e) $(3p + 4)(2p - 3)$ f) $(4p - 3)(5p + 2)$
g) $(6q - 7)(q - 3)$ h) $(3r + 2)(2r + 3)$ i) $(2a - 5)(3a - 4)$
j) $(3b + 4)(2b - 5)$ k) $(5c - 3)(3c - 2)$ l) $(6d + 5)(5d + 2)$.

4 Expand the brackets and simplify:

a) $(x + y)(x + 2y)$ b) $(x + 2y)(x - 3y)$ c) $(y + 4z)(2y - z)$
d) $(2p + q)(2p - q)$ e) $(2s + 3t)(3s - 2t)$ f) $(4u + 5v)(2u + 3v)$
g) $(3y - 4z)(4y - 5z)$ h) $(7a - 3b)(3a + 2b)$ i) $(4t + 5u)(5t - 3u)$
j) $(6b - 7c)(b - 2c)$ k) $(5f - 3g)(4g + 5f)$ l) $(7p + 5q)(6p + 7q)$.

5 Expand the brackets and simplify:

a) $(x + 1)^2$ b) $(t + 2)^2$ c) $(z - 1)^2$
d) $(y - 3)^2$ e) $(p - 2)^2$ f) $(q + 3)^2$
g) $(r - s)^2$ h) $(a - b)^2$ i) $(2x + 1)^2$
j) $(1 - 3x)^2$ k) $(2 - 3y)^2$ l) $(2x + 3y)^2$.

6 This is the plan of a hotel function suite.
All the measurements are in metres.

a) Calculate the area of the dance floor.
b) Calculate the area for the round tables.
c) Calculate the area for the square tables.
d) Calculate the area for the band.
e) Calculate the total area of the suite.
f) Calculate the area of the suite again but by a different method.
g) Hence write down an appropriate 'first, inner, outer, last' equation.

	3	2
4	dance floor	round tables
2	square tables	band

7 This is the plan of a garden.
All the measurements are in metres.

a) Calculate the area of the lawn.
b) Calculate the area for the vegetables.
c) Calculate the area for the flowers.
d) Calculate the area for the hut.
e) Calculate the total area of the garden.
f) Check your answer to (e) by writing down an appropriate 'first, inner, outer, last' equation.

	r	s
p	lawn	flowers
q	vegetables	hut

8 Expand the brackets and collect the like terms:

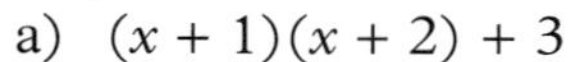

a) $(x + 1)(x + 2) + 3$ b) $(x - 1)(x + 2) - 4x$
c) $(p - 3)(p + 4) + 2p^2$ d) $(q + 1)(q - 1) + 1$
e) $(k + 2)(k - 3) + 11$ f) $(m - n)(m - 2n) - mn$
g) $(x + y)(x - 2y) + 3xy$ h) $(2x + 3y)(3x - 4y) - 6x^2$
i) $(1 - 3x)(2 + x) + x - 1$ j) $1 - (x + 1)(x - 2)$
k) $3 + (x + 4)(x + 5)$ l) $5 - (2x + 1)(x - 3)$
m) $2x^2 - (x - 2)(x + 5)$ n) $1 - (2y - 1)^2$
o) $k^2 - (k - 2)(k + 2)$ p) $(3p + 2q)(2p - 3q) - 6pq$.

9 This diagram shows the plan of a lawn L metres long and B metres broad.
A rectangular pond measuring l metres by b metres is to be constructed centrally in the lawn.

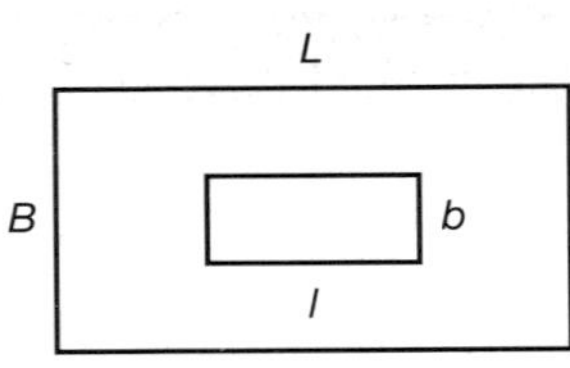

a) Show that after installing the pond, the area, A square metres, of the remaining lawn is given by $A = LB - lb$.

b) Express A in terms of x when

(i) $L = x + 5$ $B = x - 2$ $l = x$ $b = x - 4$

(ii) $L = x + 7$ $B = x + 1$ $l = x + 1$ $b = x - 1$

(iii) $L = x + 6$ $B = x + 2$ $l = x + 3$ $b = x - 1$

(iv) $L = x + 4$ $B = x - 1$ $l = x - 2$ $b = x - 3$.

10B Expand:

a) $(x^2 + 3)(x^2 + 4)$
b) $(y^2 + 2)(y^2 - 1)$
c) $(z^2 - 3)(z^2 + 2)$
d) $(w^2 - 1)(w^2 - 2)$
e) $(2x^2 + 3y^2)(3x^2 - y^2)$
f) $(3p^2 + 2q^2)(2p^2 + 3q^2)$
g) $(3a - b^3)(4a + b^3)$
h) $(2x + y^4)(x - y^4)$
i) $(a^2 - 2b^2)(a^2 + 2b^2)$
j) $(3p^2 + q^2)(3p^2 - q^2)$
k) $(x - 1)(x^3 - 1)$
l) $(x - a)(x^3 - a^3)$.

28 Common Factors

Reminder

In Exercise 26 you practised expanding brackets, e.g. $2(x + y) = 2x + 2y$.
Factorising is simply the reverse process of this, i.e. given $2x + 2y$, you can 'take out the common factor' of 2 and revert to $2(x + y)$.
Each time you take out a common factor, it is wise to expand the brackets mentally to check that it brings you back to the original expression.

To factorise an expression means to reduce it to a single term.

$2x + 2y$ is not factorised because it has two terms, $2x$ and $2y$
$2(x + y)$ is factorised because it has one term with two factors, 2 and $(x + y)$

Example Factorise:

a) $ax + ay$
b) $4x + 6y$
c) $x^2 + xy$
d) $2x + 4y - 6z$
e) $2x - 4x^2$.

Solution a) ax has four factors: $1, a, x, ax$ and ay has four factors $1, a, y, ay$
1 and a are common to both lists of factors, but 1 is of no help.
thus a is the common factor
thus $ax + ay = a(x + y)$ (Mentally expand $a(x + y)$ as a check.)

Reminder continued

b) $4x$ has factors 1, 2, 4, x, $2x$, $4x$ and
$6y$ has factors 1, 2, 3, 6, y, $2y$, $3y$, $6y$
so the common factor is 2
(Deduce this by inspection from now on.)
i.e. $4x + 6y = 2(2x + 3y)$

c) the common factor is x hence $x^2 + xy = x(x + y)$

d) the common factor is 2 hence $2x + 4y - 6z = 2(x + 2y - 3z)$

e) the highest common factor of $2x$ and $4x^2$ is $2x$
($2x \times 1 = 2x \quad 2x \times 2x = 4x^2$)
hence $2x - 4x^2 = 2x(1 - 2x)$

Note that if you only used 2 or x individually as a common factor, the answers would be $2(x - 2x^2)$ and $x(2 - 4x)$, which are not fully factorised.

Exercise 28

1 Factorise:

a) $2x + 2y$ b) $3a + 3b$ c) $3a + 6b$ d) $3a - 6b$
e) $2p + 4q$ f) $5p - 10q$ g) $4r - 8$ h) $9s - 18$
i) $6t + 3$ j) $8u - 4$ k) $9v + 3$ l) $7w + 14$
m) $3k - 12$ n) $4 - 16m$ o) $7 + 21n$ p) $32t - 8$.

2 Factorise:

a) $6x + 9$ b) $4x - 6$ c) $8x + 10$ d) $12x - 9$
e) $4 + 10y$ f) $6 - 9y$ g) $9 + 15y$ h) $10 - 15y$
i) $7x - 7y$ j) $8p + 2q$ k) $3k - 6m$ l) $25t + 5y$
m) $25p - 10q$ n) $11r + 22s$ o) $15t + 35u$ p) $30a - 40b$.

3 Factorise:

a) $x^2 + x$ b) $y^2 - zy$ c) $ax + ay$ d) $bp - pq$
e) $t^2 + 2t$ f) $t^3 + 2t^2$ g) $2y + y^2$ h) $3m + m^2$
i) $4n - n^2$ j) $4r - 2r^2$ k) $6r^2 + 3r$ l) $2t + 2t^2$
m) $2x^2 - 10x$ n) $2ar^2 + 2ar$ o) $3x^2 + 6x + 9$ p) $x^3 + 3x^2 + 4x$.

4B Factorise:

a) $\pi r - \pi s$ b) $2\pi r - 2\pi s$ c) $\pi r^2 + \pi s^2$
d) $2\pi rh + 2\pi r^2$ e) $3(x + 1) + 4(x + 1)$ f) $a(x + 1) - b(x + 1)$
g) $(x + 2)p + (x + 2)q$ h) $(a + b)x + (a + b)y$ i) $x^2 (x - 3) + (x - 3)$
j) $t^2 (t - 1) + 2(t - 1)$ k) $(u + 2)u - 3(u + 2)$ l) $(v^2 + 1)v - (v^2 + 1)$
m) $(x + 1)x - (x + 1)$ n) $(y - 2)y + 2(y - 2)$ o) $(z + 3)t + (z + 3)u$
p) $(a + 2b)c - (a + 2b)c^2$.

5B Factorise:

a) $3x + 6y - 9z$
b) $3k - 6l + 3m$
c) $5p + 10q - 15$
d) $6a - 9b - 12$
e) $x^3 + x^2 + x$
f) $k^4 + 2k^2 + 3k$
g) $2p^3 + 3p^2 + 5p$
h) $4q^4 + 3q^3 + 2q^2$
i) $2a^4 - 4a^3 + 6a^2$
j) $3b^4 + 6b^3 - 3b^2$
k) $5t^3 + 10t^2 + 15t$
l) $12t^2 + 18t^3 + 6t^4$.

29 Factorising the Difference of Two Squares

Reminder

When we apply our 'first, inner, outer, last' rule to expand $(x - y)(x + y)$, we obtain $x^2 - yx + xy - y^2$. This simplifies to $x^2 - y^2$, which is called a difference of two squares.
Reversing this process, we can factorise $x^2 - y^2$ to obtain $(x - y)(x + y)$.
Any difference of two squares can be factorised similarly,

e.g. $a^2 - b^2 = (a - b)(a + b)$
$p^2 - q^2 = (p - q)(p + q)$.

We can check that these answers are correct by (mentally) expanding the brackets again.

Note
- $(x - y)(x + y)$ and $(x + y)(x - y)$ are equally acceptable
- $x^2 + y^2$ is the sum of two squares. It does not factorise.

Example Factorise:

a) $p^2 - 4$
b) $9x^2 - 16y^2$
c) $2t^2 - 2u^2$.

Solution

a) $p^2 - 4 = p^2 - 2^2 = (p - 2)(p + 2)$
b) $9x^2 - 16y^2 = (3x)^2 - (4y)^2 = (3x - 4y)(3x + 4y)$
c) $2t^2 - 2u^2$ is not a difference of two squares, but it does have a common factor of 2
i.e. $2t^2 - 2u^2 = 2(t^2 - u^2)$
now $(t^2 - u^2)$ is a difference of two squares and factorises to give $(t - u)(t + u)$
hence $2t^2 - 2u^2 = 2(t^2 - u^2) = 2(t - u)(t + u)$

Exercise 29

1 Factorise:

a) $f^2 - g^2$
b) $y^2 - z^2$
c) $m^2 - n^2$
d) $u^2 - v^2$
e) $a^2 - 4^2$
f) $b^2 - 3^2$
g) $t^2 - 5^2$
h) $u^2 - 6^2$
i) $p^2 - 4$
j) $q^2 - 9$
k) $r^2 - 16$
l) $s^2 - 25$
m) $36 - t^2$
n) $49 - u^2$
o) $16 - v^2$
p) $9 - w^2$.

2B Factorise:

a) $1 - 4b^2$ b) $36 - 25a^2$ c) $9c^2 - 1$ d) $25d^2 - 4$
e) $81 - 16s^2$ f) $64 - 9t^2$ g) $121u^2 - 1$ h) $144v^2 - 25$
i) $4p^2 - 9q^2$ j) $16a^2 - 9b^2$ k) $25x^2 - 16y^2$ l) $4t^2 - 25u^2$
m) $100u^2 - 49v^2$ n) $81r^2 - 64s^2$ o) $36p^2 - 49q^2$ p) $25a^2 - 64b^2$.

3B Factorise fully (i.e. take out the common factor first to reveal a difference of two squares)
a) $2p^2 - 2q^2$ b) $3x^2 - 3y^2$ c) $5u^2 - 5v^2$ d) $8a^2 - 8b^2$
e) $5r^2 - 20$ f) $6s^2 - 54$ g) $7t^2 - 28$ h) $8u^2 - 200$
i) $5v^2 - 20w^2$ j) $3a^2 - 27b^2$ k) $64c^2 - 4d^2$ l) $24f^2 - 6g^2$
m) $12a^2 - 27b^2c^2$ n) $50p^2 - 72q^2r^2$ o) $80m^2t - 45n^2t$ p) $2 - 98x^2$.

4 Calculate by factorising: (e.g. $14^2 - 11^2 = (14 - 11)(14 + 11) = (3)(25) = 75$)

a) $11^2 - 9^2$ b) $12^2 - 8^2$ c) $13^2 - 7^2$ d) $11^2 - 7^2$
e) $13^2 - 9^2$ f) $14^2 - 4^2$ g) $15^2 - 5^2$ h) $17^2 - 13^2$
i) $18^2 - 12^2$ j) $19^2 - 11^2$ k) $17^2 - 3^2$ l) $999^2 - 1^2$
m) $15^2 - 12^2$ n) $13^2 - 12^2$ o) $61^2 - 60^2$ p) $25^2 - 24^2$.

5 Calculate the length of the missing side in each diagram, by writing down the appropriate statement of the Theorem of Pythagoras and applying the factorisation of a difference of two squares, for example:

169 119 x

$x^2 = 169^2 - 119^2 = (169 - 119)(169 + 119) = (50)(288) = 100 \times 144$
$\Rightarrow x = 10 \times 12 = 120$

a)
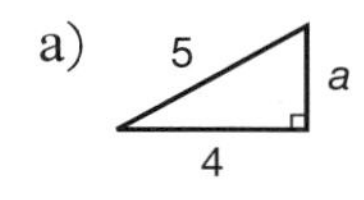

b)
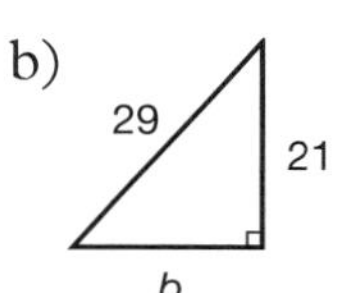

c)
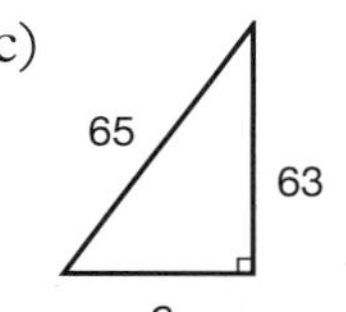

d)
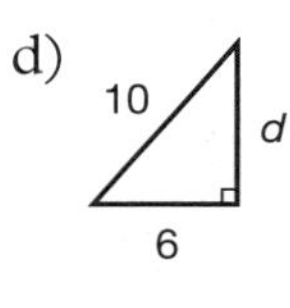

6B Calculate in terms of π, the area of each washer, by factorising a difference of squares.

a)
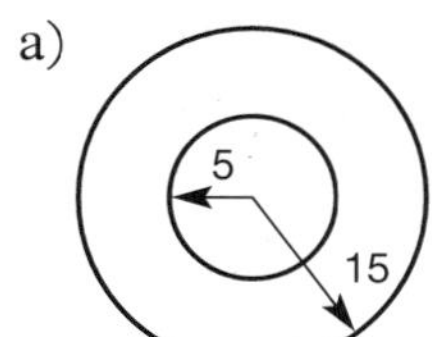

b)
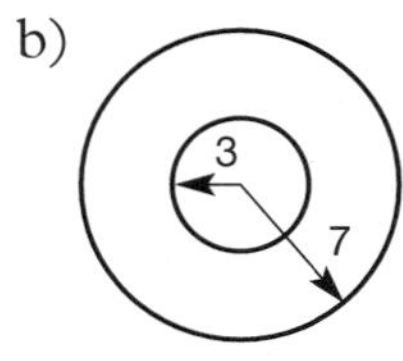

c)
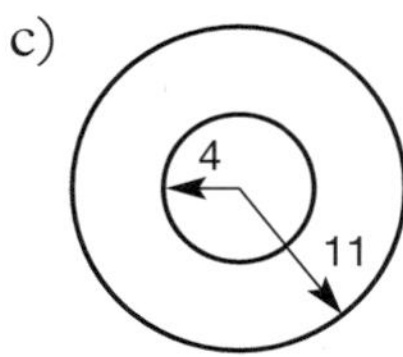

d)
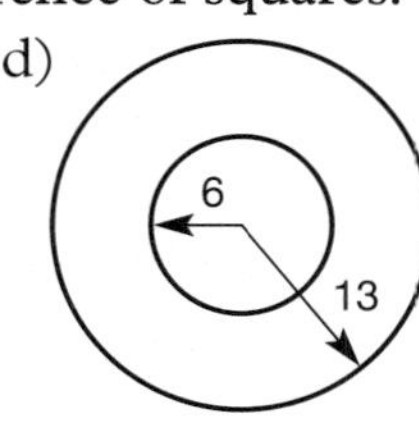

Reminder

(Some harder examples.)

Example Factorise:

a) $(x + 1)^2 - (x - 2)^2$ b) $p^2 - (m + n)^2$
c) $y^4 - 1$ d) $2z^4 - 2$.

Reminder continued

Solution (introduce a different style of bracket for clarity; watch the signs)

a) $(x+1)^2-(x-2)^2=[(x+1)-(x-2)][(x+1)+(x-2)]=3(2x-1)$
b) $p^2-(m+n)^2=[p-(m+n)][p+(m+n)]=(p-m-n)(p+m+n)$

notice the difference of squares within a difference of squares!

c) $y^4-1=(y^2)^2-1^2=[y^2-1][y^2+1]=(y-1)(y+1)(y^2+1)$
d) $2z^4-2=2(z^4-1)=2\,[z^2-1][z^2+1]=2(z-1)(z+1)(z^2+1)$

7B Factorise:

a) $(a+b)^2-c^2$ b) $(a-b)^2-c^2$ c) $a^2-(b+c)^2$
d) $a^2-(b-c)^2$ e) $(p+q)^2-r^2$ f) $25-(v-w)^2$
g) $(r+s)^2-4t^2$ h) $9a^2-4(b-c)^2$ i) $(a+b)^2-(b-c)^2$
j) $(z+2)^2-(z-3)^2$ k) $64-(2a-1)^2$ l) $(x+y-z)^2-(x-y+z)^2$.

8B Express each of these as a product of three factors:

a) x^4-1 b) y^4-16 c) x^4-y^4
d) $p^4-q^4r^4$ e) $16z^4-1$ f) u^4-81v^4.

9B Express each of these as a product of four factors:

a) $3t^4-3$ b) $2u^4-32$ c) ap^4-aq^4
d) $80v^4-5$ e) $3p^4-243q^4$ f) $162a^4-32v^4$.

30 Factorising Trinomials (x^2+bx+c)

Reminder

In Exercise 27, we practised expressing the product of two binomials as a trinomial. Factorising a trinomial is the reverse process. We finish up with a product of two binomials.
Consider these four examples of a product of binomials which expand to become a trinomial

$(x+3)(x+4)=x^2+7x+12$
$(x-3)(x-4)=x^2-7x+12$

$(x-3)(x+4)=x^2+x-12$
$(x+3)(x-4)=x^2-x-12$

Note $+12$ comes from $(3)(4)$ or $(-3)(-4)$
$7x$ comes from $3x+4x$
$-7x$ comes from $-3x-4x$

-12 comes from $(-3)(4)$ or $(3)(-4)$
x comes from $4x-3x$
$-x$ comes from $3x-4x$

Reminder continued

Hence to decide which signs appear inside the binomials (the brackets), these rules apply:

(i) the + before 12 signifies like signs
the + before the 7 means two + signs
the – before the 7 means two – signs

(ii) the – before the 12 signifies unlike signs
the + before the x means the larger factor (4) is +
the – before the x means the larger factor (4) is –

(Note: (i) 3 and 4 are factors of 12 with a *sum* of 7 (ii) 3 and 4 are factors of 12 with a *difference* of 1.)

Example

Factorise:

a) $x^2 + 5x + 6$
b) $x^2 - 6x + 8$
c) $x^2 + 2x - 15$
d) $x^2 - x - 42$.

Solution

a) $x^2 + 5x + 6 = (x \quad)(x \quad)$ is the first thing to write down
+6 means like signs (i.e. the same) +5 means both '+'
so we can write $x^2 + 5x + 6 = (x + \quad)(x + \quad)$;
now look for factors of 6 with a sum of 5 i.e. 2 & 3
hence $x^2 + 5x + 6 = (x + 2)(x + 3)$

b) $x^2 - 6x + 8 = (x \quad)(x \quad)$ since the factors of x^2 can only be x and x signs are the same, both '–', so seek factors of 8 with a sum of 6 i.e. 2 & 4
hence $x^2 - 6x + 8 = (x - 2)(x - 4)$

c) $x^2 + 2x - 15 = (x \quad)(x \quad)$
signs in the brackets are different, the larger factor (of 15) is '+'
look for factors of 15 with a difference of 2 i.e. 3 & 5
hence $x^2 + 2x - 15 = (x - 3)(x + 5)$

d) $x^2 - x - 42 = (x \quad)(x \quad)$
signs are different, larger factor is '–';
seek factors of 42 with a difference of 1 i.e. 6 & 7
hence $x^2 - x - 42 = (x + 6)(x - 7)$

Exercise 30

1 Factorise:

a) $x^2 + 3x + 2$ b) $y^2 + 9y + 18$ c) $z^2 + 7z + 10$ d) $w^2 + 10w + 21$
e) $x^2 + 9x + 20$ f) $y^2 + 5y + 6$ g) $z^2 + 11z + 28$ h) $w^2 + 11w + 30$
i) $x^2 - 3x + 2$ j) $y^2 - 8y + 15$ k) $z^2 - 4z + 3$ l) $w^2 - 7w + 10$
m) $x^2 - 6x + 5$ n) $y^2 - 9y + 14$ o) $z^2 - 9z + 18$ p) $w^2 - 12w + 35$.

2 Factorise:

a) $x^2 + x - 2$ b) $y^2 - y - 2$ c) $z^2 + 2z - 8$ d) $w^2 - 2w - 8$
e) $u^2 + u - 20$ f) $v^2 + 3v - 18$ g) $r^2 - 4r - 5$ h) $s^2 - 5s - 14$
i) $t^2 + 2t - 24$ j) $k^2 + 3k - 28$ k) $m^2 + 6m - 16$ l) $n^2 - 4n - 45$
m) $a^2 + a - 6$ n) $b^2 - b - 30$ o) $c^2 - 3c - 10$ p) $d^2 - d - 56$.

3 Factorise:

a) $x^2 + 8x + 15$ b) $y^2 + 5y - 14$ c) $z^2 - 14z + 33$ d) $w^2 - 16w + 63$
e) $u^2 + 13u + 22$ f) $v^2 + 2v - 35$ g) $r^2 - 5r - 36$ h) $s^2 - 13s + 30$
i) $t^2 + t - 42$ j) $k^2 - 17k + 72$ k) $m^2 + 2m - 35$ l) $n^2 + 6n - 27$
m) $a^2 + 4a + 3$ n) $b^2 - 17b + 70$ o) $c^2 + 2c - 99$ p) $d^2 - 5d - 24$.

4 Factorise, by first taking out the highest common factor:
[e.g. $8x^2 + 24x - 80 = 8(x^2 + 3x - 10) = 8(x - 2)(x + 5)$]

a) $3x^2 - 21x + 36$ b) $5y^2 + 25y - 70$
c) $7z^2 - 84z + 77$ d) $8w^2 + 88w + 144$.

5B Factorise, by first re-arranging the given expression in the form $x^2 + bx + c$:
[e.g. $x(x - 1) - 2(2x - 3) = x^2 - x - 4x + 6 = x^2 - 5x + 6 = (x - 2)(x - 3)$]

a) $x(x - 5) + 4(x - 3)$ b) $x(x + 5) - 2(x - 1)$
c) $x(x - 14) + 3(x + 8)$ d) $x(x + 7) - 2(2x + 5)$
e) $(x + 2)(x - 3) + 4$ f) $(x - 5)(x + 7) + 20$
g) $(x + 3)(x - 5) + 7$ h) $(x - 2)(x - 6) - 5$.

6B Factorise, noticing that these trinomials have two variables e.g. $p^2 + 3pq + 2q^2$:
[since $x^2 + 3x + 2 = (x + 1)(x + 2)$, $p^2 + 3pq + 2q^2 = (p + q)(p + 2q)$]

a) $a^2 + 5ab + 6b^2$ b) $p^2 - pq - 12q^2$
c) $t^2 - 12tu + 35u^2$ d) $u^2 + 2uv - 15v^2$
e) $k^2 - kt - 2t^2$ f) $x^2 + 3xy - 28y^2$
g) $y^2 + 4yz + 4z^2$ h) $f^2 - 9g^2$.

7B Factorise:

a) $x^2 + (a + b)x + ab$ b) $x^2 + (n - m)x - mn$
c) $x^2 - (p + q)x + pq$ d) $x^2 + (5a - 2b)x - 10ab$.

31 Factorising trinomials ($ax^2 + bx + c$)

Reminder

Consider these four examples of a product of two binomials which expand to become a trinomial:

$(2x + 5)(3x + 4) = 6x^2 + 23x + 20$

$(2x - 5)(3x - 4) = 6x^2 - 23x + 20$

$(2x + 5)(3x - 4) = 6x^2 + 7x - 20$

$(2x - 5)(3x + 4) = 6x^2 - 7x - 20$

Note $+20$ comes from $(5)(4)$ or $(-5)(-4)$
$23x$ comes from $(5 \times 3x) + (2x \times 4)$
$-23x$ comes from $((-5) \times 3x) + (2x \times (-4))$

-20 comes from $(5)(-4)$ or $(-5)(4)$
$7x$ comes from $(5 \times 3x) + (2x \times (-4))$
$-7x$ comes from $((-5) \times 3x) + (2x \times 4)$

Hence to decide which signs appear inside the binomials (the brackets), these rules apply:

(i) the + before 20 signifies like signs
the + before 23 means two + signs
the – before 23 means two – signs

(ii) the – before 20 signifies unlike signs
the + before 7 means the larger product is +
the – before 7 means the larger product is –

(Note: (i) 15 and 8 are products with a sum of 23 (ii) 15 and 8 are products with a difference of 7.)

Example Factorise:

a) $6x^2 + 17x + 12$
b) $8x^2 - 18x + 9$
c) $12x^2 + x - 6$
d) $12x^2 - 7x - 10$.

Solution a) The '+' before 12 indicates like signs. The '+' before 17 indicates that both are '+'.
The factor pairs for $6x^2$ are: x and $6x$, or $2x$ and $3x$;
The factor pairs for 12 are: 1 and 2, or 2 and 6, or 3 and 4;
There are therefore 12 possiblities which give rise to $6x^2$ and 12.
We are only interested in the arrangement in which inner and outer products add up to $17x$ however. (The working can be done using the following table:)

[factors of 6] [factors of 12 (both ways round)]

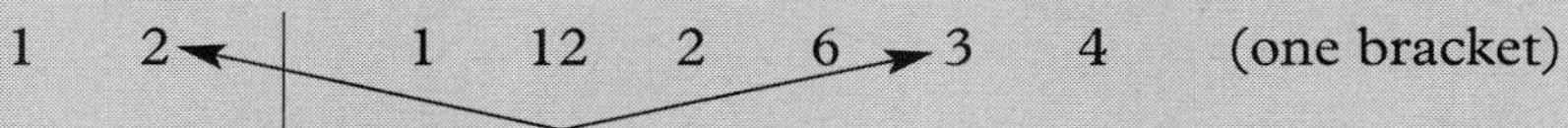

Since $(2 \times 4) + (3 \times 3) = 17$, then, $6x^2 + 17x + 12 = (2x + 3)(3x + 4)$

Reminder continued

b) signs the same, both '–'

1 2 | 1 9 3
8 4 | 9 1 3

look for inner and outer products with a sum of 18

$(2 \times 3) + (4 \times 3) = 18 \quad \Rightarrow \quad 8x^2 - 18x + 9 = (2x - 3)(4x - 3)$

c) signs different, larger product '+'

1 2 3 | 1 6 – 2
12 6 4 | 6 1 + 3

look for products with a difference of 1

$(3 \times 3) - (4 \times 2) = 1$; to get +9 and –8, we need +3 and –2, so add these to the table

$\Rightarrow \quad 12x^2 + x - 6 = (3x - 2)(4x + 3)$

d) signs different, larger product '–'

1 2 3 | 1 10 + 2
12 6 4 | 10 1 – 5

look for products with a difference of –7

$(4 \times 2) - (3 \times 5) = -7 \quad \Rightarrow \quad 12x^2 - 7x - 10 = (3x + 2)(4x - 5)$

Exercise 31

1B Factorise:

a) $2x^2 + 5x + 2$
b) $3x^2 + 10x + 3$
c) $5x^2 + 8x + 3$
d) $7x^2 + 12x + 5$
e) $5x^2 - 26x + 5$
f) $3x^2 - 10x + 7$
g) $3x^2 - 10x + 3$
h) $5x^2 - 16x + 3$
i) $2x^2 - 3x - 2$
j) $3x^2 + x - 2$
k) $2x^2 - 3x - 5$
l) $3x^2 + 5x - 2$
m) $5x^2 + 3x - 2$
n) $7x^2 - 20x - 3$
o) $5x^2 + 9x - 2$
p) $3x^2 - 4x - 7$.

2B Factorise:

a) $3x^2 + 25x + 8$
b) $6y^2 - 7y + 1$
c) $7y^2 + 23y + 6$
d) $3z^2 - 16z + 21$
e) $9z^2 + 18z + 5$
f) $5z^2 - 17z + 6$
g) $5c^2 - 21c + 4$
h) $30a^2 - 17a + 2$
i) $2x^2 - x - 10$
j) $4y^2 - y - 3$
k) $5y^2 + 19y - 4$
l) $2z^2 - z - 28$
m) $3a^2 - 2a - 8$
n) $4b^2 + 11b - 3$
o) $4x^2 + x - 5$
p) $3x^2 - 10x - 8$.

3B Factorise:

a) $6x^2 + 13x + 6$ b) $6x^2 - 23x + 20$ c) $10y^2 + 23y + 12$
d) $15y^2 - 34y + 15$ e) $4x^2 - 8x - 21$ f) $6x^2 - 7x - 20$
g) $8y^2 + 6y - 9$ h) $12y^2 - 8y - 15$ i) $4z^2 + 12z - 27$
j) $9z^2 + 9z - 28$ k) $15z^2 + 37z + 20$ l) $14z^2 - 41z + 15$
m) $21c^2 + 25c - 4$ n) $6d^2 + 23d + 15$ o) $18b^2 + 27b + 10$
p) $20d^2 - 44d - 15$.

4B Factorise, by first taking out the highest common factor:

a) $3x^2 - 3x - 6$ b) $6x^2 - 3x - 3$ c) $12x^2 + 14x - 6$
d) $24x^2 - 28x + 8$ e) $30y^2 - 65y + 30$ f) $12y^2 + 34y + 24$
g) $18y^2 + 33y - 30$ h) $56y^2 + 42y - 63$.

5B Factorise, by first re-arranging the given expression in the form $ax^2 + bx + c$:
[e.g. $(x + 2)(x - 3) + 4 = x^2 + 2x - 3x - 6 + 4 = x^2 - x - 2 = (x + 1)(x - 2)$]

a) $(x - 1)(21x - 1) - 9$ b) $(x + 1)(20x - 19) + 7$ c) $(x - 3)(6x - 11) + 2$
d) $(x + 1)(12x + 11) - 1$ e) $(6x - 1)(x + 3) + 10$ f) $(6x + 1)(x - 2) - 8$.

6B Factorise:

a) $6x^2 - 5xy - 6y^2$ b) $8x^2 + 22xy + 12y^2$ c) $12x^2 - 25xy + 12y^2$
d) $15x^2 + xy - 2y^2$ e) $12p^2 + 17pq + 6q^2$ f) $20p^2 + 13pq - 15q^2$
g) $42p^2 - 47pq + 10q^2$ h) $12p^2 + 4pq - 21q^2$ i) $6k^2 - kl - 12l^2$.

32 Miscellaneous Factors

Reminder

common factor e.g. $7x + 14y = 7(x + 2y)$
difference of two squares e.g. $x^2 - 9y^2 = (x - 3y)(x + 3y)$
common factor and difference of squares e.g. $3x^2 - 12 = 3(x^2 - 4) = 3(x^2 - 2^2)$
$= 3(x - 2)(x + 2)$
trinomial ($x^2 + bx + c$) e.g. $x^2 + 2x - 63 = (x - 7)(x + 9)$
trinomial ($ax^2 + bx + c$) e.g. $8x^2 - 22x + 15 = (2x - 3)(4x - 5)$
common factor and trinomial e.g. $2x^2 - 16x + 30 = 2(x^2 - 8x + 15)$
$= 2(x - 3(x - 5)$
not every expression factorises! e.g. $x^2 + 1$, $3x^2 + 2x + 1$

Summary of technique
- look for any common factor (take out the highest common factor)
- if there are only two terms, are they a difference of two squares?
- if there are three terms, can the trinomial be factorised?

Exercise 32

Factorise fully, where possible:

1 $ax + by$
2 $ax^2 + ax$
3 $4x^2 - 1$
4 $x^2 - x - 20$
5 $2x^2 - 18$
6 $x^2 + 2x - 15$
7 $2x^2 + 2x - 12$
8 $x^2 + x + 1$
9 $2x^2 - 9x - 5$
10 $2x^2 + 18$
11 $3x^2 - 3$
12 $x^2 - 5x - 14$
13 $9x^2 + 9x + 2$
14 $6x^2 + x - 12$
15 $6x^2 - 21x - 45$
16 $ad - bc$
17 $a^2x - axy$
18 $49x^2 - 36y^2$
19 $4x^2 - 36$
20 $4x^2 - 12x + 9$
21 $6x^2 - 5x - 21$
22 $x^2 - 16x + 63$
23 $ax + a^2y + a^3z$
24 $49 - 36t^2$
25 $200x^2 - 128z^2$
26 $5x^2 + 10x + 5$
27 $x^2 + 3x + 5$
28 $4x^2 + 2y + 8z$
29 $35x^2 + x - 6$
30 $20x^2 - 66x + 54$.

33 Revision of Algebra

Exercise 33

1 Expand the brackets:
a) $4(x + 2)$
b) $7(y - 1)$
c) $8(x^2 + 3)$
d) $5(4 - x)$
e) $3(x + y + 2z)$
f) $4(7 - 2p + 3q)$
g) $-(2a - 3b + c)$
h) $-5(2p + 3q - r)$.

2 Expand the brackets and simplify:
a) $2(x + 3) + 1$
b) $3(y + 4) - 2$
c) $5(z - 2) + 3$
d) $7(3 - s) - 4$
e) $1 + 3(x + 2)$
f) $4 + 5(2 - y)$
g) $6 - 7(3 - z)$
h) $11 - (t - 1)$.

3 Expand the brackets and simplify:
a) $(x + 1)(x - 2)$
b) $(y + 3)(y - 4)$
c) $(2z + 3)(z - 5)$
d) $(5k + 3)(k - 2)$
e) $(3k - 7)(2k + 3)$
f) $(2z - 3)(3z - 8)$
g) $(3p - q)(4p + q)$
h) $(4a + 3b)(3a - 5b)$.

4B Expand the brackets and simplify:
a) $(x + 1)(x^2 + x + 1)$
b) $(x + 3)(2x^2 - x - 1)$
c) $(2x + 3)(3x^2 + 2x + 4)$
d) $(3x - 2)(2x^2 - 4x + 5)$.

5 Factorise each of the following, using the highest common factor:
a) $3x + 6y$
b) $9t + 81$
c) $16 + 64w$
d) $14u + 49v$
e) $4x^2 + 7x$
f) $6y^2 + 9y$
g) $4z^2 + 6z$
h) $8p^2q + 12pq^2$.

6 Factorise each difference of two squares:
a) $x^2 - 81$
b) $y^2 - 49$
c) $49z^2 - 64$
d) $64p^2 - 25q^2$
e) $t^4 - 81u^2$
f) $100f^2 - 169g^2$
g) $121p^2 - 36q^2$
h) $400a^2 - 121b^2$.

7B Factorise, using a common factor then a difference of two squares:

a) $2x^2 - 2$ b) $3y^2 - 27$ c) $125 - 5z^2$
d) $196 - 4a^2$ e) $7p^2 - 7q^2$ f) $ax^2 - ay^2$
g) $300m^2 - 243n^2$ h) $175r^2 - 112s^2$.

8 Factorise these trinomials:

a) $x^2 - 2x - 35$ b) $y^2 + 6y - 27$ c) $z^2 - 13z + 36$
d) $a^2 + 9a + 14$ e) $b^2 - 13b + 22$ f) $c^2 + 15c + 36$
g) $d^2 - 12d + 32$ h) $f^2 + f - 42$.

9B Factorise these trinomials:

a) $6x^2 + x - 1$ b) $5y^2 + 3y - 2$ c) $3z^2 - 5z + 2$
d) $10t^2 + 21t - 10$ e) $30u^2 + 7u - 15$ f) $21v^2 - 16v + 3$
g) $15w^2 + 22w + 8$ h) $8k^2 + 6k - 9$.

10B Factorise fully, where possible:

a) $3p + 4q$ b) $5x^2 - 5$ c) $x^2 + 2x - 35$
d) $2y^2 - 30y + 72$ e) $3x^2 - 9x + 6$ f) $7p^2 - 343q^2$
g) $x^2 + 2x + 17$ h) $12x^2 - 10x - 8$.

11B Factorise by first re–arranging the given expression in the form $ax^2 + bx + c$:
[This question may take you some time.
A clue is that a is of the form k^2 and the factors are of the form $(kx\ .....)(kx\ .....)$]

a) $24(x - 1)(x + 2) + 25(x - 2)(x + 3)$
b) $60(x - 2)(x + 3) + 61(x - 3)(x + 4)$.

34 Test on Algebra

Allow 45 minutes for this test

1 Expand these brackets:

a) $3(x + 2)$ b) $4(2y + z)$ c) $3(x + y - 2z)$ d) $x(x + 2y)$.

2 Expand and simplify:

a) $1 + 2(x + 2)$ b) $2 - 3(y - 1)$ c) $3(2z - 3) + 5$ d) $2(3p + 4q) - p$.

3 Express each product as a trinomial:

a) $(x + 3)(x + 5)$ b) $(y + 4)(y - 5)$ c) $(2z - 1)(3z - 2)$ d) $(4t - 3)(2t + 5)$.

4 Factorise:

a) $2x + 6y$ b) $x^2 + 3x$ c) $3x^2 + 51$ d) $6r^2s + 4rs^2$.

5 Factorise:

a) $100 - x^2$ b) $p^2 - 4q^2$ c) $25a^2 - 16b^2$ d) $49u^2 - 4v^2$.

6B Factorise:

a) $2x^2 - 8$ b) $5y^2 - 45$
c) $3 - 75z^2$ d) $20p^2 - 125q^2$.

7 Factorise:

a) $x^2 + 4x + 3$

b) $y^2 - 15y + 36$

c) $z^2 - 4z - 32$

d) $t^2 + 5t - 14$.

8B Factorise:

a) $6x^2 + x - 1$

b) $15y^2 + 14y - 8$

c) $15z^2 - 11z - 12$

d) $30u^2 - 37u + 10$.

9B Expand:

a) $(x + 1)(x^2 + 2x + 3)$

b) $(x - 1)(3x^2 - 2x + 1)$

c) $(2x + 3)(x^2 + x - 1)$

d) $(3x - 2)(2x^2 - 3x + 4)$.

CIRCLE PROPERTIES

35 The Length of an Arc

Reminder

The length of the arc AB is proportional to the angle at the centre $A\hat{C}B$.

[We sometimes say that the arc AB **subtends** the angle $A\hat{C}B$ at the centre.]

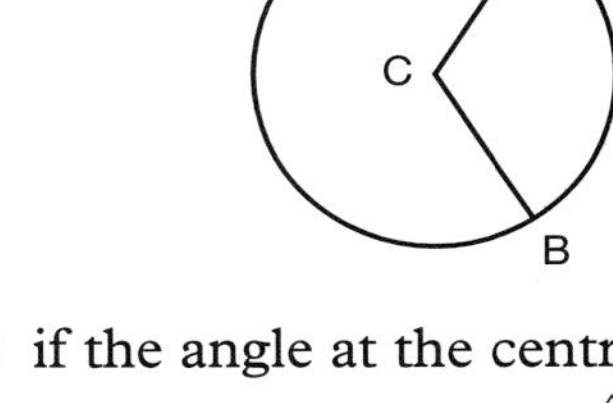

i.e.

(i) if the angle at the centre is $\frac{1}{2}$ of 360° then ACB is a diameter and the arc AB is $\frac{1}{2}$ of the circumference i.e. half the angle gives half the arc

(ii) if the angle at the centre $x°$ is doubled to $2x°$, i.e. $A\hat{C}B_1 = x°$ and $A\hat{C}B_2 = 2x°$ then the length of arc $AB_2 = 2l$ and arc $AB_1 = l$ i.e. double the angle gives double the arc.

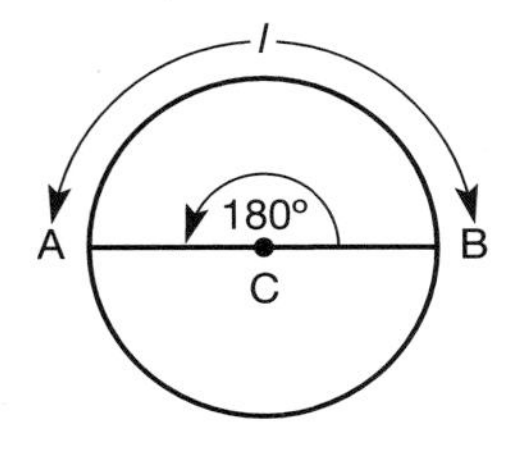

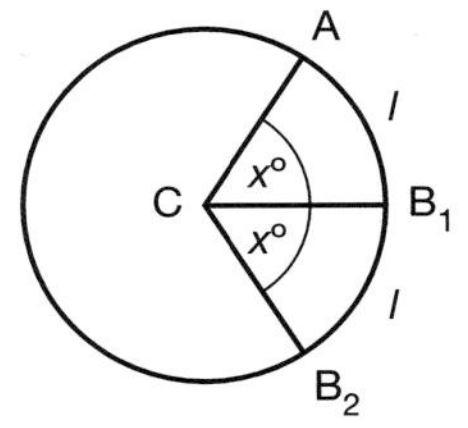

Remember that the circumference of the circle of radius r is $C = 2\pi r$ or that the circumference of the circle of diameter d is $C = \pi d$.

Example In a circle with centre C and radius 45 mm, an arc AB subtends an angle of 160° at the centre.

Calculate the length of the arc AB.

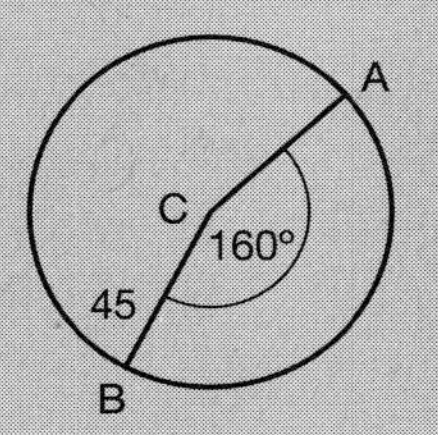

Reminder continued

Solution (Method 1) $\frac{160}{360} = \frac{16}{36} = \frac{4}{9}$

circumference $= 2\pi r = 2 \times \pi \times 45 = 90\pi$

arc AB $= \frac{4}{9} \times$ circumference $= \frac{4}{9} \times 90\pi = 40\pi = 125{\cdot}7$ mm.

(Method 2) $\frac{\text{arc length}}{\text{circumference}} = \frac{\text{angle at centre}}{360}$

$\Rightarrow \quad \frac{l}{2\pi r} = \frac{160}{360} = \frac{16}{36} = \frac{4}{9}$

$\Rightarrow \quad l = \frac{4}{9} \times 2\pi r = \frac{4}{9} \times 2\pi \times 45 = 40\pi = 125{\cdot}7$ mm.

Exercise 35

Give each answer correct to the nearest mm, inch, or degree.

1 A circle with centre C has radius 100 mm.
Calculate the length of the arc AB when $\hat{ACB}$ is

a) 180° b) 60° c) 45°
d) 135° e) 300°.

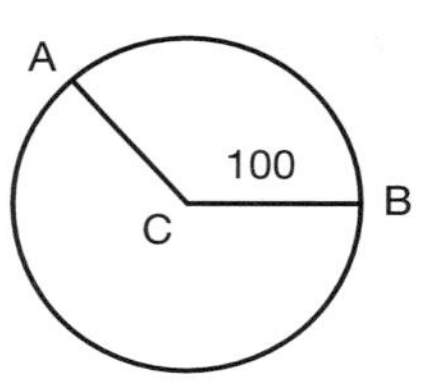

2 A circle with centre Q has radius 70 mm.
Calculate the length of the arc PR when $\hat{PQR}$is

a) 30° b) 150° c) 45°
d) 225° e) 120° f) 330°.

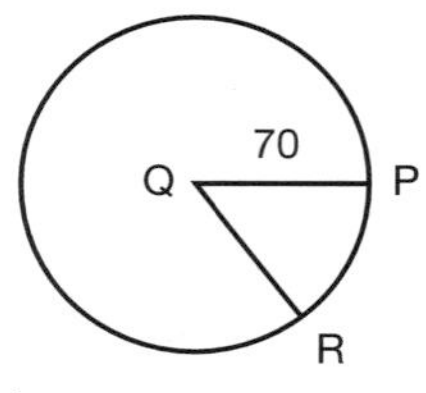

3 PQRS is a square inscribed in a circle with centre O and radius 120 mm.
Calculate a) the size of $\hat{POQ}$
b) the length of the arc PQ.

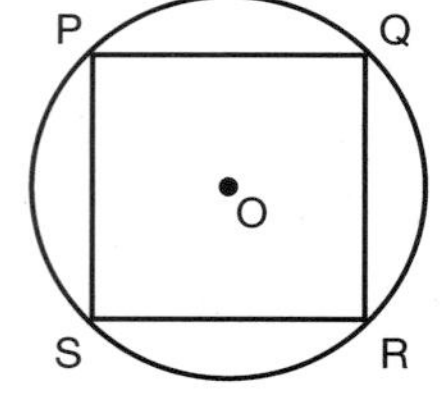

4 ABCDEF is a regular hexagon inscribed in a circle with centre O and radius 80 mm.
Calculate a) the size of $\hat{AOB}$
b) the length of the arc AB.

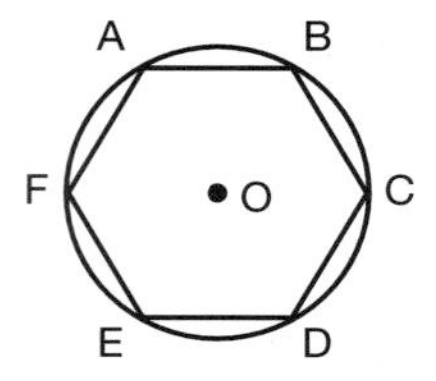

5 JKLMN is a regular pentagon inscribed in a circle with centre O and radius 5 m.
Calculate the length of the arc MN.

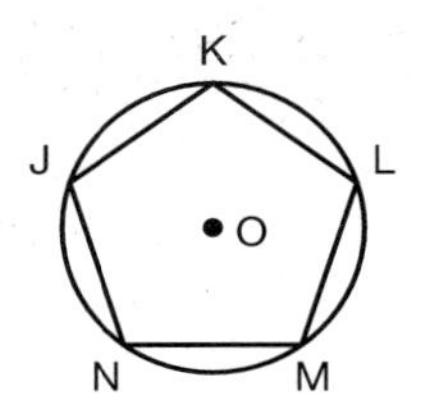

6 The diagram shows a road bending through 30°.
The road is 10 m wide and the kerbs at the bend form arcs of circles. The radius of curvature of the inner kerb is 30 m.
Calculate the length of a) the inner kerb
b) the outer kerb.

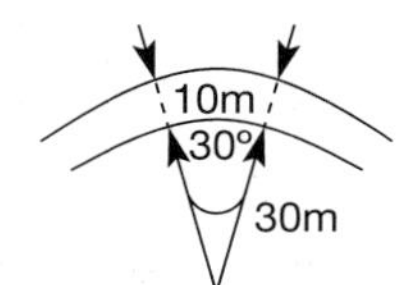

7 The diagram shows a table in a café.
The shape of the table is a circle with centre C and radius 750mm with a minor segment AB removed and $A\hat{C}B = 120°$.
The side AB is secured to the wall and plastic edging is glued to the curved edge.
Calculate the length of plastic edging required.

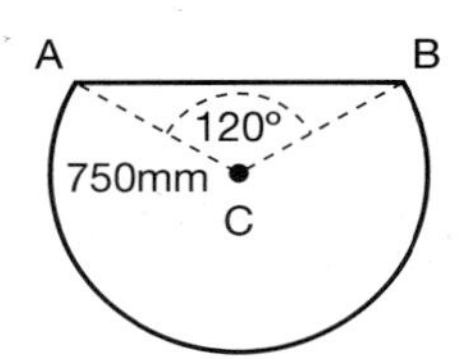

8 Tony's racing cycle has 26 inch wheels (i.e. the diameter of the wheel is 26 inches).
When the bike is rolled forward and the wheels rotate through 216°, how far has the cycle moved?

9 PQ is an arc of length 210 mm in a circle with centre C and radius 200 mm.
Calculate a) the circumference of the circle
b) the size of $P\hat{C}Q$
c) the perimeter of the sector PCQ.

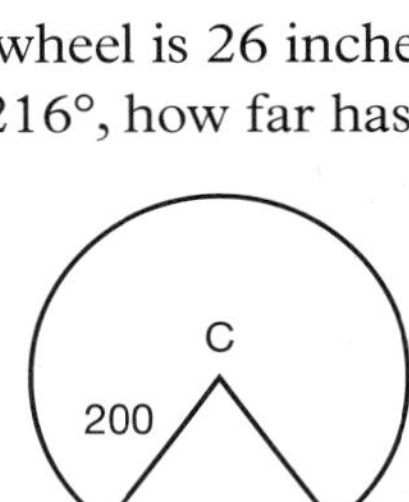

10 The points A, B, C, D, E, F lie on the circumference of a circle with centre O and radius 110 mm.
Calculate the size of
a) $A\hat{O}B$ if the arc AB has length 50 mm
b) $A\hat{O}C$ if the arc AC has length 140 mm
c) $A\hat{O}D$ if the arc AD has length 200 mm
d) $A\hat{O}E$ if the arc AE has length 300 mm
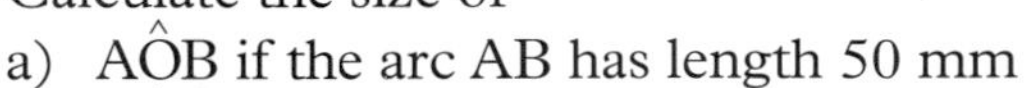
e) $A\hat{O}F$ if the arc AF has length 450 mm.

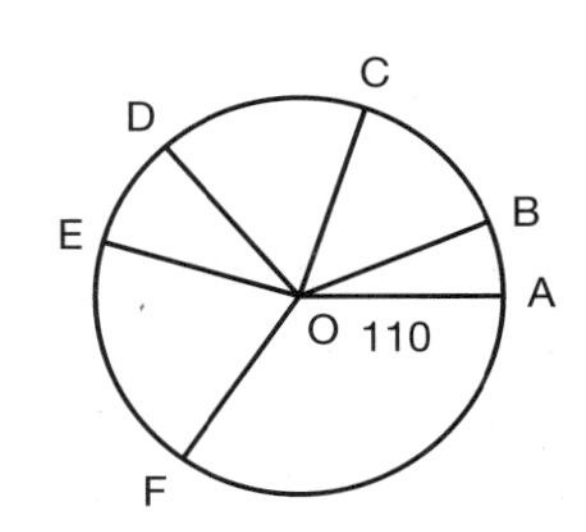

11 The diagram shows five concentric circles with centre O.
The radii have these lengths (in mm):
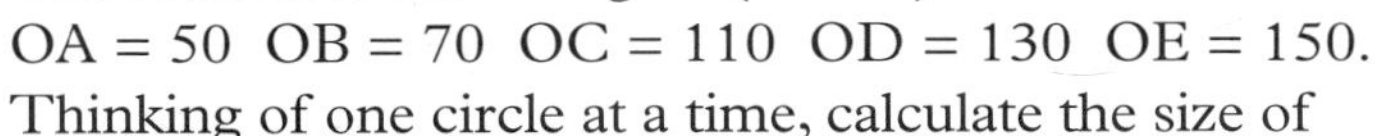
OA = 50 OB = 70 OC = 110 OD = 130 OE = 150.
Thinking of one circle at a time, calculate the size of

a) $A\hat{O}F$ if the arc AF has length 30 mm
b) $B\hat{O}G$ if the arc BG has length 110 mm
c) $C\hat{O}H$ if the arc CH has length 200 mm
d) $D\hat{O}I$ if the arc DI has length 350 mm
e) $E\hat{O}J$ if the arc EJ has length 500 mm.

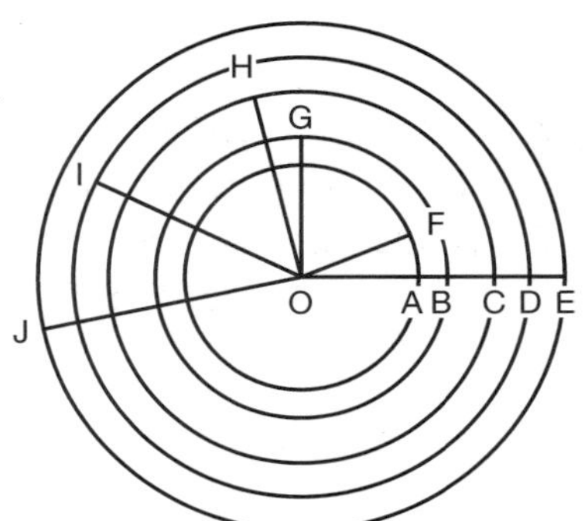

12 A cycle with 26 inch diameter wheels is rolled forward 91 inches. Through what angle do the wheels rotate?

13 SV is an arc of length 250 mm in a circle with centre T and $S\hat{T}V = 60°$
Calculate the length of a) the circumference
b) the radius TV
c) the perimeter of the sector STV.

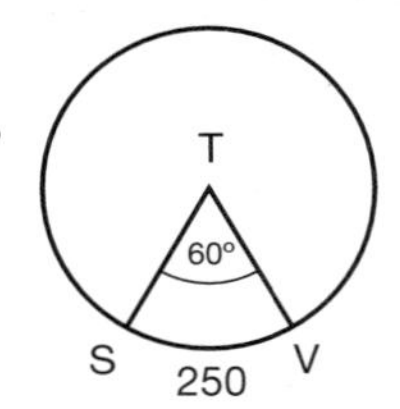

14 When I roll my wheel-barrow forward by 1 m, the wheel rotates through 310°. Calculate the diameter of the wheel.

36 The Area of a Sector

Reminder

The area of the sector ACB is proportional to the angle at the centre, $A\hat{C}B$.

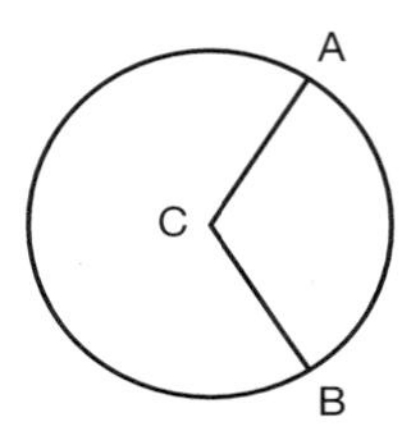

i.e.

(i) if the angle at the centre is $\frac{1}{2}$ of 360° then ACB is a diameter and the sector ACB is a semi-circle
i.e. half the angle gives half the area

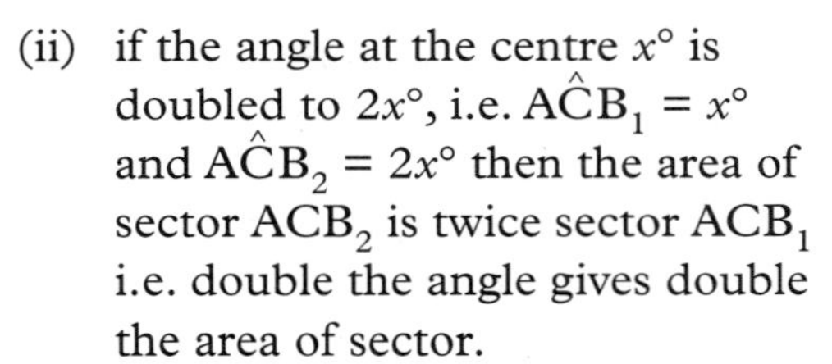

(ii) if the angle at the centre $x°$ is doubled to $2x°$, i.e. $A\hat{C}B_1 = x°$ and $A\hat{C}B_2 = 2x°$ then the area of sector ACB_2 is twice sector ACB_1
i.e. double the angle gives double the area of sector.

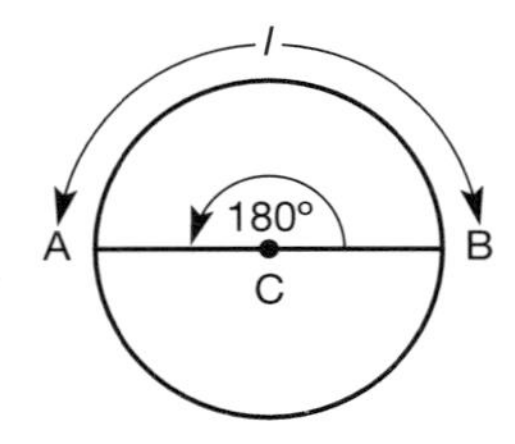

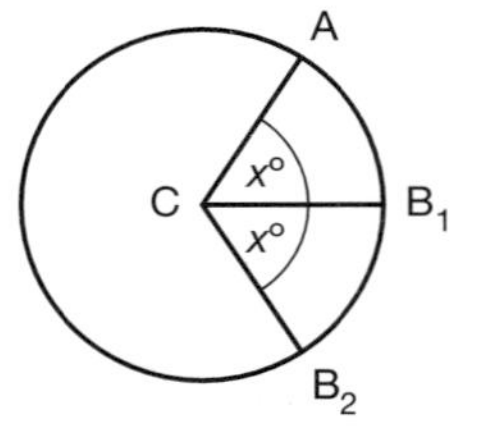

Example In a circle with centre C and radius 750 mm, the major arc AB subtends an angle of 200° at the centre.

Calculate the area of the major sector ACB.

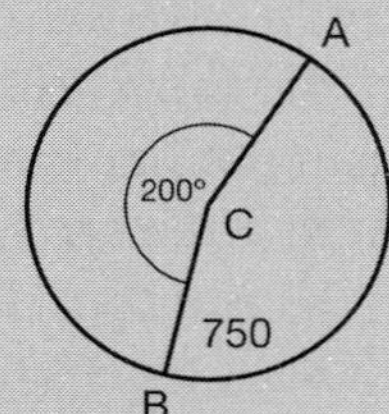

Reminder continued

Solution (Method 1) area of the circle $= \pi r^2 = \pi\,(750)^2$

$$\frac{200}{360} = \frac{20}{36} = \frac{5}{9}$$

$$\text{area of sector} = \frac{5}{9} \times \text{area of the circle} = \frac{5}{9} \times \pi(750)^2$$

$$= 981\,748\ \text{mm}^2$$

(Method 2) $$\frac{\text{area of sector}}{\text{area of circle}} = \frac{\text{angle at centre}}{360}$$

$$\Rightarrow \quad \frac{A}{\pi(750)^2} = \frac{200}{360} = \frac{20}{36} = \frac{5}{9}$$

$$\Rightarrow \quad A = \frac{5}{9} \times \pi(750)^2 = 981\,748\ \text{mm}^2.$$

Exercise 36

Express each answer correct to the nearest mm^2.

1 A circle with centre C has radius 120 mm.
Calculate the area of the sector ACB when $\hat{ACB}$ is:

a) 180° b) 30° c) 45°
d) 225° e) 330°.

2 In a circle with centre Q and radius 80 mm, calculate the area of the sector PQR when $\hat{PQR}$ is:

a) 60° b) 90° c) 120°
d) 150° e) 210° f) 240°.

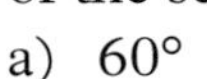

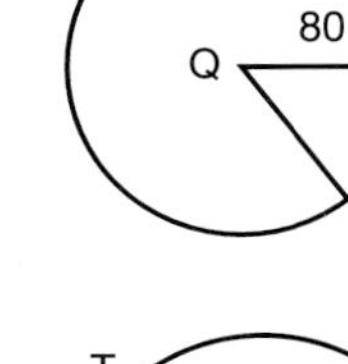

3 STUV is a square inscribed in a circle with centre O and radius 170 mm.
Find a) the size of $\hat{SOT}$
b) the area of sector SOT.

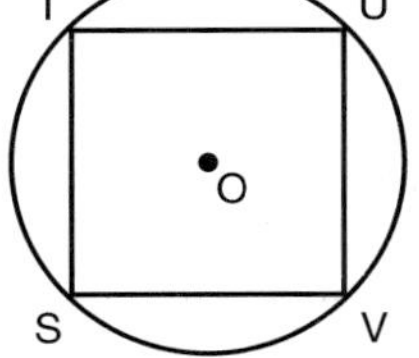

4 ABCDEF is a regular hexagon inscribed in a circle with centre O and radius 1 m.
Find a) the size of $\hat{AOB}$
b) the area of sector AOB.

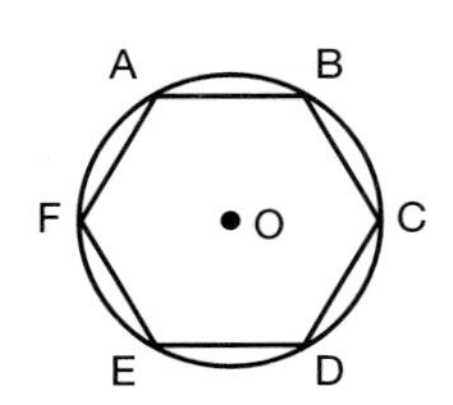

5 PQRST is a regular pentagon inscribed in a circle with centre O and radius 2 m.
Calculate the area of sector POQ.

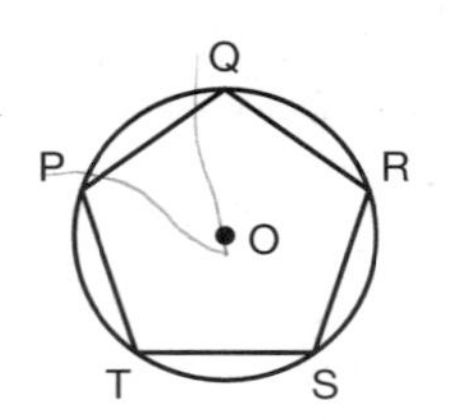

6B This diagram shows two straight pieces of 00 gauge model railway track joined by what is known as a No. 1 radius standard curve, whose rails are arcs of circles.
The track is 16·5 mm wide.
The radius of curvature of the inner rail is 371 mm and the track bends through $22\frac{1}{2}°$.
Calculate the area between the two curved sections of rail.

7 Two walls of an ice cream parlour meet at 120°.
A table secured to these walls is in the shape of a circle with a minor sector removed.
The radius of the table is 760 mm.
It is recommended that each customer has 300 000 mm^2 of table top space.
The café owner plans to fix five stools around this table.
Would his plan comply with the recommendations?
[Remember that the answer to this question is not just yes or no. Calculations must be shown.]

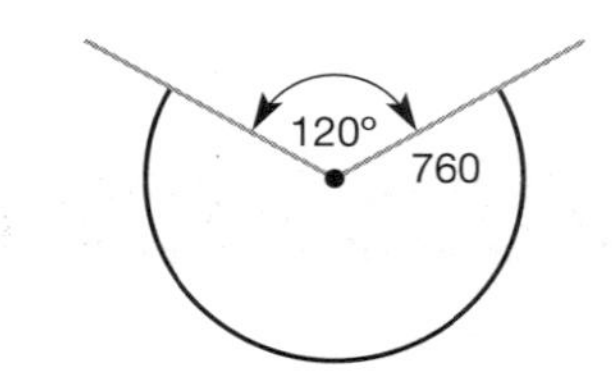

8 A lever for changing points in a signal box is 1·2 m long and moves through 75° from the main line position to the siding position.
a) Calculate the area swept out by the lever.
b) If it takes the signalman 2 seconds to move the lever, state the areal speed of the lever (i.e. the area swept out per second).

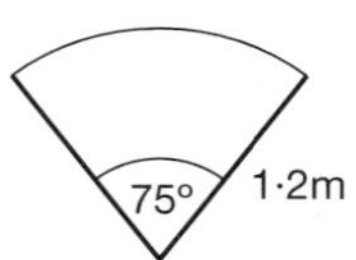

9 POQ is a sector of a circle with centre O and radius 150 mm.
The area of sector POQ is 50 000 mm^2.
Find: a) the area of the circle
b) the size of $P\hat{O}Q$.

10 The points A, B, C, D, E lie on the circumference of a circle with centre O and radius 250 mm.
Calculate the size of:
a) $A\hat{O}B$ if the area of sector AOB is 30 000mm^2
b) $A\hat{O}C$ if the area of sector AOC is 50 000mm^2
c) $A\hat{O}D$ if the area of sector AOD is 90 000mm^2
d) $A\hat{O}E$ if the area of sector AOE is 125 000mm^2.

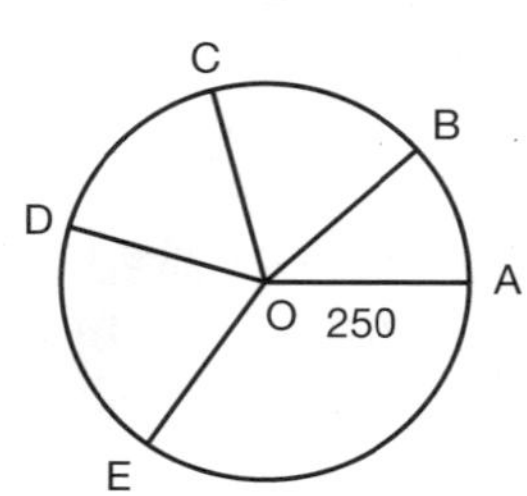

11 PKQ is a sector of a circle with centre K.
The area of the sector PKQ is 3500 mm^2 and $P\hat{K}O = 40°$.
Calculate a) the area of the circle
b) the length of the radius.

12B The rotating 'arm' on a radar screen sweeps out an area of 17 000 mm^2 per second. Each object identified by the scanner remains visible on the screen for one second and a sector with an angle of 50° at the centre is visible behind the 'arm' as it sweeps round.
Calculate the size of the diameter of the radar screen.

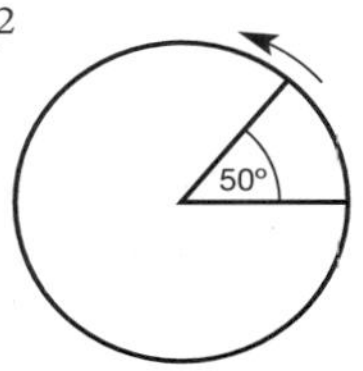

37 Tangent and Radius

Note: Since the trigonometry of right-angled triangles features prominently in the rest of this topic, you might find it advisable at this stage to do Exercise 42 from Unit 2 which revises this work.

Reminder

This diagram shows a family of parallel lines.
The lowest line cuts the circle in two points.
The top line does not cut the circle at all.
The middle line touches the circle at exactly one point and is known as a tangent.

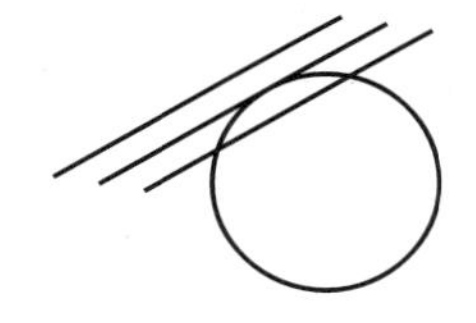

The tangent to a circle is perpendicular to the radius at its point of contact.
[Any point P other than T on the tangent is outside the circle, so CP > CT.
So CT is the shortest distance from C to the tangent, hence the perpendicular.]

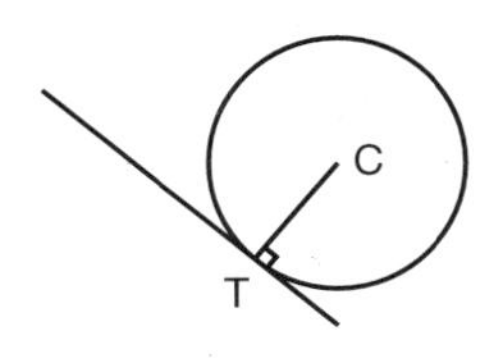

Example PT is a tangent to a circle with centre O and radius 5 m.
OP cuts the circle at Q. The length of PT is 12 m.
Calculate: a) the length of PQ
b) the size of (i) $T\hat{P}O$
(ii) $P\hat{O}T$.

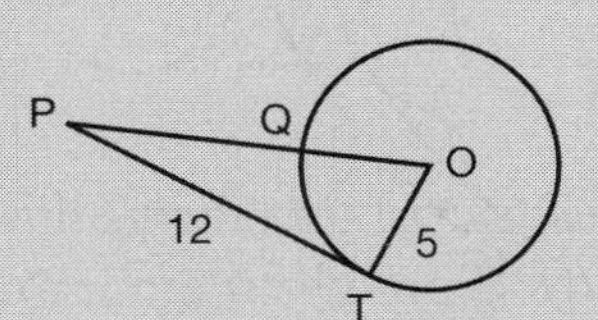

Reminder continued

Solution

a) $P\hat{T}O = 90°$ (radius is perpendicular to tangent)

$\Rightarrow PO = \sqrt{5^2 + 12^2} = \sqrt{25 + 144} = \sqrt{169} = 13$

$\Rightarrow PQ = PO - OQ = 13 - 5 = 8$ m

b) (i) $\tan T\hat{P}O = \frac{5}{12} \Rightarrow T\hat{P}O = 22{\cdot}6°$

(ii) $P\hat{O}T = 90° - T\hat{P}O = 67{\cdot}4°$

[Remember that in a right-angled triangle the two acute angles add up to 90°.]

Exercise 37

1 Calculate the size of $P\hat{O}T$.

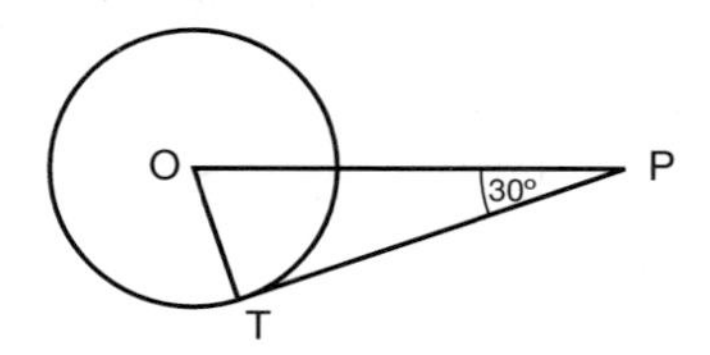

2 In each diagram, C is the centre of the circle and AB is the tangent at A. Calculate the sizes of p, q, r, s and t.

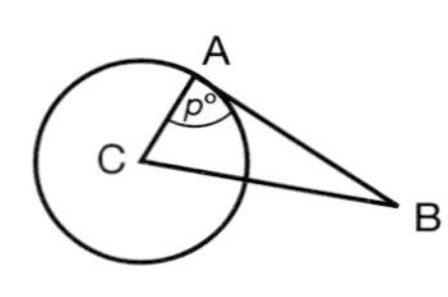

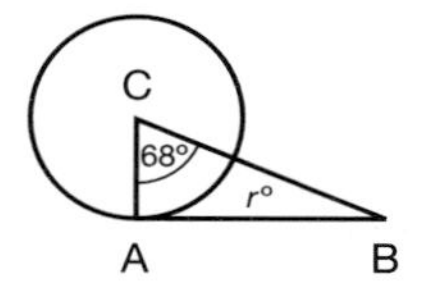

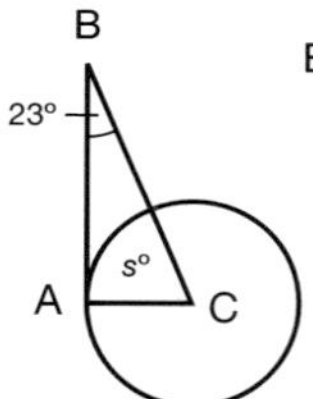

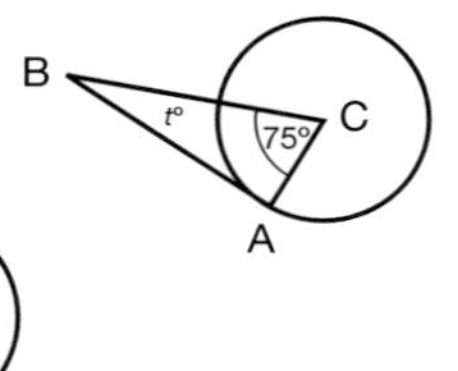

3 In each diagram, the length of the tangent PT and the distance of P from the centre C of the circle are given in metres. Calculate the length of the radius in each case.

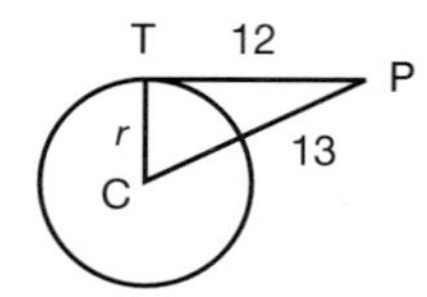

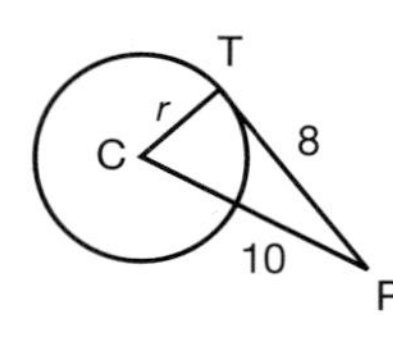

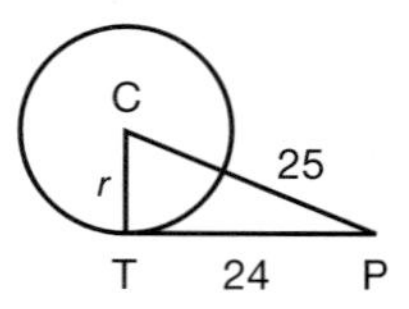

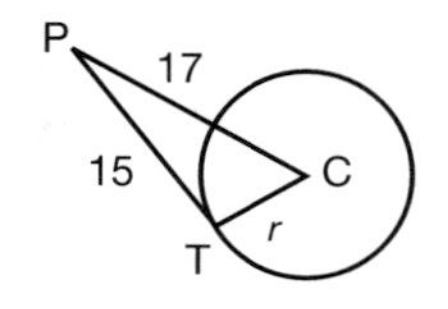

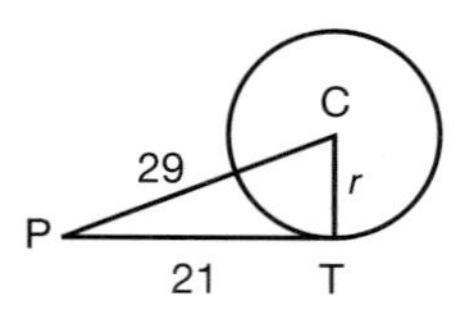

4 In each diagram, the radius and the length of the tangent PT at T to the circle with centre C are given in metres. Calculate the length of PC in each case.

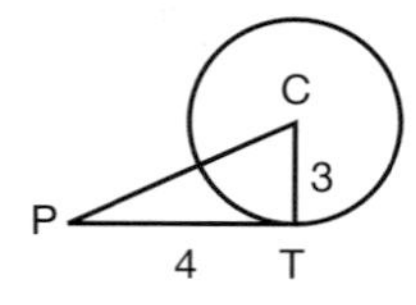

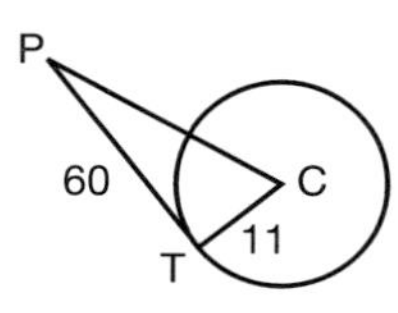

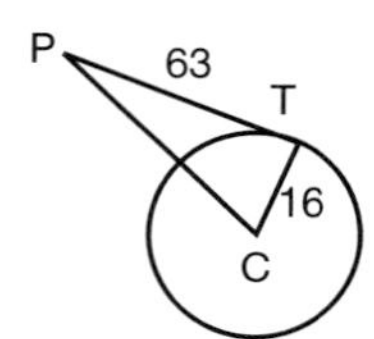

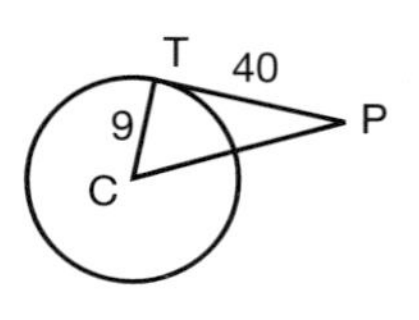

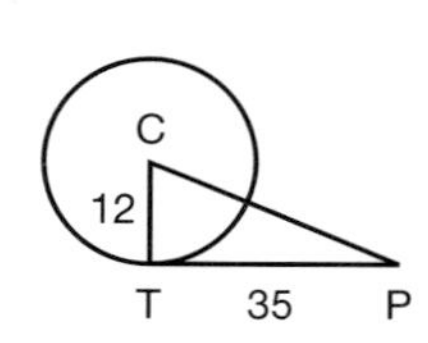

5 In each diagram, FG is the tangent at G to the circle with centre H. The lengths are given in metres. Calculate the length of the tangent FG in each case.

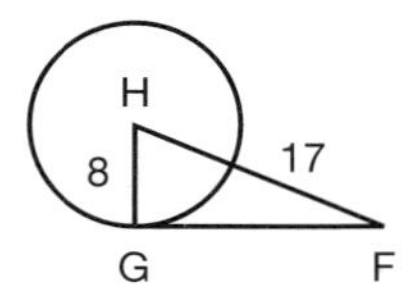

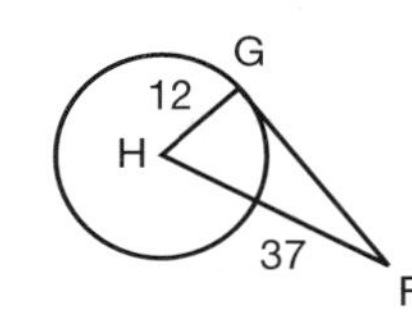

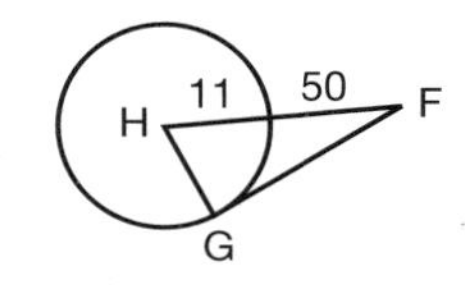

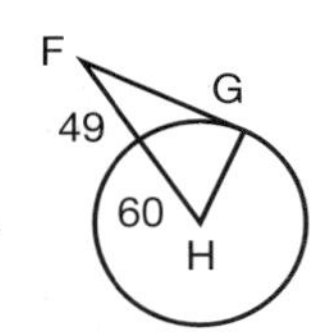

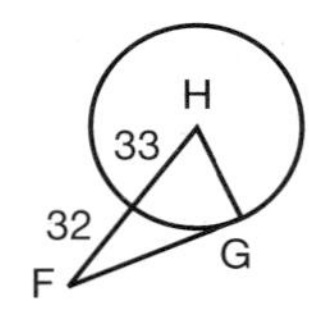

Reminder

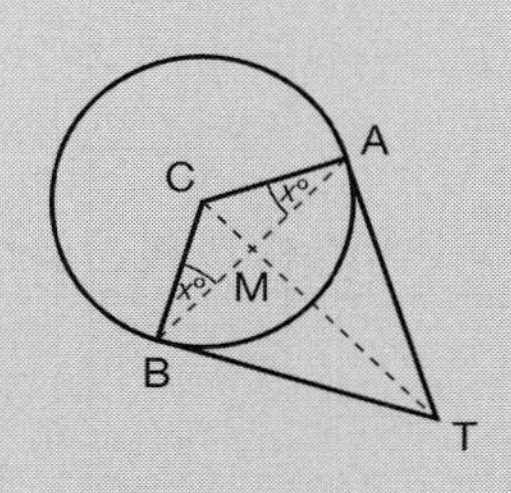

CA = CB (radii of the same circle)

$\Rightarrow$ ΔACB is isosceles $\Rightarrow$ $\hat{CBA} = \hat{CBA} = x°$ (base angles)

$\Rightarrow$ $\hat{BAT} = 90° - x° = \hat{ABT}$ (tangent $\perp^r$ radius)

$\Rightarrow$ ΔABT is isosceles

$\Rightarrow$ AT = TB

i.e. tangents to a circle from an external point are equal;
also ACBT is a kite, known as the tangent kite;
M is the midpoint of AB; AB $\perp^r$ CT; CT is the axis of symmetry.

6 QP and QR are tangents to the circle with centre O, and $\hat{POQ} = 65°$.
Calculate the size of $\hat{PQR}$, showing each step of your working.

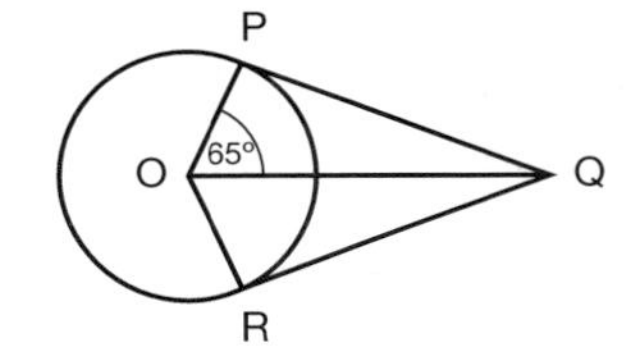

7 LK and LM are tangents to the circle with centre N, and $\hat{KLN} = 28°$.
Calculate the size of $\hat{KNM}$, showing each step of your working.

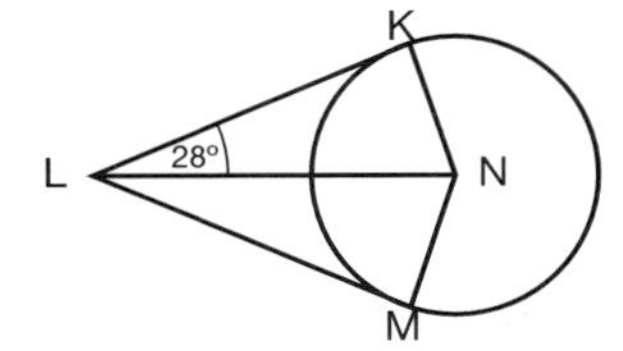

8B AB is the tangent at B to the circle with centre C.
ACD is a straight line, and $\hat{BAC} = 32°$.
Calculate the size of $\hat{DBC}$, showing each step of your working.

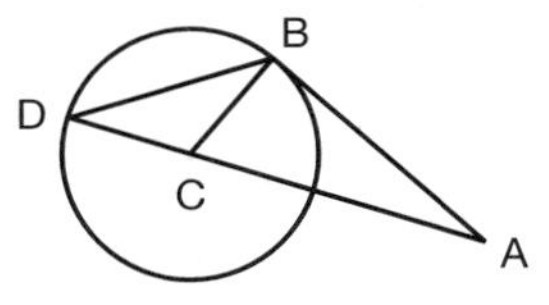

9B PQ is the tangent to the circle with centre R.
PRT and QRS are straight lines, and $\hat{TPQ} = 40°$.
Calculate the size of $\hat{STP}$, showing each step of your working.

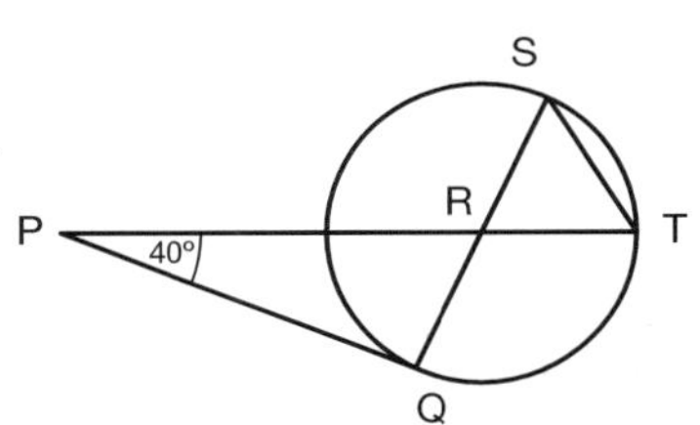

10B WX is the tangent at X to the circle with centre U, and $X\hat{V}W = 130°$.
Calculate the size of $V\hat{W}X$, showing each step of your working.

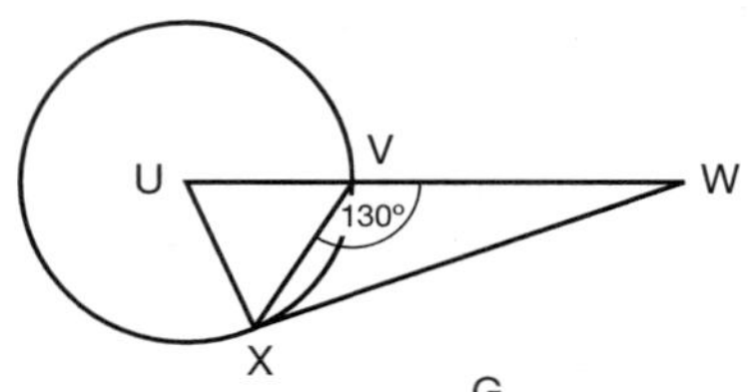

11H FG is the tangent at G to the circle with centre H.
GHL and FKHM are straight lines, and $M\hat{L}H = 50°$.
Calculate the size of $F\hat{G}K$, showing each step of your working.

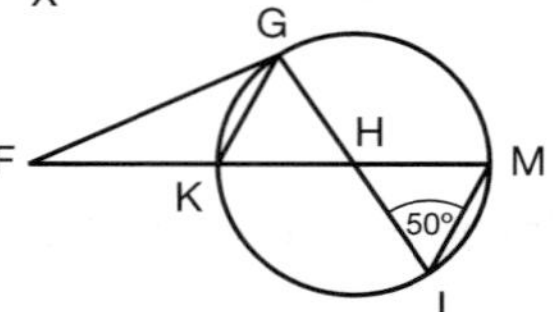

12H ABC is the tangent at B to the circle with centre E.
AED is a straight line, and $C\hat{B}D = 65°$.
Calculate the size of $B\hat{A}E$, showing each step of your working.

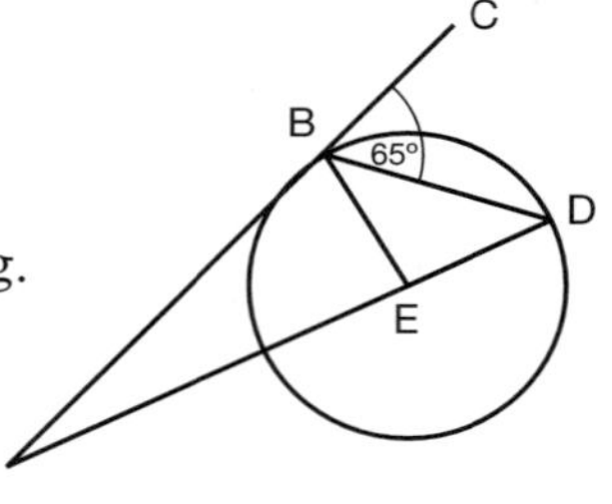

13H The line PQR is the tangent at Q to the circle with centre T.
$Q\hat{P}T = 40°$ and $Q\hat{R}S = 50°$.
Prove that RS is the tangent at S, showing each step of your working.
[Hint: join QS.]

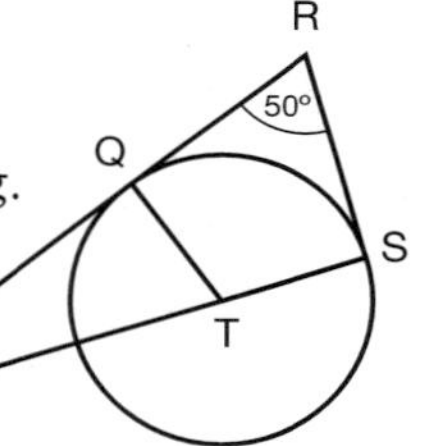

14 In each diagram, QT is the tangent at T to the circle with centre A.
Calculate, correct to one decimal place, the size $A\hat{Q}T$ of in each case.

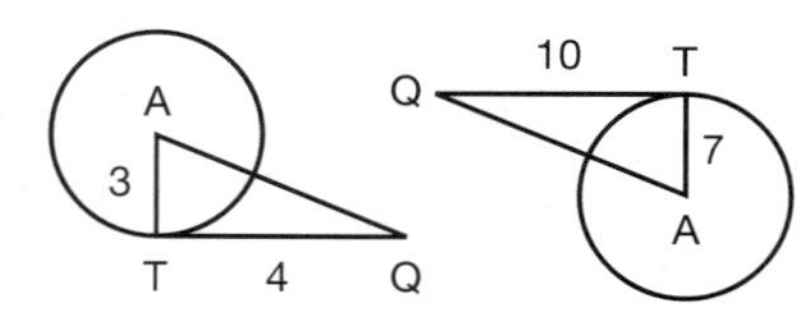

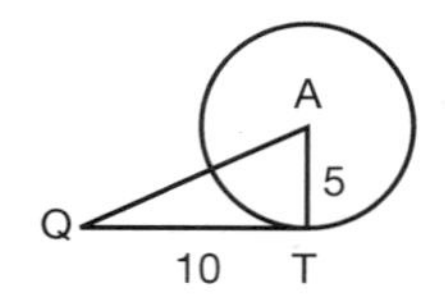

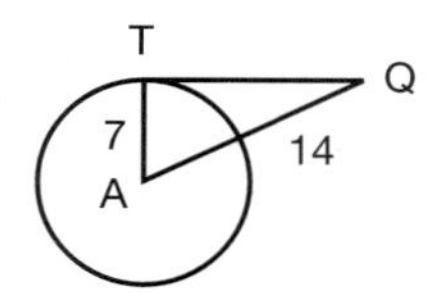

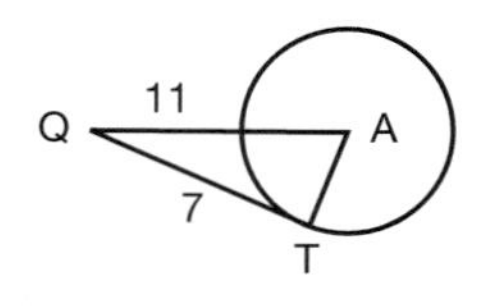

15 In each diagram, AB is the tangent at B to the circle with centre C.
All measurements given are in metres.
Calculate, correct to 2 significant figures, the length of the radius in each case.

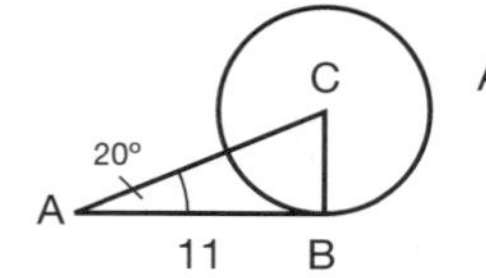

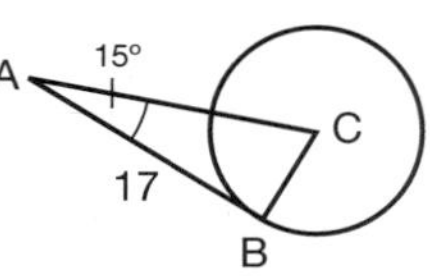

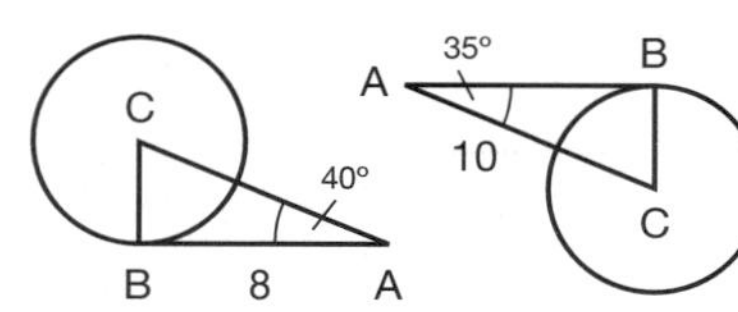

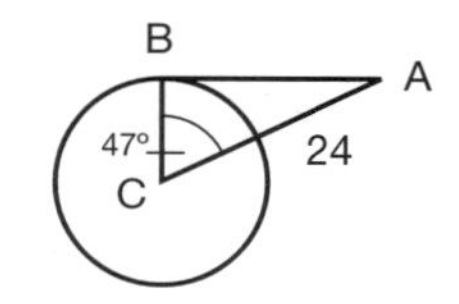

16B Calculate, correct to the nearest metre, the length of the tangent in each diagram. All lengths are given in metres.

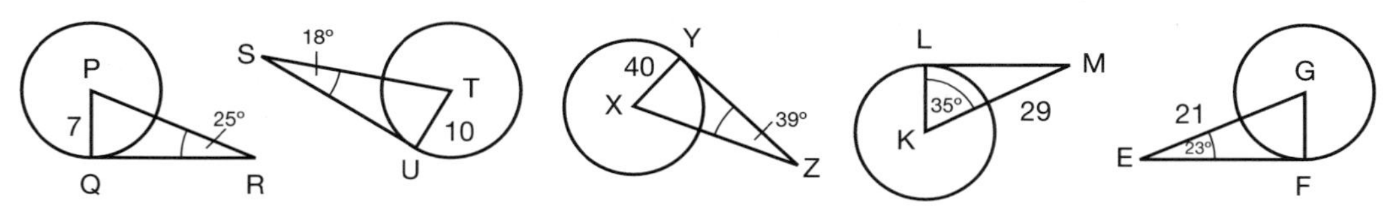

38 The Angle in a Semi-circle

Reminder

If AB is a diameter of a circle and C lies on the circumference, then $A\hat{C}B = 90°$.
This is often abbreviated to the statement that 'an angle in a semi-circle is a right angle'.

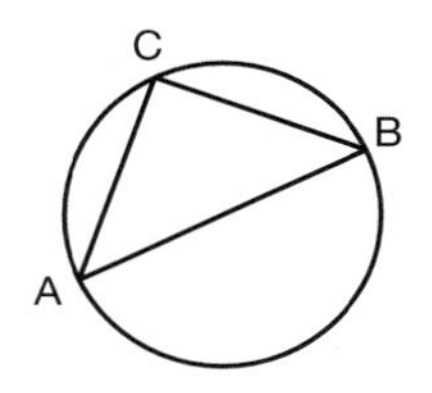

Example C lies on the circumference of a circle with diameter AB.
AC = 45 mm and BC = 28 mm
Calculate a) the length of AB
b) the size of (i) $C\hat{A}B$
(ii) $A\hat{B}C$.

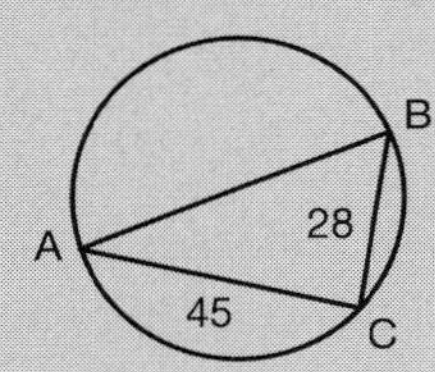

Solution a) $A\hat{C}B = 90°$ (angle in a semi-circle)
Therefore using the Theorem of Pythagoras:

$$AB = \sqrt{45^2 + 28^2} = \sqrt{2025 + 784} = \sqrt{2809} = 53 \text{ mm}$$

b) (i) $\tan C\hat{A}B = \frac{28}{45} \Rightarrow C\hat{A}B = 31{\cdot}9°$
(ii) $A\hat{B}C = 90° - C\hat{A}B = 90° - 31{\cdot}9° = 58{\cdot}1°$

Exercise 38

1 In each diagram, PQ is a diameter of the circle which passes through R.
Find the sizes of *a*, *b*, *c*, *d* and *e*.

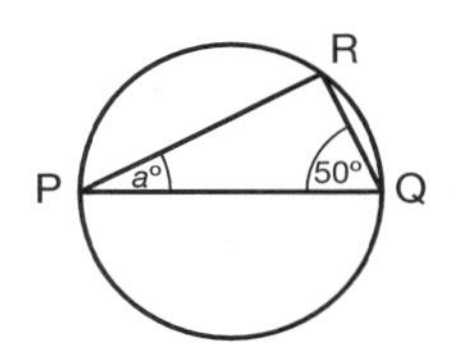

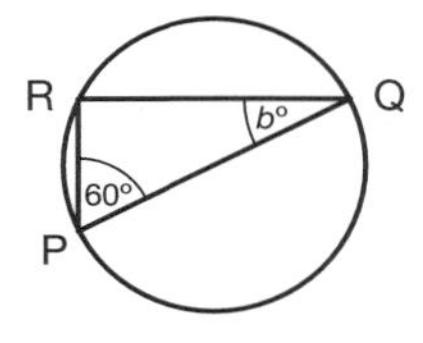

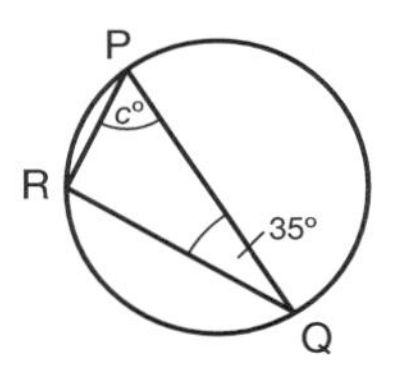

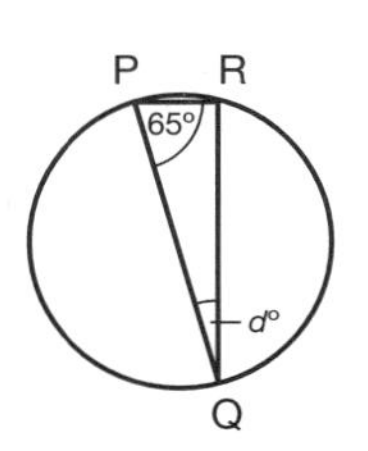

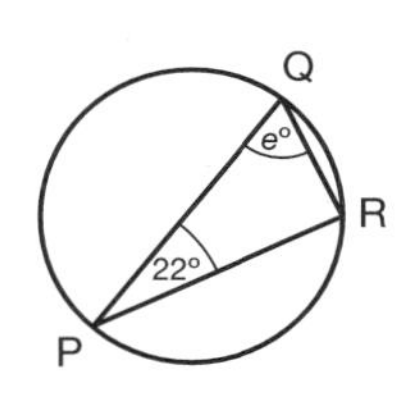

2 Calculate the length of the diameter of each circle. All lengths are given in millimetres.

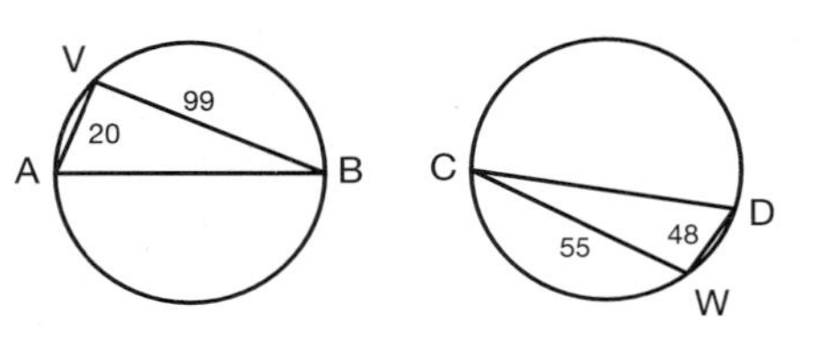

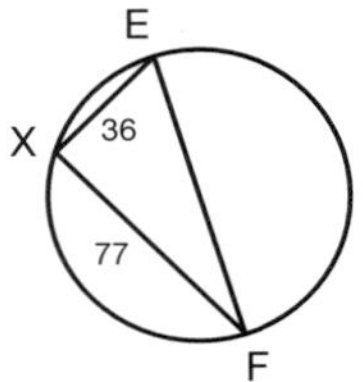

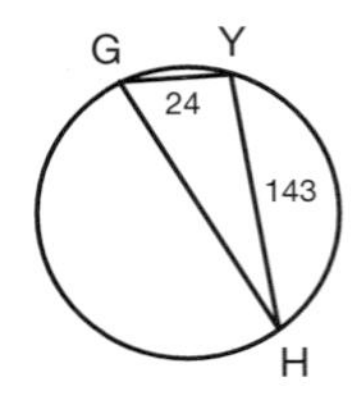

3 In each diagram, PQ is a diameter of the circle. All lengths are given in millimetres. Calculate the length of the unknown side in each triangle.

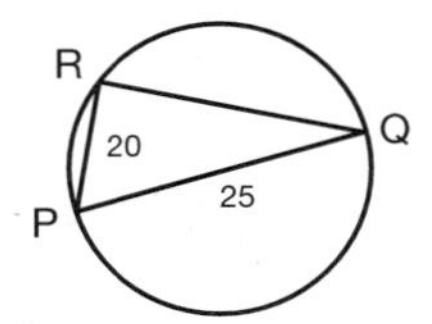

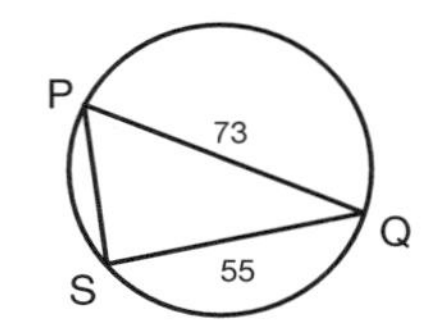

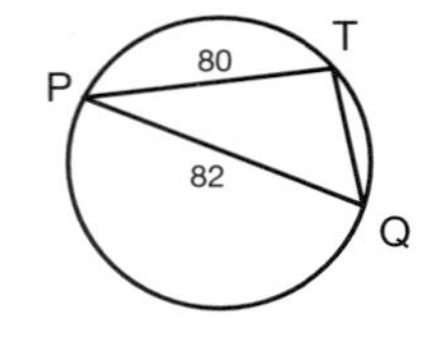

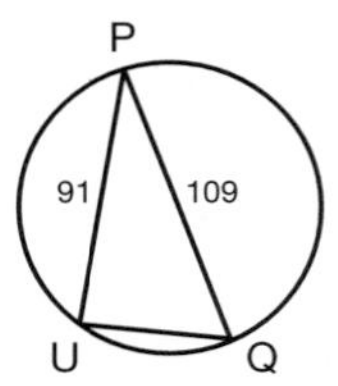

4 In this circle AB is a diameter.
All lengths are given in millimetres.
Calculate the length of BD.

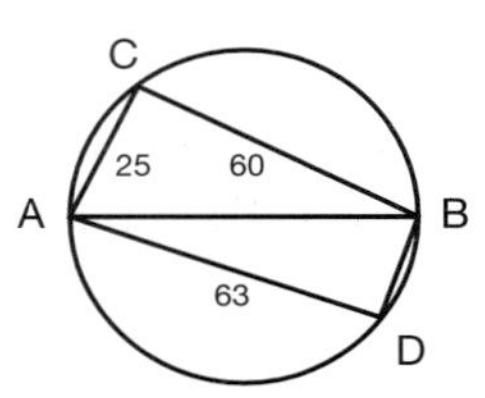

5 In this circle KL is a diameter.
Find the sizes of the three angles of triangle KLM, showing each step of your working.

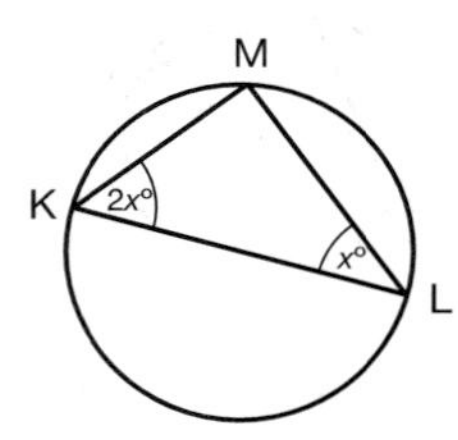

6 Repeat question 5 if $\hat{L} = 2x°$ and $\hat{K} = 3x°$.

7 In this circle KL is a diameter, MN is perpendicular to KL, and $M\hat{K}L = 25°$.
Find the size of $M\hat{N}L$, showing each step of your working.

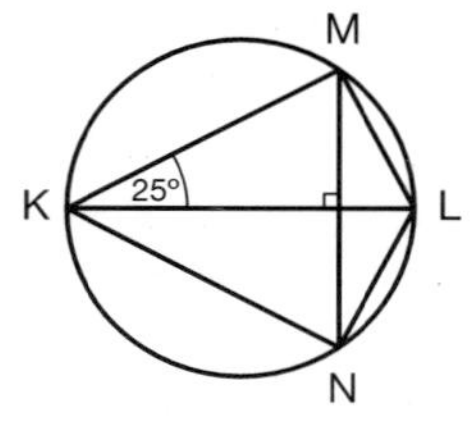

8 In this circle RS is a diameter and $V\hat{R}T = 20°$.
Find the size of $V\hat{S}T$, showing each step of your working.

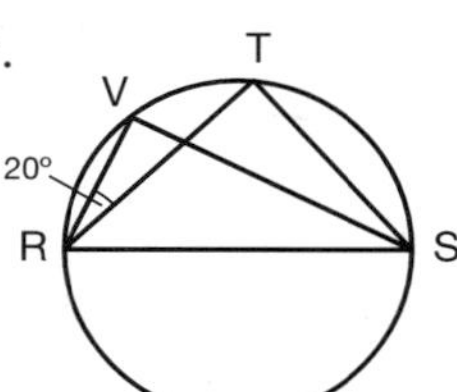

9 LN is a diameter of the circle with centre M, and $K\hat{M}N = 80°$. Find the size of $K\hat{L}M$, showing each step of your working.

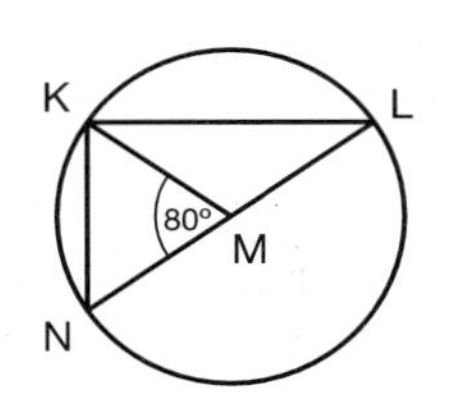

10 In this circle DC is a diameter and AB is parallel to DC. Also, $A\hat{D}B = 28°$ and $B\hat{D}C = 31°$.
Calculate the size of $B\hat{A}C$, showing each step of your working.

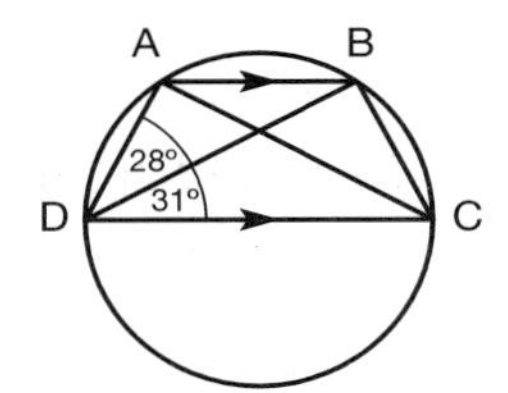

11 The line TAN is the tangent at A to the circle with centre O; POQ is a diameter and $Q\hat{A}N = 26°$.
Calculate the size of $A\hat{P}Q$, showing each step of your working.

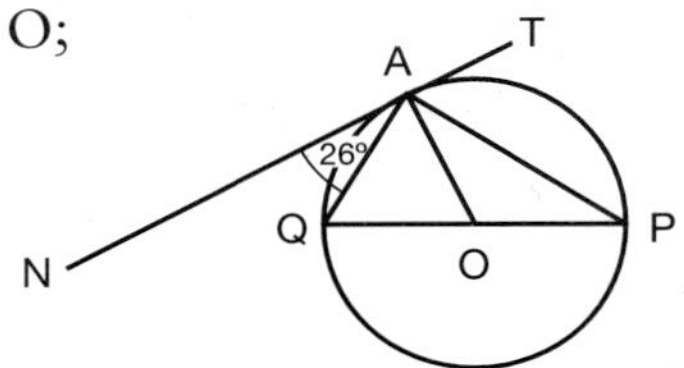

12 In each diagram, PQ is a diameter of the circle which passes through R.
All the lengths are given in metres. Calculate, correct to one decimal place, the sizes of a, b, c, d and e.

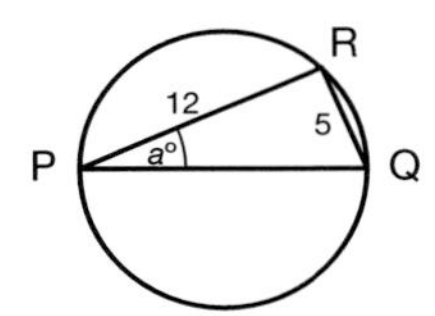

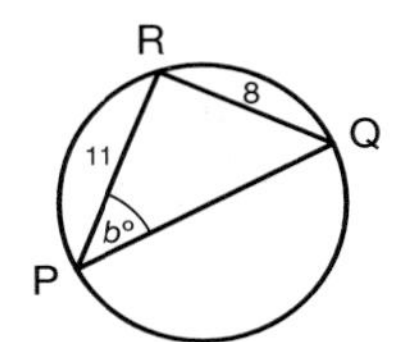

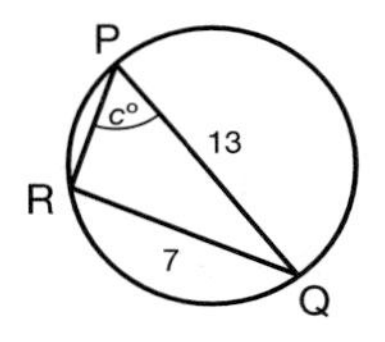

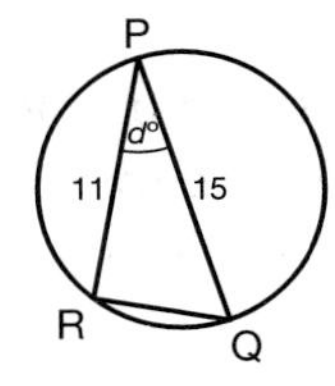

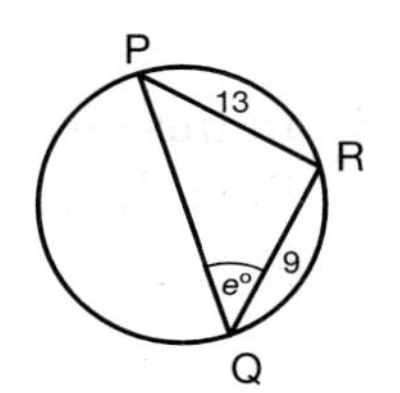

13 In each diagram, AB is a diameter of the circle. All the lengths are given in metres.
Calculate, correct to 2 significant figures, the lengths of BC, AD and AE.

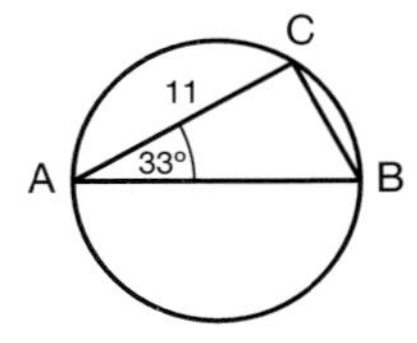

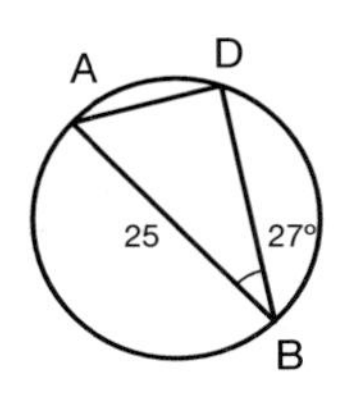

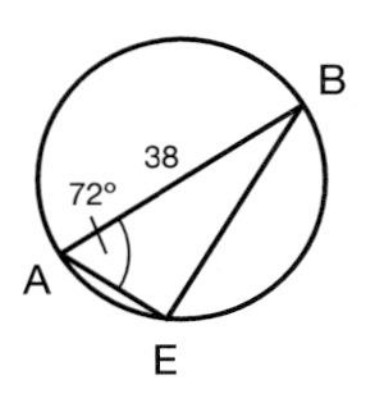

14B In these circles XY, PQ and AB are diameters. All the lengths are given in metres.
Calculate, correct to 3 significant figures, the lengths of XZ, PQ and AB.

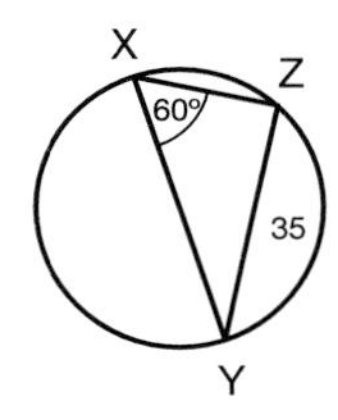

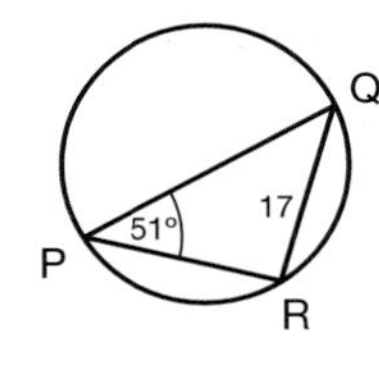

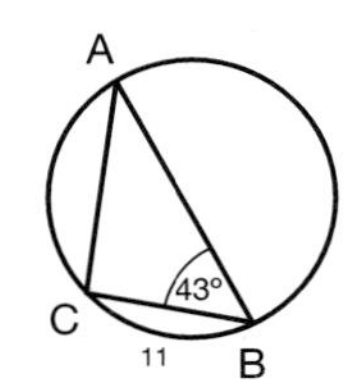

39 The Centre of a Circle and the Perpendicular Bisector of a Chord

Reminder

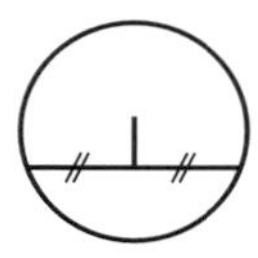

a line from the centre of a circle to the midpoint of a chord **is perpendicular to the chord**

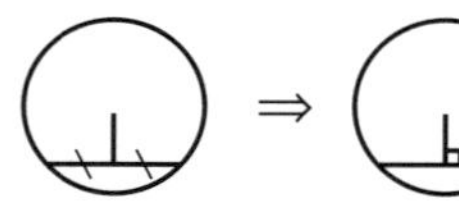

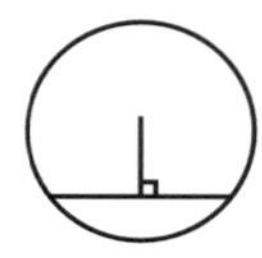

a line from the centre of a circle perpendicular to a chord **bisects the chord** i.e. passes through the midpoint

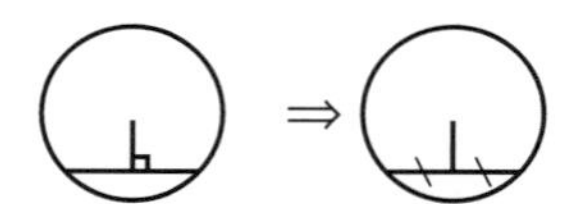

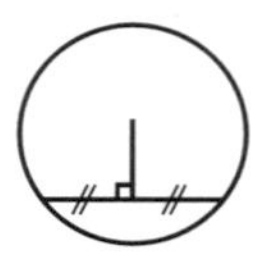

the perpendicular bisector of a chord of a circle **passes through the centre** of the circle

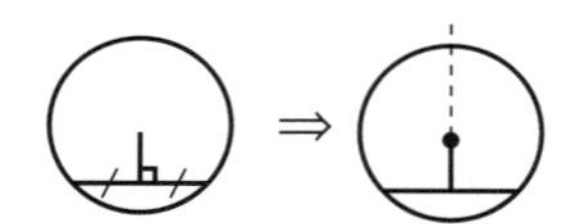

Example Find the centre of the circle which passes through the points A (4, 0), B (0, 4) and C (10, 4).

Solution Plot the points to help to understand the problem.
AB, BC and AC are chords of this circle, so the centre is where the perpendicular bisectors of these chords meet (AB and BC are the easiest two to deal with)
The midpoint of AB is (2, 2),
so the perpendicular bisector of AB has equation $y = x$.
The midpoint of BC is (5, 4),
so the perpendicular bisector of BC has equation $x = 5$.
These lines meet at (5, 5), so (5, 5) is the centre of the circle

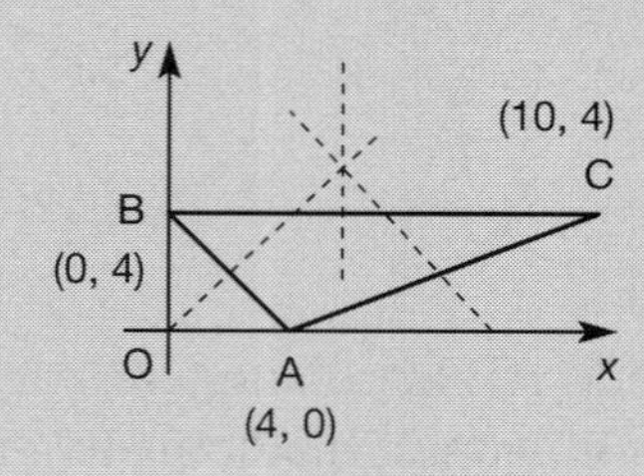

Exercise 39

1 Find the coordinates of C, the centre of the circle which passes through the points O (0, 0), A (6, 0) and B (0, 8).

2 Find the coordinates of C, the centre of the circle which passes through the points P (0, 39), Q (0, 9), R (13, 0) and S (27, 0), and verify that the radii CP, CQ, CR, CS are all equal in length.

3 AB is a chord of length 240 mm in a circle with centre C and radius 130 mm. Calculate the length of CM, the distance of the chord from the centre of the circle.

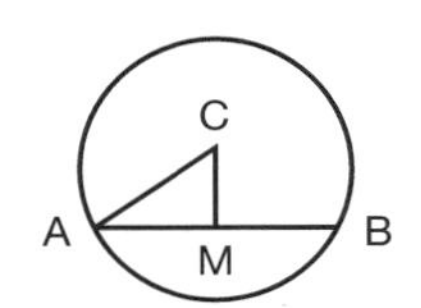

4 A chord KL of length 30 m is 8 m away from the centre C of the circle shown. Calculate the length of the radius of the circle.

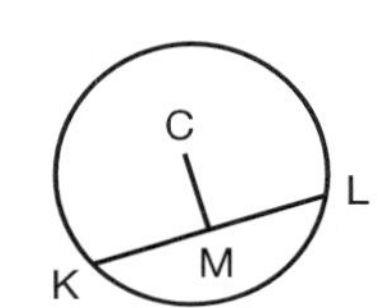

5 A chord PQ of a circle of radius 25 mm is 7 mm away from the centre of the circle. Calculate the length of the chord PQ.

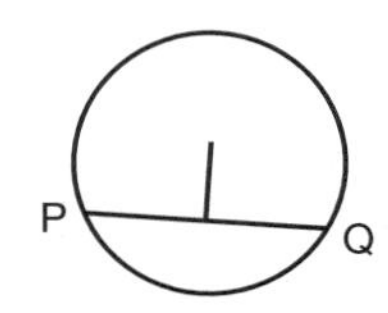

6 PQ is a diameter of a circle with centre O. R is the midpoint of the chord SQ. Prove that PS is parallel to OR.

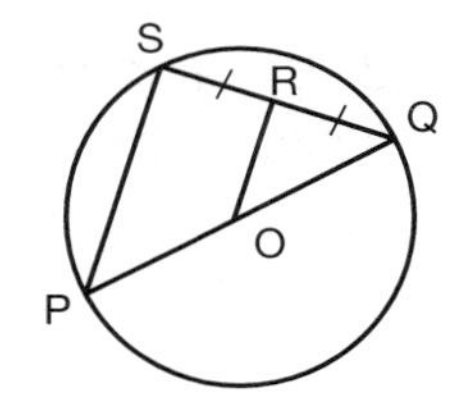

7 The line TAN is the tangent at A to the circle with centre C, and AE is a diameter. D is the midpoint of the chord AB, and $N\hat{A}D = 68°$. Find the size of $D\hat{C}E$, showing each step of your working.

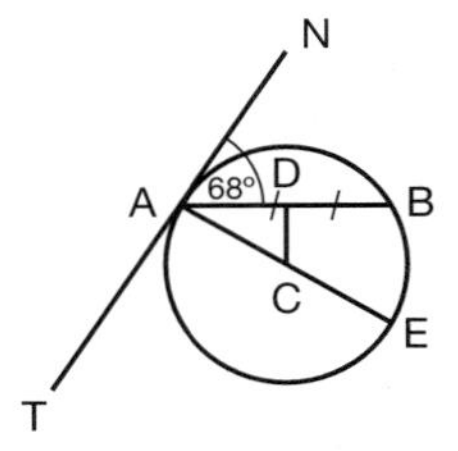

8 AF is the tangent at A to the circle with centre C. AB is a diameter, E is the midpoint of the chord AD, and $C\hat{F}A = 36°$. Find the size of $A\hat{B}D$, showing each step of your working.

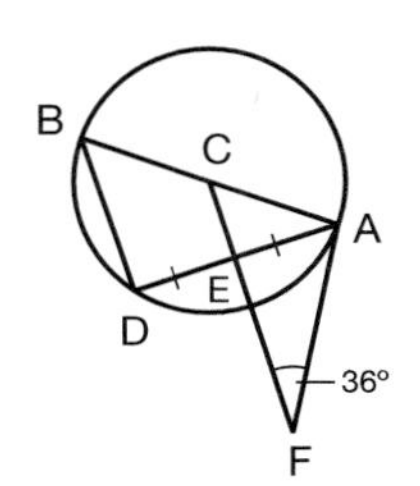

9B The diagram shows the circular cross-section of an oil tank of radius 1·5 m.
PQ represents the surface of the oil and AB is the dip stick used to measure the depth of the oil.
Calculate the width of the surface of the oil when it is 0·6 m deep.

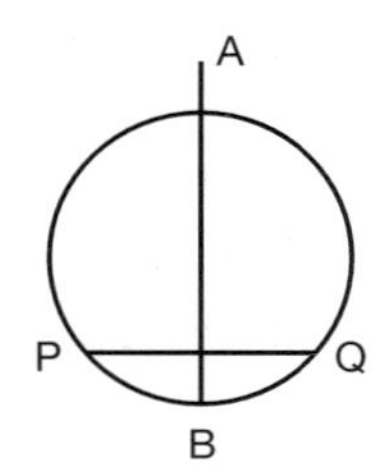

10B In this diagram PQCR is a tangent kite.
The line PC cuts the circle at A (and when produced, at B).
QC = 12 m and PQ = 35 m.
Calculate a) the length of AP
b) the size of $C\hat{B}R$.

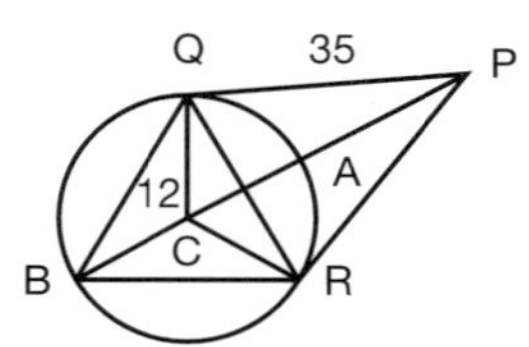

40 Revision of Circle Properties

Exercise 40

Round each answer to 1 decimal place.

1 AB is an arc of the circle with centre C and radius 15 mm, as shown.
AB subtends an angle of 40° at C.
Calculate: a) the length of the arc AB
b) the area of the sector ACB.

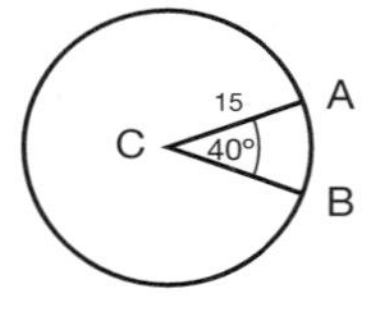

2 In the circle with centre O and radius 17·3 mm, the arc PQ has length 15·1 mm.
Calculate: a) the size of $P\hat{O}Q$
b) the area of sector POQ.

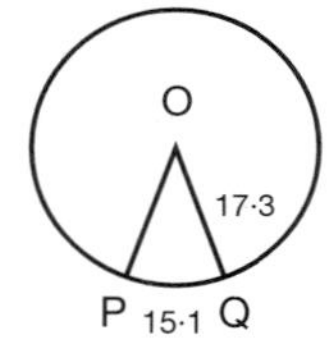

3 In the circle with centre L and radius 59 mm, the sector KLM has an area of 4253 mm^2.
Calculate: a) the size of $K\hat{L}M$
b) the length of the arc KM.

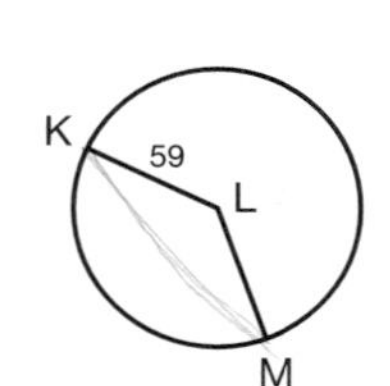

4 A silver pendant is in the shape of the shaded region shown.
If the inner and outer radii of the circles are 56mm and 82mm, and the angle at the centre is 35°, calculate the area of one side of this pendant.

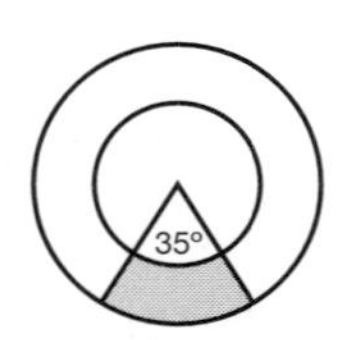

5 Acorn Investments plc wish to adopt a new logo based on concentric circles as shown. The angle at the centre is 60°.
The areas shaded horizontally and vertically are equal.
The inner radius is 12mm.
Calculate the length of the outer radius.

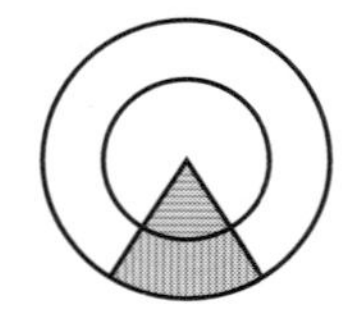

6 The diagram shows what is known as a No. 2 radius double curve for an N gauge railway set.
Eight of these curves fit together to make a complete circle.
The length of the inner rail is 207 mm, and N gauge rails are 9 mm apart.
Calculate, to the nearest millimetre, the length of the outer rail.

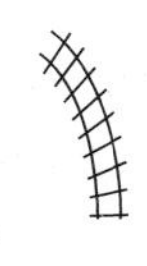

7B In the diagram FOG is a sector of a circle with centre O with $F\hat{O}G = 100°$.
The length of the arc FG is 34·9 m.
Calculate, correct to the nearest square metre, the area of the sector FOG.

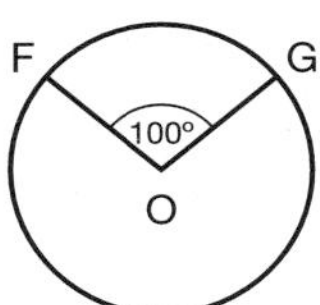

8B RST is a sector of a circle with centre S and radius 6·98 mm.
The area of sector RST is 34 mm^2.
Calculate the length of the arc RT.

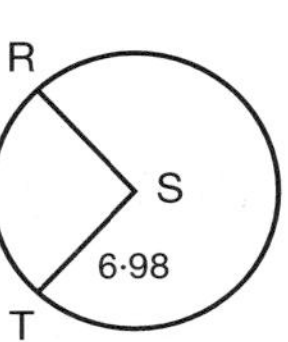

9 AB is a diameter of the circle with centre C, and AT is the tangent at A to the circle.
The line TC produced meets the circle at P, and $P\hat{A}B = 24°$.

a) Calculate the size of $A\hat{T}C$.
b) Given that AT = 12 units, calculate the length of the diameter of the circle.

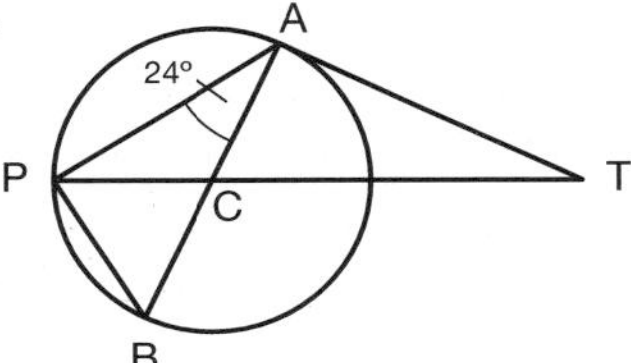

10 AB is a diameter of the circle with centre O. The point T lies on AB produced and PT is the tangent at P to the circle.

a) If $B\hat{T}P = 32°$, calculate the size of (i) $B\hat{P}T$
(ii) $O\hat{P}A$.
b) If the radius of the circle is 33·1 m, calculate the length of the tangent PT correct to 3 significant figures.

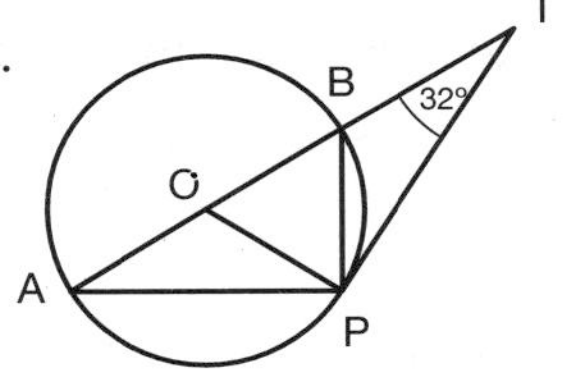

11 AB is a diameter of the circle with centre O and radius 5 units.
OM is the perpendicular bisector of the chord BC,
AC = 8 units, $C\hat{A}B = x°$, and $M\hat{O}B = y°$.

a) What is the relationship between x and y?
b) Calculate the value of x.
c) Find the length of OM.
d) What is the relationship between AC and OM?

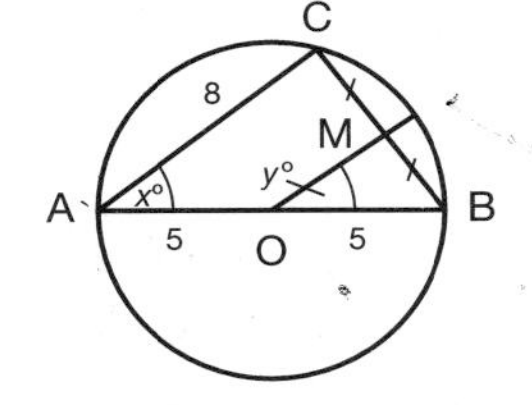

12 The diagram shows two concentric circles with centre O and radii 35 mm and 53 mm.
The chord AB of the outer circle is 28 mm away from O and cuts the inner circle at C and D.
Calculate the length of AC.

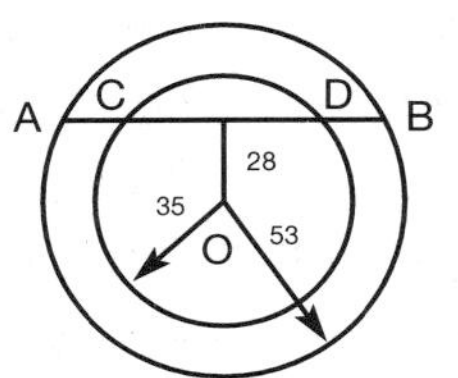

13 Two circles, one with centre P and radius 53 mm, and the other with centre Q and radius 75 mm intersect in R and S. The common chord RS has length 90 mm. Calculate the length of PQ.

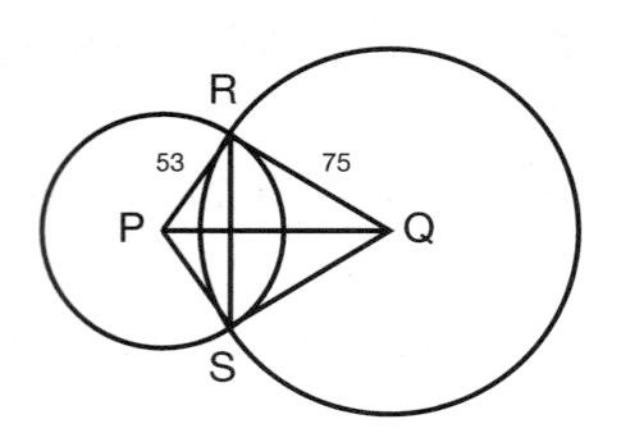

14 AB is a diameter of the circle with centre O and radius 97 mm. The parallel chord CD and the diameter AB are 65 mm apart. Calculate the length of CD.

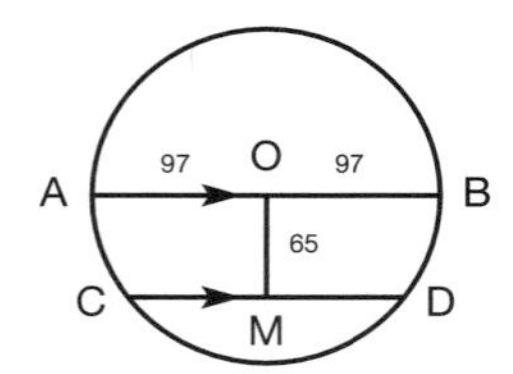

15 AB and CD are two equal parallel chords of a circle of radius 101 mm.
They are 40 mm apart.
Calculate the length of each chord.

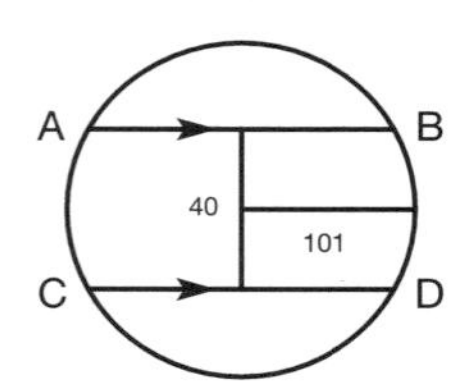

16B XY is a chord of length 30 mm in the circle with centre O. The perpendicular from O to XY meets XY in W and the circle at Z.
If WZ = 9 mm, calculate the length of the radius.

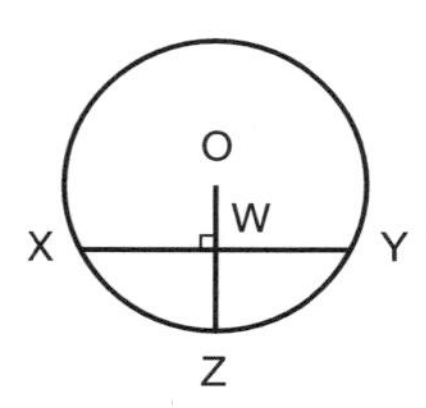

17H KL and MN are parallel chords (on opposite sides of the centre) of a circle.
KL = 102 mm and MN = 72 mm.
They are 145 mm apart.
Calculate the length of the radius of the circle.

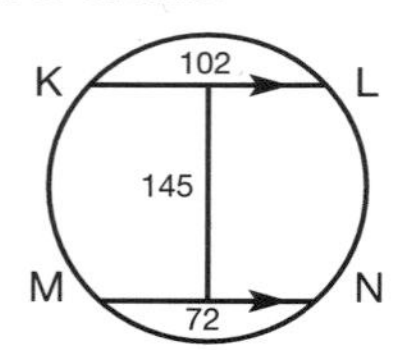

18B The diagram shows the doorway into an alcove in the Altvennie Turkish Baths. It consists of part of a circle of diameter 10 feet and three sides of a square of side 6 feet.
Calculate:

a) the height of the centre of the circle above the top of the square
b) the perimeter of this doorway (including the floor line)
c) the area of this doorway.

19B A circular sheet of card of radius 150 mm has a 120° sector removed, and the remaining card is made into a cone to form a party hat.

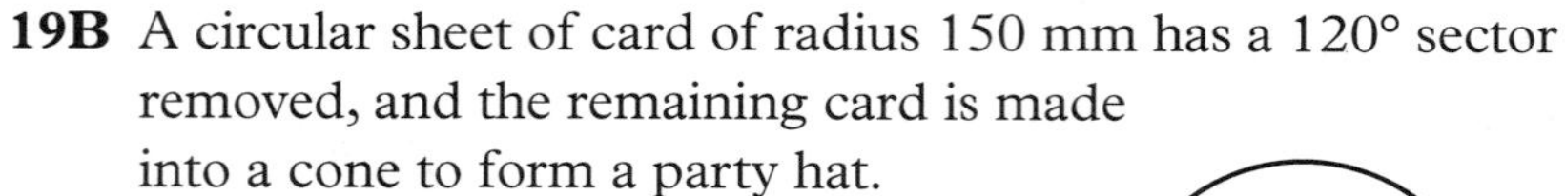

Calculate: a) the radius of the hat
b) the height of the hat.

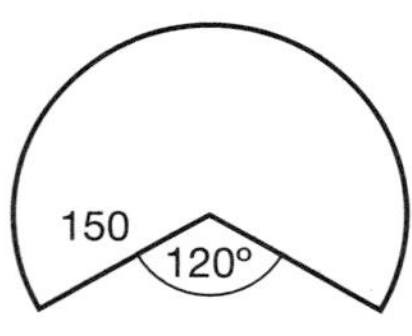

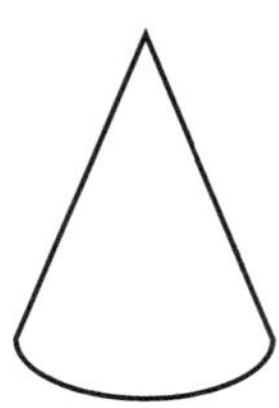

41 Test on Circle Properties

Allow 45 minutes for this test

1 Calculate the length of the arc PQ.

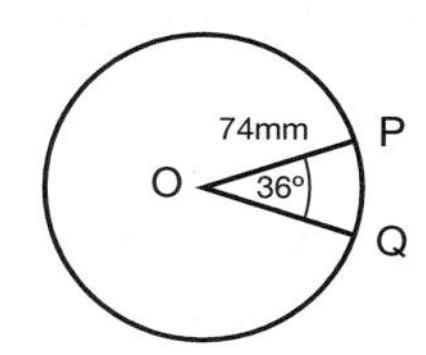

2 Calculate the area of the sector ACB.

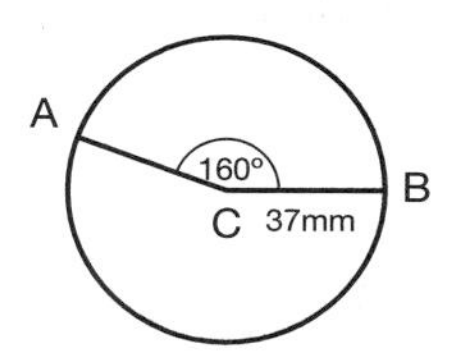

3 CD is the tangent at C to the circle with centre E; $C\hat{D}E = 27°$.
Find the size of $C\hat{E}D$.

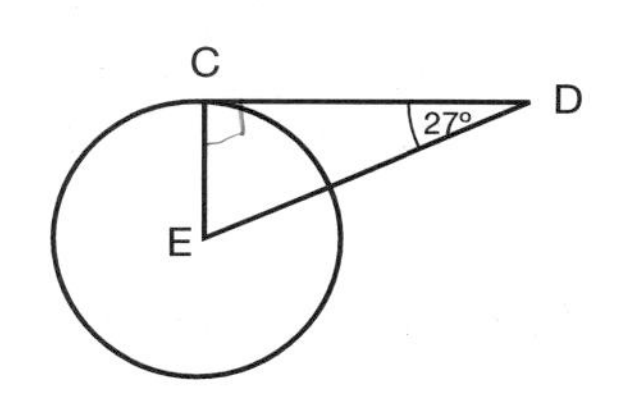

4 KM is a diameter of the circle and $L\hat{K}M = 38°$.
Find the size of $L\hat{M}K$.

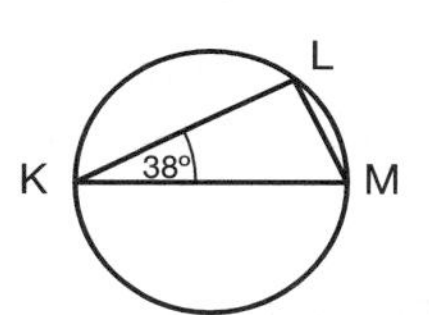

5 W is the midpoint of the chord XY of the circle with centre V; $W\hat{Y}V = 41°$.
Calculate the size of $X\hat{V}W$.

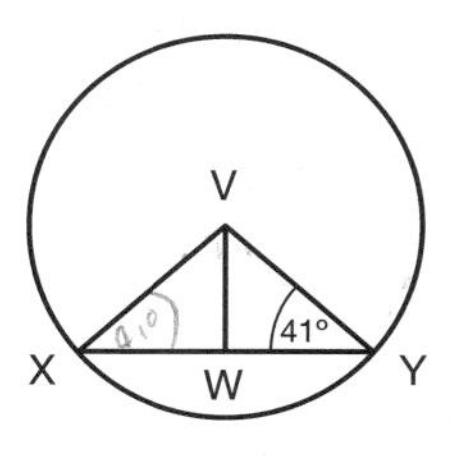

6 The length of the arc CE is 69·1 mm.
The length of the radius DE is 27·5 mm.
Calculate the size of $C\hat{D}E$.

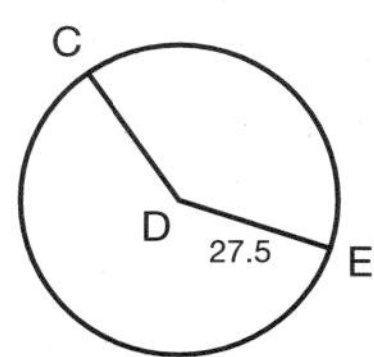

7 The area of the sector POQ is 736 mm^2.
The radius OQ is 75 mm.
Calculate the size of $P\hat{O}Q$.

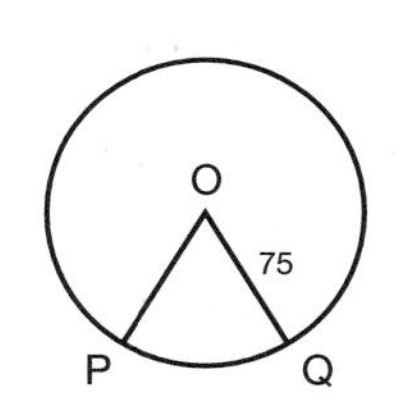

8 Line PQR is the tangent at Q to the circle with centre S. The lengths shown are in mm.
Find: a) the length of the radius
b) the size of $Q\hat{R}S$.

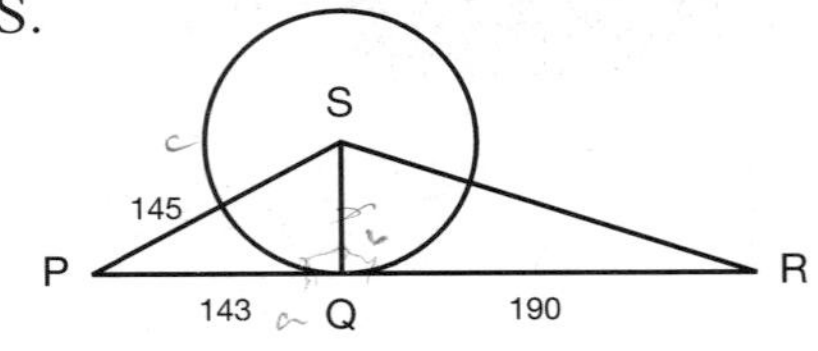

9 AC is a diameter of the circle shown, and AB = 36 mm, BC = 77 mm, $C\hat{A}D$= 37°.
a) Calculate the length of the diameter.
b) Calculate the length of the chord CD.

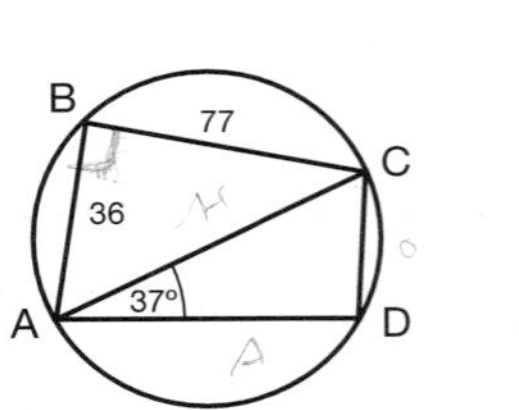

10 TA and TB are the tangents at A and B to the circle with centre C. TC cuts the circle at D and E, and $A\hat{E}C$ = 25°.
Calculate the size of $D\hat{B}T$.

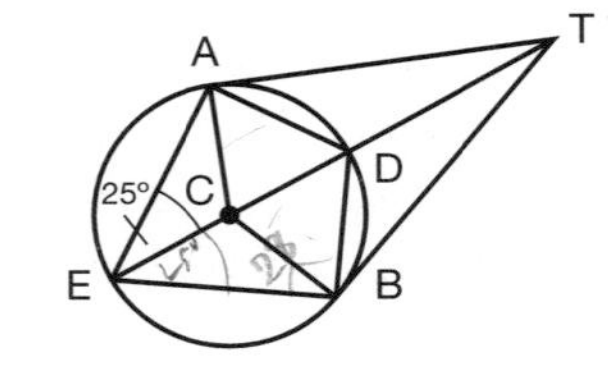

TRIGONOMETRY

42 Revision of Right-Angled Triangle Trigonometry

Reminder

There are nine types of calculation involved, three for each of sine, cosine and tangent. You have to be able to calculate an angle, or the 'easy' side, or the 'awkward' side. The third type requires one more line of working and is part of the Credit syllabus, but not the General.
First of all, ensure that your calculator is in degree mode.

	angle	'easy' side	'awkward' side
sine	(*all lengths are in metres*) 13, 5, $x°$ $\sin x° = \frac{5}{13}$ $\Rightarrow x = 22{\cdot}6°$	73, p, 41° $\sin 41° = \frac{p}{73}$ $\Rightarrow p = 73 \times \sin 41°$ $= 47{\cdot}9$	a, 14, 37° $\sin 37° = \frac{14}{a}$ $\Rightarrow a \times \sin 37° = 14$ $\Rightarrow a = \frac{14}{\sin 37°} = 23{\cdot}3$
cosine	14, 11, $y°$ $\cos y° = \frac{11}{14}$ $\Rightarrow y = 38{\cdot}2°$	52, q, 49° $\cos 49° = \frac{q}{52}$ $\Rightarrow q = 52 \times \cos 49°$ $= 34{\cdot}1$	b, 42, 55° $\cos 55° = \frac{42}{b}$ $\Rightarrow b \cos 55° = 42$ $\Rightarrow b = \frac{42}{\cos 55°} = 73{\cdot}2$

Reminder continued

	angle	'easy' side	'awkward' side
tangent	(*all lengths are in metres*) 13, 11, $z°$ $\tan z° = \frac{11}{13}$ $\Rightarrow z = 40{\cdot}2°$	27, r, 53° $\tan 53° = \frac{r}{27}$ $\Rightarrow r = 27 \times \tan 53°$ $= 35{\cdot}8$	c, 84, 67° $\tan 67° = \frac{84}{c}$ $\Rightarrow c \times \tan 67° = 84$ $\Rightarrow c = \frac{84}{\tan 67°} = 35{\cdot}7$

Example Find the value of t.
(The lengths are in millimetres.)

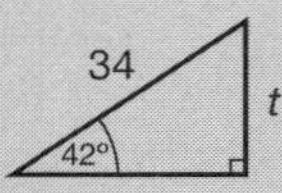

Solution Decide first which trig ratio (sine, cosine or tangent) is appropriate.
How are 't' and '34' related to the given acute angle, 42°?
't' is *opposite* and '34' is the *hypotenuse*
Recall that $\sin(\angle) = \frac{\text{opposite}}{\text{hypotenuse}}$ and hence use sine. i.e. $\sin 42° = \frac{t}{34}$
now multiply both sides by 34 $34 \times \sin 42° = t$
hence $t = 22{\cdot}75$

Note: To use 'SOHCAHTOA', you must remember how to spell it.

Exercise 42

1 Write down the value of:
a) $\sin p$ b) $\cos p$ c) $\tan p$.

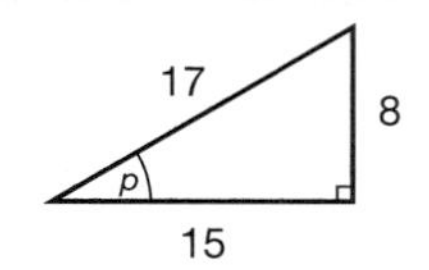

2 Write down the value of:
a) $\sin q$ b) $\cos q$ c) $\tan q$.

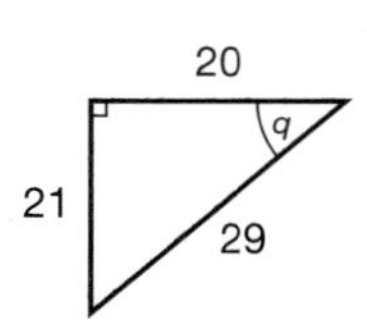

3 Write down the value of:

a) $\sin r$ b) $\cos r$ c) $\tan r$.

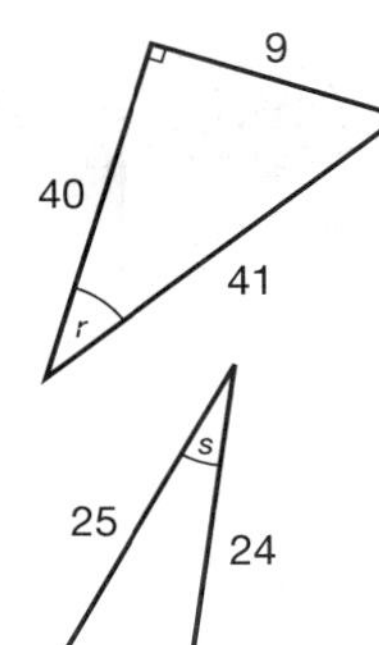

4 Write down the value of:

a) $\sin s$ b) $\cos s$ c) $\tan s$.

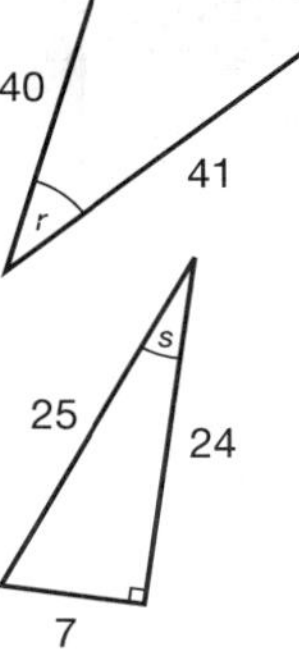

In each of the questions 5 to 8, write down a ratio (in terms of the letters used for the lengths of the sides) for: a) sin A b) cos A c) tan A.

5

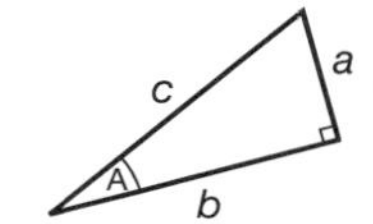

6

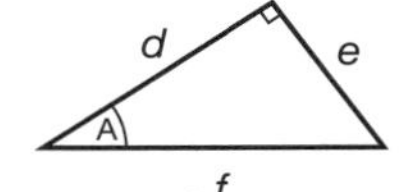

7

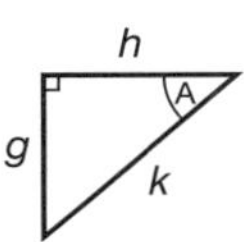

8

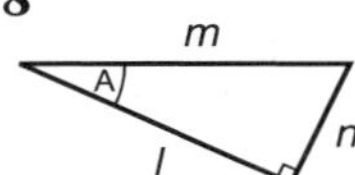

9 Find, correct to one decimal place, the sizes of a, b and c.

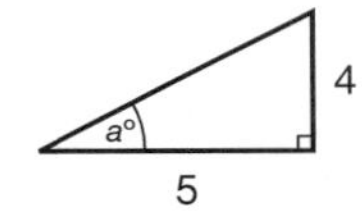

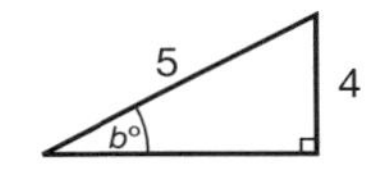

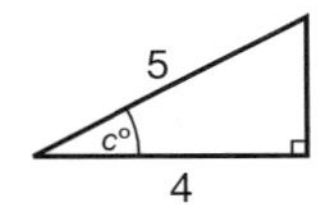

10 Find, correct to three significant figures, the values of p, q and r.

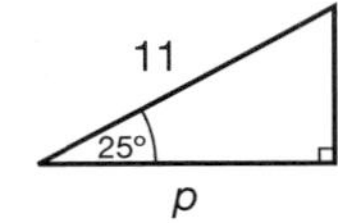

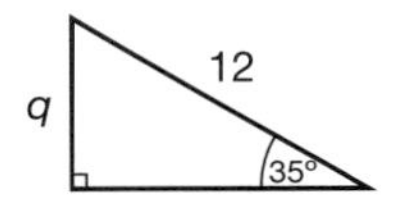

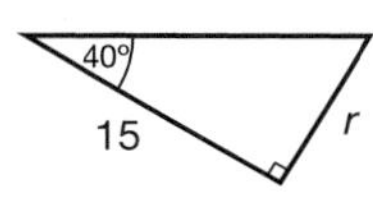

11 Find, correct to three significant figures, the values of x, y and z.

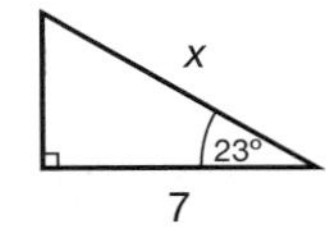

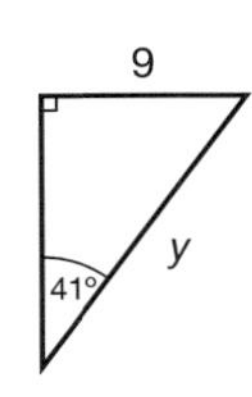

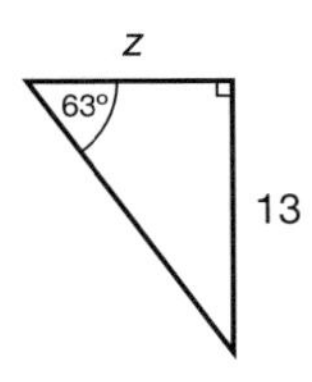

12 A ship sails 10 km from a port P on a bearing of 065° to a point S. [Remember that bearings are measured clockwise from North.]

a) How far east of P is S?

b) How far north of P is S?

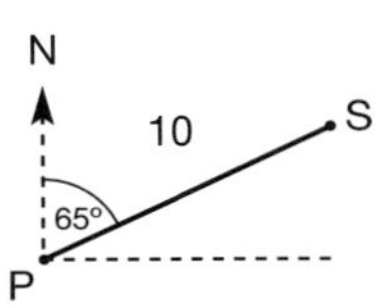

13 A ladder of length 4 m standing on level ground leans at an angle of 70° to the horizontal against a vertical wall.

a) How far up the wall does the ladder reach?

b) How far from the wall is the foot of the ladder?

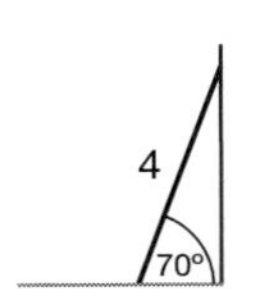

14 On level ground, a javelin rests as shown on a medicine ball which has a diameter of 300 mm.
The tip of the javelin (on the ground) is 1·6 m from where it touches the ball.
Calculate the angle between the javelin and the ground.

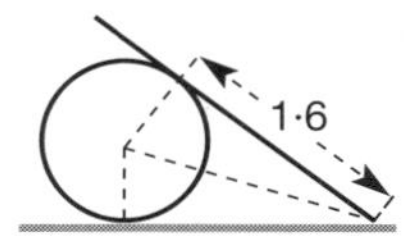

15 A technician is creating a clock face for a new town clock. T, F and S are the points on the circumference where the hours twelve, four and six appear, and C is the centre.
a) Write down the size of (i) $S\hat{C}F$ (ii) $C\hat{T}F$
b) If TF = 170 mm and SF = 95 mm, does this confirm that the clockface has been made accurately?

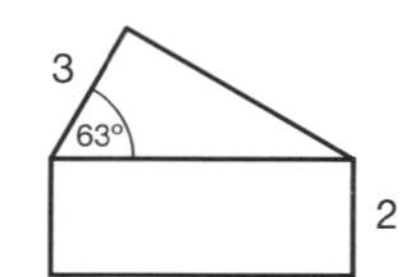

16 A holiday chalet is built with an upstairs lounge.
The south facing windows are 3 m tall and are inclined at 63° to the lounge floor which is 2 m above the ground floor.
Calculate the overall height of the chalet.

17 A landscape gardener plans to instal a circular fish pond of diameter 4 m in the corner of a rectangular garden and set it in a triangular area of crazy paving with angles of 30°, 60° and 90°, as shown.

If C represents the corner of the garden and AB the edge of the crazy paving, calculate the lengths of AC and BC and hence the area of crazy paving required.

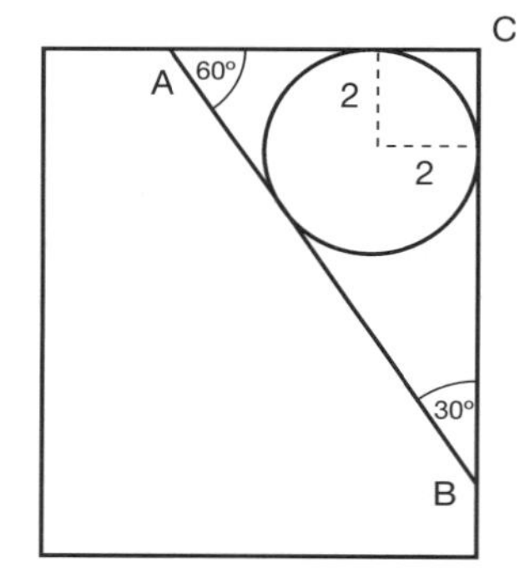

Reminder

For $0 \leqslant x \leqslant 90$, these are the graphs of $y = \sin x°$, $y = \cos x°$ and $y = \tan x°$:

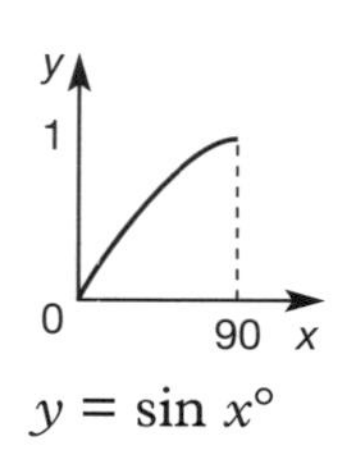

$y = \sin x°$

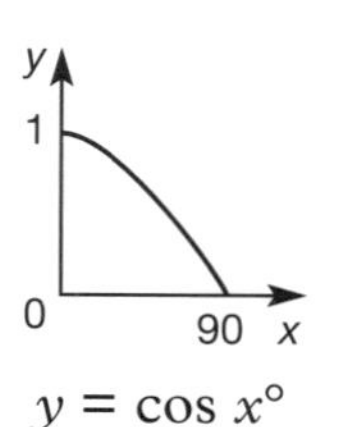

$y = \cos x°$

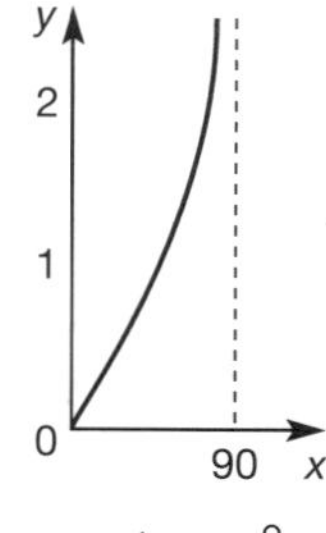

$y = \tan x°$

The graphs of sine and tangent are increasing, but the graph of cosine is decreasing.
The graph of tangent has an 'asymptote' at $x = 90$.

18 Without referring to your calculator, write down the value of:

a) $\sin 0°$ b) $\cos 0°$ c) $\tan 0°$ d) $\cos 90°$
e) $\sin 90°$ f) $\tan 90°$.

19 Without referring to your calculator, solve these equations:

a) $\sin x° = 1$ b) $\cos x° = 0$ c) $\tan x° = 0$ d) $\cos x° = 1$.

43 Angles greater than 90°

Reminder

For non-acute angles, it is inadequate to use the simple acute angle definitions of sine, cosine, and tangent (i.e. opposite over hypotenuse, adjacent over hypotenuse, and opposite over adjacent).

We need to use the more formal definitions: $\sin A = \frac{y}{r}$

$\cos A = \frac{x}{r}$

$\tan A = \frac{y}{x}$

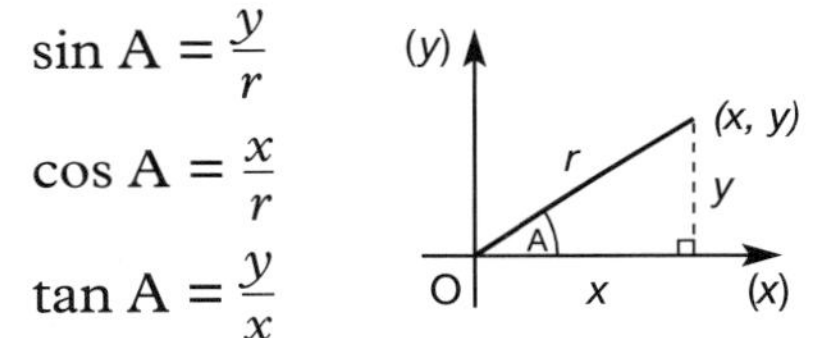

The x, y, r definitions are clearly equivalent to the opp., adj., hyp. definitions, but they have the advantage that they can also be applied to angles of any size; 'r' is always positive.

($\hat{A}$ is a second quadrant angle)

$$\left.\begin{array}{l} x < 0 \\ y > 0 \\ r > 0 \end{array}\right\} \Rightarrow \left\{\begin{array}{l} \sin A > 0 \\ \cos A < 0 \\ \tan A < 0 \end{array}\right.$$

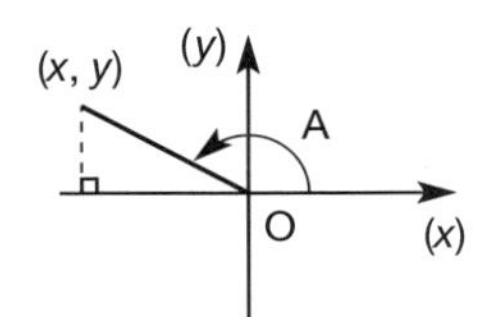

($\hat{A}$is a third quadrant angle)

$$\left.\begin{array}{l} x < 0 \\ y < 0 \\ r > 0 \end{array}\right\} \Rightarrow \left\{\begin{array}{l} \sin A < 0 \\ \cos A < 0 \\ \tan A > 0 \end{array}\right.$$

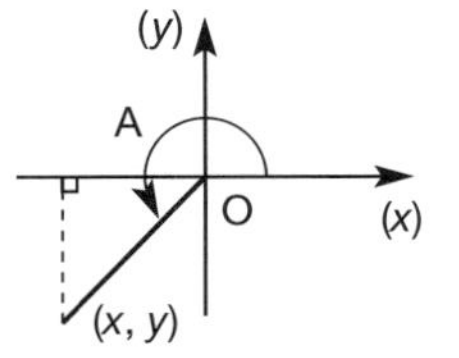

($\hat{A}$is a fourth quadrant angle)

$$\left.\begin{array}{l} x > 0 \\ y < 0 \\ r > 0 \end{array}\right\} \Rightarrow \left\{\begin{array}{l} \sin A < 0 \\ \cos A > 0 \\ \tan A < 0 \end{array}\right.$$

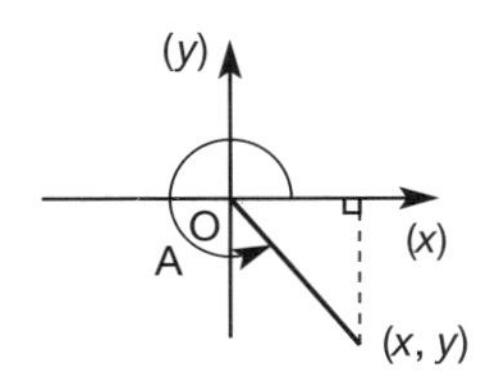

Reminder continued

Combining these results leads to the familiar $\frac{S \mid A}{T \mid C}$ diagram, which is a summary of:

sin(∠) > 0 in the 1st and 2nd quadrants	sin(∠) < 0 in the 3rd and 4th quadrants
cos(∠) > 0 in the 1st and 4th quadrants	cos(∠) < 0 in the 2nd and 3rd quadrants
tan(∠) > 0 in the 1st and 3rd quadrants	tan(∠) < 0 in the 2nd and 4th quadrants

Although in practice your calculator will tell you the sin, cos, or tan of any angle directly, in order to work backwards, you need to relate every non-acute angle to an appropriate acute angle, and find the sign from 'all, sin, tan, cos' (or a graph).

The appropriate acute angle: in the 2nd quadrant is (180° – A),
in the 3rd is (180° + A), and
in the 4th is (360° – A).

(e.g. $\sin 130° = \sin (180 - 50)° = + \sin 50° = 0{\cdot}766$
$\cos 220° = \cos (180 + 40)° = - \cos 40° = -0{\cdot}766$
$\tan 340° = \tan (360 - 20)° = - \tan 20° = -0{\cdot}364$)

And working backwards:

$\sin X = 0{\cdot}700 \Rightarrow X = 44{\cdot}4°$ or $(180° - 44{\cdot}4°) = 44{\cdot}4°$ or $135{\cdot}6°$
$\cos Y = 0{\cdot}200 \Rightarrow Y = 78{\cdot}5°$ or $(360° - 78{\cdot}5°) = 78{\cdot}5°$ or $281{\cdot}5°$
$\tan Z = 0{\cdot}300 \Rightarrow Z = 16{\cdot}7°$ or $(180 + 16{\cdot}7°) = 16{\cdot}7°$ or $196{\cdot}7°$

Take care with negative values. First determine the relevant quadrants, then find the appropriate acute angle A from your calculator, ignoring the negative sign.

e.g. $\sin P = -0{\cdot}700 \Rightarrow P = (180° + A)$ or $(360° - A)$ and $A = 44{\cdot}4°$
$\Rightarrow P = (180° + 44{\cdot}4°)$ or $(360° - 44{\cdot}4°) = 224{\cdot}4°, 315{\cdot}6°$
$\cos Q = -0{\cdot}200 \Rightarrow Q = (180° - A)$ or $(180° + A)$ and $A = 78{\cdot}5°$
$\Rightarrow Q = (180° - 78{\cdot}5°)$ or $(180° + 78{\cdot}5°) = 101{\cdot}5°, 258{\cdot}5°$
$\tan R = -0{\cdot}300 \Rightarrow R = (180° - A)$ or $(360° - A)$ and $A = 16{\cdot}7°$
$\Rightarrow R = (180° - 16{\cdot}7°)$ or $(360° - 16{\cdot}7°) = 163{\cdot}3°, 343{\cdot}3°$.

Exercise 43

1 Use your calculator to evaluate, correct to three decimal places:

a) sin 132° b) cos 132° c) tan 132°
d) sin 222° e) cos 222° f) tan 222°
g) sin 301° h) cos 301° i) tan 301°.

(Do not use your calculator for questions 2, 3 and 4.)

2 a) given that $\sin 47° = 0{\cdot}731$, find an obtuse angle whose sine is $0{\cdot}731$
b) given that $\sin 83° = 0{\cdot}993$, find an obtuse angle whose sine is $0{\cdot}993$
c) given that $\sin 11{\cdot}5° = 0{\cdot}199$, find an obtuse angle whose sine is $0{\cdot}199$.

3 a) given that $\cos 23° = 0{\cdot}921$, find an obtuse angle whose cosine is $-0{\cdot}921$
b) given that $\cos 48° = 0{\cdot}669$, find an obtuse angle whose cosine is $-0{\cdot}669$
c) given that $\cos 84° = 0{\cdot}105$, find an obtuse angle whose cosine is $-0{\cdot}105$.

4 a) given that $\tan 17° = 0{\cdot}306$, find an obtuse angle whose tangent is $-0{\cdot}306$
b) given that $\tan 77° = 4{\cdot}331$, find an obtuse angle whose tangent is $-4{\cdot}331$
c) given that $\tan 52° = 1{\cdot}280$, find an obtuse angle whose tangent is $-1{\cdot}280$.

5 Find two angles (between 0° and 360°) which satisfy the equations:
a) $\sin x° = 0{\cdot}578$ b) $\sin y° = 0{\cdot}901$ c) $\sin z° = 0{\cdot}309$.

6 Find the obtuse angle which satisfies the equation:
a) $\cos p° = -0{\cdot}545$ b) $\cos q° = -0{\cdot}967$ c) $\cos r° = -0{\cdot}738$.

Reminder

The graphs of $y = \sin x°$, $y = \cos x°$ and $y = \tan x°$ for $0 \leqslant x \leqslant 90$ were given in Exercise 42. Making use of the connection between any angle and its appropriate acute angle allows the graphs to be extended:

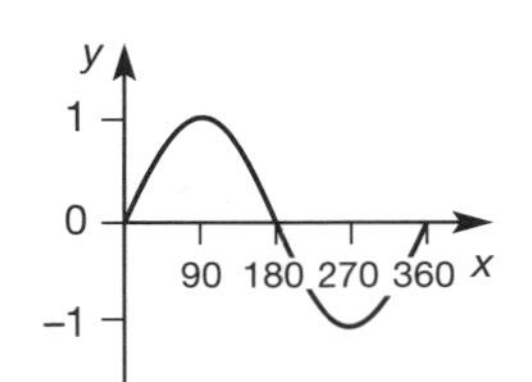

$y = \sin x°$

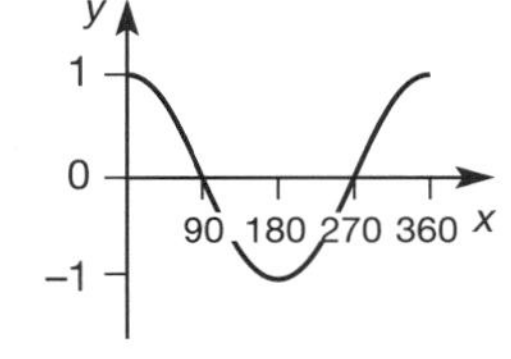

$y = \cos x°$

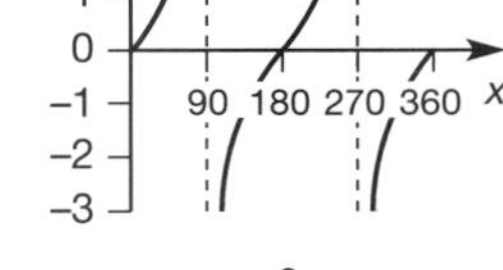

$y = \tan x°$

These graphs show how any angle is related to its appropriate acute angle, a.

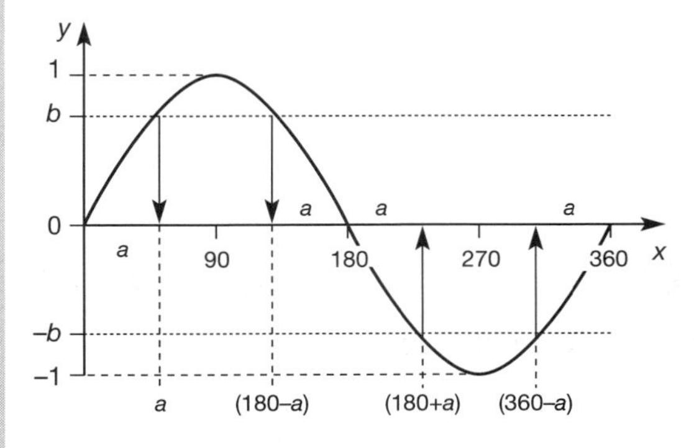

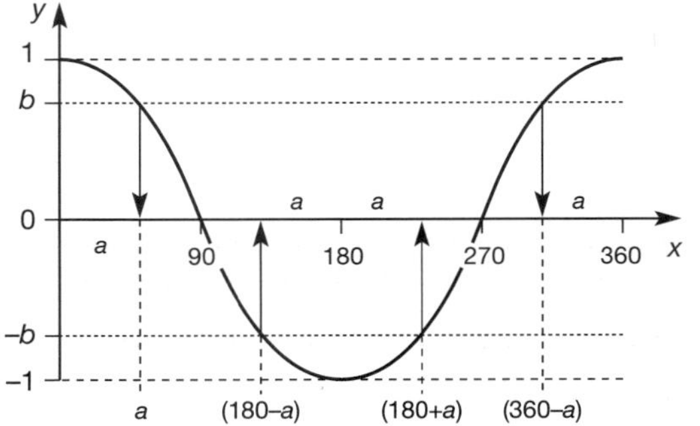

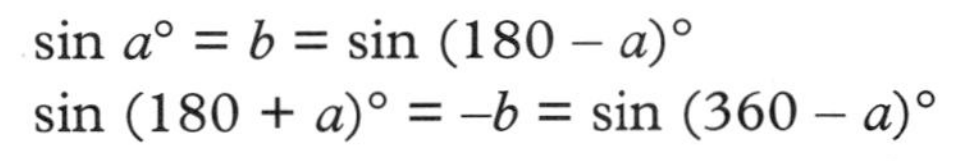

$\sin a° = b = \sin (180 - a)°$
$\sin (180 + a)° = -b = \sin (360 - a)°$

$\cos a° = b = \cos (360 - a)°$
$\cos (180 - a)° = -b = \cos (180 + a)°$

Reminder continued

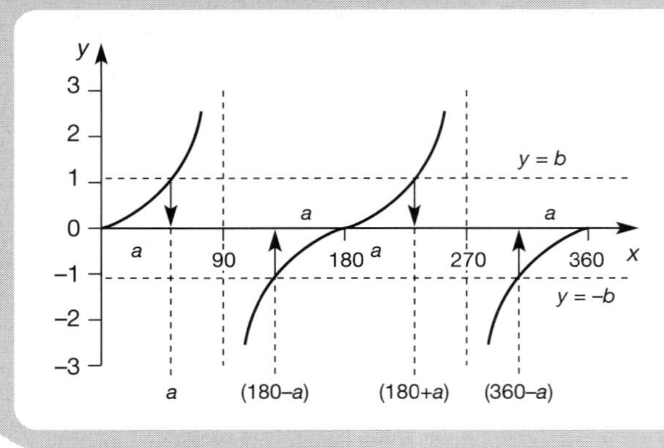

$\tan a^\circ = b = \tan (180 + a)^\circ$

$\tan (180 - a)^\circ = -b = \tan (360 - a)^\circ$

(Do not use your calculator for questions 7, 8 and 9.
The answers required for questions 7 to 12 all lie within the range 0° to 360°.)

7 Given that cos 72° = 0·309, write down:
a) another angle whose cosine is 0·309
b) two angles whose cosine is –0·309.

8 Given that sin 83° = 0·993, write down:
a) another angle whose sine is 0·993
b) two angles whose sine is –0·993.

9 Given that tan 22° = 0·404, write down:
a) another angle whose tangent is 0·404
b) two angles whose tangent is –0·404.

10 a) Find two angles whose sine is 0·530.
b) Find two angles whose sine is –0·530.

11 a) Find two angles whose cosine is 0·682.
b) Find two angles whose cosine is –0·682.

12 a) Find two angles whose tangent is 2·145.
b) Find two angles whose tangent is –2·145.

(You will get further practice with this kind of work when you do unit 3.)

44 The Area of a Triangle

Reminder

The area of a triangle = $\frac{1}{2}$(base) × (height)

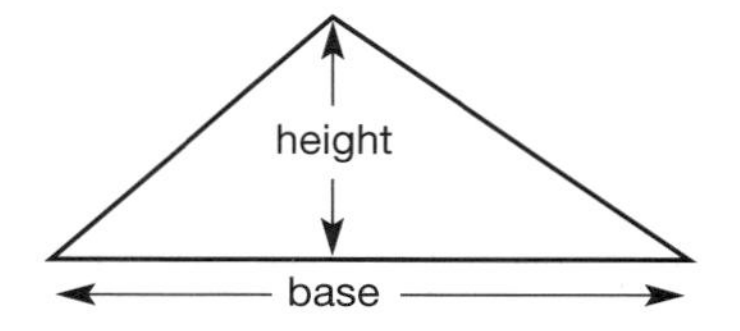

You will remember this result easily if you can recall how it is proved.

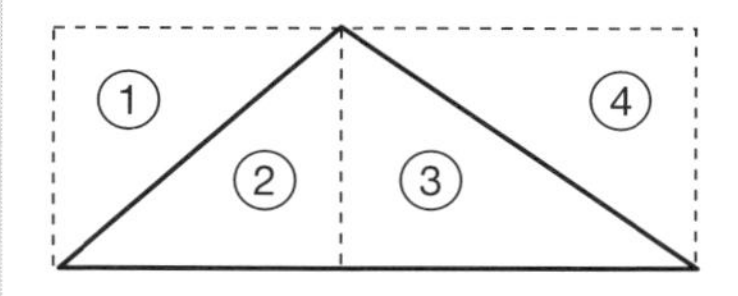

area ① = area ②
area ③ = area ④
hence the triangle is half of the surrounding rectangle.

The formula works in all three possible cases:

(i) the altitude lies within the triangle

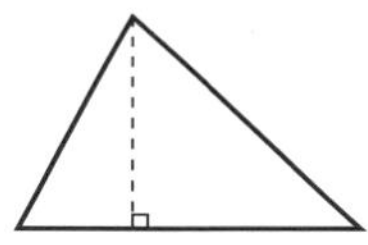

(ii) the altitude lies outside the triangle

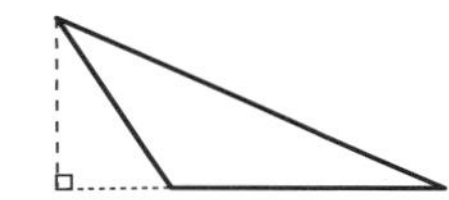

(iii) the altitude is a side of the triangle

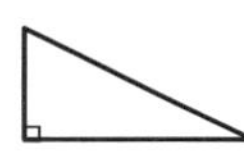

Any side may be considered to be the base.
The corresponding altitude is at right angles to that.

The area of this triangle could be calculated using $\frac{1}{2}$(BC) × (AD) or $\frac{1}{2}$(AB) × (FC) or $\frac{1}{2}$(AC) × (BE).

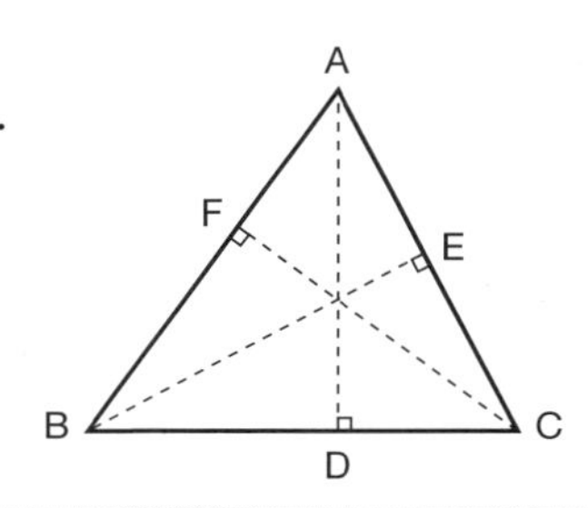

Example Calculate the area of triangle PQR.
(The lengths are given in millimetres.)

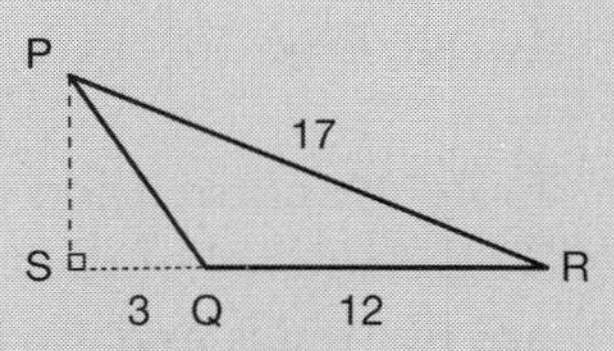

Solution $PS^2 = PR^2 - SR^2 = 17^2 - 15^2 = (17 - 15)(17 + 15) = 2 \times 32 = 64$

$\Rightarrow PS = 8$

$\Rightarrow$ area of triangle PQR = $\frac{1}{2}(QR) \times (PS) = \frac{1}{2} \times 12 \times 8 = 48\text{ mm}^2$.

Exercise 44

1 Calculate the area of each triangle. (All lengths are given in millimetres.)

a)

b) 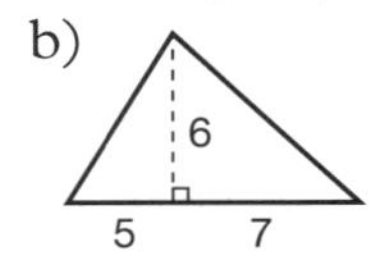

c) 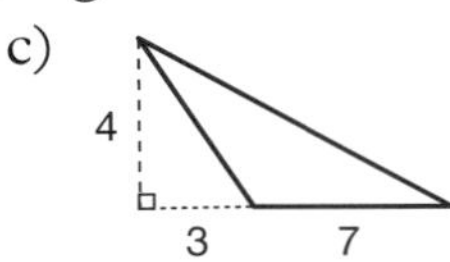

d) 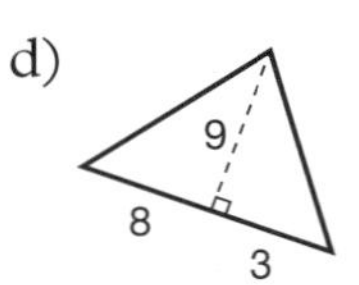

2 For each triangle, use trigonometry to calculate the height and hence find the area. (All lengths are given in millimetres.)

a)

b) 

c) 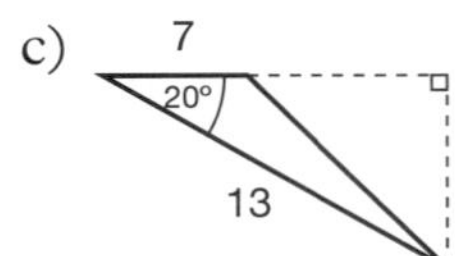

d) 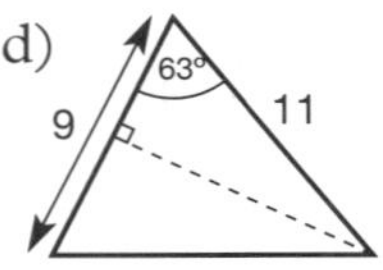

3 Apply the technique used in question 2, to

a) express the height in terms of b and $\hat{C}$

b) obtain a formula involving $\sin C$ for the area of triangle ABC.

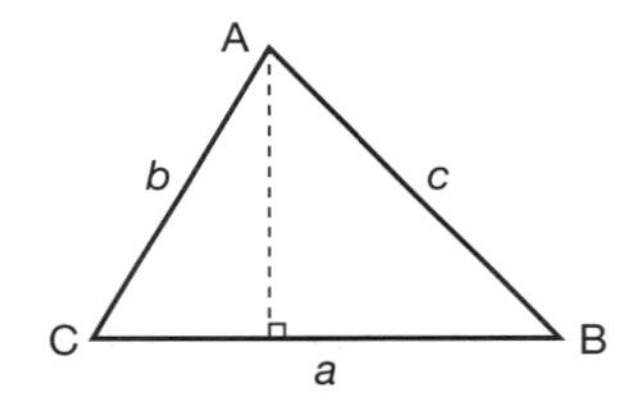

Reminder

The area of triangle ABC $= \frac{1}{2}ab\sin C$

The same answer can be obtained from

$\frac{1}{2}ac\sin B$ or $\frac{1}{2}bc\sin A$.

Note the convention that BC, the side opposite $\hat{A}$, is called a, etc.
The elements of the triangle required to be known to be able to use this formula are two sides and *the included angle.*

Example Calculate the area of this triangle.
(All lengths are given in millimetres.)

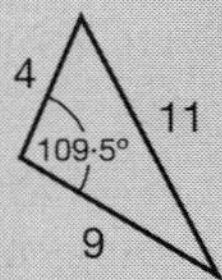

Solution Note that there is more information given here than is required. (It is all consistent, of course.) The sides which include the known angle are the lengths 4 and 9, hence

$$\text{area} = \frac{1}{2} \times 4 \times 9 \times \sin 109{\cdot}5° = 16{\cdot}967 \approx 17{\cdot}0\ \text{mm}^2.$$

4 Use the $\frac{1}{2}ab \sin C$ formula to calculate the area of each triangle. (All lengths are in mm.)

a)

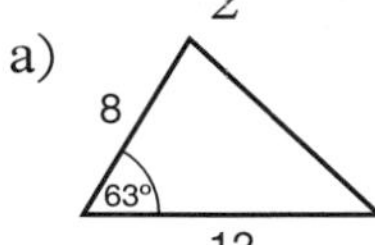

b)

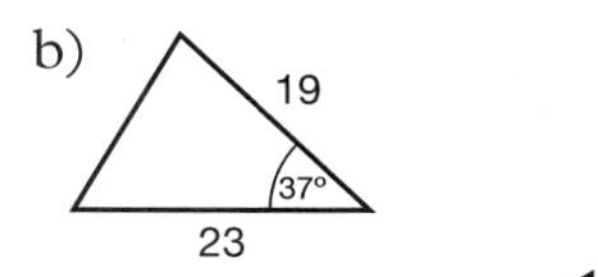

c) 22 29° 18

d)

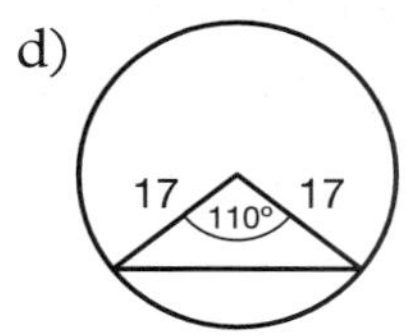

5 Calculate the total area of sail being used by this sailing boat. (All lengths are given in metres.)

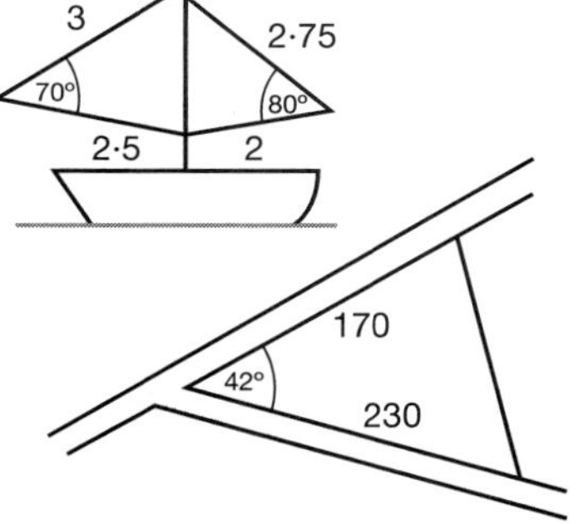

6 Find the area of this field bounded by two roads and a hedge. (All lengths are given in metres.)

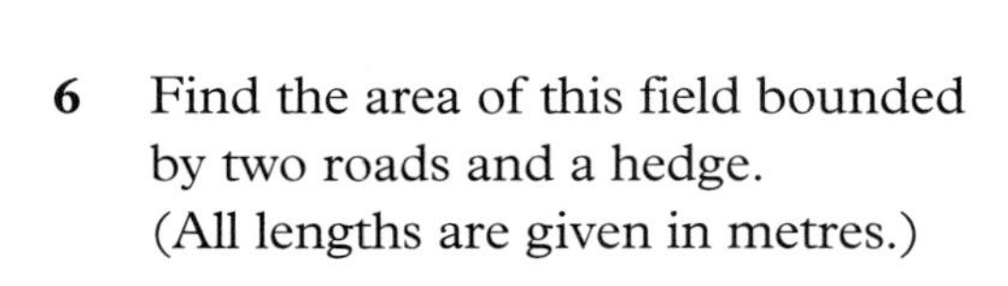

7 A fisherman makes triangular brackets to screw to the wall of his shed to support his rods. The lengths of the sides of these brackets are 200 mm, 300 mm and 450 mm. The top edge of each bracket is at 52·8° to the vertical, to ensure no rod rolls off. Calculate the area of one side of one of these brackets.

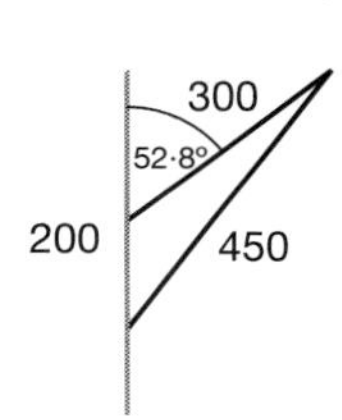

8B Two sides of a triangle measure 4 m and 3 m and the area of the triangle is 3 m^2. What size must the angle between these two sides be? There are two possible answers; make a sketch of the two possible triangles.

9B Two sides of a triangle measure 98 mm and 103 mm and the area of the triangle is 5000 m^2. What size must the angle between these two sides be? [Two answers.]

10B A chord AB, of a circle of radius 130 mm subtends an angle of 54° at the centre O.
Calculate the area of a) the minor sector AOB
b) triangle AOB
c) the minor segment AB.

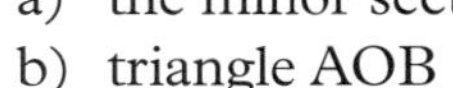

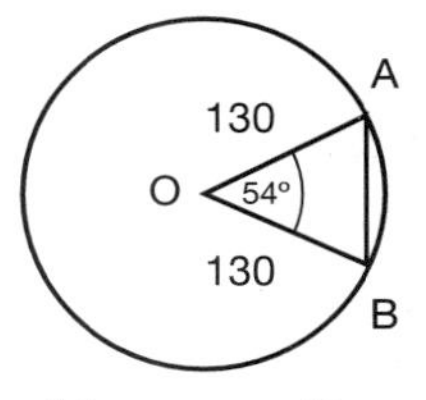

11B A regular hexagon ABCDEF is inscribed in a circle with centre O and radius 12 m. Calculate the area of triangle AOB and hence the area of the hexagon.

12H Calculate the area of these regular polygons:
a) a pentagon inscribed in a circle of radius 100 mm
b) an octagon inscribed in a circle of radius 1 m
c) a 36-sided polygon inscribed in a circle of radius 1 m.

13H Find the area of the parallelogram HKUV where HK = 68·1 mm, HV = 79·3 mm and $\hat{K}$ = 65·3°.

14H Triangle PQR and parallelogram ABCD are equal in area. PQ = 53·6 mm, PR = 93·2 mm, and $\hat{P}$ = 74·3°; AB = 82·1 mm, AD = 64·9 mm. Calculate the possible sizes of $\hat{A}$.

45 The Sine Rule (to find a side)

Reminder

If you are given two angles of a triangle (e.g. $\hat{B}$ and $\hat{C}$) and a side opposite one of them (e.g. c), then you can calculate b, the length of the side opposite the other given angle by first drawing the altitude through A, as shown.

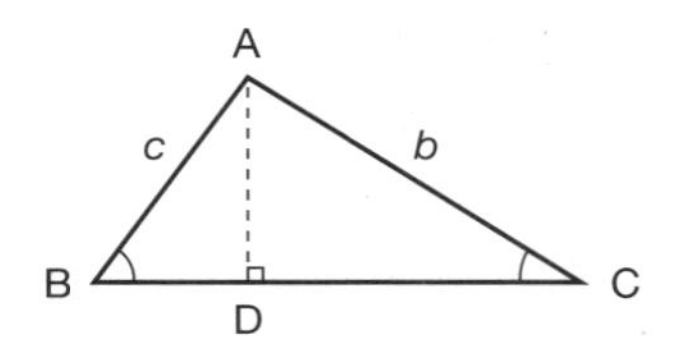

Example Find the length of b when $c = 7$, $\hat{B} = 60°$ and $\hat{C} = 40°$.

Solution In triangle ABD

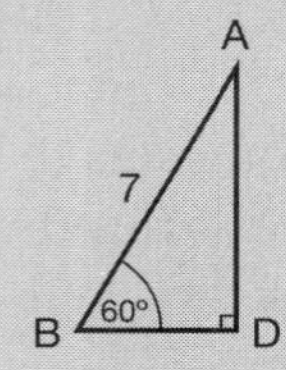

$$\sin 60° = \frac{AD}{7}$$

$$\Rightarrow AD = 7 \times \sin 60° = 6{\cdot}06$$

In triangle ACD

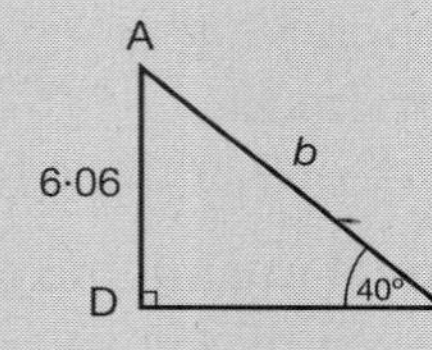

$$\sin 40° = \frac{6{\cdot}06}{b}$$

$$\Rightarrow b = \frac{6{\cdot}06}{\sin 40°} = 9{\cdot}43$$

Note that if you are given two angles of a triangle, then the third angle is also easily found because all three angles add up to 180°.

Exercise 45

1 By drawing the appropriate altitudes, calculate, to the nearest whole number, the values of x, y and z.

a)

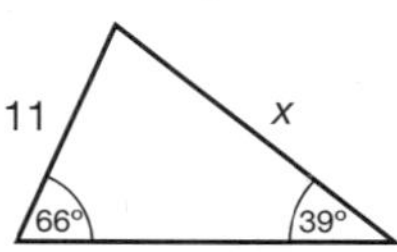

b)

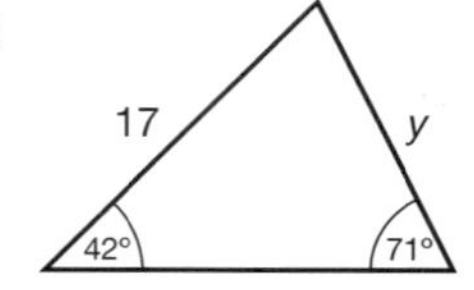

c)

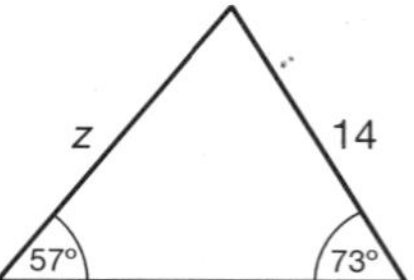

2 Triangle ABC has altitude AD.

a) Use triangle ABD to express AD in terms of c and $\hat{B}$.

b) Use triangle ADC to obtain an expression for b in terms of AD and $\hat{C}$.

c) Combine these results to express b in terms of c, $\hat{B}$, and $\hat{C}$.

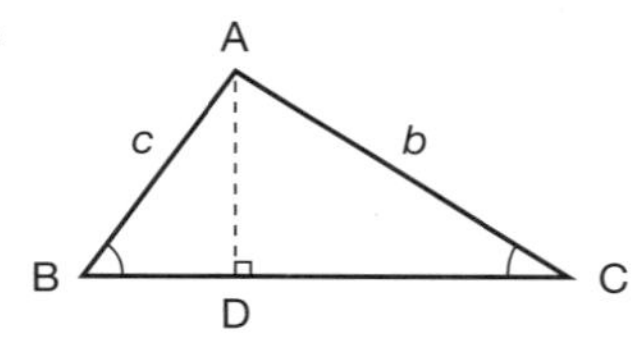

Reminder

The Sine Rule: $\dfrac{a}{\sin \hat{A}} = \dfrac{b}{\sin \hat{B}} = \dfrac{c}{\sin \hat{C}}$

Each fraction is the length of a side over the sine of the opposite angle.
The Sine Rule is used in scalene triangles (these have no right angles).

Example Find the value of k in the diagram given.
(All the lengths are given in metres.)

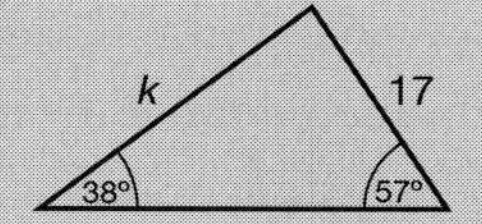

Solution k is opposite the 57° angle, and 17 is opposite the 38° angle, so

$$\frac{k}{\sin 57°} = \frac{17}{\sin 38°}$$

now multiply both sides by sin 57°, to make k the subject of the formula; $k = \dfrac{17}{\sin 38°} \times \sin 57° = 23{\cdot}2$ m

(a single calculation by calculator)

3 All the following lengths are given in metres. Calculate, correct to the nearest millimetre, the values of x, y and z.

a) 

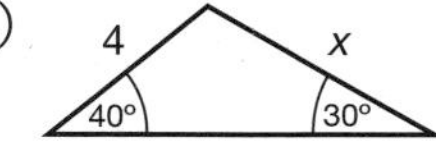

b)

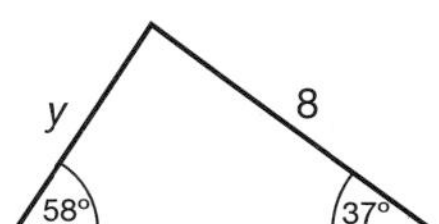

c)

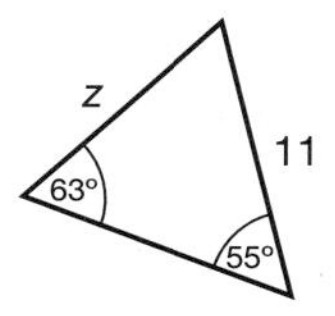

4 In traingle ABC,

a) $a = 8$ mm,	$\hat{A} = 49°$,	$\hat{B} = 59°$;	calculate b
b) $b = 2{\cdot}8$ m	$\hat{A} = 53°$,	$\hat{B} = 61°$;	calculate a
c) $a = 25$ mm,	$\hat{A} = 100°$,	$\hat{C} = 28°$;	calculate c
d) $a = 5{\cdot}15$ m,	$\hat{B} = 66°$,	$\hat{C} = 53{\cdot}8°$;	calculate b. [Find $\hat{A}$ first]

5 State the Sine Rule for triangles:

a) PQR b) XYZ c) KLM.

6 A surveyor measures a base line AB to be 440 m long, and takes bearings of a landmark C from A and B. He finds that $B\hat{A}C = 48°$ and $A\hat{B}C = 75°$. Calculate the distances of C from A and B.

7 D and E are landmarks, E being 15 km east of D. A third landmark F bears 126° from D (i.e. N126°E) and 226° from E. Calculate the distances of F from D and E.

8B From a point A level with the foot of a hill the angle of elevation of the top of the hill is 15°. From a point B, 1·2 km nearer the hill, the angle of elevation is 35°. Find the height of the hill. [Hint: First make a copy of the diagram and calculate the size of every angle in it.]

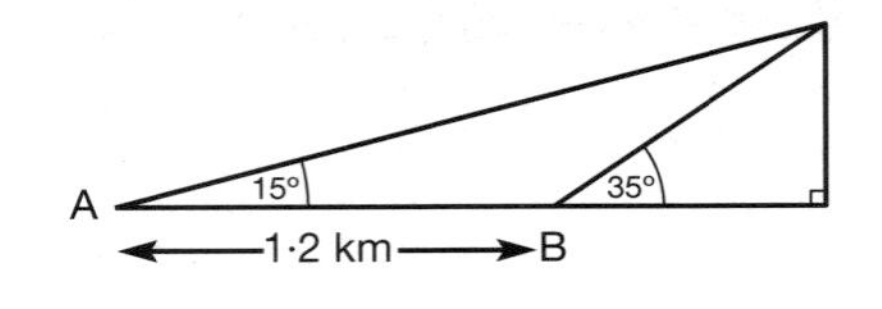

9B A ship sails due east at 12 km/h for an hour from a point P to a point Q. At P the bearing of a lighthouse from the ship was 055°. At Q it is 295°. Calculate the distance from the ship to the lighthouse:

a) at P
b) at Q
c) when the ship is closest to the lighthouse.

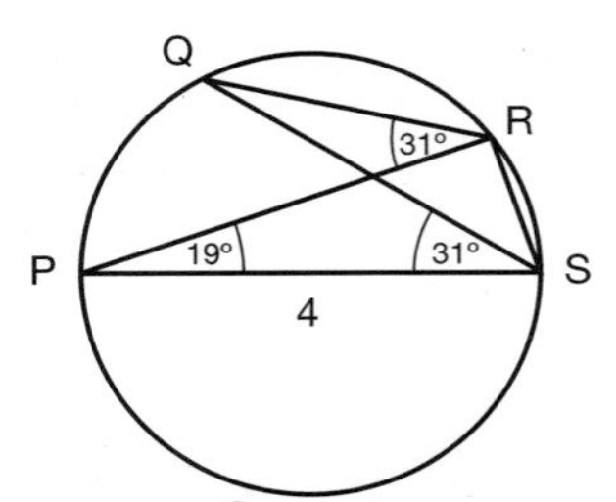

10B The diameter of the circle, PS, is 4 m in length. Also, $\hat{P} = 19°$, and $Q\hat{R}P = Q\hat{S}P = 31°$.
Find the lengths of RS and QR.

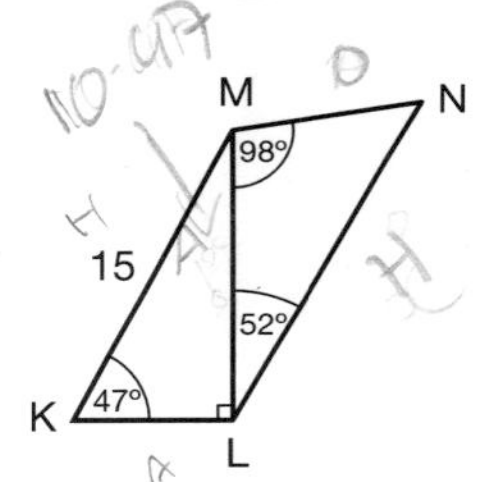

11B The diagram represents the framework of a simple crane, in which
KM = 15 m, $\hat{K} = 47°$, $L\hat{M}N = 98°$, and $M\hat{L}N = 52°$.
Calculate the length of LN.

46 The Sine Rule (to find an angle)

Reminder

The Sine Rule: $\dfrac{a}{\sin \hat{A}} = \dfrac{b}{\sin \hat{B}} = \dfrac{c}{\sin \hat{C}}$.

Example A bag of nuts for garden birds is tied to two branches of a tree at the same height by strings of length 0·8 m and 1·5 m. The shorter string is at an angle of 47° to the horizontal. Find the inclination of the longer string to the horizontal.

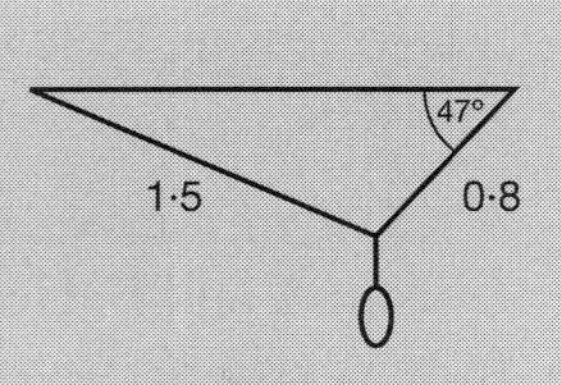

Reminder continued

Solution

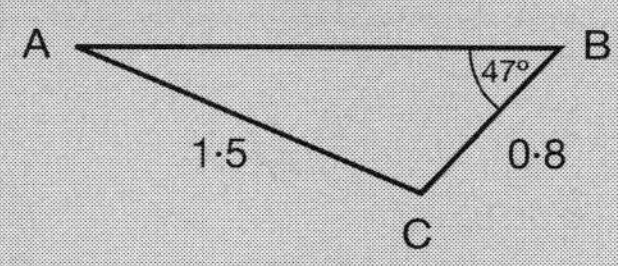

By the Sine Rule: $\dfrac{0{\cdot}8}{\sin \hat{A}} = \dfrac{1{\cdot}5}{\sin 47°}$

cross multiply: $0{\cdot}8 \times \sin 47° = 1{\cdot}5 \times \sin \hat{A}$

divide both sides by 1·5: $\sin \hat{A} = \dfrac{0{\cdot}8 \times \sin 47°}{1{\cdot}5}$

hence: $\hat{A} = 23{\cdot}0°$

Remember that sin 157° = sin 23°, but the 157° can be rejected by common sense.
Take care: sometimes the obtuse answer will be the correct one.

Exercise 46

1 In triangle ABC, $b = 60$ mm, $c = 53$ mm, $\hat{B} = 42°$, calculate $\hat{C}$.

2 In triangle ABC, $a = 17$ mm, $b = 9$ mm, $\hat{A} = 55°$, calculate $\hat{B}$.

3 In triangle ABC, $b = 21$ mm, $c = 16$ mm, $\hat{B} = 43°$, calculate $\hat{C}$.

4 In triangle PQR, $\hat{P} = 100°$, PR = 45 mm, QR= 79 mm, calculate $\hat{Q}$.

5 In triangle DEF, DE = 123 m, EF = 88 m, $\hat{F} = 114°$, calculate $\hat{E}$.

6 The leg (with the steps) on a set of step ladders is 2·3 m long, and when fully extended is inclined at 65° to the horizontal. The supporting leg is 2·1 m long. Calculate its inclination to the horizontal.

7 An extension to a house has a pitched roof. The rafters on one side of the roof are 3·5 m long and are inclined at 49° to the horizontal. The rafters on the other side are 3·1 m long. Calculate their inclination to the horizontal.

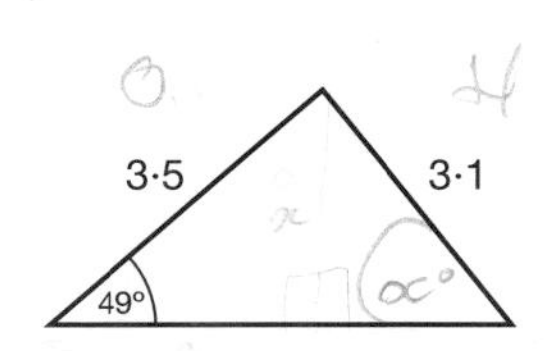

8B ABCD is a parallelogram with AB = 56 mm, AC = 78 mm and $B\hat{A}D = 49{\cdot}6°$. Calculate the size of $C\hat{A}D$.

9B P lies 25 km from Q on a bearing of 330°, and R lies due east of Q and is 37 km from P. Calculate the three figure bearing of P from R.

10B The diagram represents the framework for a simple crane. AB = 7 m, BD = 11 m, $C\hat{A}B = 50°$, and $D\hat{C}B = 105°$. Calculate the size of $C\hat{B}D$.

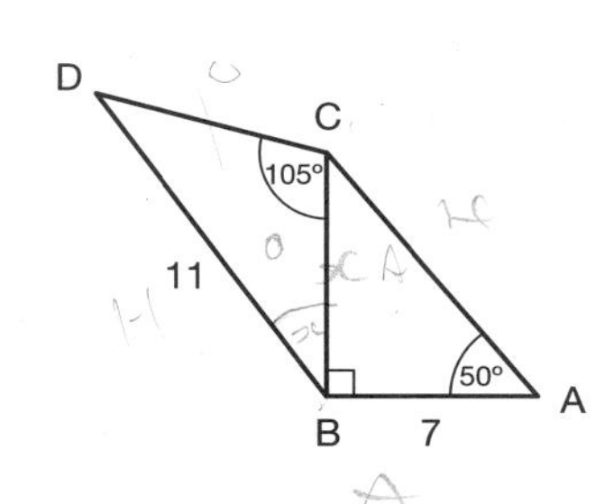

47 The Cosine Rule (to find a side)

Reminder

If you are given two sides of a triangle (e.g. b and c) and the angle included by these sides (i.e. $\hat{A}$), then you can find the length of the third side (a) by drawing the altitude through B (or C), as shown.

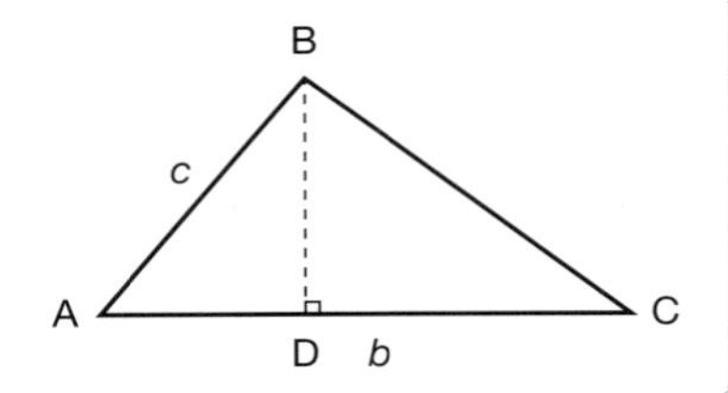

Example Find the length of BC when $c = 7$ m, $b = 9$ m and $\hat{A} = 39°$.

Solution In triangle ABD

$\sin 39° = \frac{BD}{7}$
$\Rightarrow BD = 7 \times \sin 39°$
$= 4{\cdot}405$

$\cos 39° = \frac{AD}{7}$
$\Rightarrow AD = 7 \times \cos 39°$
$= 5{\cdot}440$

So DC = AC – AD = 9 – 5·440 = 3·560

In triangle BDC

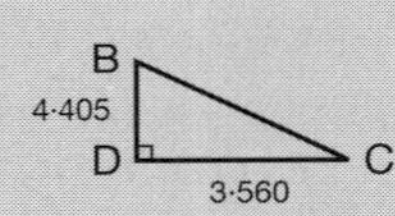

$BC^2 = 4{\cdot}405^2 + 3{\cdot}560^2 = 32{\cdot}0776$
$\Rightarrow BC = 5{\cdot}6637 \approx 5{\cdot}7$ m

Exercise 47

1 By drawing an appropriate altitude, calculate the values of x, y and z:

a) 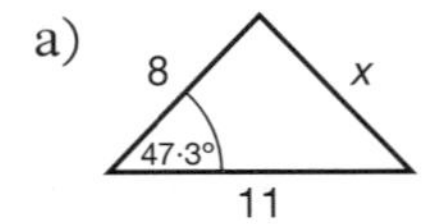

b) 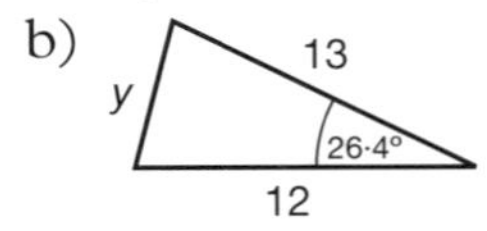

c) 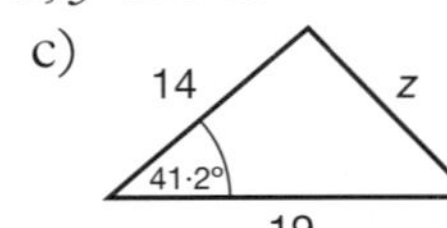

2 Triangle ABC has altitude BD. Let AD = x and BD = y.

a) Write down an expression for c^2 in terms of x and y.

b) Write down an expression for DC in terms of x and b.

c) Write down an expression for a^2 in terms of
   (i) y and DC
   (ii) x, y and b
   (iii) b, c and x.

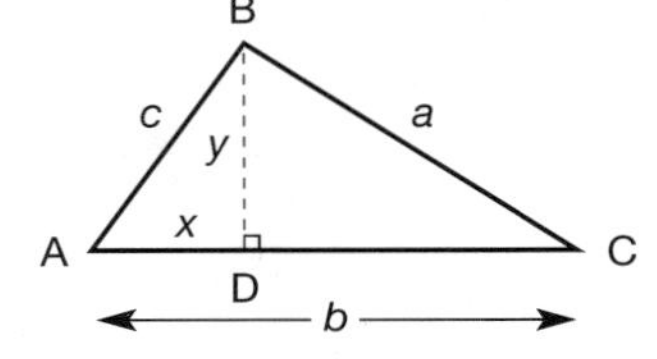

d) Use triangle ABD to express x in terms of c and $\hat{A}$, and hence express a^2 in terms of b, c, and $\hat{A}$.
Check your answer. This is the Cosine Rule.

Reminder

The Cosine Rule: $a^2 = b^2 + c^2 - 2bc\cos\hat{A}$
or $b^2 = a^2 + c^2 - 2ac\cos\hat{B}$
or $c^2 = a^2 + b^2 - 2ab\cos\hat{C}$

Example Find the value of x.
(All lengths are given in metres.)

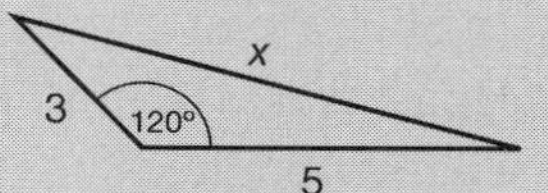

Solution

$x^2 = 3^2 + 5^2 - (2 \times 3 \times 5 \times \cos 120°)$
$= 49$ (all one calculator operation)
$\Rightarrow x = 7$

Note that it is not necessary to letter the diagram and use the formula as a piece of algebra. It can be easier, given two sides *and the included angle*, to think:

'(third side)^2 = (first side)^2 + (second side)^2 – 2(first side) × (second side) × (cosine of angle between)'

3 State the cosine rule for a) p^2 in triangle PQR
b) m^2 in triangle KLM.

4 Which of the following is a correct cosine rule formula for something in triangle UVW? If so what does the expression equal?

a) $v^2 + u^2 - 2vu\cos\hat{W}$
b) $u^2 + w^2 + 2uw\cos\hat{V}$
c) $v^2 + w^2 - vw\cos\hat{U}$
d) $w^2 + u^2 - 2uw\cos U\hat{V}W$

5 Town B is 20 km north of town A, and town C is 15 km north west of A. Calculate the distance between B and C.

6 Two ships leave port at noon, one sailing on a bearing of 045° at 9 km/h, the other due east at 12 km/h. How far apart will they be at 14:30?

7 The radar screen on a battleship shows two other ships, A, 30 km away on a bearing of 040°, and B, 50 km away at 123°. Calculate the distance between A and B.

8 A fishing boat steams 3 km NE from its home port and then 2 km north. How far is it from port?

9 Telecommunication cable is being laid alongside the B215. The lochan shown is too close to the road to allow the cable to be laid by the roadside between A and B, so it has been re–routed via C.
AB = 100 m, AC = 40 m, $B\hat{A}C = 38°$.
How much more cable did this detour require?

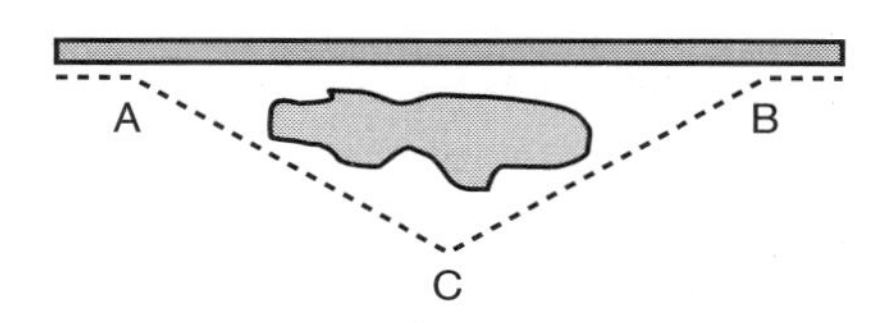

10 Hamish and Catriona live in a croft (C on the 'map'), and there is a path across the moor to the shop S. Catriona cycles the 5 km to the shop along the roads at a speed of 15 km/h. If Hamish leaves the croft at the same time, at what speed will he need to travel across the moor path to be at the shop just before Catriona?

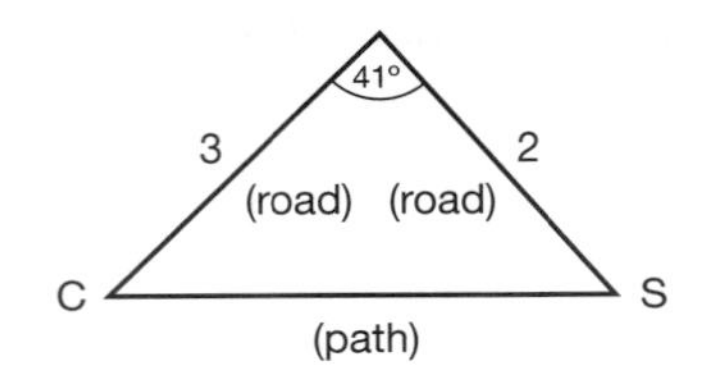

11 The pencil in a pair of compasses is wrongly inserted, so that the lengths of the legs of the compasses are 80 mm and 90 mm. If the legs are set at 26° apart, what radius of circle will be obtained?

48 The Cosine Rule (to find an angle)

Reminder

By changing the subject of a cosine rule formula for a side, a formula for the cosine of an angle can be obtained.

$\cos \hat{A} = \dfrac{b^2 + c^2 - a^2}{2bc}$ $\qquad \cos \hat{B} = \dfrac{a^2 + c^2 - b^2}{2ac}$ $\qquad \cos \hat{C} = \dfrac{a^2 + b^2 - c^2}{2ab}$

Example Calculate the value of x.
(All the lengths are given in metres.)

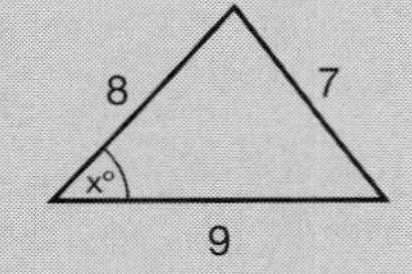

Solution $\cos x° = \dfrac{8^2 + 9^2 - 7^2}{2 \times 8 \times 9} = \dfrac{2}{3}$,

so $x = 48{\cdot}2$

Note again that it is not necessary to label all the vertices and sides and substitute into an algebraic formula.

Exercise 48

1 State the cosine rule for a) $\cos \hat{P}$ in triangle PQR
b) $\cos \hat{Z}$ in triangle XYZ.

2 Which of the following is a correct cosine rule formula for something in triangle UVW? If so what does the expression equal?

a) $\dfrac{u^2 + v^2 - w^2}{2uv}$ b) $\dfrac{u^2 + w^2 - v^2}{2wu}$ c) $\dfrac{v^2 + w^2 - u^2}{2uv}$ d) $\dfrac{u^2 + v^2 - 2uv}{w^2}$.

3 In triangle ABC, $a = 10$ mm, $b = 8$ mm, and $c = 11$ mm. Calculate $\hat{C}$.

4 In triangle DEF, $d = 10$ mm, $e = 8$ mm, and $f = 15$ mm. Calculate $\hat{F}$.

5 In triangle KLN, $k = 4$ m, $l = 5$ m, and $n = 6$ m. Calculate $\hat{L}$ and $\hat{N}$.

6 In triangle PQR, $p = 3$ m, $q = 5$ m, and $r = 6$ m. Calculate $\hat{P}$ and $\hat{Q}$.

7 Find the size of the largest angle of a triangle whose sides measure 3 m, 5 m, and 7 m.

8 As part of a larger structure, a welder joins three rods of lengths 200 mm, 210 mm and 290 mm together to form a triangle. Calculate the size of the smallest angle of this triangle.

49 Miscellaneous Triangles

Reminder

In practice, you have to decide which is the most appropriate trigonometrical strategy to use: right angles, sine rule, cosine rule, or a mixture.

Example In the quadrilateral ABCD, BC = 5 m, CD = 7 m
$\hat{B} = 65°$, and $A\hat{C}D = 40°$.
Calculate the length of AD.

Solution AD is in triangle ADC.
Can I use the sine rule in triangle ADC?
(No, I need another angle.)
Can I find another angle? (No.)
Can I use the cosine rule in triangle ADC?
(No, I need another side.)
Can I find another side? Yes. AC, from triangle ABC, by the sine rule.

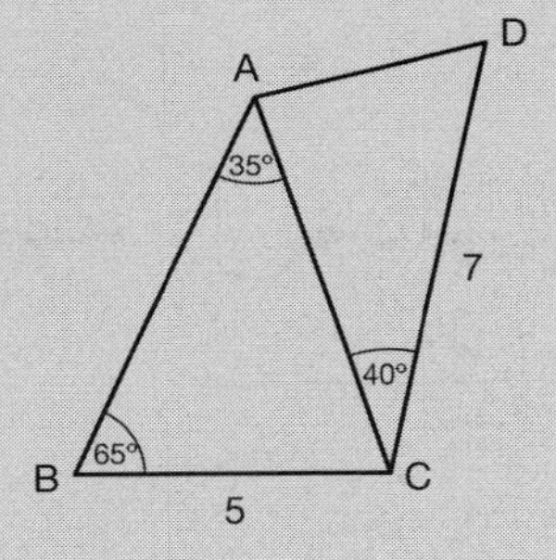

In triangle ABC, $\dfrac{5}{\sin 35°} = \dfrac{AC}{\sin 65°} \Rightarrow AC = \dfrac{5 \times \sin 65°}{\sin 35°} = 7{\cdot}9$

In triangle ADC, $AD^2 = 7{\cdot}9^2 + 7^2 - 2 \times 7 \times 7{\cdot}9 \times \cos 40° = 26{\cdot}685$

$\Rightarrow AD = 5{\cdot}16579 \approx 5{\cdot}2$ m

Exercise 49

1 From its home harbour a fishing boat steams 8 km NW, then 10 km north. Calculate its distance and bearing from the harbour.

2 At 14:45, an aircraft flying at 420 km/h on a bearing of 075° passes over a castle. At 15:15, the pilot changes course to 120°. Find the aircraft's distance and bearing from the castle at 15:35.

3 Triangle ABC represents the jib of a crane with BC vertical. The lengths are given in metres.
Calculate the vertical height of A above C.

4 A ship is 5 km NW of a coastguard station and steaming on a bearing of 075°. Forty minutes later the bearing of the ship from the coastguard station is 040°. Calculate her speed.

5 To 'solve a triangle' means to find all three sides and all three angles from whatever information is given (usually three pieces). Solve triangle ABC, given that $a = 5$ m, $c = 7$ m and $\hat{B} = 60°$.

6 In quadrilateral ABCD, AB = 8 m, BC = 4 m, $\hat{A} = 30°$, $\hat{C} = 60°$, and $A\hat{D}B = 70°$. Calculate the length of DC.

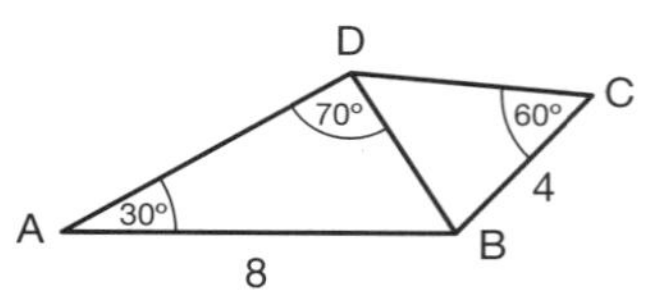

50 Revision of Trigonometry

Exercise 50

In each of questions 1 to 8, calculate the unknown quantity (side or angle) indicated.

1

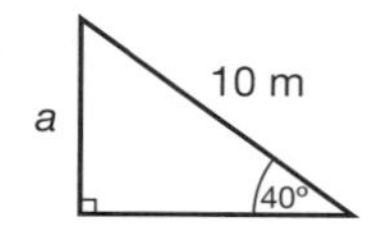

2

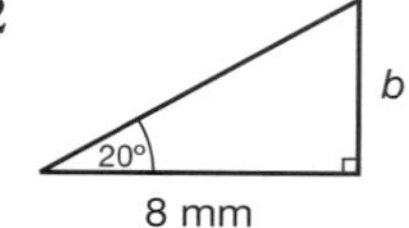

3

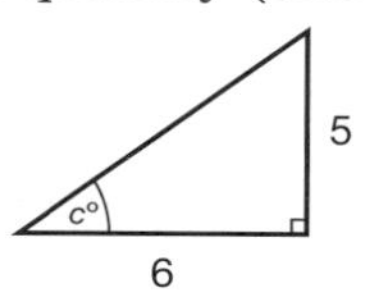

4

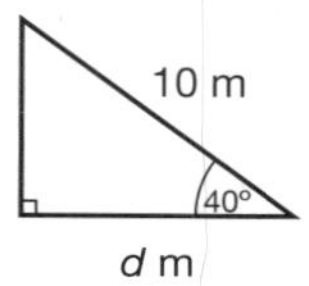

5

e
8 m
75°

6

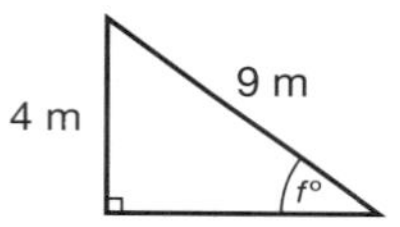

7

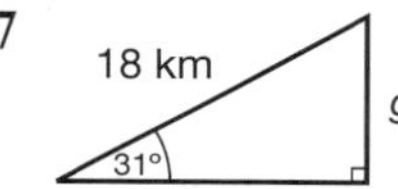

8

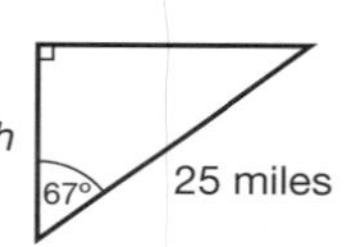

9 For each of the following equations, find, without using your calculator, two solutions for x ($0 \leqslant x \leqslant 360$)

a) $\sin x° = \sin 30°$	b) $\cos x° = \cos 60°$	c) $\tan x° = \tan 45°$
d) $\sin x° = \sin 72°$	e) $\cos x° = \cos 85°$	f) $\tan x° = \tan 21°$
g) $\sin x° = \sin 172°$	h) $\cos x° = \cos 330°$	i) $\tan x° = \tan 200°$
j) $\sin x° = -\sin 20°$	k) $\cos x° = -\cos 50°$	l) $\tan x° = -\tan 80°$
m) $\sin x° = \sin 210°$	n) $\cos x° = \cos 250°$	o) $\tan x° = \tan 100°$.

10 a) Make rough sketches of the graphs of $y = \sin x°$, $y = \cos x°$ and $y = \tan x°$ for $0 \leqslant x \leqslant 360$.

b) Use your graphs to *write down* the value of

(i) sin 180°	(ii) cos 270°	(iii) tan 180°	(iv) sin 270°
(v) cos 360°	(vi) tan 360°	(vii) sin 90°	(viii) cos 0°
(ix) tan 0°	(x) sin 0°	(xi) cos 90°	(xii) tan 270°.

c) Use your graphs to solve these equations:

(i) $\sin x° = 1$	(ii) $\cos x° = -1$	(iii) $\sin x° = -1$	(iv) $\sin x° = 0$
(v) $\cos x° = 1$	(vi) $\tan x° = 0$	(vii) $\cos x° = 0$.	

11 In triangle PQR, $p = 110$ mm, $r = 130$ mm and $\hat{Q} = 51{\cdot}3°$.
Calculate: a) the area of triangle PQR
b) the length of PR.

12 In triangle KLM, KL = 15 m, $\hat{L} = 47{\cdot}3°$, and $\hat{M} = 33{\cdot}4°$.
Calculate: a) the length of KM
b) the area of triangle KLM.

13 In triangle XYZ, $z = 17$ m, $y = 23$ m, and $\hat{Y} = 37{\cdot}3°$.
Calculate the size of $\hat{Z}$.

14 In triangle FGH, $h = 18$ m, $g = 20$ m and $f = 23$ m.
Calculate the size of $\hat{G}$.

51 Test on Trigonometry

Allow 45 minutes for this test

1 Use your calculator to find the value of
a) sin 304° b) cos 227° c) tan 139°.

2 Find two angles in the range 0° to 360° which satisfy:
a) $\sin x° = 0{\cdot}27$ b) $\sin x° = -0{\cdot}27$ c) $\cos x° = 0{\cdot}48$
d) $\cos x° = -0{\cdot}48$ e) $\tan x° = 1{\cdot}6$ f) $\tan x° = -1{\cdot}6$.

3 ABCDE is a regular pentagon inscribed in a circle with centre O and radius 75 mm.
Calculate the area of the pentagon.

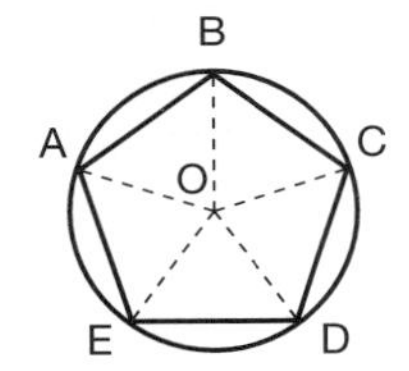

4 In quadrilateral ABCD, BD = 7 m, DC = 11 m, $\hat{A} = 41°$, $\hat{C} = 37°$ and $\text{A}\hat{\text{D}}\text{B} = 53°$.
Calculate: a) the length of AB
b) the size of $\text{C}\hat{\text{B}}\text{D}$.

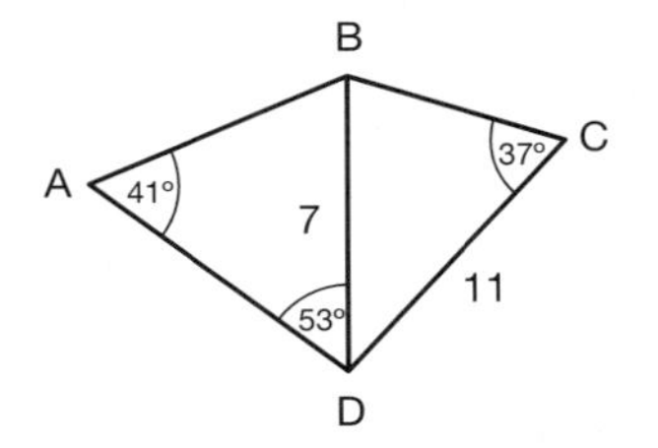

5 In quadrilateral PQRS, PS = 12 m, PR = 9 m, PQ = 10 m, QR = 11 m and $\text{S}\hat{\text{P}}\text{R} = 38°$.
Calculate: a) the length of SR
b) the size of $\hat{Q}$.

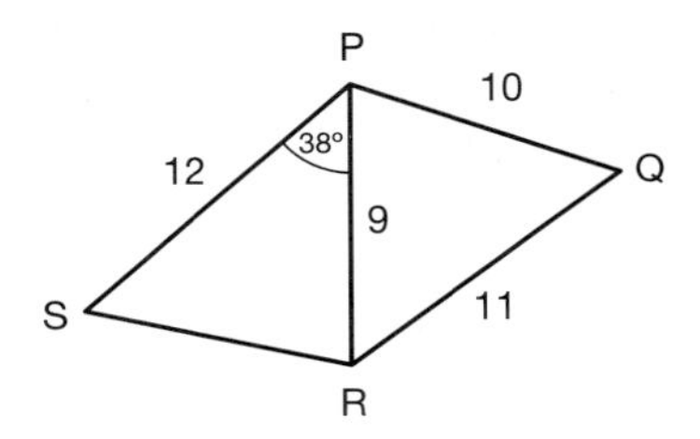

SIMULTANEOUS EQUATIONS

52 Linear Relationships

Reminder

Near the end of Exercise 23, we looked at linear relationships and their graphs.

Recall that the equation $y = mx + c$ represents the straight line with gradient m and intercept c (on the y-axis). The expression $mx + c$ (or $ax + b$) is a linear function of x.

Example Entry to Henblime Palace grounds costs £10 per car (and driver) plus £3 per passenger. Draw a graph of the cost against the number of passengers.

Solution First make up a table of values, starting with no passengers (i.e. the driver only), up to 6. More than 6 passengers would need a minibus.

number of passengers	0	1	2	3	4	5	6
cost (£)	10	13	16	19	22	25	28

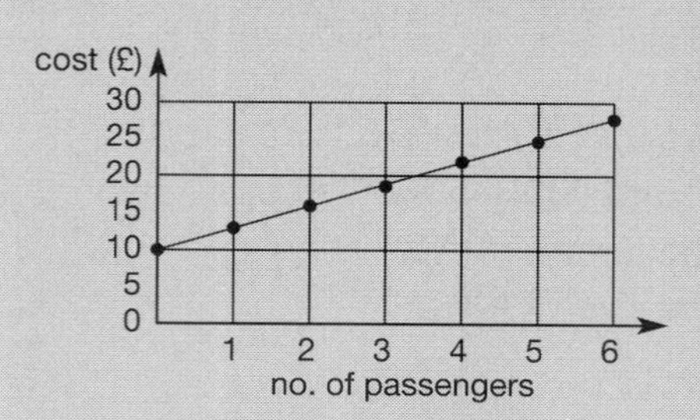

Note that the graph in this case consists strictly of only the seven dots, but joining them up shows more clearly that they lie in a straight line. This line does not allow *interpolation*; you cannot find the cost for 2·75 passengers.

Exercise 52

1 An oven has already been heated to 100°C. The thermostat is set higher and the temperature is rising by 5°C every minute.

a) Complete this table of values:

time (minutes)	0	5	10	15	20	25	30
temperature (°C)	100						

b) Draw a graph to show the increase in temperature over the first half hour.
c) Can you use this graph to estimate the temperature after 12 minutes?

2 The washing machine repairer from *@ureservice* has a call-out charge of £40, plus a labour charge of £30 per hour.
a) Draw a graph of the labour charges against the duration of the call (for up to 3 hours).
b) Use the graph to estimate the labour charge for a 1 hour 20 minute visit.
c) Check this answer by direct calculation.

3 Granny gave Calum a personal stereo for his birthday and £10 to buy replacement batteries when they were required. Each set of replacement batteries costs £2·25.
a) Draw a graph of the 'battery money' that Calum has left against the sets of replacement batteries.
b) From your graph, find how many sets of replacement batteries the £10 covered.
c) How much of Calum's own money did he have to use to buy the next set of batteries?

4 Mike and Angelo have been awarded the contract to paper and paint the ceilings of St. Magnus and St. Margaret's Churches in Auchenturpie. They based their estimate for the job on £500 for the hire of scaffolding, £200 for paper and paint, plus £300 per day.
a) Draw a graph of the cost against the number of days required to do the job.
b) If their tender for the job was £2000, in how many complete days do you think they intended to complete the work?

5 A joiner is nailing sarking boards to the rafters on the roof of a new house, and is using up nails at the rate of 7 per minute. He started with 250 nails in the pouch on his tool belt.
a) Draw a graph to show the number of nails left in the pouch against time.
b) Use your graph to estimate how many nails he had left in his pouch after half an hour.
c) Check this estimate by direct calculation.

6 *Kittbuild* bought a field and obtained permission to erect 65 houses on the site. They completed the houses at the rate of 3 per month.
a) Draw a graph of the number of houses built against time.
b) Draw a graph of the number of houses left to be built, against time.
c) Was there any month at the end of which there were as many houses left to build as there were already built?

7B Madge washes the 15 *Dunnikier Athletic* football strips every week. When she hangs the shirts out to dry, it takes 3 minutes to put out the rope and stretchers, and 40 seconds to peg each shirt to the line.

a) Draw a graph of the time taken (t minutes) to hang out a given number (n) of shirts.

b) Find the gradient of this line and hence the equation of the line, in terms of t and n.

c) Madge can peg up a pair of shorts in 30 seconds. Draw a graph of the number of pairs of shorts still to be hung out against time.
(Remember that this number will be 15 until all the shirts are hung out.)

d) Use your second graph to estimate how long it takes before all the shirts and shorts are hung out, and check your answer by direct calculation.

53 The Point of Intersection of Two Graphs

Reminder

At the point of intersection of two graphs, both graphs are at the same height. To the left of this point, one graph is below the other, but to the right it is the other way round. This has to be interpreted in context.

Example

I sail to my holiday destination and hire a car for a week.
Two firms on the island hire cars; *Selflimo* charges £20 per day plus 2p per mile,
Hirewheels charges £15 per day plus 3p per mile.

a) Draw a graph to illustrate the cost against the number of miles driven for both firms.

b) How should I decide which firm to use?

Solution

a) First make up a table of values:

no. of miles	0	1000	2000	3000	4000
Selflimo	140	160	180	200	220
Hirewheels	105	135	165	195	225

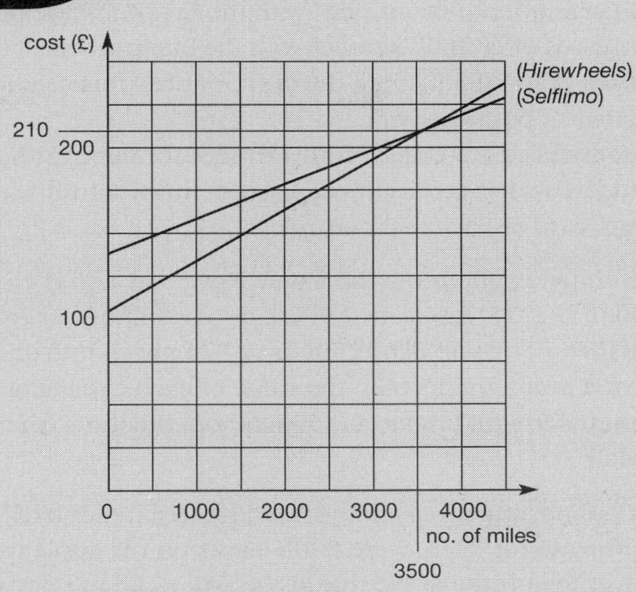

b) The point of intersection is (3500, 210).
Hence, if I drive exactly 3500 miles, the cost is the same with either company.
If I drive less than 3500 miles, the cost is less if I use *Hirewheels*.
If I drive more than 3500 miles, the cost is less if I use *Selflimo*.

Exercise 53

1 A couple fly to Jersey and hire a car for their week's holiday.
They have been given leaflets for two hire firms:
Flashcars charges £26 per day plus 4p per mile.
Mymotors charges £22 per day plus 5p per mile.

a) Draw a graph to illustrate the cost against the number of miles driven, for both firms.
b) What advice would you give the couple to ensure that they obtained the better deal?

2 Harry bought a new flat and was trying to choose one of two phone companies.
Rapicall charge a monthly line rental of £30 plus 2p per minute for all local or national calls.
Budgtel charge a monthly line rental of £25 plus 3p per minute for all local or national calls.

a) Draw a graph to illustrate the monthly cost for each phone company against the number of minutes used.
b) What advice would you give Harry regarding which phone company to choose?

3 Dorothy decided to change to a credit card which gave her airmiles as an incentive. *Bluebell* cards offer 4000 airmiles with the first purchase plus 1 airmile for each £1 spent. *Primrose* cards offer 2000 airmiles with the first purchase plus 2 airmiles for each £1 spent.
   a) Draw a graph to illustrate the number of airmiles gained on each card against the number of pounds spent.
   b) Dorothy needs to collect 7000 airmiles for her next holiday. Which credit card would provide her with these sooner, and what value of purchases would be needed on the card to earn these airmiles?

4 Hugh is thinking about buying a new Rover car and is considering the '45' model at £12 000 or the '25' model at £9000. By consulting car magazines, he has concluded that the Rover '45' will depreciate by £170 per month and the Rover '25' by £110.
   a) Draw a graph to illustrate the value of each car against time for 5 years.
   b) From the graph, how would Hugh hope to obtain the higher trade-in value for each model?

5 Charlie has an apple tree and a sycamore tree in her back garden. Last autumn, at the start of the month, there were 2000 leaves on the apple tree and 3000 on the sycamore. The leaves fell off the apple tree at the rate of 50 per day and off the sycamore at 100 per day.
   a) Draw a graph to illustrate the number of leaves left on each tree against the number of days since the start of the month.
   b) Was there any time when there were less leaves on the sycamore tree?

6 Laura and Craig are looking for a photographer for their wedding. Nicol Harvey will charge £400 plus £5 per photograph in the album. Lewis Johns will charge £250 plus £8 per photograph in the album.
   a) Draw a graph to illustrate the cost for each firm against the number of photographs in the album.
   b) What advice would you give the couple regarding which photographer to choose?

54 Graphical Solution of Simultaneous Equations

Reminder

Example Ed and Barry are twins who are about to start secondary school. Their dad offers them a choice of one of two schemes for payment of their weekly pocket money while at secondary school.
Ed chose scheme 1, and Barry chose scheme 2.
Scheme 1: £3 more than the year of secondary school which he is in.
Scheme 2: £5 more than half of the year of secondary school which he is in.

Reminder continued

a) Find an equation to express the weekly pocket money, £y, in terms of the school year, x, which the twins are in, for both schemes.
b) Sketch the lines which have these equations.
c) Use your sketch to find if there is any school year in which both boys receive the same amount of money.

Solution

a) scheme 1: $y = x + 3$ scheme 2: $y = \frac{1}{2}x + 5$

b)

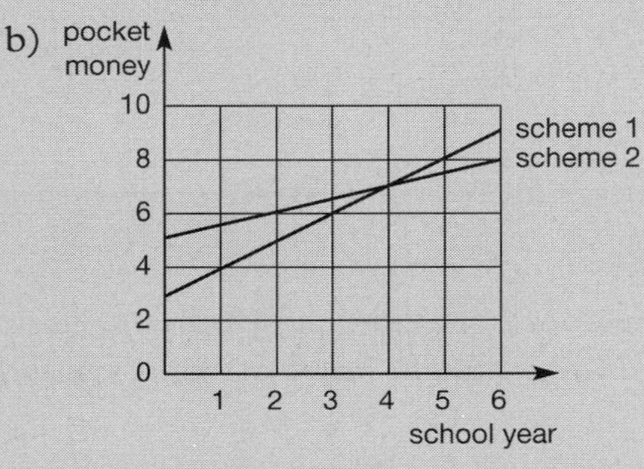

c) From the graph it can be seen that in 4th year, each twin receives £7 per week.

Exercise 54

1 Sam and Amy have to decide from which firm to take their new gas supply. *Scotgas* has a standing charge of £30 per quarter and charges 2p per therm. *Fifegas* has a standing charge of £20 per quarter and charges 3p per therm.
a) Find the equation for the quarterly cost of gas from each company in terms of the number of therms used.
b) Draw a graph to find the coordinates of the point of intersection of the lines which have these equations.
c) What advice would you give Sam and Amy regarding their choice of gas supplier.

2 A Croatian pine and a leylandi are planted in a garden. The pine is 1·5 m tall and grows at the rate of 0·25 m per annum. The leylandi is 0·5 m tall and grows 0·45 m every year.
a) Find the equation for the height of each tree in terms of the number of years since they were planted.
b) Draw a graph to find the coordinates of the point of intersection of the lines which have these equations.
c) When will the leylandi be taller than the pine?

By making a sketch of the lines which have these equations, find the coordinates of the point of intersection of each of the following pairs.

3 $\begin{cases} y = x + 2 \\ y = 2x + 1 \end{cases}$

4 $\begin{cases} y = x + 5 \\ y = 3x + 1 \end{cases}$

5 $\begin{cases} y = 3x + 2 \\ y = 2x + 3 \end{cases}$

6 $\begin{cases} y = 3x - 2 \\ y = 7 \end{cases}$

7 $\begin{cases} y = \frac{1}{3}x + 5 \\ y = \frac{1}{2}x + 4 \end{cases}$

8 $\begin{cases} y = x + 1 \\ y = \frac{1}{2}x + 3 \end{cases}$

9 $\begin{cases} y = 8 - x \\ y = 4 - \frac{1}{3}x \end{cases}$

10 $\begin{cases} y = 9 - x \\ y = \frac{3}{5}x + 1 \end{cases}$

11 $\begin{cases} y = -x \\ y = 2x + 6 \end{cases}$

Reminder

The equation of a straight line does not always appear in the form $y = mx + c$. Sometimes it is necessary to draw a straight line by finding out where it crosses the axes.

Example

Find the coordinates of the point of intersection of the lines with equations
$x + 3y = 9$ and $2x + 3y = 12$.

Solution

$x + 3y = 9$ crosses the y-axis where $x = 0 \Rightarrow 0 + 3y = 9 \Rightarrow (0, 3)$
$x + 3y = 9$ crosses the x-axis where $y = 0 \Rightarrow x + 0 = 9 \Rightarrow (9, 0)$
(hence)

$2x + 3y = 12$ crosses the y-axis where $x = 0 \Rightarrow 0 + 3y = 12 \Rightarrow (0, 4)$
$2x + 3y = 12$ crosses the x-axis where $y = 0 \Rightarrow 2x + 0 = 12 \Rightarrow (6, 0)$
(hence)

Putting both together gives:

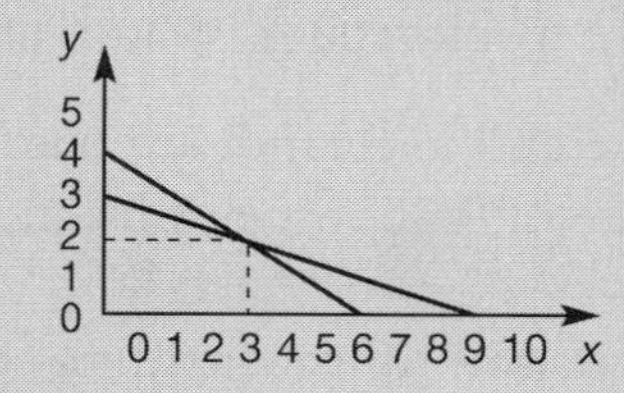

hence (3, 2)

By making a sketch of the lines which have these equations, find the coordinates of the point of intersection of each of the following pairs.

12 $\begin{cases} x + 4y = 8 \\ x + 2y = 6 \end{cases}$

13 $\begin{cases} y = \frac{1}{2}x + 3 \\ x + y = 9 \end{cases}$

14 $\begin{cases} 5y = 3x + 15 \\ y = x + 1 \end{cases}$

15 $\begin{cases} 3x + 4y = 36 \\ y = 3 \end{cases}$

16 $\begin{cases} x + y = 7 \\ y = x + 3 \end{cases}$

17 $\begin{cases} x + 3y = 6 \\ 5y = x + 2 \end{cases}$

18 $\begin{cases} y = 3x + 6 \\ x + y = 2 \end{cases}$

19 $\begin{cases} 5x + 7y = 35 \\ y = x - 1 \end{cases}$

20 $\begin{cases} y = 2x - 3 \\ y = \frac{1}{2}x + 2 \end{cases}$

55 Algebraic Solution of Simultaneous Equations

Reminder

The graphical solution of simultaneous equations has disadvantages; it needs squared paper, it is time consuming, and it is usually inaccurate where the answers involve fractions. Using algebra gets round these problems.

Example Find the coordinates of the point of intersection of the lines with equations $y = 140 + 0{\cdot}02x$ and $y = 105 + 0{\cdot}03x$.

Solution Equate the two expressions for y: $140 + 0{\cdot}02x = 105 + 0{\cdot}03x$

$$\Rightarrow \quad 140 - 105 = (0{\cdot}03 - 0{\cdot}02)x$$

$$\Rightarrow \quad 35 = 0{\cdot}01x$$

$$\Rightarrow \quad x = 3500$$

then either $y = 140 + 0{\cdot}02 \times 3500 = 210$

or $y = 105 + 0{\cdot}03 \times 3500 = 210$ hence $(3500, 210)$

[These equations could have been used in the reminder at the start of Section 53 about *Hirewheels* and *Selflimo*.]

Exercise 55

Solve the following pairs of simultaneous equations algebraically:

1 $\begin{cases} y = 4x + 5 \\ y = 5x + 4 \end{cases}$

2 $\begin{cases} y = 3x + 7 \\ y = 4x + 6 \end{cases}$

3 $\begin{cases} y = 2x + 5 \\ y = 5x - 1 \end{cases}$

4 $\begin{cases} y = 7x - 2 \\ y = 3x + 6 \end{cases}$

5 $\begin{cases} y = 8x + 3 \\ y = 5x + 12 \end{cases}$

6 $\begin{cases} y = 4x - 3 \\ y = 8x - 19. \end{cases}$

Reminder

A common type of example contains equations of the form $\begin{cases} x + y = \dots \\ x - y = \dots. \end{cases}$

Example

Solve $\begin{cases} x + y = 17 \\ x - y = 13. \end{cases}$

Solution

Adding these two equations will eliminate y, so $2x = 30 \Rightarrow x = 15$

EITHER
subtract the equations to eliminate x

$$y - (-y) = 17 - 13$$
$$\Rightarrow \quad 2y = 4$$
$$\Rightarrow \quad y = 2$$

OR
substitute in either equation: e.g. $x + y = 17$
becomes $15 + y = 17 \quad \Rightarrow \quad y = 2$

hence $x = 15, y = 2$

Solve the following pairs of simultaneous equations algebraically:

7 $\begin{cases} x + y = 11 \\ x - y = 9 \end{cases}$

8 $\begin{cases} x + y = 8 \\ x - y = 4 \end{cases}$

9 $\begin{cases} x + y = 19 \\ x - y = 9 \end{cases}$

10 $\begin{cases} x + y = 5 \\ x - y = 7 \end{cases}$

11 $\begin{cases} x + y = 10 \\ x - y = 18 \end{cases}$

12 $\begin{cases} x + y = 2 \\ x - y = -6. \end{cases}$

Reminder

Where one of the four coefficients of x or y is 1, the method of *substitution* can be useful.

Example

Solve $\begin{cases} y - 2x = 1 \\ 3x + 4y = 15. \end{cases}$

Solution

First make y the subject of the first equation by adding $2x$ to both sides
i.e. $y = 2x + 1$

Now replace y in the second equation by $(2x + 1)$
i.e. $3x + 4(2x + 1) = 15$

$$\Rightarrow \quad 3x + 8x + 4 = 15$$
$$\Rightarrow \quad 11x = 11$$
$$\Rightarrow \quad x = 1$$

now use $y = 2x + 1$ with $x = 1$ to calculate that $y = 3$

hence $x = 1, y = 3$

Solve the following pairs of simultaneous equations by the method of substitution:

13 $\begin{cases} y - 3x = 4 \\ 2x + 3y = 1 \end{cases}$

14 $\begin{cases} y - 4x = 2 \\ 3x - 2y = 6 \end{cases}$

15 $\begin{cases} y - 5x = 3 \\ 7x - 4y = 14 \end{cases}$

16 $\begin{cases} x + 4y = 7 \\ 3x + 2y = 11 \end{cases}$

17 $\begin{cases} x - 3y = 4 \\ 4x - 5y = 2 \end{cases}$

18 $\begin{cases} x + 2y = 1 \\ 3x + 4y = 9. \end{cases}$

Reminder

Where none of the four coefficients is 1, it is usually easier to use the method of *elimination*. This means multiplying one or both equations to obtain the same coefficient of x or y and then adding or subtracting to eliminate that variable. Basically it is done the same way as questions 7 to 12 but with an extra step at the beginning. You will finish up with the same equation to solve as you would obtain by the method of substitution.

Example Solve $\begin{cases} 3x + 5y = 11 \text{ ....(a)} \\ 4x - 3y = 5 \text{ ....(b).} \end{cases}$

Solution Notice that it is useful to label the equations. This helps you if you make an error and have to retrace your working. It helps a marker to follow your working.

Here we have a choice between arranging to have either $12x$ or $15y$ in each equation.

(Method 1)

$$\begin{array}{ll} \text{(a)} \times 4 & 12x + 20y = 44 \\ \text{(b)} \times 3 & 12x - 9y = 15 \\ \hline \text{subtracting} & 29y = 29 \\ \Rightarrow & y = 1 \end{array}$$

substituting in (a)

$$\begin{aligned} &\Rightarrow 3x + (5 \times 1) = 11 \\ &\Rightarrow 3x = 6 \\ &\Rightarrow x = 2 \end{aligned}$$

hence $x = 2, y = 1$

(Method 2)

$$\begin{array}{ll} \text{(a)} \times 3 & 9x + 15y = 33 \\ \text{(b)} \times 5 & 20x - 15y = 25 \\ \hline \text{adding} & 29x \qquad = 58 \\ \Rightarrow & x = 2 \end{array}$$

substituting in (a)

$$\begin{aligned} &\Rightarrow (3 \times 2) + 5y = 11 \\ &\Rightarrow 5y = 5 \\ &\Rightarrow y = 1 \end{aligned}$$

hence $x = 2, y = 1$

Note that questions 13 to 18 could also have been solved by this method. Ability to apply both methods is desirable.

Solve the following pairs of simultaneous equations by the method of elimination:

19 $\begin{cases} 3x + 2y = 13 \\ 4x - 3y = 6 \end{cases}$ **20** $\begin{cases} 2x + 3y = 16 \\ 3x - 4y = 7 \end{cases}$ **21** $\begin{cases} 5x - 7y = 13 \\ 3x + 4y = 16 \end{cases}$

22 $\begin{cases} 7x + 3y = 11 \\ 5x - 4y = 14 \end{cases}$ **23** $\begin{cases} 6x + 5y = 8 \\ 5x + 3y = 9 \end{cases}$ **24** $\begin{cases} 5x - 4y = 6 \\ 4x - 3y = 4. \end{cases}$

Solve the following pairs of simultaneous equations by what you consider to be the most appropriate method:

25 $\begin{cases} y = 2x + 5 \\ y = 3x - 4 \end{cases}$ **26** $\begin{cases} p + q = 13 \\ p - q = 7 \end{cases}$ **27** $\begin{cases} y = 2x + 1 \\ 3x + 5y = 18 \end{cases}$

28 $\begin{cases} 3x + 5y = 1 \\ x = 7 \end{cases}$ **29** $\begin{cases} 5x - 2y = 19 \\ 3x + 4y = 1 \end{cases}$ **30** $\begin{cases} 2x + 3y = 20 \\ 5x - y = 33. \end{cases}$

31 Bill, who drives 800 miles per week, buys a second hand car with 25 000 miles on the clock. At the same time, his wife Susan, who drives 500 miles a week, buys a car with 40 000 miles on the clock. Find an equation for the mileage of each car in terms of the number of weeks since purchased by Bill and Susan, and hence find how long it will be before Bill's car has done the same mileage as Susan's.

32 Morag is heating up two pots of soup for a large number of guests. The smaller gas ring on her cooker raises the soup temperature by 3°C per minute, and the larger by 5°C per minute. When the pot on the smaller ring has reached 40°C and the other 60°C, Morag swaps them over. Form an equation for the temperature of each pot of soup in terms of time from the swap-around and determine how long it will be till both pots of soup are equally hot.

56 Problem Solving with Simultaneous Equations

Reminder

Information can often be expressed in the form of a pair of simultaneous equations. Take care to explain the meaning of any variables you introduce.

Example Six tourists visited a National Trust castle and in the forenoon they bought 4 cups of tea and two coffees, which cost them £7·60. In the afternoon they were charged £8 for 2 cups of tea and 4 coffees. Find the costs of a tea and a coffee.

Solution Let a cup of tea cost x pence and a coffee cost y pence.

in the morning: $4x + 2y = 760$ .... (a)
in the afternoon: $2x + 4y = 800$ .... (b)

Reminder continued

solve (a) & (b) simultaneously: (a) × 2 ⇒ $8x + 4y = 1520$
(b) ⇒ $2x + 4y = 800$
(subtracting) $6x = 720$
⇒ $x = 120$

substituting in (a): $4(120) + 2y = 760$
⇒ $2y = 760 - 480 = 280$
⇒ $y = 140$

hence tea costs £1·20 per cup and coffee £1·40

Exercise 56

1 The sum of two whole numbers is 40 and their difference is 6.
Let the numbers be x and y and suppose $x > y$. Write down two equations involving x and y. Solve these equations simultaneously to find the numbers.

2 Two angles are complementary. One angle is 30° larger than the other. Find the sizes of the angles.

3 Two angles are supplementary. One is 62° smaller than the other. Find the sizes of the angles.

4 Sally puts £50 in an account and adds £2 per week. Teresa puts £70 in her account and adds £1 per week. After how many weeks do Sally and Teresa have equal amounts in their accounts?

5 The perimeter of a rectangle is 174 mm. The length is 21 mm longer than the breadth. Calculate the area of the rectangle.

6 Bob the baker charged Mrs Smith £4·10 for 6 scones and 4 pancakes, and charged Mrs Jones £4·70 for 5 scones and 7 pancakes. Find the cost of 7 scones and 5 pancakes for Mr Bell.

7 Mr and Mrs McPherson paid £26 for themselves and their three children to get into the pantomime. The show cost £25 for Mr and Mrs Niven, their adult daughter, and her child. Find the costs of the adult and child tickets.

8 A batch of the alloy *argolite* contains 3 tonnes of tin and 5 tonnes of copper and costs £370. A batch of the alloy *bargon* contains 5 tonnes of tin and 4 tonnes of copper and costs £400. Find the cost of a batch of another alloy, *chiprok*, which contains 2 tonnes of tin and 3 tonnes of copper.

9 A group of five Italian tourists bought 3 fish suppers and 2 haggis suppers in Johnny's chip shop in Pittenweem and it cost them £12·90. A group of six Dutch tourists later bought 5 fish suppers and 1 haggis supper and were charged £15·90. Find the price of a fish supper and a haggis supper.

10 Years ago, grocers used to create their own blends of tea. Jimmy Webster's 'Golden Blend' was a mixture of 3 kg of Indian tea and 2 kg of Ceylon tea and sold at 26p per kg. Let Indian tea cost x pence per kg and Ceylon tea y pence per kg and show that the above information leads to the equation $3x + 2y = 130$.
His 'Silver Blend' was a mixture of 2 kg of Indian tea and 3 kg of Ceylon tea and sold at 24p per kg. Form a second equation in x and y and hence find the selling price for the 'Bronze Blend' which was a mixture of 1 kg of Indian tea and 4 kg of Ceylon tea.

57 Revision of Simultaneous Equations

Exercise 57

1 A cellar was flooded and contained 800 litres of flood water. This was pumped out at the rate of 50 litres per minute. Draw a graph of the amount of water left in the cellar against time.

2 Nancy had £100 in her savings account and added £1 every week.
a) Draw a graph of Nancy's savings against time for a year, ignoring any interest due.
b) Use your graph to estimate how many weeks it took for her to accumulate £130 in savings, and check your answer by direct calculation.

3 Mr and Mrs Nairn have won £2000 in a charity raffle. They put their winnings in a separate bank account and withdrew £50 every week to spend on a gourmet dinner at the 'Old Rectory' Hotel. Their teenage son Donny stacks shelves part-time in Teskways and has a savings account with £240 in it. He puts £60 from his earnings into this account every week.
a) Draw graphs on the same diagram to illustrate how the parents' money diminishes and Donny's grows.
b) When will Donny have more money in his account than his parents do?

4 Solve graphically: a) $\begin{cases} y = x + 4 \\ 3x + 2y = 18 \end{cases}$ b) $\begin{cases} y = 2x \\ y = \frac{1}{2}x + 5. \end{cases}$

5 Solve algebraically: a) $\begin{cases} 5x + 4y = 1 \\ 3x + 5y = 11 \end{cases}$ b) $\begin{cases} 7x + y = 10 \\ 2x - 3y = 16. \end{cases}$

6 Jim and Claire are on honeymoon on the tropical island of Fajima, where they can hire a *4×4* for a week from *Dirt Jeep* for £200 plus 5p per km, or from *Gumshoe* for £150 plus 7p per km.
a) Draw a graph to show cost against distance travelled for both firms.
b) What advice would you give the happy couple to minimise their expense?

7B Sandra has two leylandi cuttings of heights 160 mm and 120 mm.
She feeds the larger one on 'Rapigro' and it grows 5 mm every week. She feeds the smaller one on 'Spurt' and it grows 7 mm every week.
Find algebraically if and when they will ever be of the same height.

8B Grandad is currently 15 times Rupert's age. In 4 years he will be 8 times Rupert's age. How old was Grandad when Rupert was born?

58 Test on Simultaneous Equations

Allow 45 minutes for this test

1 Tony the plumber charges £60 for a call-out plus £40 per hour labour.
a) Draw a graph to illustrate his charges against the time spent on the job.
b) Use your graph to estimate how long Tony spent at a house where his total charge was £160.

2 Margaret and Sheila work for a market research company.
Margaret has a weekly wage of £340 plus £30 for each client that she interviews.
Sheila has a weekly wage of £400 plus £25 for each client that she interviews.
a) Draw a graph to show earnings against number of clients interviewed for both women.
b) For how many clients does (i) Sheila earn more than Margaret
(ii) Margaret earn more than Sheila?

3 Solve graphically: a) $\begin{cases} x + y = 9 \\ y = \frac{1}{2}x \end{cases}$ b) $\begin{cases} y = 2x + 8 \\ 4x + y = 2. \end{cases}$

4 Solve algebraically: a) $\begin{cases} 5x + y = 13 \\ 3x - 4y = 17 \end{cases}$ b) $\begin{cases} 3x + 4y = 32 \\ 5x - 3y = 5. \end{cases}$

5 Two phone firms are available to houses in a new estate on the outskirts of Dunfermline.
Allcall charges a monthly line rental of £15, and 2p per phone unit.
Callwyde charges a monthly line rental of £10, and 3p per phone unit.
What advice would you give to a home buyer in this estate wishing to choose between these two firms? [Do not use a graph.]

6 In an old set of primary school counting rods, all rods of one colour are the same length. When laid in line, 6 green rods and 4 purple ones have a total length of 620 mm. Also, 4 green rods and 5 purple ones have a total length of 530 mm.
How long would a line of 2 green rods and 3 purple rods be?

7B A bag of 15 coins worth £5·40 contains only 20p and 50p pieces. How many of each are there?

8B If the numerator and denominator of a certain fraction are both increased by 1, then the fraction is equal to $\frac{3}{4}$. If they are both decreased by 1, then the fraction is equal to $\frac{1}{2}$. Find the fraction.

GRAPHS, CHARTS AND TABLES

59 Revision of Interpreting Graphs

Reminder

You have to be able to extract data from bar graphs, line graphs, pie charts, and stem and leaf diagrams, and interpret the information gathered.

Example The temperature in Mr. Napier's maths classroom was recorded every hour, and the next morning his first year class produced the following graph.

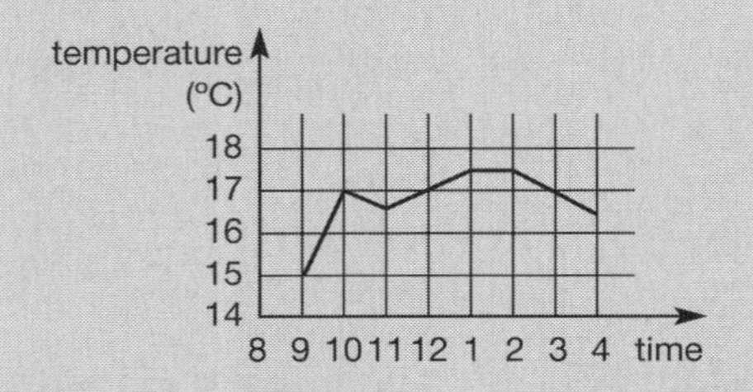

a) When was the room temperature at its lowest?
b) When was the room temperature at its highest?

Solution a) 9 a.m. [This is where the graph is lowest, but we do not really know if the temperature fell below 15°C before it rose to 17°C at 10 a.m.]

b) 1 p.m. to 2 p.m.
[Again we do not know what happened in between. This could be the lunch hour. Was the heating turned off then? Was the door left open? Was there a lunchtime activity in the room?]

Example Mr. Napier's favourite saying appears to be 'Yes, correct, well done.' One day some likeable rogues from the school newspaper organised a count of how many times he used this phrase each period, and published this bar graph.

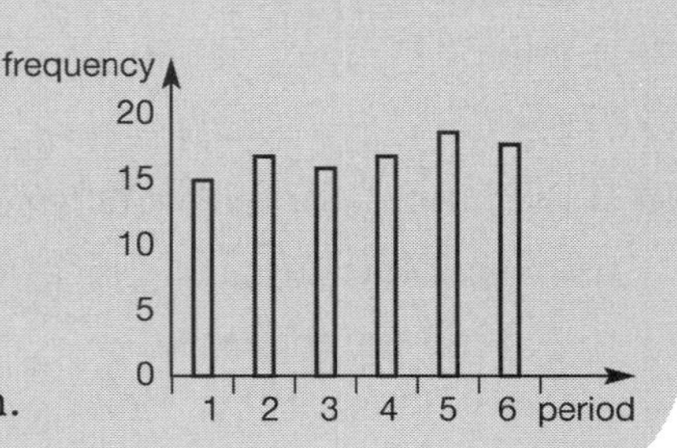

Reminder continued

a) How many times did he use the phrase during period 3?
b) During which period did he use the phrase most often?
c) Suggest reasons why he used the phrase less often during period 1.

Solution

a) 16
b) 5
c) more wrong answers given; fewer pupils in the class; doing a different activity; etc

Example

This pie chart shows the number of brothers and sisters of each of the 24 pupils in Mr. Napier's Intermediate 2 class.

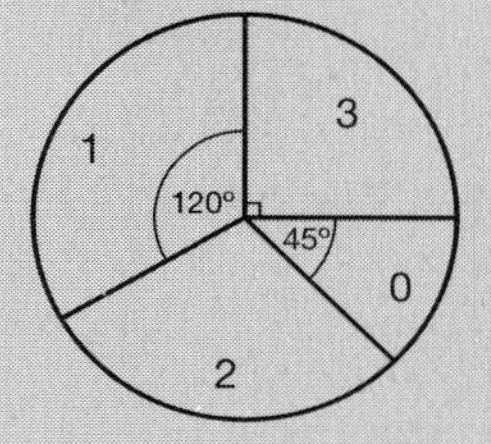

a) How many pupils have no brothers or sisters?
b) What proportion of the class were in a family of two children?
c) How many pupils had two other siblings?

Solution

a) $\frac{45}{360} \times 24 = 3$
b) $\frac{120}{360} = \frac{1}{3}$
c) $360 - (90 + 120 + 45) = 105;\ \frac{105}{360} \times 24 = 7$

Example

This stem and leaf diagram shows the percentage marks of Mr. Napier's Int. 2 class in the prelim exam on Units 1 and 2.

Stem	Leaf
4	7
5	0 1 3 7
6	2 2 5 9 9
7	1 1 4 5 7 7 8 9
8	0 2 5 8
9	3 6

6 | 3 denotes 63

a) Write down the highest and lowest percentage marks.
b) How many pupils scored less than half marks?
c) How many pupils would you expect to gain an 'A' pass at the end of the course?

Reminder continued

Solution a) 96, 47

b) 1

c) You need a scaled mark of 70% for an 'A' pass. The top 14 pupils might be successful.

Example This back to back stem and leaf diagram shows the percentage marks of the same class in their final prelim exam on all three units, girls on the left, boys on the right.

(girls)	Stem	(boys)
1	4	3 8
1 3 7	5	2
6 9	6	4 6 7
3 5 7	7	2 3
2 3	8	1 6 8
2	9	5

Comment This shows the girls' scores ranging from 41% to 92% and the boys' from 43% to 95%. One girl and two boys scored less than half marks. Six boys and six girls scored over 70%; two less than last time, which is quite realistic.

Exercise 59

1 A class of sixth year pupils were asked how many children were in each of their families. The results are shown in this bar graph:

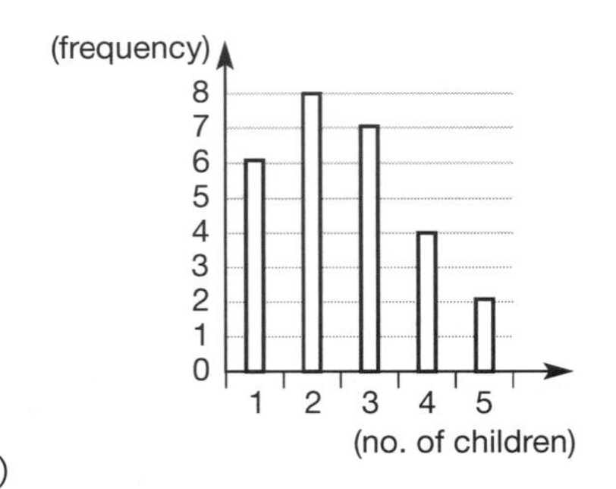

a) How many pupils were in the class ?

b) How many families had four children?

c) How many pupils had no brothers or sisters?

d) How many families had more than three children?

e) State the modal number of children. (The mode is the value with the highest frequency.)

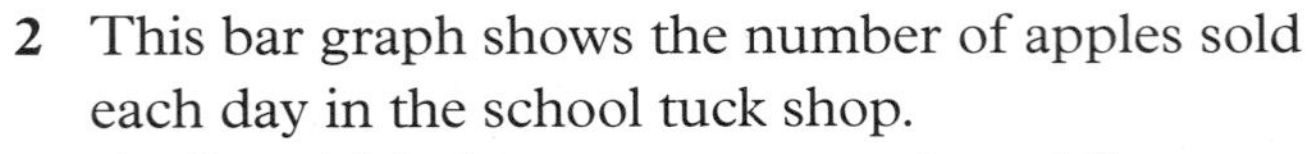

2 This bar graph shows the number of apples sold each day in the school tuck shop.

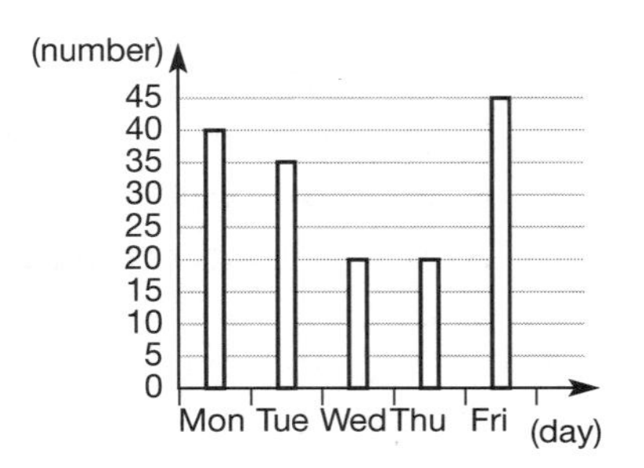

a) On which day were most apples sold?

b) On which two days were the sales equal?

c) How many apples were sold on Tuesday?

d) What perecentage of the apples sold that week were sold on Friday?

3 One hundred people were asked for their preferred type of meat. The results are shown in this bar graph. (The initials denote l: lamb p: pork s: steak v: veal g: gammon c: chicken k: liver or kidney)

a) Which type of meat was most popular?
b) Which was the least popular?
c) How many people prefer chicken?
d) How many more people chose steak than gammon?
e) How many less chose pork than chose lamb?

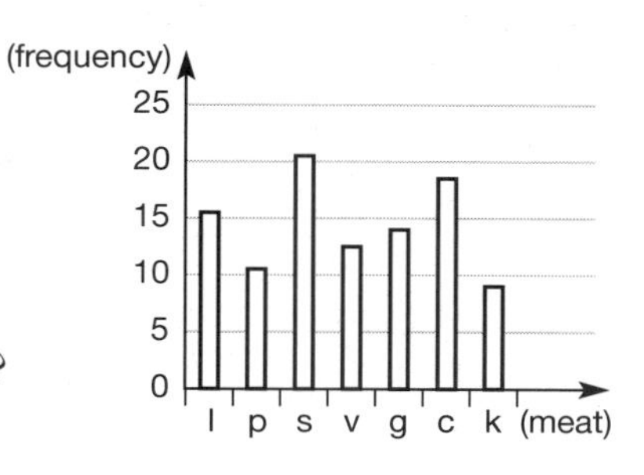

4 A sample of pupils was asked by the organisers of their Christmas party what their favourite flavour of ice cream was. The results were recorded in this bar graph.

a) Which flavour was chosen by 8 pupils?
b) How many chose chocolate?
c) Which flavour was most popular?
d) How many more people chose chocolate than chose strawberry?
e) Which flavour was least popular?
f) How many pupils were in this sample?

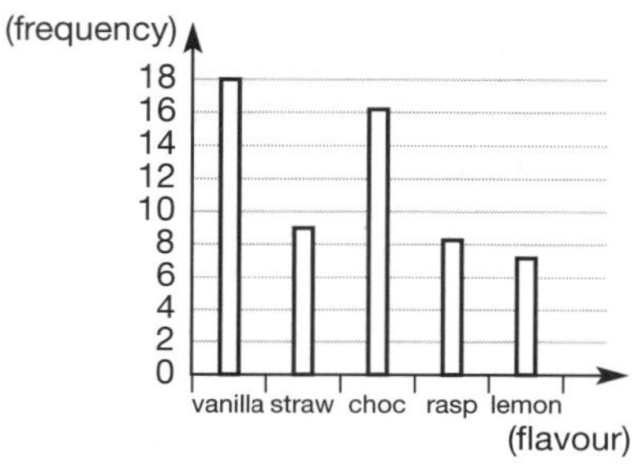

5 The line graph records the height (in millimetres) of a boy on his birthday each year during primary school.

a) How tall was he when he became 6?
b) How tall was he on his 10th birthday?
c) At what age was he 1 m tall?
d) How much did he grow in height between his 10th and 11th birthdays?
e) How much taller was he when he was 9 than when he was 6?

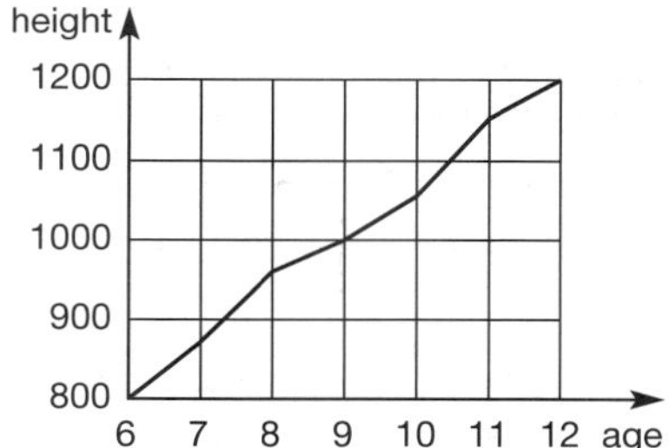

6 The amount of petrol used by a car depends on the speed at which it is driven. The distance covered per litre of petrol (in km per litre) is shown against speed (in km/h) for a particular model, by the line graph below.

a) What is the distance per litre at
   (i) 30 km/h
   (ii) 90 km/h?
b) Between what speeds is the distance per litre more than 3 km/l?
c) At what speeds is the petrol consumption equal to 2 km/l?
d) For this car what is
   (i) the most economical speed
   (ii) the least economical speed?

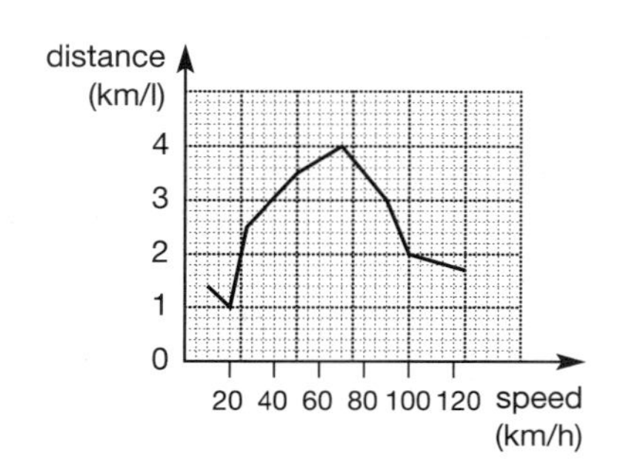

7 This line graph shows the annual rainfall in inches at Sandy Cove during the 1930's.

a) In which year did most rain fall?
b) In which year did least rain fall?
c) How many inches of rain fell in 1935?
d) In which year did 46 inches of rain fall?
e) Between which two consecutive years did the greatest difference in rainfall occur?

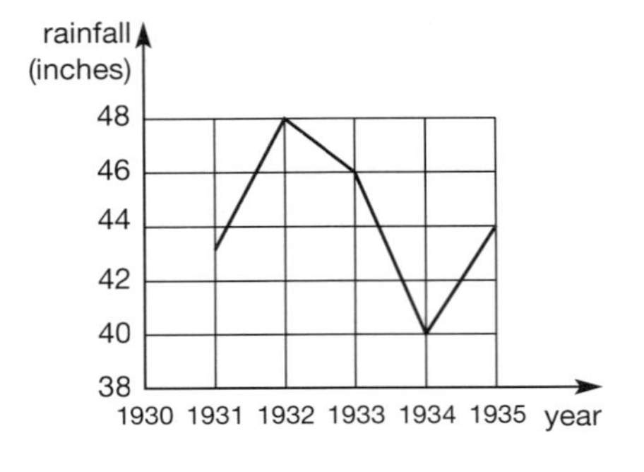

8 This line graph shows how the price of petrol (in pence per litre) varied over a seven month period.

a) What was the highest price during the period?
b) What was the lowest price during the period?
c) What price was it in February?
d) When did it stay unchanged for longest?
e) When did the biggest increase occur?

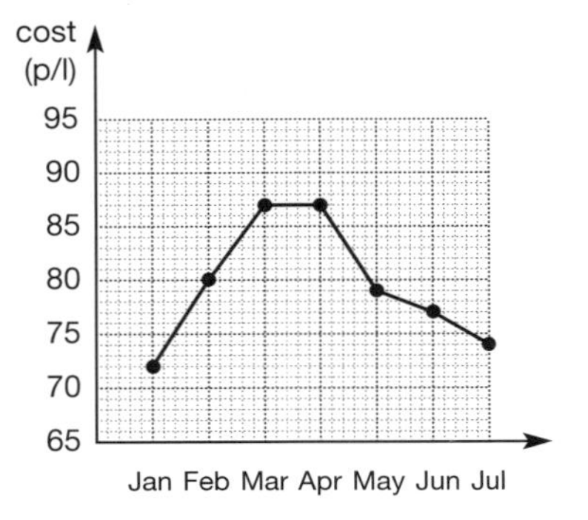

9 This pie chart shows the colour of hair of 100 fifth year boys.

a) Which is the most common colour?
b) Which colour occurs least often?
c) How many boys had blonde hair?
d) Which colours occur equally often?
e) Did more boys have brown hair or blonde hair?

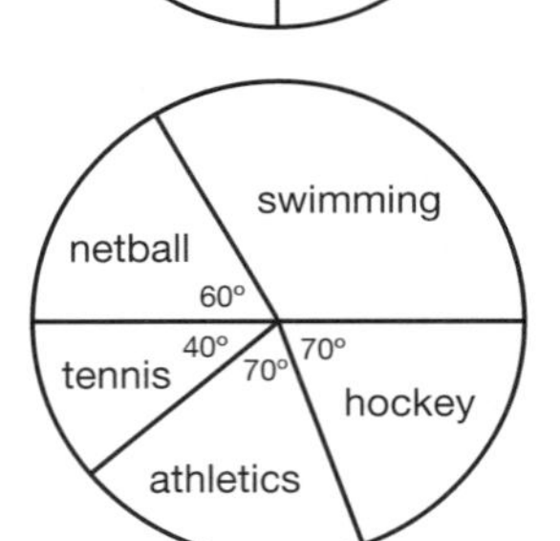

10 This pie chart shows the favourite sports of all the fifth year girls.

a) Which sport is most popular?
b) Which sport is least popular?
c) Which two sports are equally popular?
d) Is netball more popular than hockey?
e) Find the swimming sector angle.

11 This pie chart shows the colour of ring binder chosen by 120 fifth year maths pupils.

a) Name two pairs of colours which were equally well liked.
b) Which was the least popular colour?
c) How many pupils chose a yellow folder?
d) What is the red sector angle.

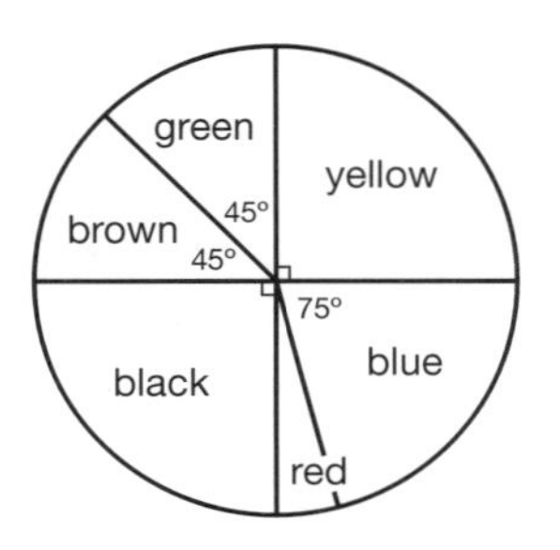

12 This pie chart shows the type of drink chosen by 360 senior pupils at a ceilidh.

a) What was the most popular drink?
b) What was the least popular drink?
c) Which drinks were equally popular?
d) How many chose ginger beer?
e) Calculate the size of the angle in the Irn Bru sector.

13 The ages of the first fifteen cars to arrive at a veteran and vintage car rally are shown in this stem and leaf diagram.

4	3 4 7
5	0 1 2 5 8
6	2 3 6 9
7	3 7 8

4 | 3 denotes 43

a) What was the age of the oldest car to arrive?
b) Were more or less of these cars over 60 years old?
c) Were any of these cars made and sold in the same year?

14 The 25 pupils in Mr Napier's S4 maths class measured their heights and recorded them in this stem and leaf diagram.

15	1 2 3 5 6 7 9
16	0 1 2 3 5 7 8 9
17	0 1 1 3 4 5 6 6 8
18	1

15 | 1 denotes 1·51 m

a) Write down the heights of the tallest and the shortest pupils in the class.
b) What percentage of pupils were less than 1·60 m tall?
c) What percentage of pupils were between 1·60 m and 1·70 m tall (inclusive)?

15 These 25 pupils in Mr Napier's S4 maths class discovered their verbal reasoning quotients (VRQ's) and recorded them in this stem and leaf diagram.

10	3 5 8 9
11	1 3 4 5 7 8 9
12	0 1 2 4 5 6 8 8 9
13	1 3 4 6 7

12 | 3 denotes 123

a) Write down the highest and lowest VRQ's in the class.
b) Do the majority of pupils have a VRQ of less than, or more than 120?
c) Write down the mode of this distribution.

16 Last Tuesday the practice nurse in Dr Finlay's surgery took the blood pressure of 25 patients. The diastolic blood pressures observed are displayed in this stem and leaf diagram.

Stem	Leaves
7	0 1 3 4
7	5 7 8 9
8	0 0 1 2 4 4
8	6 7 7 8
9	0 1 2 3
9	5 6 8

8 | 3 denotes 83

a) Write down the highest and lowest blood pressures observed in this sample.
b) What percentage of these patients had a diastolic blood pressure above 90?
c) How many fewer people had a diastolic blood pressure over 85 than under 85?

17 Construct an ordered back to back stem and leaf diagram to illustrate the weekly earnings of Mr. Napier's Intermediate 2 Maths class from the data given:

Boys	£24	£17	£28	£35	£19	£38	£45	£12	£33	£27	£39	£0
Girls	£25	£30	£13	£31	£19	£34	£0	£17	£40	£29	£35	£0.

60 Revision of Constructing Graphs

Reminder

Example Draw a bar graph to illustrate the number of bananas sold in the school dining room last week, from this data:

Monday 24 Tuesday 31 Wednesday 29 Thursday 27 Friday 30

Solution Mark the days on the 'x' axis and the frequency on the 'y' axis. (This is not a coordinate diagram.)

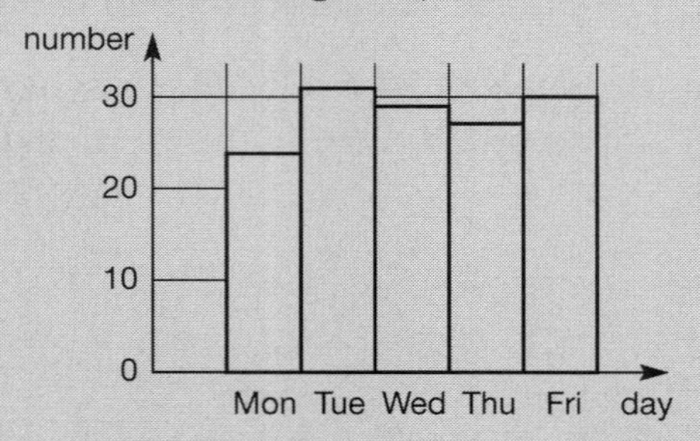

Example From the following data draw a line graph to illustrate the temperature in the dining room around lunchtime.

time	12:30	12:45	13:00	13:15	13:30	13:45
temperature (°C)	15	15	17	18	16	14

Reminder continued

Solution Plot these points as on a coordinate diagram, and join successive points.

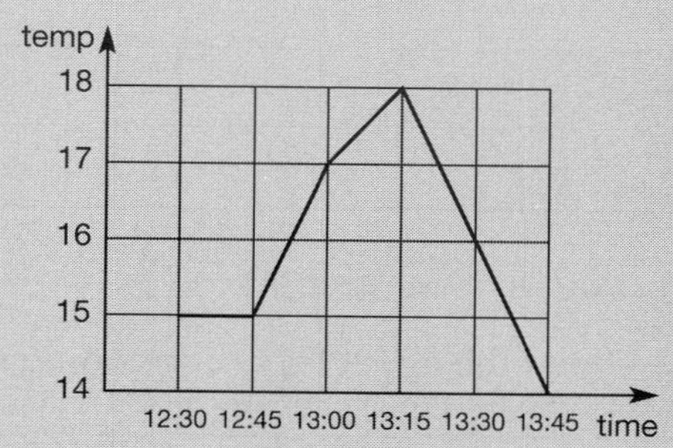

Example The number of grammes of mashed potato that a sample of boys received with their school lunches were as follows:

103	106	110	104	112	116	102	108	95	109
113	105	99	107	115	98	109	96	111	112

Draw a stem and leaf diagram to illustrate this data.

Solution Look through the data and note the range, from 95 to 112.
So we need 90, 100 and 110 as stems.
(Now go through the data systematically, e.g. by columns, so for 103 put a 3 in the row with the stem 10 and for 113, put a 3 in the row with stem 11.)

9	9 8 6 5
10	3 6 5 4 7 2 9 8 9
11	3 0 2 5 6 1 2

(Now put the 'leaves' in ascending order.)

9	5 6 8 9
10	2 3 4 5 6 7 8 9 9
11	0 1 2 2 3 5 6

(Finally, include the key to ensure your notation is understood.)

9 | 3 denotes 93

Exercise 60

1 This frequency table shows the number of punctures sustained by the 18 members of a cycling club over the past three years:

number of punctures	0	1	2	3	4
frequency	1	3	5	7	2

Illustrate this information on a bar graph.

2 This frequency table shows the ages of those attending a junior disco at a holiday camp:

age	7	8	9	10	11	12	13
frequency	5	11	16	23	14	2	1

a) Illustrate this information on a bar graph.
b) What was the modal age of those in attendance?

3 This frequency table shows the number of times the pupils in one class requested to see their guidance teacher last term:

no. of visits	0	1	2	3	4
frequency	2	5	6	4	1

Illustrate this information on a bar graph.

4 This frequency table shows the number of pens carried by the pupils in one class last Friday.

no. of pens	0	1	2	3	4	5	6
frequency	4	5	7	3	2	2	1

a) Illustrate this information on a bar graph.
b) What was the modal number of pens carried?

5 The price of a packet of 40 tea bags varied as follows:

Jan	Feb	Mar	Apr	May	Jun	Jul	Aug	Sep	Oct	Nov	Dec
89p	92p	91p	87p	88p	86p	88p	89p	90p	89p	91p	92p.

a) Illustrate this data by a line graph.
b) When was tea
(i) cheapest,
(ii) most expensive?

6 Last January, Ferrybridge High School suffered a flu epidemic. For the first ten days of term the numbers of absentees were as follows:

Day	1	2	3	4	5	6	7	8	9	10
no. of absentees	31	34	59	77	88	83	69	43	30	29

a) Illustrate this data by a line graph.
b) For what percentage of this time were there more than 60 pupils absent?

7 Baby Anna was weighed at the clinic every fortnight for her first four months, with these results:

age in weeks	0	2	4	6	8	10	12	14	16
mass (kg)	3·1	3·2	3·6	4·0	4·1	4·3	4·3	4·8	5·1

a) Illustrate this data by a line graph.
b) At what age do you think Anna had a mass of 4·5 kg?

8 Arthur measured the length of each of his tropical fish and obtained these lengths (in mm):

61 52 49 57 51 48 53 45 48 58 58 60
53 55 60 54 52 50 49 66 65 56 57 53

Construct a stem and leaf diagram to illustrate this data.

9 After two rounds of a golf competition the leading twenty scores were:

155 151 149 153 158 144 158 147 165 149
146 166 158 160 153 156 141 161 168 158

Construct a stem and leaf diagram to illustrate this data.

10 A trading standards inspector recorded the volume of milk in 20 cartons (allegedly holding one litre). His results (in ml) were as follows:

1002 1010 1005 1011 995 988 1000 1012 1002 1008
1003 1009 989 1010 1001 1000 998 1008 996 1007

Construct a stem and leaf diagram to illustrate this data.

61 Scattergraphs

Reminder

We can construct a scatter diagram when we have two pieces of information about each member of the sample. Some scattergraphs are only qualitative.

Example This scattergraph plots the value of a car against its brake horse power.
A car P is shown.
Plot a point for another car Q, which is more powerful than A but cheaper.

value
• P
brake horse power

Solution Q is more powerful than P,
so Q must be to the right of P;
Q is cheaper than P
so Q must be below the level of P.

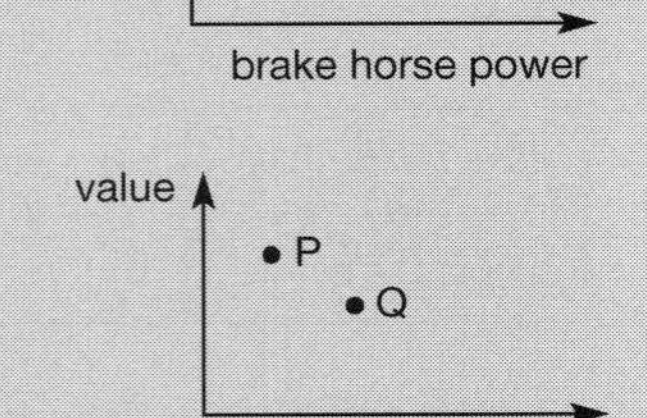

Example In Drumlithy High School, 14 pupils do both Maths and Physics at Advanced Higher level. Their prelim percentages are shown:

Pupil	Alf	Bob	Dot	Eve	Fay	Ian	Jim	Ken	May	Pam	Sam	Tom	Una	Val
physics	51	56	44	46	52	38	39	59	29	36	24	21	42	48
maths	82	91	73	76	85	64	67	97	52	61	43	37	71	79

a) Illustrate this data on a scattergraph (maths against physics).
b) Draw the best fitting straight line through this data.
c) Use your line to estimate what a pupil might have scored in his Maths exam if he had scored 54% for physics.

Solution a) b)

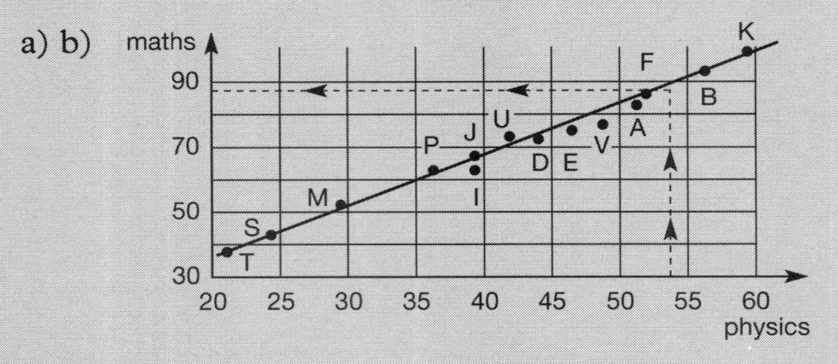

It is not usual to label the points.

c) follow the arrows from 54 on the x-axis to obtain 88%.

Exercise 61

1 Consider this scattergraph showing altitude against temperature.
Two statements can be made: D is higher than C,
D is warmer than C

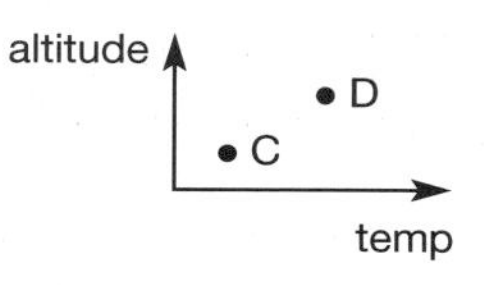

Similarly, write down two statements for each of the following:

a) 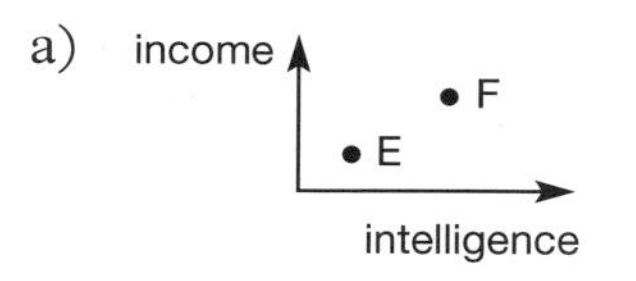

b) 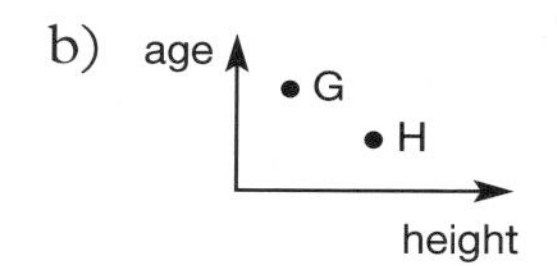

c) 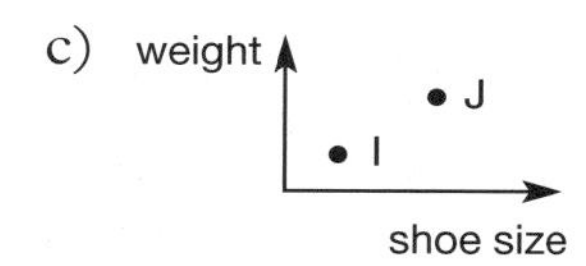

d) 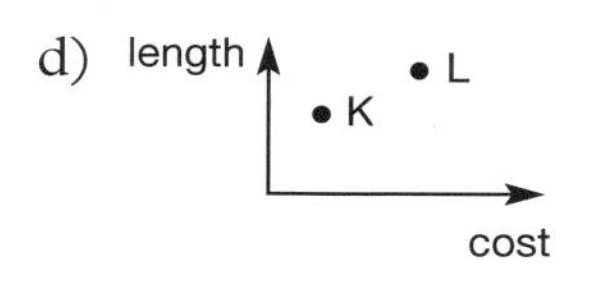

e) 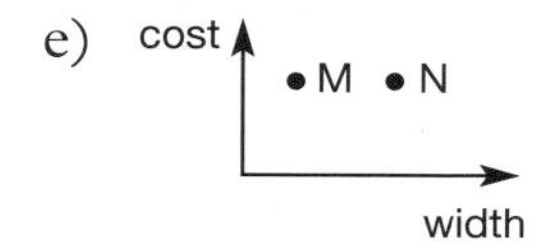

f) 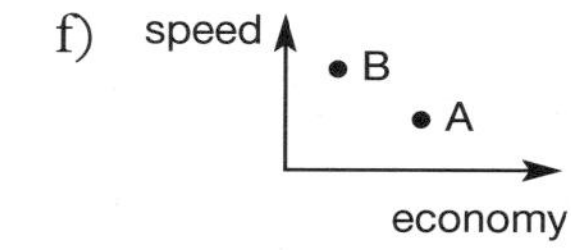

g) 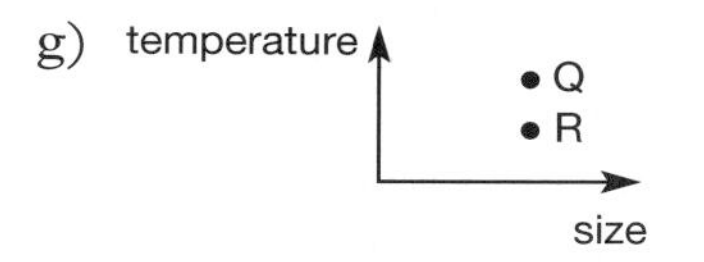

h) 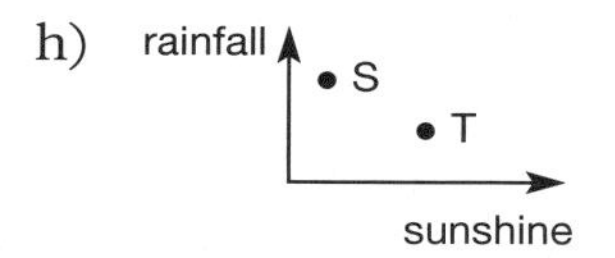

i) 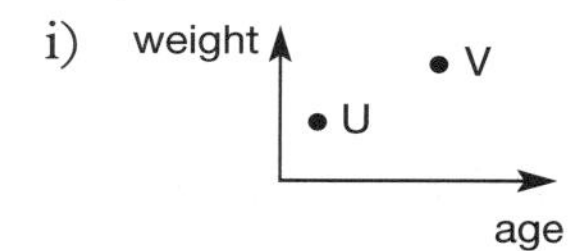

j) 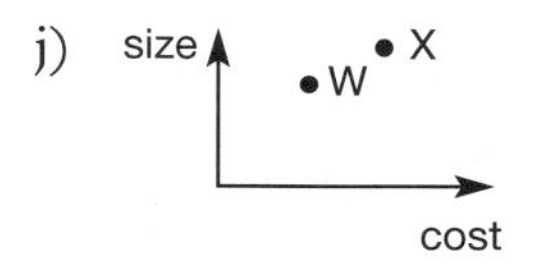

k) 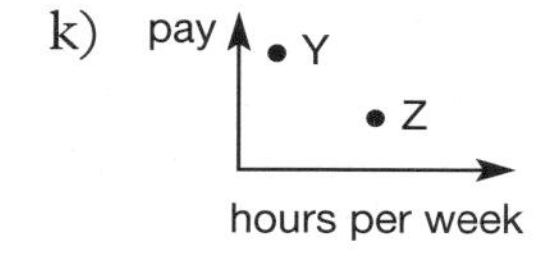

l) 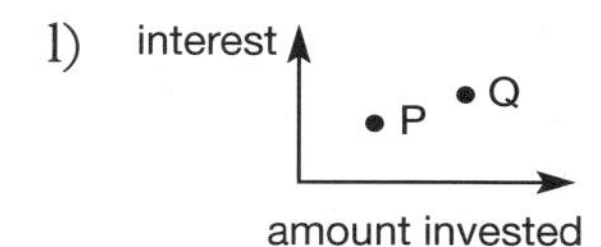

2 Here are the front views of five local hotels:

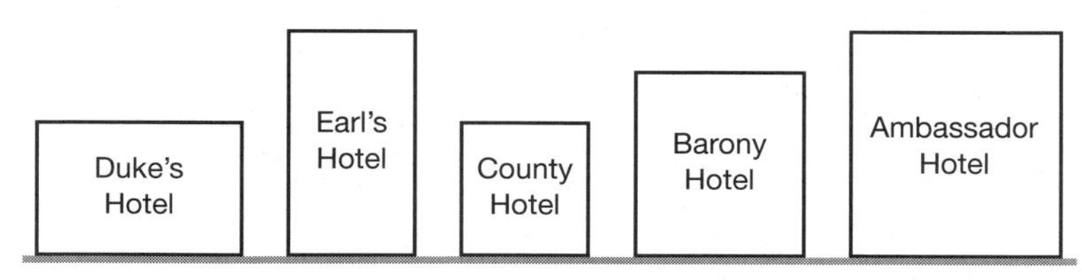

On this scattergraph, the point B represents the Barony Hotel.
Copy the diagram and add points A, C, D, E for the other four hotels.

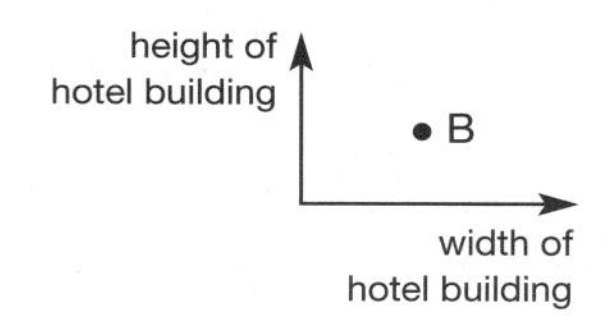

3 This map shows Kirkcaldy, West Wemyss, Thornton, Glenrothes, Auchtertool, Kinghorn, and Burntisland. Make a copy of the accompanying scatter diagram and label each point.

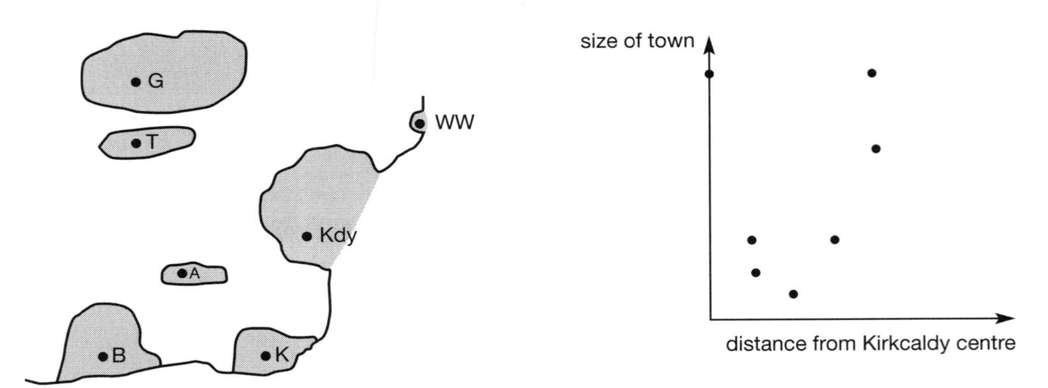

4 Here are the school reports for four boys:
Arthur has not done a stroke this term, and his marks reflect his laziness.
Billy is a very capable pupil, as his marks show, but he does not concentrate in class.
Charles is a very hard worker and deserves these excellent marks.
Dennis is a reliable worker, but did not do well in his exams.

a) Copy the accompanying scattergraph and label four points A, B, C and D for these four boys.
b) Write a suitable report for Ewan, the boy represented by the fifth point on the graph.

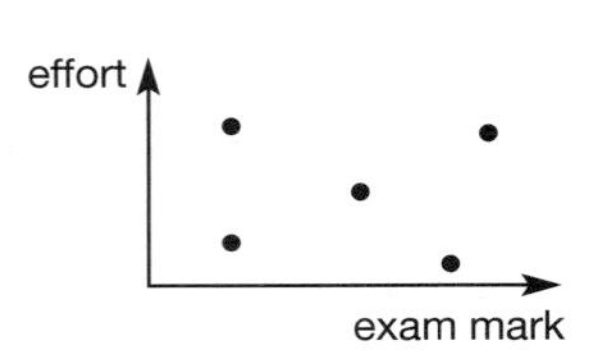

5 Fifteen children between 3 and 9 years had their heights (in mm) and their ages (in months) recorded as follows:

child	A	B	C	D	E	F	G	H	I	J	K	L	M	N	O
age	61	86	36	64	93	48	96	77	53	45	69	87	82	39	99
height	605	777	333	581	841	515	958	697	484	415	691	865	783	422	892

a) Illustrate this data on a scattergraph, plotting height against age.
b) Draw the line of best fit through the points on your graph.
c) Use your line to estimate the height of one of these children at age 7 years.

6 An artist works five days per week in his studio producing paintings.
He generally completes 12 works per month. When he has to spend time out of the studio he completes less paintings per month, as shown:

no. of days out of the studio	0	2	3	5	7	9	10	12
works completd that month	12	11	10	9	8	7	6	5

a) Illustrate this data on a scattergraph, plotting number of works against days out.
b) Draw the line of best fit through the points on your graph.
c) Can you use your line to predict how many paintings he would manage to complete in a month in which he had to be out of his studio for 15 days?
d) Why is your answer to part c) less reliable than estimating a number of paintings completed in a month with 8 days spent out of his studio?

7 The 24 pupils in an Int. 2 Maths class recorded their heights (in metres) and masses (in kg) as follows:

pupil	A	B	C	D	E	F	G	H	I
height (m)	1·58	1·89	1·41	1·73	1·67	1·49	1·81	1·70	1·54
mass (kg)	68·7	80·5	63·6	74·5	71·7	64·5	77·5	73·3	66·9

pupil	J	K	L	M	N	O	P	Q	R
height (m)	1·85	1·47	1·64	1·78	1·59	1·69	1·63	1·92	1·52
mass (kg)	79·2	62·4	70·4	75·8	68·4	72·8	70·6	80·3	65·3

a) Illustrate this data on a scattergraph, plotting mass against height.
b) Draw the line of best fit through the points on your graph.
c) Use your line to estimate the mass of a pupil of height 1·75 m.

62 Constructing Pie Charts

Reminder

In a pie chart the area of each sector is proportional to the size of the quantity it represents.
Hence the angles at the centre are in proportion to the sizes of quantities they represent.

Example This frequency table shows the number of punctures sustained by the 18 members of a cycling club over the past three years.

number of punctures	0	1	2	3	4
frequency	1	3	5	7	2

Illustrate this information in a pie chart. (The same distribution as Ex. 60 Q.1)

Reminder continued

Solution

0 punctures, frequency 1 $\Rightarrow \frac{1}{18} \times 360° = 20°$

1 puncture, frequency 3 $\Rightarrow \frac{3}{18} \times 360° = 60°$

2 punctures, frequency 5 $\Rightarrow \frac{5}{18} \times 360° = 100°$

3 punctures, frequency 7 $\Rightarrow \frac{7}{18} \times 360° = 140°$

4 punctures, frequency 2 $\Rightarrow \frac{2}{18} \times 360° = 40°$

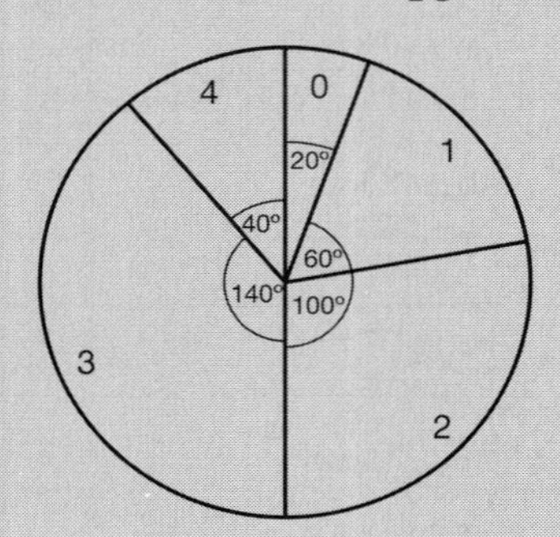

Exercise 62

1 At a wedding meal, the guests chose their main courses as follows:

32 salmon   24 gammon   40 roast beef

Illustrate this information in a pie chart.

2 One day 600 senior citizens collected their pensions from Clamers Street Post Office. They were asked how they got there, and the answers were as follows:

250 walked   175 came by bus   100 came by car   75 cycled

Illustrate this information in a pie chart.

3 This frequency table shows the number of times the pupils in one class requested to see their guidance teacher last term.

no. of visits	0	1	2	3	4
frequency	2	5	6	4	1

Illustrate this information in a pie chart. (The same distribution as Ex. 60 Q. 3, page 132)

4 This frequency table shows the number of pens carried by the pupils in one class last Friday.

no. of pens	0	1	2	3	4	5	6
frequency	4	5	7	3	2	2	1

Illustrate this information in a pie chart. (The same distribution as Ex. 60 Q. 4)

5 This frequency table shows the ages of those attending a junior disco at a holiday camp.

age	7	8	9	10	11	12	13
frequency	5	11	16	23	14	2	1

Illustrate this information in a pie chart. (The same distribution as Ex. 60 Q. 2)

6 Dalburnie Small Bore Rifle Club held a vote to determine the colour to repaint the clubhouse. The results are shown in this bar graph. Illustrate this data on a pie chart.

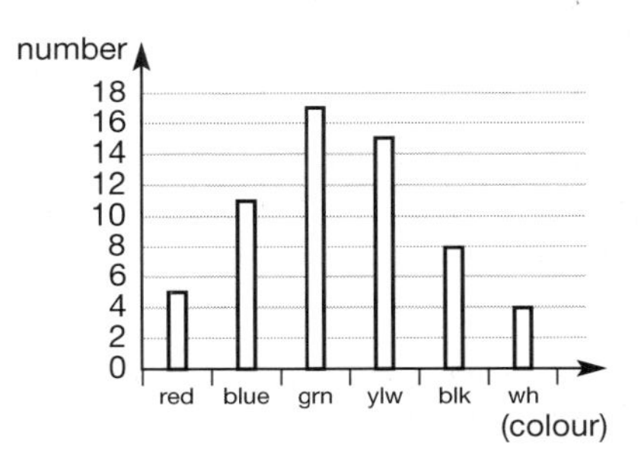

63 Dot Plots

Reminder

Dot plots provide a simple way to illustrate small numbers of data. They are really simplified bar graphs, and show quickly the shape of a distribution.

Example A woman working in her garden became aware of a blackbird with a white feather on one wing. She counted the numbers of worms he caught on each visit to her garden. These were:

2 3 0 1 5 4 2 4 3
0 2 1 7 4 0 2 3

Illustrate this data on a dot plot.

Solution (Enter one dot above the appropriate point on the scale.)

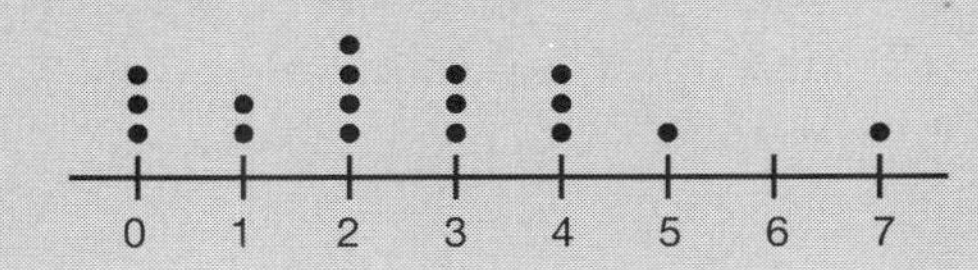

Exercise 63

1 A garden centre has a stock of *gentico aspirillium*. An assistant counts the number of leaves on each stalk and obtains:

3 3 4 5 2 4 5 3 2
3 3 5 4 6 2 4 3 5.

Illustrate this data by a dot plot.

2 Mrs. Watt recorded the number of days between consecutive mowings of her lawn, obtaining:

9 7 8 7 6 9 7 10 8

7 9 9 8 7 10 11 9 7.

Illustrate this data by a dot plot.

3 When Mr. Scott is in his back garden, his neighbour's dog comes barking for a biscuit. He counted the number of doggy biscuits he needed each day to keep the dog quiet, obtaining:

3 0 1 2 1 3 4 1
3 0 5 2 1 2 0 1.

Illustrate this data by a dot plot.

4 William counted the number of weeks between successive haircuts, obtaining:

5 7 6 6 7 8 5 5
6 7 8 6 5 7 8.

Illustrate this data by a dot plot.

5 Evelyn noted the number of clothes pegs she used on the washing line with each batch of washing, with these results:

15 11 12 14 15 13 16 18 17 14
13 15 17 11 13 15 17 16 14 13.

Illustrate this data by a dot plot.

6 When Tracey came home from her holiday, her lawn required cutting, and lots of daisies were growing in it. She noticed that they were growing in clumps, and counted the number of daisies in each clump, with these results:

21 23 18 25 19 22 21 20 23 19 24
21 25 17 18 23 20 24 19 18 20.

Illustrate this data by a dot plot.

64 Quartiles and Box Plots

Reminder

The **median** is the middle value of a data set arranged in order.
The **lower quartile** is the median of the lower half of the data set.
The **upper quartile** is the median of the upper half of the data set.
A **five figure summary** of a data set consists of the median, quartiles, and lowest and highest values.
A **box plot** is a way of conveying the five figure summary pictorially.

Example

Scotland used 13 players in a particular football international. The number of goals that each had scored for his club during that season were:

0	2	5	8	3	9	14
16	21	6	12	1	18.	

Find the median and quartiles and construct a box plot

Solution

First arrange the data in ascending order:

0	1	2	3	5	6	8	9	12	14	16	18	21

Since we have 13 players, the middle one is the 7th (since $\frac{1}{2}(13+1) = 7$), so the median is 8 (goals).

The lower quartile is the median of [0 1 2 3 5 6], and

2 and 3 are the middle numbers here, so the lower quartile is $2\frac{1}{2}$.

The upper quartile is the median of [9 12 14 16 18 21], and

14 and 16 are the middle numbers here, so the upper quartile is 15.

Thus the box plot looks like this:

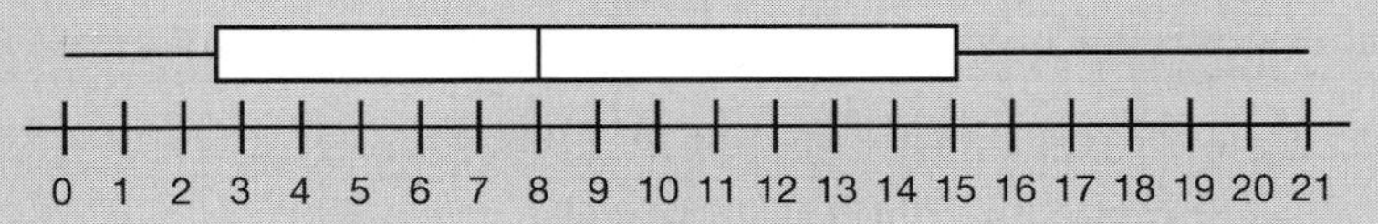

Exercise 64

1 Find the median and the quartiles, and construct a box plot for each of the following data sets:

a) 1 3 5 7 9
b) 7 7 8 9 9 10 11 11 11 12 16
c) 14 16 19 23 25 27 41
d) 85 85 86 87 87 89 91 95 100 103 4000
e) 80 75 87 73 84 71 88

2 After a day on the High Street, some student charity fund raisers had these amounts in their collecting cans:

£25·01	£19·25	£14·23	£17·36	£21·35	£27·81	£39·36
£17·32	£13·27	£15·81	£16·34	£20·00	£28·50	£12·98
£18·41	£15·95	£17·18	£22·09	£15·15	£19·08	£17·47

Obtain a five figure summary for this data set and construct a box plot to illustrate it.

3 The ages of the members of a primary school chess club are:

5 6 7 8 8 9 10 11.

Find the mode, the median and the mean, and construct a box plot.

4 The ages of the members of a secondary school chess club are:

12 13 14 14 14 15 16 16 16 16 17.

Find the mode, the median and the mean, and construct a box plot.

5 The number of new tyres that had to be fitted during the lifetime of a batch of 5-year old cars were:

8 9 7 8 5 6 9 7 9 10 9.

Find the mode, the median and the mean, and construct a box plot.

6 The number of runners entered by Spartans Athletic Club in the Christian Aid Forth Road Bridge cross in successive years was:

20 22 18 21 22 16 14 19.

Find the mode, the median and the mean, and construct a box plot.

7 The number of visits to the dentist in the last year by the members of a hockey club were:

1 5 4 2 1 1 3 5 4 3 2 4 6 4.

Find the mode, the median and the mean, and construct a box plot.

8 The ages of the members of a university rugby team are displayed on the dot plot below.

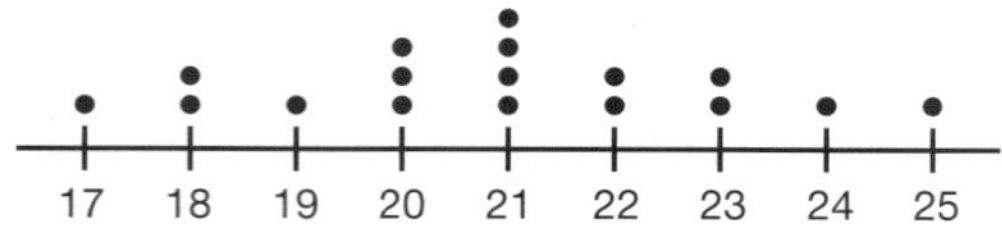

Illustrate this information in a box plot.

9 After a storm, slaters were invited to tender for repairs to the roofs of all the council houses in Munro Street. In order to obtain a suitable estimate for the work, George counted the number of broken slates on each roof, obtaining these results:

15	9	6	10	14	17	16	21	44	25	5
23	8	11	32	23	12	17	10	21	9	36
18	7	8	11	18	22	7	19	29.		

Arrange this data in an ordered stem and leaf diagram and construct a box plot.

65 Cumulative Frequency

Reminder

The cumulative frequency is the sum of all the frequencies so far, i.e. the number of the population with a score not exceeding a given score.

Example A window cleaner had a note of the number of window panes in each house in a housing estate.

no. of panes	20	21	22	23	24	25	26	27	28	29
frequency	2	4	7	16	34	39	24	13	9	2

Write out this frequency table, add a cumulative frequency column, and use it to calculate the median and quartiles.

Solution

no. of panes	f	cum. f
20	2	2
21	4	6 (2 + 4)
22	7	13 (6 + 7)
23	16	29 (13 + 16)
24	34	63
25	39	102
26	24	126
27	13	139
28	9	148
29	2	150

150 houses means that the median is the mean of 75th and 76th
Each the 64th to the 102nd has a value of 25.
Hence the median is 25.

The median of the lower 75 is the $\frac{1}{2}(75 + 1)$th;
i.e. the 38th.
Each of the 30th to 63rd has a value of 24
Hence the lower quartile is 24.

The upper quartile is therefore the (75 + 38)th or 113th value i.e. 26.

Exercise 65

1 A first year class did a mental arithmetic test and the results are shown in this frequency table:

score	3	4	5	6	7	8	9	10
frequency	2	3	5	4	6	7	2	1

a) Write out this frequency table (vertically) and add a cumulative frequency column.
b) Use this to find the median and quartiles, and construct a box plot.

2 After some more practice, the class in question 1 did a similar test and the results were:

score	4	5	6	7	8	9	10
frequency	2	2	3	5	7	6	5

a) Write out this frequency table (vertically) and add a cumulative frequency column.
b) Use this to find the median and quartiles and construct a box plot.
c) Compare the two box plots and comment on the performance of the class.

3 Draw the cumulative frequency graph for each of questions 1 and 2.

4 A church organised a sponsored 'Songs of Praise' and after consultation decided on a selection of hymns. The numbers of verses in each of these hymns were as follows:

no. of verses	2	3	4	5	6	7	8
frequency	2	5	7	3	2	1	1

Add a cumulative frequency column to this frequency table and construct a box plot.

5 The number of teams scoring a particular number of goals on Saturday was:

no. of goals	0	1	2	3	4	5	6	7
frequency	11	10	8	6	3	1	0	1

Add a cumulative frequency column to this frequency table and construct a box plot.

6 A sample of pupils were asked how many text books they were carrying, and the results were as follows:

no. of text books	0	1	2	3	4	5	6
frequency	4	7	17	23	16	12	5

Find the median and quartiles from a cumulative frequency column.

7 All first year pupils in a school were asked how often they had had chips last week, giving:

no. of times chips eaten	0	1	2	3	4	5	6	7
frequency	23	37	53	29	25	14	13	11

Find the median and quartiles from a cumulative frequency column.

(There is no revision exercise or Topic Test for this topic because there is so much overlap with the following one.)

STATISTICS

66 Revision of Probability

Reminder

Probability is a measure of chance between 0 and 1.
Events with probability 0 cannot happen, e.g. choosing a banana from a bowl of 5 apples and 3 pears.
Events with probability 1 are certain to happen, e.g. obtaining a total of less than 15 on rolling two standard dice.
Many probability calculations are based on situations with equal likely outcomes, e.g. in tossing a coin it is equally likely to obtain a head or a tail; in rolling a die a score of 1, 2, 3, 4, 5, 6 are all equally likely; in choosing a card from a standard pack, each of the 52 cards is equally likely to be picked. The formula for the probability of one of these equal likely events is

$$\left[\text{P(event)} = \frac{\text{no. of favourable outcomes}}{\text{total no. of outcomes}}\right]$$

Example A bag contains 4 red marbles and 3 blue ones. A marble is chosen at random from the bag.

a) What is the probability that the marble chosen is red?
b) If you repeated this 350 times, how many times would you expect to choose a red marble?
c) If a blue marble is obtained on the first draw and it is not replaced in the bag, what is the probability of obtaining a red marble on the second draw?

Solution For small numbers like this, drawing your own picture helps. Use coloured pens or just shading.

a) $\text{P(red)} = \dfrac{\text{no. of red marbles}}{\text{total no. of marbles}} = \dfrac{4}{7}$

b) expected no. of red marbles = (probability of red)×(no. of trials)
$= \dfrac{4}{7} \times 350 = 200$

c) In your sketch, put your thumb over one of the blue marbles, so that you are now looking at 4 red and 2 blue marbles.

Hence $\text{P(red)} = \dfrac{\text{no. of red marbles}}{\text{total no. of marbles}} = \dfrac{4}{6} = \dfrac{2}{3}$.

Exercise 66

1 a) What is the probability of obtaining a head on one toss of a coin?
b) How many heads would you expect to get on 100 tosses of a coin?

2 a) What is the probability of scoring six with one throw of a die?
b) How many sixes would you expect to get if you threw 216 dice?

3 a) If the probability of a primary school pupil in Aberforth catching flu this winter is one fifth, what is the probability of avoiding the flu?
b) In an Aberforth primary school of 550 pupils, how many do you expect to catch flu this winter?

4 A bag contains 4 red, 3 green and 2 blue counters. If a counter is drawn at random from the bag, write down the probability of obtaining:
a) a red counter b) a green counter c) a blue counter.

5 An urn contains 3 red balls and 4 white ones. A ball is drawn at random.
a) What is the probability that it is red?
b) If the drawn ball is not red and is not replaced in the urn and a second ball is drawn at random, what is the probability that the second ball is red?

6 There are 9 cream filled sweets in a paper bag; 3 coffee, 3 lime and 3 strawberry flavoured.
One is eaten at random.
a) What is the probability that it is coffee flavoured?
b) If it was coffee, what is the probability that the next sweet chosen at random is coffee?
c) If the first sweet was not coffee, what is the probability of the second one being coffee?

7 A scrabble player holds the tiles necessary to spell the word PARALLEL, which means that he has an extra tile and must return one to the pool. The tiles are shuffled face down and an opponent chooses a tile at random. What is the probability that it is
a) an R b) an A c) an L d) a vowel e) not a P?

8 Make an array of all the possible outcomes of throwing a red die and a blue die.
What is the probability of obtaining:
a) a double
b) a total of 12
c) a total of 7
d) no sixes
e) an even total
f) a total less than 4
g) 2 scores which multiply to give 13
h) two scores whose difference is less than 6
i) different red and blue scores
j) a red score greater than the blue score
k) an even red score and an odd blue score
l) the red score two less than the blue score?

9 A number is formed at random from the digits 2, 3, 4, 5 without repetition. If one of these numbers is selected at random, what is the probability that it is:

a) even
b) divisible by 5
c) divisible by 3
d) greater than 2000?

10 During an epidemic, it is estimated that the probability of a school pupil catching the infection is 0·14.

a) In a school of 750 pupils, how many pupils would you expect to become infected?
b) What is the probability of the head boy NOT catching it?

11 Bertie uses a zebra crossing on his way to and from school. Often he crosses the road with no moving traffic in sight and other times the traffic has to stop for him. Bertie counted that out of 100 journeys to and from school, traffic had to stop for him on 45 occasions.

a) What is Bertie's best estimate for the probability that traffic has to stop to let him use the zebra crossing?
b) If Bertie makes 4 journeys per day, how many times in a school week would you expect the traffic to have to stop for him?

12 In a trivia quiz, the category of the question to be asked is determined by a spinner with five unequal differently coloured sectors. In 100 spins the frequencies of the colours were:

colour	red	blue	orange	yellow	green
frequency	24	17	14	15	30

a) What is your best estimate for the probability of being asked a green category question?
b) How many green category questions would you expect to be asked in a game which had 70 questions?

67 Revision of Mean, Median, Mode, and Range, of a Data Set

Reminder

The **mean** score is the sum of all the scores divided by the number of scores.
The **median** score is the middle score when the scores are arranged in order.
(There is only a middle score if the number in the population is odd;
when it is even we use the mean of the two middle scores.)
The **mode** (or modal score) is the score which occurs most often, i.e. the mode has the highest frequency. (There could be no mode or more than one.)
The **range** is the difference between the highest and lowest scores.

Reminder continued

Example In a televised snooker contest, there were 11 century breaks, with values:

121 126 119 101 126 110 114 147 104 135 117.

Calculate:

a) the mean b) the median c) the mode d) the range

of this data set.

Solution

a) mean = (121 + 126 + 119 + 101 + 126 + 110 + 114 + 147 + 104 + 135 + 117) ÷ 11 = 1320 ÷ 11 = 120

b) (first write the scores in order)

101 104 110 114 117 119 121 126 126 135 147

the middle one is $\frac{1}{2}(11 + 1) = 6$th, which is 119

c) 126 occurs twice and no other score is repeated, so the mode is 126

d) range = 147 − 101 = 46

Exercise 67

1 This data set shows net golf scores in the annual staff handicap competition:

80 77 83 86 86 75 87.

a) Find (i) the mean (ii) the median (iii) the mode (iv) the range.

b) Find the probability of randomly selecting the winner.

2 This data set is the number of points scored by each member of the school basketball team in the first term matches:

41 75 59 52 59 23 6.

a) Find (i) the mean (ii) the median (iii) the mode (iv) the range.

b) Find the probability of randomly selecting a basketball player who had scored less than 50 points.

3 This data set shows the collar sizes (in inches) of members of a rugby scrum:

16 17 17 16 16 15 15 16.

a) Find (i) the mean (ii) the median (iii) the mode (iv) the range.

b) Find the probability of randomly selecting a rugby player with a 16 inch collar size.

4 This data set shows the skirt waist sizes (in inches) for a hockey team:

24 30 26 26 24 22 28 20 26 28 18.

a) Find (i) the mean (ii) the median (iii) the mode (iv) the range.

b) Find the probability of randomly selecting a hockey player with the modal waist size.

5 This data set shows the depth (in feet) of coal pits in one county:

535 320 690 455 605 575.

a) Find (i) the mean (ii) the median (iii) the mode (iv) the range.
b) Find the probability of randomly selecting a pit less than 500 feet deep.

6 This data set shows heights (in metres) of apple trees in a small orchard:

3·0 4·2 4·1 3·9 4·2 2·5 3·8 2·1 3·5 3·7.

a) Find (i) the mean (ii) the median (iii) the mode (iv) the range.
b) Find the probability of randomly selecting an apple tree taller than 4 m.

7 This data set shows the numbers of nuts in similar packets of assorted nuts:

63 60 57 69 53 59 61 50 61 65.

a) Find (i) the mean (ii) the median (iii) the mode (iv) the range.
b) Find the probability of randomly selecting a packet with less than 60 nuts in it.

8 This data set shows the numbers of points scored by a school 1st XV in their matches up till Christmas:

68 13 33 11 36 21 8 77 28 33 15 71.

a) Find (i) the mean (ii) the median (iii) the mode (iv) the range.
b) Find the probability of randomly choosing to watch a match where the 1st XV scored less than 20 points.

68 Revision of Mean, Median, Mode, and Range, from a Frequency Table

Reminder

When several scores occur with greater frequency, data is more usefully conveyed by a frequency table.

Example The ages of a youth club football team pool are:

17 14 19 16 17 20 18 20 17 18 18 18 18.

a) Find
(i) the mean (ii) the median (iii) the mode (iv) the range.
b) Find the probability of selecting a player aged 18 at random.

Reminder continued

Solution

age (x)	tallies	frequency (f)	xf	cum f
14	|	1	14	1
15		0	0	1
16	|	1	16	2
17	|||	3	51	5
18	~~||||~~	5	90	10
19	|	1	19	11
20	||	2	40	13
		13	230	

a) (i) $\bar{x} = \dfrac{\sum xf}{\sum f} = \dfrac{230}{13} = 17{\cdot}7$

(ii) the middle score is $\frac{1}{2}(13+1)$th, i.e. the 7th; the 6th to the 10th are aged 18, so median = 18

(iii) 18 has the highest frequency (5), so the mode is 18

(iv) range = 20 – 14 = 6

b) $P(18) = \dfrac{\text{no. of players aged 18}}{\text{total no. of players}} = \dfrac{5}{13}$

Exercise 68

1 In this frequency distribution x represents the number of pairs of trousers owned by the fifth year girls:

x	3	4	5	6	7	8	9
f	4	11	41	25	7	1	1

a) Find (i) the mean (ii) the median (iii) the mode (iv) the range.

b) Find the probability of selecting at random a girl with more than 5 pairs of trousers.

2 In this frequency distribution x represents the number of occupants of cars passing the school.

x	1	2	3	4	5	6
f	37	46	35	27	12	23

a) Find (i) the mean (ii) the median (iii) the mode (iv) the range.

b) Find the probability of selecting at random a car with only the driver in it.

3 In this frequency distribution x represents the number of days lost through illness in a small firm:

x	1	2	3	5	9	22
f	8	5	7	6	1	1

a) Find (i) the mean (ii) the median (iii) the mode (iv) the range.
b) Find the probability of selecting at random an employee with less than 5 days off.

4 In this frequency distribution x represents the number of goals scored per game:

x	0	1	2	3	4	5	8	11
f	10	13	14	8	1	2	1	1

a) Find (i) the mean (ii) the median (iii) the mode (iv) the range.
b) Find the probability of selecting at random a match where more than 5 goals were scored.

5 In this frequency distribution x represents the duration (in minutes) of the headteacher's phone calls:

x	2	3	4	6	7	8	9	10	11	14	15
f	1	2	3	1	3	4	1	2	1	1	1

a) Find (i) the mean (ii) the median (iii) the mode (iv) the range.
b) On a random phone call, what is the probability that the headteacher will be engaged for more than 10 minutes?

6 This bar graph shows the number of teeth missing from the mouths of 40 army recruits.

Calculate the mean number of teeth missing.

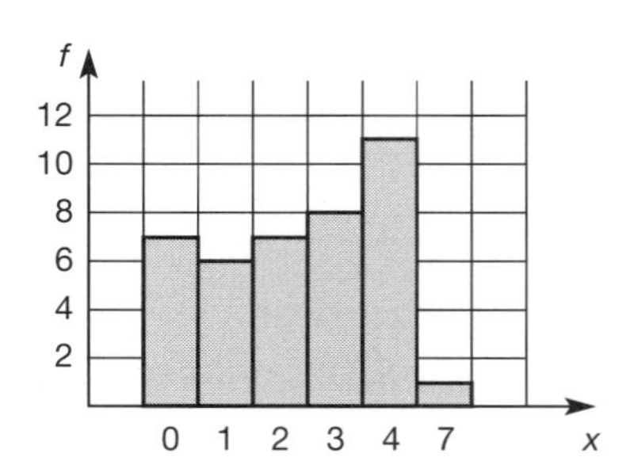

7 This bar graph shows the sizes of the football boots of the Primary Schools Champion team pool.

Calculate the mean size of boot.

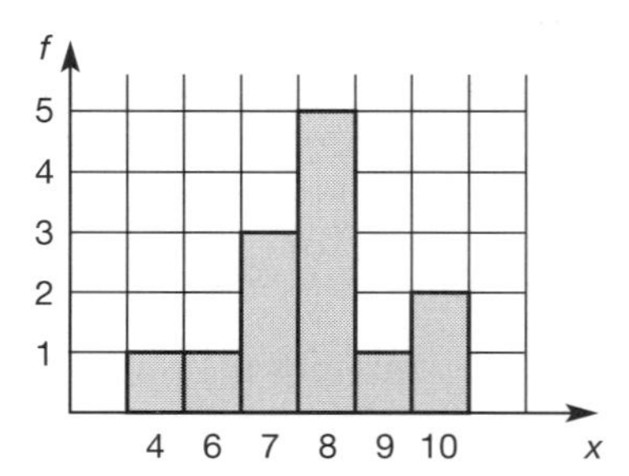

8 This bar graph shows the number of previous convictions of a group of criminals.

Calculate the mean number of previous convictions.

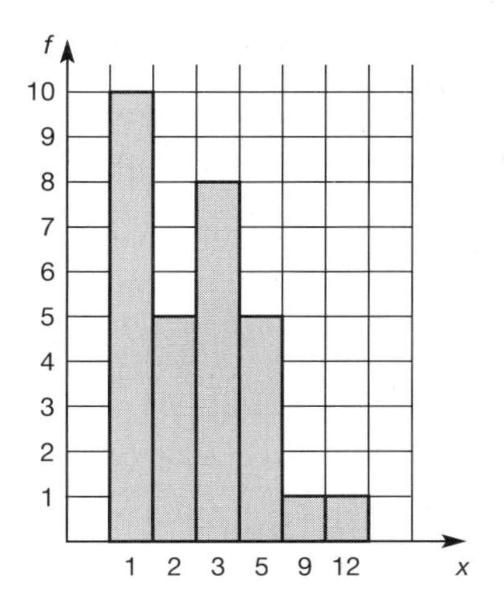

9 In 36 successive periods of Maths, Robin made the following numbers of mistakes:

1 3 1 1 0 2 5 1 2 3 1 4 1 2 2 3 2 1
2 0 2 4 3 1 0 4 3 1 2 3 0 1 2 1 3 0.

a) Construct a tally chart and hence a frequency table.
b) State the mode and the range of this distribution.
c) Add a cumulative frequency column and illustrate by a box plot.
d) Calculate the mean number of mistakes per period.
e) If an inspector visited the school one day at random, what is the probability that Robin made no mistakes on that day?
f) [Optional] Illustrate this data in
(i) a bar graph (ii) a pie chart.

10 An office staff of 40 reported that the number of cups of coffee each had consumed on Monday was:

3 2 1 3 4 3 1 2 4 7 3 2 4 1 3 0 3 3 4 3
1 4 0 2 2 2 4 3 0 2 2 3 2 2 8 4 1 7 2 4

a) Construct a tally chart and hence a frequency table.
b) State the mode and the range of this distribution.
c) Add a cumulative frequency column and illustrate by a box plot.
d) Calculate the mean number of cups of coffee consumed.
e) If one of the staff is chosen at random, what is the probability that he or she had more than two cups of coffee on the Monday?
f) [Optional] Illustrate this data in
(i) a bar graph (ii) a pie chart.

69 Quartiles and Semi-Interquartile Range

Reminder

The semi-interquartile range is half the difference between the quartiles.

Example The number of stops travelled by each passenger on one run of the number 53 bus was as follows:

no. of stops	1	2	3	4	5	6	7	8	9	10
frequency	1	3	5	8	6	9	10	8	5	7

Find the range and semi-interquartile range for this distribution.

Solution

no. of stops	f	cum f
1	1	1
2	3	4
3	5	9
4	8	17
5	6	23
6	9	32
7	10	42
8	8	50
9	5	55
10	7	62

range = 10 – 1 = 9

since 62 passengers,

median = mean of 31st and 32nd = 6

lower quartile = median of lower 31 = 16th = 4

upper quartile = median of upper 31 = 47th = 8

semi-interquartile range $= \frac{1}{2}(8 - 4) = 2$

Exercise 69

1 In this frequency distribution x represents the numbers of replica football shirts owned by a sample of 40 boys:

x	0	1	2	3	4	5
f	7	6	8	7	11	1

a) Find the semi-interquartile range.
b) If a boy is selected at random, estimate the probability that he has no such shirts.

2 In this frequency distribution x represents the number of goals scored in each game by Dysart Rob Roy last season:

x	0	1	2	3	4	5	6	7
f	7	9	7	8	1	2	1	1

a) Find the semi-interquartile range.
b) If a game is chosen at random, state the probability that the team scored more than 4 goals.

3 In this frequency distribution x represents the number of coins in the pockets of a sample of first year pupils:

x	0	1	2	3	4	5	6	7
f	3	5	16	10	3	1	1	1

Find the semi-interquartile range.

4 In this frequency distribution x represents the number of children in the households in Anstruther Avenue:

x	0	1	2	3	4	5	6
f	6	9	39	27	7	1	1

a) Find the semi-interquartile range.
b) If a house is chosen at random, state the probability that only one child lives there.

5 In this frequency distribution x represents the number of people in cars passing through a set of traffic lights:

x	1	2	3	4	5	6
f	32	29	15	12	10	2

a) Find the semi-interquartile range.
b) If a car is chosen at random, state the probability that it contained no passengers.

6 In this frequency distribution x represents the duration (in minutes) of the mobile phone calls of an electrical repair engineer last Tuesday:

x	2	3	4	6	7	8	9
f	4	3	2	3	4	3	1

Find the semi-interquartile range.

7 In this frequency distribution x represents the number of absences by workers in a factory to attend funerals last year.

x	0	1	2	3	4	5	6
f	11	8	5	7	6	2	1

a) Find the semi-interquartile range.
b) If a worker is chosen at random, state the probability that he or she attended less than 3 funerals last year.

70 Standard Deviation

Reminder

The range and semi-interquartile range are both measures of the spread of a distribution. They are limited in the information that they convey about the distribution. For example, the two data sets shown below have the same range, 80 – 1, and the same semi-interquartile range, $\frac{1}{2}(37 - 19)$, but they are clearly quite different.

1 18 18 19 20 21 22 25 34 35 36 37 71 73 80
1 6 8 19 20 21 22 23 24 25 27 37 38 39 80

The standard deviation is a measure of spread about the mean which uses every piece of data. It is the *root mean square deviation*, which is short for saying the square root of the average (mean) of the squares of the differences between each piece of data and the average (mean) of the data. (However when we are dealing with a sample of size n, we divide by $(n - 1)$ instead of n when finding the average of the squares of the deviations.)

Example Calculate the standard deviation of the heights (in metres) of the five members of a basketball team:

1·63 1·65 1·68 1·73 1·76.

Solution To find the deviation from the mean, we must first calculate the mean ($\bar{x}$).

We then add a column for the deviations, $x - \bar{x}$, and then another for $(x - \bar{x})^2$.

Reminder continued

x	$x - \bar{x}$	$(x - \bar{x})^2$
1·63	−0·06	0·0036
1·65	−0·04	0·0016
1·68	−0·01	0·0001
1·73	0·04	0·0016
1·76	0·07	0·0049
8·45		0·0118

$$\bar{x} = \frac{\sum x}{n} = \frac{8{\cdot}45}{5} = 1{\cdot}69 \quad \text{S.D} = \sqrt{\frac{\sum(x-\bar{x})^2}{n-1}} = \sqrt{\frac{0{\cdot}0118}{4}} = 0{\cdot}0543$$

As a simple way of interpreting standard deviation as a measure of spread, note that the three middle values in the data set all fall into the interval 1·69 ± 0·0543, as do the upper and lower quartiles.

Exercise 70

Calculate the standard deviation for each of the data sets given in questions 1 to 9.

1 The number of empty tables in a restaurant for the 8 p.m. sitting for dinner from Monday to Friday last week:

5 6 4 3 2.

2 The number of lamp posts in each street in a small housing estate:

7 8 12 15 32 34.

3 The number of patients who failed to keep their doctor's appointment each month from January to June last year at one particular surgery:

104 110 113 120 128 133.

4 The number of weeks that the ink cartridge lasted on Sam's PC:

7 8 10 13 16 20 24.

5 On a single journey across Edinburgh, a bus driver counted the number of consecutive sets of traffic lights that he encountered at green:

1 1 2 2 2 2 3 3 4 4 5 7.

6 All the neighbours in a row planted a ground cover rhododendron bush in their front gardens. The numbers of flowering heads on these bushes one May morning were:

18 19 25 26 29 31 35 41.

7 The number of distinct showers of rain each day during the first week in April:

1 5 7 3 6 8 5.

8 The number of paper clips in boxes marked as containing 1000:

990 996 1001 1003 1007 1009 1015.

9 The number of bars of music (excluding repeat bars) printed on each page of Joseph Haydn's *Allegro in C*:

22 28 21 23 19 17.

10 For the data set in question 8,
a) construct a box plot;
b) state the range and semi-interquartile range;
c) if one of these boxes were chosen at random, what would the probability be of finding more than 1000 paper clips?

11 Compare the median, mean and standard deviation of the two data sets given in the 'reminder' before the start of this exercise:

1 18 18 19 20 21 22 25 34 35 36 37 71 73 80
1 6 8 19 20 21 22 23 24 25 27 37 38 39 80.

Reminder

Suppose you were asked to calculate the standard deviation of a data set where the mean was not a convenient number, e.g. the data: 6·1, 6·3, 6.9, 7·2, 7·6, 8·3. yields a sum of 42·4, and so the mean is 7·07. Applying the formula to this set of data is extremely tedious and susceptible to casual errors, so it is better to use the alternative version:

$$\text{S.D.} = \sqrt{\frac{\sum x^2 - \frac{(\sum x)^2}{n}}{n-1}}$$

The working here would be:

x	x^2
6·1	37·21
6·3	39·69
6·9	47·61
7·2	51·84
7·6	57·76
8·3	68·89
42·4	303·00

$$\text{so S.D.} = \sqrt{\frac{303 - \frac{(42{\cdot}4)^2}{6}}{5}} = 0{\cdot}821$$

12 Apply this formula to the data sets in questions 1 to 5 and check that you get the same answers as before.

71 Regression

Reminder

Earlier, you drew the best fitting *regression* line on a scattergraph, and used it to estimate the y-coordinate for a given x-coordinate. Here we refine this by obtaining the equation of the regression line.

Example In Glencorthie Academy 17 pupils do both Maths and Physics at Advanced Higher. The prelim percentage marks for 16 of them are shown below:

Pupil	A	B	C	D	E	F	G	H	I	J	K	L	M	N	O	P
Maths	10	15	25	35	40	45	50	55	60	65	70	75	75	80	85	100
Physics	25	25	35	35	40	45	45	50	50	50	55	55	60	60	65	65

a) Illustrate this data on a scattergraph (physics against maths).
b) Draw the best fitting straight line through this data.
c) Find the equation of the line you have just drawn.
d) Use the equation of the line to estimate a physics mark for the 17th pupil who missed the physics exam but who scored 50% in the maths prelim.

Solution a) b)

c) (10, 25) and (80, 60) lie on the line, so the gradient is $\frac{1}{2}$, and the intercept is 20, hence $y = \frac{1}{2}x + 20$

[or (physics) $= \frac{1}{2}$(maths) + 20]

d) $x = 50 \Rightarrow y = \frac{1}{2}(50) + 20 = 45$ i.e. 45% for physics

Exercise 71

1 The coach of the basketball team totalled the number of points scored by each of his players over the last five games of the season. He also recorded the height of each player and produced this table:

player	A	B	C	D	E	F	G	H	I	J
points	30	34	38	40	42	46	50	52	54	56
height	1·30	1·35	1·40	1·50	1·60	1·65	1·70	1·80	1·80	1·90

a) Plot the height against the points scored on a scattergraph.
b) Draw the best fitting straight line through the data and determine its equation.
c) Use your equation to determine what height you need to be to hope to score more than 44 points.

2 The number of ducks sitting on the surface of the water in a park pond at noon was recorded for a week and the temperature of the water noted too. The results were as follows:

day	Sun	Mon	Tue	Wed	Thu	Fri	Sat
temperature (°C)	5	8	4	2	7	3	6
no. of ducks	39	33	43	46	37	45	38

a) Plot the number of ducks against the temperature on a scattergraph.
b) Draw the best fitting straight line through the data and determine its equation.
c) Use your equation to estimate how many ducks you might expect when the water temperature is 1°C.
d) Does this equation help you to determine how many ducks there would be when the temperature is –1°C?
e) Does the equation help you to determine how many ducks there might be in the unlikely event of water temperature rising to 26°C?
f) What do we call the process of trying to estimate values outwith the range of the given data?

3 During one week in summer, the owner of an ice cream van in the park recorded the number of litres of ice cream sold each day, and the temperature at 3 p.m.:

day	Sun	Mon	Tue	Wed	Thu	Fri	Sat
temperature (°C)	19	22	17	13	21	15	20
litres of ice cream	65	80	58	44	72	48	71

a) Plot the volume of ice cream against the temperature on a scattergraph.
b) Draw the best fitting straight line through the data and determine its equation.
c) Use your equation to estimate how many litres of ice cream the van might expect to sell on a day when the forecast is for a mid-afternoon temperature of 18°C.
d) How do we describe the difference between the correlation in this question and the previous one?
e) Give some reasons why this question is an oversimplification.

4 A group of men were discussing the cost of motor insurance and generally believed that the insurance premium was closely related to the value of the car. They produced these results:

man	A	B	C	D	E	F	G	H	I
value of car (× £1000)	10	5	13	9	7	11	15	6	14
insurance premium (£)	360	225	380	270	260	320	415	235	405

a) Plot the premium (in £) against the value of the car (in £1000s) on a scattergraph.
b) Draw the best fitting straight line through the data and determine its equation.
c) Using your equation, determine what premium you might expect to pay for a car worth (i) £8000 (ii) £12 000.

5 Malcolm is in the middle of preparing for a hill race in which he has to carry a 10 kg rucksack, by running a six mile course with different weights in his rucksack.
He compiled the following table of results:

weight (kg) in rucksack	2	3	4	5	6	7	8	9
average speed (km/h)	$11\frac{1}{2}$	$11\frac{3}{4}$	11	$10\frac{3}{4}$	$9\frac{3}{4}$	$9\frac{1}{4}$	9	$8\frac{3}{4}$

a) Plot the average speed against the weight carried on a scattergraph.
b) Draw the best fitting straight line through the data and determine its equation.
c) From your equation, at what average speed can Malcolm run when he is carrying 10 kg? How confident can you be about this prediction?
d) From your equation, at what average speed can Malcolm run when he is running without a rucksack?
Are you more or less confident about this prediction than the previous one?

6 At the end of the football season, the points gained in the league by each team were compared with the goal difference, giving these results:

team	A	B	C	D	E	F	G	H	I	J
points gained	72	70	62	59	52	50	36	35	27	23
goal diff.	24	33	17	12	10	−1	−22	−18	−32	−41

a) Plot the goal difference against the points gained on a scattergraph.
b) Express in words any relationship between these two quantities which is obvious from the scattergraph.
c) Draw the best fitting straight line through the data and determine its equation.
d) If, over the season, a team scored as many goals as it conceded, whereabouts in the league would you expect it to finish?

72 Revision of Graphs, Charts, Tables and Statistics

Exercise 72

1 The first dot plot shows the pulse rates of a group of teachers in the staff room at lunch time. The second dot plot shows the pulse rates of a class of pupils just before lunch.

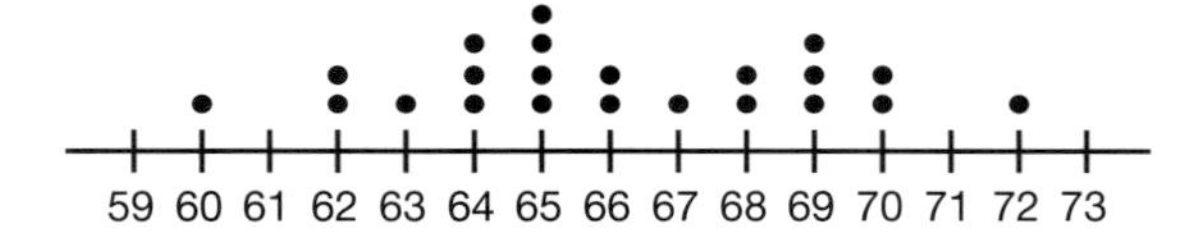

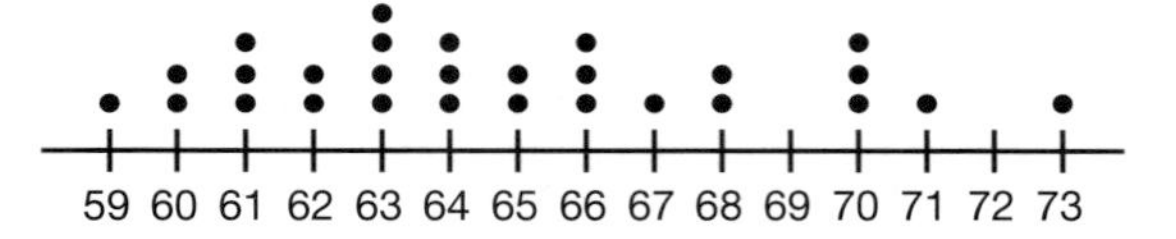

a) Draw a box plot for each distribution and make some comparisons.
b) If a teacher and a pupil are selected at random from each of these groups, find the probabilities that each has a pulse rate less than 63.

2 Over the season Priory Primrose won 17 games, drew 10, and lost 9.
Construct a pie chart to illustrate this.

3 Find the range, median, lower and upper quartiles, and semi-interquartile range of these data sets:

a) 2 3 4 6 8 9 11
b) 2 3 4 6 8 9
c) 2 3 4 6 8 9 11 14.

4 The number of litres of petrol sold to each customer at the Corner Garage one morning was:

52	35	37	40	21	19	42	45	55	26	41
32	15	50	25	63	50	31	16	32	40	30
42	34	40	47	20	59	38	43	51	44	30
43	49	31	42	60	20	28	23	20	22.	

a) Construct an ordered stem and leaf diagram from this data.
b) Obtain the five figure summary for this distribution.
c) Calculate the mean number of litres sold to customers that morning.

5 Draw a dot plot to illustrate these council tax bands in an area of mixed housing:

C	A	D	C	E	A	C	H	D	E	G	F	D
C	E	A	B	B	D	E	F	G	G	D	H	F.

6 Find the range, median, lower and upper quartiles, and semi-interquartile range of these data sets:

a) 104	105	108	109	110	111	111					
b) 14	19	22	23	25	27	28	29	30	31	33	
c) 70	72	74	75	77	78	79	85	86			
d) 91	92	93	93	97	98	99	99	106			
e) 28	29	31	35	37	42	68	92				
f) 82	56	20	59	11	7	57	88	90	51	91	12.

7 This data set gives the ages of the staff football team members:

23 24 25 25 25 25 26 26 27 35 42.

a) Find (i) the mean (ii) the median (iii) the mode (iv) the range.
b) Find the probability of randomly selecting a teacher over 30.

8 New gas ovens were tested for the accuracy of their thermostats. They were set at gas mark 4 for 30 minutes and the temperature recorded. The results (in °F) were:

341 347 352 357 360 361.

Calculate the mean and the standard deviation.

9 This data set gives the ages of the members of a ladies slimming club:

43	46	47	44	45	47	47	43	48	50
47	46	44	47	43	49	49	49	51	43.

a) Find (i) the mean (ii) the median (iii) the mode (iv) the range.
b) Find the probability of randomly selecting a 47-year old.

10 Hugh was organising the sweepstake at the annual curling club dance. He had a theory that the luckiest people had their birthdays closest to the middle of the month (which he took as the 15th of the month), so he asked each prizewinner for the date of their birthday and obtained these results:
(D denotes the birth date (day) and V the value of the prize (£))

D	19	10	14	19	9	8	22	23	27	6	29	4
V	12	11	10	9	8	7	6	5	4	3	2	1

a) Plot the values of V against the number of days away from the middle of the month (e.g. the 14th and 16th become 1) on a scattergraph.
b) Draw the best fitting straight line through this data and determine its equation.
c) Use your equation to calculate what value of prize Hugh would expect someone born on the 20th of a month to win.
d) Is Hugh likely to obtain the same regression line from next year's data?

11 Tickets at a hospital charity fayre for the wheel of fortune (shown here) cost £2 each.

a) Calculate the probability of winning
(i) £5
(ii) £3
(iii) £1.

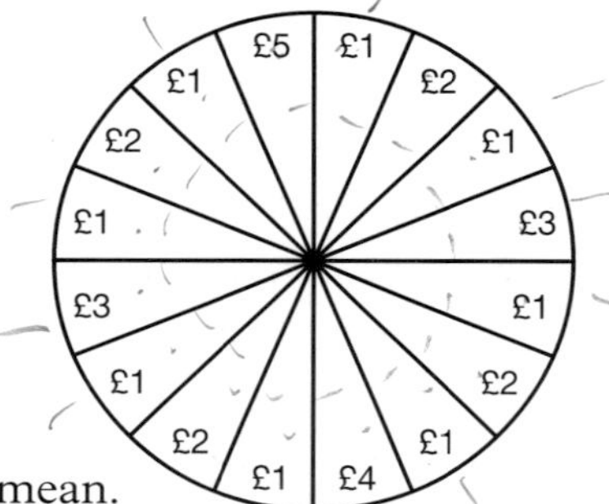

b) How many times would you expect to win £4 if you bought 80 tickets for a spin of the wheel?
c) Draw a bar graph to show how often each prize appears on the wheel.
d) Express this data in a frequency table and calculate the mean.
e) Comment on your value of the mean and the price of a ticket.

73 Test on Graphs, Charts, Tables, and Statistics

Allow 45 minutes for this test

1 A card is drawn at random from a standard pack of cards. Find the probability that the card is:
(a) the ace of spades (b) a king (c) red
(d) a heart (e) a black queen.

2 A card is drawn at random from a standard pack of cards and not replaced.
a) Find the probability that it is a spade.
b) If it is a spade, find the probability that a second card drawn at random is also a spade.
c) If it is not a spade, find the probability that a second card drawn at random is also not a spade.

3 Sarah asked the other pupils in her class how many times they had been to the High Street in the last month, and obtained the following data set:

6 0 6 4 2 4 3 4 4 4
1 5 7 5 1 3 4 0 3 8
3 3 5 8 2 8 5 0 1 0.

a) Construct a frequency table and calculate the mean.
b) Illustrate the data on a bar graph.

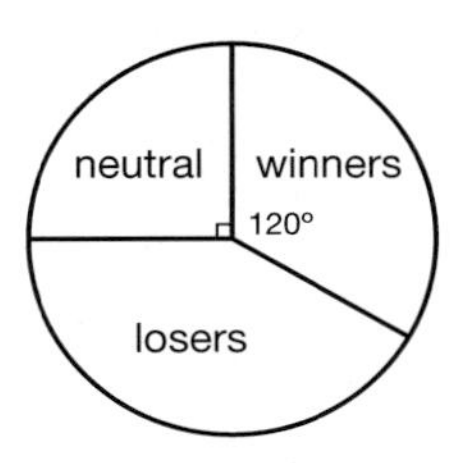

4 This pie chart shows the support of the crowd at the cup final. There were 12 000 neutral supporters.
Calculate the number of: a) winning supporters
b) losing supporters.

5 The ages of the supporters on a bus on the way to the cup final were:

37 10 48 37 29 38 11 42 62 23 17 33 55
39 44 57 14 39 46 21 64 53 51 9 60 48
67 30 17 23 66 63 10 28 51 58 34 18 10
12 38 46 29 48 9 57 25 19 26 41 66.

a) Construct a stem and leaf diagram and use it to construct a box plot for this data.
b) After the game, the passengers arrive back at the bus in random order. What is the probability that the first passenger back to the bus is under 20 years of age?

6 At the rifle club competition, each competitor was allowed 10 shots and the number of bullseyes recorded as shown:

6 7 10 0 8 9 3 2 4 2 3 4 6
5 6 5 5 8 1 6 9 7 3 6 5 6
4 1 3 8 6 7 7 7 4 4 5 7.

a) Construct a frequency table and state the range.
b) Add a cumulative frequency table and calculate the semi-interquartile range.

7 The retiral ages of a group of insurance agents were:

55 56 59 59 61 64.

Write down the semi-interquartile range and calculate the standard deviation.

8 The number of wins and the numbers of points gained for the 10 teams in a football league at the end of the season were:

no. of wins	19	19	15	15	12	13	7	8	5	6
no. of points	68	66	57	54	49	48	35	33	28	26

a) Plot the number of points against the number of wins on a scattergraph.
b) Draw the best fitting straight line through this data and determine its equation.
c) Use your equation to determine how many points you would expect a team to accumulate if they achieved 11 wins during the season.

FURTHER ALGEBRA

74 Simplifying Algebraic Fractions

Reminder

The rules for simplifying algebraic fractions are no different to those for reducing numerical fractions to their lowest terms. It is, therefore, worth taking another look at numerical fractions before we embark on algebraic fractions.

Example Find three equivalent fractions for $\frac{3}{5}$.

Solution Equivalent fractions are found by multiplying the numerator (i.e. the top line, 3 in this case) and the denominator (i.e. the bottom line, 5 in this case) by the same quantity.

e.g. multiplying top and bottom by 4 gives $\frac{3 \times 4}{5 \times 4} = \frac{12}{20}$

multiplying top and bottom by 7 gives $\frac{21}{35}$

multiplying top and bottom by 11 gives $\frac{33}{55}$.

We need to use equivalent fractions when we add and subtract fractions, but it is more common to be asked to reduce fractions to their lowest terms. This is where we divide top and bottom by the same number.

Example Express each of these vulgar fractions in their lowest terms:

a) $\frac{8}{12}$ b) $\frac{48}{84}$ c) $\frac{34}{85}$ d) $\frac{68}{100}$ e) $\frac{48}{1024}$.

Solution

a) The highest common factor of 8 and 12 is 4, so divide top and bottom by 4

i.e. $\frac{8}{12} = \frac{8 \div 4}{12 \div 4} = \frac{2}{3}$ (or divide top and bottom by 2 twice).

b) The highest common factor of 48 and 84 is 12, so divide top and bottom by 12

i.e. $\frac{48}{84} = \frac{4}{7}$.

Reminder continued

c) The highest common factor of 34 and 85 is 17, so divide top and bottom by 17

i.e. $\frac{34}{85} = \frac{2}{5}$.

d) The highest common factor of 68 and 100 is 4, so divide top and bottom by 4.

i.e. $\frac{68}{100} = \frac{17}{25}$.

e) The highest common factor of 48 and 1024 is 16, but you are unlikely to be able to divide by 16 without a calculator, so divide top and bottom by 4 and 4 again

i.e. $\frac{48}{1024} = \frac{12}{256} = \frac{3}{64}$.

Exercise 74

Do not use your calculator

1 Arrange these six vulgar fractions in three pairs of equivalent fractions:

$\frac{12}{21}, \frac{15}{27}, \frac{7}{28}, \frac{20}{35}, \frac{11}{44}, \frac{25}{45}$.

2 Which, if any, of these fractions is equal to $\frac{13}{18}$?

$\frac{3}{8}, \frac{14}{19}, \frac{23}{28}, \frac{39}{54}, \frac{52}{72}, \frac{65}{95}, \frac{130}{180}$.

3 Copy and fill the gaps: $\frac{}{5} = \frac{8}{} = \frac{}{15} = \frac{24}{30} = \frac{36}{} = \frac{72}{} = \frac{}{270}$.

4 Express these common fractions in their lowest terms:

a) $\frac{10}{12}$ b) $\frac{7}{42}$ c) $\frac{12}{28}$ d) $\frac{22}{26}$

e) $\frac{14}{91}$ f) $\frac{85}{100}$ g) $\frac{16}{36}$ h) $\frac{15}{21}$

i) $\frac{18}{27}$ j) $\frac{49}{63}$ k) $\frac{18}{28}$ l) $\frac{60}{132}$.

5 Express these common fractions in their lowest terms:

a) $\frac{26}{65}$ b) $\frac{16}{44}$ c) $\frac{56}{84}$ d) $\frac{20}{48}$

e) $\frac{20}{55}$ f) $\frac{36}{66}$ g) $\frac{60}{96}$ h) $\frac{60}{69}$

i) $\frac{64}{96}$ j) $\frac{63}{91}$ k) $\frac{48}{84}$ l) $\frac{78}{91}$.

Reminder

We apply the same rules to the simplification of algebraic fractions; we can only cancel a factor of the whole of the top line with a factor of the whole of the bottom line. To identify these factors, you may need to factorise an expression by taking out a common factor, or factorise a difference of two squares or a trinomial.

Example Simplify a) $\frac{6x+18y}{3}$ b) $\frac{(x+2)^2}{(x+2)^3}$ c) $\frac{x^2+2x}{x^2-4}$ d) $\frac{x^2+x-2}{2x^2-2x}$

Solution

a) Take out the common factor on the top line, then divide top and bottom by 3:

$$\frac{6x+18y}{3} = \frac{\overset{2}{\cancel{6}}(x+3y)}{\underset{1}{\cancel{3}}} = 2(x+3y).$$

b) Until we have done Indices again, we will simply write out the factors and cancel:

$$\frac{(x+2)^2}{(x+2)^3} = \frac{\overset{1}{\cancel{(x+2)}}\overset{1}{\cancel{(x+2)}}}{\underset{1}{\cancel{(x+2)}}\underset{1}{\cancel{(x+2)}}(x+2)} = \frac{1}{(x+2)}.$$

Strictly speaking, we should say $x \neq -2$, or the bottom line would be zero and the fraction therefore infinite, but we can consistently choose to omit such references.

c) Use the common factor on the top and the difference of two squares underneath:

$$\frac{x^2+2x}{x^2-4} = \frac{x\overset{1}{\cancel{(x+2)}}}{(x-2)\underset{1}{\cancel{(x+2)}}} = \frac{x}{x-2}.$$

Note: there is no more cancelling.

d) Factorise the trinomial on the top and take out the common factor on the bottom.

$$\frac{x^2+x-2}{2x^2-2x} = \frac{(x+2)\overset{1}{\cancel{(x-1)}}}{2x\underset{1}{\cancel{(x-1)}}} = \frac{x+2}{2x}.$$

Note: there is no more cancelling.

6 Simplify:

a) $\frac{xy^2}{xy^3}$ b) $\frac{a^2b}{ab^2}$ c) $\frac{(pq)^2}{p^2q}$ d) $\frac{r^2s}{(rs)^2}$ e) $\frac{(xy)^3}{x^2y}$.

7 Simplify:

a) $\dfrac{4x + 6y}{2}$ b) $\dfrac{12p + 9q}{3}$ c) $\dfrac{4}{8u + 12v}$ d) $\dfrac{5x}{5x + 10y}$ e) $\dfrac{2k}{4k + 8kl}$.

8 Simplify:

a) $\dfrac{2p - 2q}{2p - 2q}$ b) $\dfrac{3x - 3y}{3y - 3x}$ c) $\dfrac{4m - 8n}{8n - 4m}$ d) $\dfrac{3x - 6}{8 - 4x}$ e) $\dfrac{x^2 - 4x}{3x - 12}$.

9 Simplify:

a) $\dfrac{(p + q)^2}{(p + q)^3}$ b) $\dfrac{m^2 + mn}{n^2 + nm}$ c) $\dfrac{x + xy}{2 + 2y}$ d) $\dfrac{6 - 3x}{8 - 4x}$ e) $\dfrac{5x + 10}{5y - 5}$.

10 Simplify:

a) $\dfrac{x^2 - 4}{3x + 6}$ b) $\dfrac{x^2 - 1}{4x - 4}$ c) $\dfrac{y^2 - 9}{xy - 3x}$ d) $\dfrac{z^2 - 16}{4z + 16}$ e) $\dfrac{p^2 - q^2}{2p - 2q}$.

11 Simplify:

a) $\dfrac{x^2 - 5x + 6}{6x - 18}$ b) $\dfrac{y^2 - y - 2}{2y + 2}$ c) $\dfrac{p^2 + 4p + 3}{3p + 3}$ d) $\dfrac{k^2 + 3k - 4}{3k + 12}$ e) $\dfrac{z^2 + 3z - 10}{5z - 10}$.

12 Simplify:

a) $\dfrac{4x - 12}{x^2 + x - 12}$ b) $\dfrac{p^2 + 2p - 3}{p^2 - 1}$ c) $\dfrac{x^2 - 9}{x^2 - 2x - 3}$ d) $\dfrac{x^2 - 3x}{x^2 - x - 6}$ e) $\dfrac{3x - 3}{x^3 - x}$.

75 The Four Rules for Algebraic Fractions

Reminder

It is again useful to be reminded of the four rules (add, subtract, multiply, divide) for numerical fractions.

Example Evaluate a) $\dfrac{5}{12} + \dfrac{4}{9}$ b) $\dfrac{3}{10} - \dfrac{2}{11}$ c) $\dfrac{6}{7} \times \dfrac{14}{27}$ d) $\dfrac{8}{9} \div \dfrac{16}{15}$.

Solution

a) To add or subtract fractions they must be expressed with the same denominator.
The least common multiple of 12 and 9 is 36, so express both fractions in 36ths.

$$\frac{5}{12} + \frac{4}{9} = \underbrace{\frac{5 \times 3}{12 \times 3} + \frac{4 \times 4}{9 \times 4}}_{A} = \underbrace{\frac{15}{36} + \frac{16}{36}}_{B} = \underbrace{\frac{15 + 16}{36}}_{C} = \frac{31}{36}$$

Step A (and even step B) may be omitted after sufficient practice.

Step C helps to avoid the dreadful error of writing $\dfrac{(15 + 16)}{(36 + 36)}$.

Reminder continued

b) The least common multiple of 10 and 11 is 110, so express both fractions in 110ths

$$\frac{3}{10} - \frac{2}{11} = \frac{3 \times 11}{10 \times 11} - \frac{2 \times 10}{11 \times 10} = \frac{33}{110} - \frac{20}{110} = \frac{33 - 20}{110} = \frac{13}{110}$$

Before considering multiplication, you may prefer to complete Exercise 75 questions 1 to 4.

c) There is NO NEED to find equivalent fractions in order to carry out multiplication (or division). Essentially, you multiply the top lines and multiply the bottom lines, but in practice, to avoid large numbers, we cancel first, as if it were a single fraction.

$$\frac{6}{7} \times \frac{14}{27} = \frac{6 \times 14}{7 \times 27} = \frac{{}^{2}\cancel{6} \times \cancel{14}^{2}}{{}_{1}\cancel{7} \times \cancel{27}_{9}}$$ (Divide 7 and 14 by 7.) (Divide 6 and 27 by 3.)

$$= \frac{2 \times 2}{1 \times 9} = \frac{4}{9}$$

d) Division by $\frac{a}{b}$ is the same as multiplication by $\frac{b}{a}$, so the rule for dividing fractions is to turn the second one upside down and multiply.

$$\frac{8}{9} \div \frac{16}{15} = \frac{8}{9} \times \frac{15}{16} = \frac{{}^{1}\cancel{8} \times \cancel{15}^{5}}{{}_{3}\cancel{9} \times \cancel{16}_{2}}$$ (Divide 8 and 16 by 8.) (Divide 9 and 15 by 3.)

$$= \frac{1 \times 5}{3 \times 2} = \frac{5}{6}$$

Note that $\frac{16}{15}$ is the same as the mixed number $1\frac{1}{15}$, but we do not use mixed numbers with algebraic fractions, so we can ignore here the refinements needed for coping with mixed number fractions.

Exercise 75

1 Add these fractions expressing your answers in their lowest terms:

a) $\frac{1}{5} + \frac{2}{5}$ b) $\frac{2}{15} + \frac{3}{15}$ c) $\frac{2}{10} + \frac{3}{10}$ d) $\frac{5}{12} + \frac{7}{12}$ e) $\frac{9}{16} + \frac{3}{16}$.

2 Subtract these fractions expressing your answers in their lowest terms:

a) $\frac{9}{10} - \frac{3}{10}$ b) $\frac{17}{20} - \frac{11}{20}$ c) $\frac{23}{24} - \frac{7}{24}$ d) $\frac{19}{25} - \frac{4}{25}$ e) $\frac{23}{30} - \frac{17}{30}$.

3 Add:

a) $\frac{1}{4} + \frac{3}{8}$ b) $\frac{2}{5} + \frac{1}{6}$ c) $\frac{2}{7} + \frac{3}{5}$ d) $\frac{3}{7} + \frac{4}{9}$ e) $\frac{7}{12} + \frac{3}{8}$.

4 Subtract:

a) $\frac{9}{10} - \frac{3}{5}$ b) $\frac{7}{8} - \frac{1}{4}$ c) $\frac{5}{9} - \frac{1}{6}$ d) $\frac{11}{12} - \frac{5}{8}$ e) $\frac{5}{7} - \frac{3}{5}$.

5 Multiply:

a) $\frac{4}{9} \times \frac{3}{8}$ b) $\frac{5}{6} \times \frac{9}{10}$ c) $\frac{6}{7} \times \frac{7}{9}$ d) $\frac{8}{9} \times \frac{15}{16}$ e) $\frac{12}{25} \times \frac{5}{18}$.

6 Divide:

a) $\frac{6}{25} \div \frac{9}{10}$ b) $\frac{4}{15} \div \frac{6}{5}$ c) $\frac{4}{35} \div \frac{32}{7}$ d) $\frac{11}{42} \div \frac{55}{28}$ e) $\frac{12}{25} \div \frac{32}{15}$.

Reminder

Do not invent any new rules of your own. Apply the same rules to the algebraic fractions as you did for the numerical ones.

Example Simplify:

a) $1 + \frac{1}{x}$ b) $\frac{x}{y} - \frac{y}{x}$ c) $\frac{x+2}{x-2} + \frac{2}{x}$

d) $\frac{x^2}{x^2 - 9} \times \frac{x+3}{x}$ e) $\left(\frac{y^2 - 4}{y^2 + 2y}\right) \div \left(\frac{y+3}{y}\right)$.

Solution

a) Make 1 into a fraction with denominator x, i.e. $1 = \frac{x}{x}$.

Thus $1 + \frac{1}{x} = \frac{x}{x} + \frac{1}{x} = \frac{x+1}{x}$

b) The lowest common denominator is xy, so multiply $\frac{x}{y}$ by $\frac{x}{x}$ and $\frac{y}{x}$ by $\frac{y}{y}$.

Thus $\frac{x}{y} - \frac{y}{x} = \frac{x}{y} \times \frac{x}{x} - \frac{y}{x} \times \frac{y}{y} = \frac{x^2}{xy} - \frac{y^2}{xy} = \frac{x^2 - y^2}{xy}$

c) The lowest common denominator is $x(x - 2)$,

so multiply $\frac{x+2}{x-2}$ by $\frac{x}{x}$ and $\frac{2}{x}$ by $\frac{x-2}{x-2}$.

$$\text{Thus } \frac{x+2}{x-2} + \frac{2}{x} = \frac{x+2}{x-2} \times \frac{x}{x} + \frac{2}{x} \times \frac{x-2}{x-2}$$

$$= \frac{x^2 + 2x}{x(x-2)} + \frac{2x - 4}{x(x-2)} = \frac{x^2 + 2x + 2x - 4}{x(x-2)}$$

$$= \frac{x^2 + 4x - 4}{x(x-2)}$$

Reminder continued

d) Factorise and cancel:

$$\frac{x^2}{x^2-9} \times \frac{x+3}{x} = \frac{\overset{x}{\cancel{x^2}} \times \cancel{(x+3)}^1}{(x-3)\cancel{(x+3)}_1\cancel{x}^1} = \frac{x}{x-3}$$

e) To divide, turn the second fraction upside down and multiply, then factorise and cancel:

$$\left(\frac{y^2-4}{y^2+2y}\right) \div \left(\frac{y+3}{y}\right) = \left(\frac{y^2-4}{y^2+2y}\right) \times \left(\frac{y}{y+3}\right)$$

$$= \frac{(y-2)\overset{1}{\cancel{(y+2)}}}{\underset{1}{\cancel{y}}\underset{1}{\cancel{(y+2)}}} \times \frac{\overset{1}{\cancel{y}}}{y+3} = \frac{y-2}{y+3}$$

7 Add:

a) $1 + \frac{1}{k}$ b) $2 + \frac{3}{p}$ c) $p + \frac{1}{q}$

d) $2x + \frac{x}{3}$ e) $1 + \frac{1}{1+x}$.

8 Subtract:

a) $2 - \frac{1}{y}$ b) $p - \frac{2}{q}$ c) $3x - \frac{4}{x}$

d) $3x - \frac{4}{y}$ e) $z^2 - \frac{1}{z}$.

9 Add:

a) $\frac{1}{x} + \frac{1}{x+1}$ b) $\frac{2}{y} + \frac{3}{y-1}$ c) $\frac{z}{2} + \frac{3z}{4}$

d) $\frac{1}{a+1} + \frac{1}{a-1}$ e) $\frac{3}{b-1} + \frac{4}{b+1}$.

10 Subtract:

a) $\frac{1}{x} - \frac{1}{x+1}$ b) $\frac{2}{y} - \frac{5}{y+2}$ c) $\frac{3}{z-2} - \frac{4}{z}$

d) $\frac{1}{a-1} - \frac{1}{a+1}$ e) $\frac{4}{b-1} - \frac{3}{b+1}$.

11 Simplify:

a) $\frac{3}{a+b} + \frac{4}{a-b}$ b) $\frac{a}{a+b} - \frac{b}{a-b}$ c) $p - \frac{1}{2p}$

d) $1 + q + \frac{1}{3q}$ e) $\frac{m}{n} - \frac{n}{m}$.

12B Multiply:

a) $\dfrac{x^2+xy}{xy-y^2}\times\dfrac{y}{x}$ b) $\dfrac{x-1}{x^2-4}\times\dfrac{x-2}{x^2-1}$ c) $\dfrac{x+3}{x^2-16}\times\dfrac{x+4}{x^2-9}$

d) $\dfrac{3y^2-27}{2y+8}\times\dfrac{y+4}{y-3}$ e) $\dfrac{x+4}{x^2-6x+5}\times\dfrac{x-5}{x^2+3x-4}$.

13B Divide:

a) $\dfrac{p^2-pq}{q^2-pq}\div\dfrac{p}{q}$ b) $\dfrac{u^2-uv}{v^2+uv}\div\dfrac{u}{v}$ c) $\dfrac{a^2+2ab}{2ab-b^2}\div\dfrac{a}{b}$

d) $\dfrac{x+1}{x^2+x-6}\div\dfrac{x^2-1}{x-2}$ e) $\dfrac{y+3}{y^2+y-2}\div\dfrac{y^2+2y-3}{y+2}$.

14B Simplify:

a) $\dfrac{3}{x+1}+\dfrac{2}{x^2-1}$ b) $\dfrac{x+1}{x^2-4}-\dfrac{2}{x+2}$ c) $\dfrac{p+q}{p-q}-\dfrac{p-q}{p+q}$

d) $\dfrac{1}{k^2-2k+1}-\dfrac{1}{k^2-1}$ e) $\dfrac{2}{x^2+x-2}+\dfrac{3}{x^2-3x+2}$.

76 Change of Subject of a Formula

Reminder

Changing the subject of a formula involves the same processes as solving an equation, so it is worthwhile reviewing the processes available for solving simple equations.

Example Solve:

a) (i) $x-5=0$ (ii) $x-a=0$

b) (i) $x+7=0$ (ii) $x+b=0$

c) (i) $2x=16$ (ii) $kx=16$ (iii) $kx=t$

d) (i) $\dfrac{x}{3}=4$ (ii) $\dfrac{x}{m}=4$ (iii) $\dfrac{x}{m}=n$

e) (i) $2x+3=11$ (ii) $px+3=11$ (iii) $px+q=r$.

Solution

a) (i) add 5 to both sides $\Rightarrow x-5+5=0+5 \Rightarrow x=5$

(ii) add a to both sides $\Rightarrow x-a+a=0+a \Rightarrow x=a$

b) (i) subtract 7 from both sides $\Rightarrow x+7-7=0-7 \Rightarrow x=-7$

(ii) subtract b from both sides $\Rightarrow x+b-b=0-b \Rightarrow x=-b$

c) (i) divide both sides by 2 $\Rightarrow \dfrac{2x}{2}=\dfrac{16}{2} \Rightarrow x=8$

(ii) divide both sides by k $\Rightarrow \dfrac{kx}{k}=\dfrac{16}{k} \Rightarrow x=\dfrac{16}{k}$

(iii) divide both sides by k $\Rightarrow \dfrac{kx}{k}=\dfrac{t}{k} \Rightarrow x=\dfrac{t}{k}$

Reminder continued

d) (i) multiply both sides by 3 $\Rightarrow 3\left(\frac{x}{3}\right) = 3 \times 4 \quad \Rightarrow x = 12$

(ii) multiply both sides by m $\Rightarrow m\left(\frac{x}{m}\right) = m \times 4 \quad \Rightarrow x = 4m$

(iii) multiply both sides by m $\Rightarrow m\left(\frac{x}{m}\right) = m \times n \quad \Rightarrow x = mn$

e) (i) subtract 3, divide by 2

$\Rightarrow 2x + 3 - 3 = 11 - 3$

$\Rightarrow 2x = 8$

$\Rightarrow \frac{2x}{2} = \frac{8}{2}$

$\Rightarrow x = 4$

(ii) subtract 3, divide by p

$\Rightarrow px + 3 - 3 = 11 - 3$

$\Rightarrow px = 8$

$\Rightarrow \frac{px}{p} = \frac{8}{p}$

$\Rightarrow x = \frac{8}{p}$

(iii) subtract q, divide by p

$\Rightarrow px + q - q = r - q$

$\Rightarrow px = r - q$

$\Rightarrow \frac{px}{p} = \frac{r - q}{p}$

$\Rightarrow x = \frac{r - q}{p}$

Example Change the subject of the formula a) $v = \frac{1}{2}(u + gh)$ to h

b) $p(q + r) = s$ to r.

Solution

a) Clear the fraction first, by doubling both sides: $2v = u + gh$.
Isolate the term which includes h by subtracting u from both sides:

$\Rightarrow gh = 2v - u$

$\Rightarrow \frac{gh}{g} = \frac{2v - u}{g}$

i.e. $h = \frac{2v - u}{g}$

b) Expand the brackets and isolate the term which includes r:

i.e. $pq + pr = s$

$\Rightarrow pq + pr - pq = s - pq$

$\Rightarrow pr = s - pq$

$\Rightarrow \frac{pr}{p} = \frac{s - pq}{p}$

i.e. $r = \frac{s - pq}{p}$

Exercise 76

Change the subject of the formula:

1	$P = 3a$	to a,	**2**	$V = IR$	to R,
3	$A = lb$	to b,	**4**	$P = mf$	to f,
5	$c = np$	to n,	**6**	$PV = RT$	to V,
7	$E = mc^2$	to m,	**8**	$H = \frac{Tv}{550}$	to v,
9	$I = \frac{PTR}{100}$	to P,	**10**	$A = \frac{1}{2}bh$	to b,
11	$E = \frac{1}{2}mv^2$	to m,	**12**	$g = c + p$	to p,
13	$P = 2x + y$	to x,	**14**	$P = 2(l + b)$	to l,
15	$s = \frac{1}{2}(u + v)t$	to u,	**16**	$q = a(r + s)$	to s,
17	$y = ax + b$	to a,	**18**	$A = \frac{1}{2}(a + b)h$	to h,
19	$A = 2\pi r(r + h)$	to h,	**20**	$mu + nv = c$	to u,
21	$p = q + 2fg$	to g,	**22**	$t = u(1 + v)$	to v,
23	$k = \frac{1}{3}(p + q)t$	to q,	**24**	$s = ut + \frac{1}{2}ft^2$	to u,
25	$s = ut + \frac{1}{2}ft^2$	to f,	**26**	$v^2 = u^2 + 2fs$	to f,
27B	$A = \pi r^2$	to r,	**28B**	$v^2 = u^2 + 2fs$	to u,
29B	$T = 2\pi\sqrt{\frac{l}{g}}$	to g,	**30B**	$L = \frac{b}{1 - a}$	to a,
31H	$x^2 + y^2 = r^2$	to y,	**32H**	$n(p - a) = m(b - p)$	to p,
33H	$x = \frac{1 - t^2}{1 + t^2}$	to t.			

77 Simplification of Surds

Reminder

A surd is an algebraic irrational number, i.e. it is a solution of an algebraic equation, e.g. $x^2 - 5 = 0 \Rightarrow x^2 = 5 \Rightarrow x = \pm\sqrt{5}$.

So $\sqrt{5}$ is a surd.

$\sqrt{4}$ looks like a surd but it is not, since $\sqrt{4} = 2 = \frac{2}{1}$, which is rational, i.e. the ratio of two integers (2 and 1).

$\sqrt[3]{4}$ and $\sqrt[4]{7}$ are also surds, but we can confine our attention to quadratic surds, i.e. square roots.

Reminder continued

Example Express: (i) $\sqrt{40}$ (ii) $\sqrt{108}$ in its simplest form.

Solution (i) $\sqrt{40} = \sqrt{4 \times 10} = \sqrt{4} \times \sqrt{10} = 2\sqrt{10}$

(ii) $\sqrt{108} = \sqrt{36 \times 3} = \sqrt{36} \times \sqrt{3} = 6\sqrt{3}$

Note that 36 was chosen as one of the factors of 108 because it is the largest perfect square which divides 108.

Example Express: (i) $3\sqrt{2}$ (ii) $4\sqrt{7}$ as an entire surd.

Solution (i) $3\sqrt{2} = \sqrt{9} \times \sqrt{2} = \sqrt{9 \times 2} = \sqrt{18}$

(ii) $4\sqrt{7} = \sqrt{16} \times \sqrt{7} = \sqrt{16 \times 7} = \sqrt{112}$

Example Simplify:

(i) $4\sqrt{3} + 3\sqrt{3}$

(ii) $2\sqrt{3} + 3\sqrt{2} + 3\sqrt{3}$

(iii) $\sqrt{45} + \sqrt{12} + \sqrt{20}$

Solution Like surds are collected in the same way as like terms, and unlike surds are treated in the same way as unlike terms.

(i) $4\sqrt{3} + 3\sqrt{3} = (4 + 3)\sqrt{3} = 7\sqrt{3}$

(ii) $2\sqrt{3} + 3\sqrt{2} + 3\sqrt{3} = 5\sqrt{3} + 3\sqrt{2}$

(Surds may have to be expressed in their simplest form to find which are like surds.)

(iii) $\sqrt{45} + \sqrt{12} + \sqrt{20} = (\sqrt{9} \times \sqrt{5}) + (\sqrt{4} \times \sqrt{3}) + (\sqrt{4} \times \sqrt{5})$

$= 3\sqrt{5} + 2\sqrt{3} + 2\sqrt{5} = 5\sqrt{5} + 2\sqrt{3}$

Exercise 77

1 Which of the following are surds?

a) $\sqrt{9}$ b) $\sqrt{12}$ c) $\sqrt{16}$ d) $\sqrt{21}$ e) $\sqrt{27}$

f) $\sqrt{0{\cdot}9}$ g) $\sqrt{0{\cdot}09}$ h) $\sqrt{1{\cdot}6}$ i) $\sqrt{0{\cdot}16}$ j) $\sqrt{121}$

k) $\sqrt{1{\cdot}21}$ l) $\sqrt{12{\cdot}1}$.

2 If a and b represent the sizes of the two shorter sides of a right-angled triangle, and c the hypotenuse, in which of the following is the number represented by c a surd?

a) $a = 2, b = 3$ b) $a = 7, b = 8$ c) $a = 4, b = 3$

d) $a = 2, b = 4$ e) $a = 5, b = 13$ f) $a = 7, b = 24$.

3 Express each of the following in its simplest form:

a) $\sqrt{8}$ b) $\sqrt{12}$ c) $\sqrt{28}$ d) $\sqrt{27}$
e) $\sqrt{18}$ f) $\sqrt{50}$ g) $\sqrt{75}$ h) $\sqrt{20}$
i) $\sqrt{24}$ j) $\sqrt{45}$ k) $\sqrt{32}$ l) $\sqrt{54}$
m) $\sqrt{98}$ n) $\sqrt{200}$ o) $\sqrt{72}$ p) $\sqrt{147}$
q) $\sqrt{242}$ r) $3\sqrt{125}$ s) $10\sqrt{40}$ t) $2\sqrt{63}$
u) $3\sqrt{44}$ v) $10\sqrt{1000}$ w) $7\sqrt{48}$ x) $3\sqrt{450}$.

4 Express each of the following as an entire surd:

a) $2\sqrt{7}$ b) $3\sqrt{5}$ c) $3\sqrt{3}$ d) $2\sqrt{2}$
e) $5\sqrt{3}$ f) $4\sqrt{2}$ g) $3\sqrt{7}$ h) $10\sqrt{5}$
i) $4\sqrt{3}$ j) $7\sqrt{2}$ k) $5\sqrt{5}$ l) $3\sqrt{11}$.

5 Collect the like surds and hence simplify:

a) $5\sqrt{2} + 3\sqrt{2}$ b) $9\sqrt{5} - 4\sqrt{5}$
c) $8\sqrt{3} + 4\sqrt{3}$ d) $\sqrt{10} + 5\sqrt{10} - 6\sqrt{10}$
e) $2\sqrt{7} - 5\sqrt{7}$ f) $\sqrt{2} + \sqrt{2} - 2\sqrt{2}$
g) $7\sqrt{11} - 9\sqrt{11} + 3\sqrt{11}$ h) $2\sqrt{5} + 5\sqrt{3} + 3\sqrt{5} - \sqrt{3}$
i) $15\sqrt{3} + 3\sqrt{15} - 9\sqrt{15}$ j) $7\sqrt{2} - 3\sqrt{7} - 2\sqrt{7} - \sqrt{2}$
k) $3\sqrt{3} + 4\sqrt{5} - 2\sqrt{3} - 2\sqrt{5}$ l) $5(1 + 3\sqrt{6}) - 2(7\sqrt{6} - 3)$.

6 Simplify:

a) $\sqrt{8} + \sqrt{18} + \sqrt{20}$ b) $\sqrt{27} + \sqrt{24} - \sqrt{12}$
c) $\sqrt{45} + \sqrt{75} - \sqrt{20} + \sqrt{12}$ d) $\sqrt{54} - \sqrt{80} + \sqrt{24} - \sqrt{45}$
e) $\sqrt{112} - \sqrt{50} - \sqrt{63} - \sqrt{18}$ f) $\sqrt{125} - \sqrt{28} + \sqrt{180} - \sqrt{175}$
g) $\sqrt{27} + \sqrt{96} - \sqrt{12} - \sqrt{150}$ h) $\sqrt{98} + \sqrt{147} - \sqrt{72} - \sqrt{48}$
i) $\sqrt{294} + \sqrt{252} - \sqrt{24} - \sqrt{112}$.

7 This diagram shows two right-angled triangles.
AB = 2 cm, BC = 1 cm = DC.
Calculate the lengths of AC and AD (as surds).

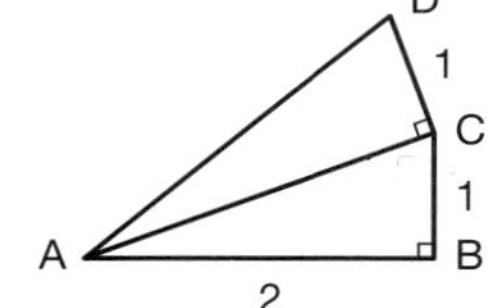

8 Each edge of this cube is 2 cm long.
Calculate the length of:

a) a face diagonal
b) a space diagonal, expressing your answers as surds in their simplest form.

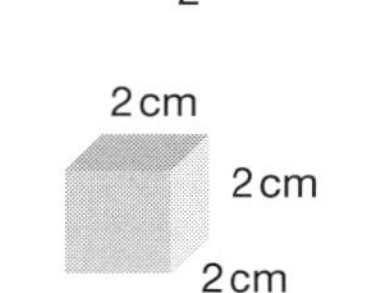

9 A cuboid measures 6 cm × 3 cm × 2 cm.
Calculate, expressing your answers as surds in their simplest form:

a) the lengths of the different face diagonals
b) the length of a space diagonal.

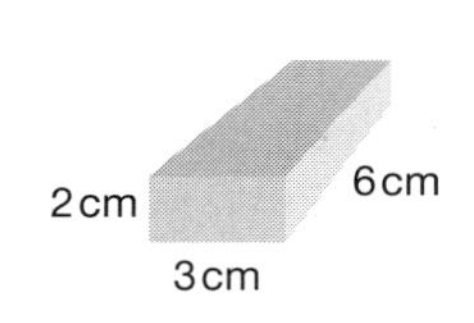

78 The Algebra of Surds

Reminder

Example If $p = 2 + \sqrt{3}$, evaluate p^2.

Solution [Recall $(a + b)^2 = a^2 + 2ab + b^2$]
[Recall $(\sqrt{3})^2$ or $= 3$ or $(\sqrt{a})^2 = a$]

$$p^2 = (2 + \sqrt{3})^2 = 2^2 + 2 \times 2 \times \sqrt{3} + (\sqrt{3})^2$$
$$= 4 + 4\sqrt{3} + 3$$
$$= 7 + 4\sqrt{3}$$

Example Evaluate $\sqrt{48} \times \sqrt{75}$.

Solution

$$\sqrt{48} \times \sqrt{75} = \sqrt{16 \times 3} \times \sqrt{25 \times 3}$$
$$= \sqrt{16} \times \sqrt{3} \times \sqrt{25} \times \sqrt{3}$$
$$= 4\sqrt{3} \times 5\sqrt{3}$$
$$= 20 \times 3 = 60$$

Exercise 78

1 Simplify:

a) $\sqrt{7} \times \sqrt{7}$ b) $\sqrt{2} \times \sqrt{2}$ c) $\sqrt{5} \times \sqrt{5}$
d) $\sqrt{13} \times \sqrt{13}$ e) $\sqrt{a} \times \sqrt{a}$ f) $\sqrt{3} \times \sqrt{2}$
g) $\sqrt{4} \times \sqrt{5}$ h) $\sqrt{6} \times \sqrt{a}$ i) $\sqrt{9} \times \sqrt{a}$
j) $\sqrt{x} \times \sqrt{5}$ k) $\sqrt{2} \times \sqrt{8}$ l) $\sqrt{5} \times \sqrt{20}$
m) $\sqrt{6} \times \sqrt{3}$ n) $\sqrt{9} \times \sqrt{18}$ o) $2\sqrt{2} \times \sqrt{2}$
p) $2\sqrt{3} \times 3\sqrt{3}$ q) $3\sqrt{2} \times 5\sqrt{3}$ r) $4\sqrt{5} \times 2\sqrt{3}$
s) $\sqrt{8} \times \sqrt{12}$ t) $2\sqrt{27} \times 3\sqrt{75}$.

2 Expand the brackets and simplify:

a) $\sqrt{3}(\sqrt{3} + 1)$ b) $\sqrt{5}(\sqrt{5} - 1)$ c) $\sqrt{2}(1 - \sqrt{2})$
d) $\sqrt{5}(2 + \sqrt{5})$ e) $\sqrt{2}(3 + 4\sqrt{2})$ f) $(\sqrt{5} + \sqrt{2})(\sqrt{5} - \sqrt{2})$
g) $(\sqrt{2} + 1)(\sqrt{2} - 1)$ h) $(\sqrt{7} - 3)(\sqrt{7} + 3)$ i) $(\sqrt{2} + 1)^2$
j) $(\sqrt{7} - \sqrt{11})(\sqrt{7} + \sqrt{11})$ k) $(\sqrt{3} + \sqrt{5})^2$ l) $(\sqrt{2} - \sqrt{3})^2$
m) $\sqrt{5}(\sqrt{2} + \sqrt{6})(\sqrt{2} - \sqrt{6})$ n) $\sqrt{3}(\sqrt{8} + 2)^2$ o) $(\sqrt{8} + \sqrt{12})(\sqrt{48} - \sqrt{18})$.

3 Given that $x = 1 + \sqrt{2}$ and $y = 1 - \sqrt{2}$, evaluate:

a) $5x + 5y$ b) $2xy$ c) $x^2 + y^2$.

4 Given that $p = \sqrt{7} + \sqrt{3}$ and $q = \sqrt{7} - \sqrt{3}$, evaluate:

a) $3p - 3q$ b) $4pq$ c) $p^2 - q^2$ d) $(2p - 3q)^2$.

5 A rectangle has length $(2 + \sqrt{2})$cm and breadth $(2 - \sqrt{2})$cm.
Calculate: a) the area of the rectangle
b) the length of a diagonal.

6 The point $P(\sqrt{2}, k)$ lies on the curve with equation $y = 2 - \frac{1}{2}x^2$.
Find the value of k and the distance from P to the origin.

Reminder

Some division of surds is straightforward, e.g. $\frac{\sqrt{15}}{\sqrt{3}} = \sqrt{5}$.

Other examples can be more awkward, but we can use a result that we saw earlier, namely $(\sqrt{a})^2 = a$.

When $\sqrt{a}$ appears as the denominator, we multiply top and bottom by $\sqrt{a}$.

e.g. $\sqrt{3} \div \sqrt{5} = \frac{\sqrt{3}}{\sqrt{5}} = \frac{\sqrt{3}}{\sqrt{5}} \times \frac{\sqrt{5}}{\sqrt{5}} = \frac{\sqrt{15}}{5}$ (which is a single surd).

Example Express (i) $\frac{6}{3\sqrt{2}}$; (ii) $\frac{18}{\sqrt{12}}$

with a rational denominator in its simplest form.

Solution

(i) $\frac{6}{3\sqrt{2}} = \frac{6}{3\sqrt{2}} \times \frac{\sqrt{2}}{\sqrt{2}} = \frac{6\sqrt{2}}{3 \times 2} = \frac{6\sqrt{2}}{6} = \sqrt{2}$

(ii) Ensure that the surd is in its simplest form before rationalising in order to minimise cancelling.

$$\frac{18}{\sqrt{12}} = \frac{18}{2\sqrt{3}} = \frac{9}{\sqrt{3}} = \frac{9}{\sqrt{3}} \times \frac{\sqrt{3}}{\sqrt{3}} = \frac{9\sqrt{3}}{3} = 3\sqrt{3}.$$

7 Simplify:

a) $\frac{\sqrt{12}}{\sqrt{3}}$ b) $\frac{\sqrt{50}}{\sqrt{2}}$ c) $\frac{\sqrt{27}}{\sqrt{3}}$ d) $\frac{\sqrt{40}}{\sqrt{8}}$

e) $\frac{\sqrt{60}}{\sqrt{15}}$ f) $\frac{2\sqrt{6}}{3\sqrt{2}}$ g) $\frac{4\sqrt{10}}{2\sqrt{5}}$ h) $\frac{5\sqrt{20}}{\sqrt{5}}$

i) $\frac{\sqrt{14}}{\sqrt{14}}$ j) $\frac{\sqrt{75}}{\sqrt{12}}$ k) $\frac{\sqrt{125}}{\sqrt{180}}$ l) $\frac{\sqrt{39}}{\sqrt{156}}$.

8 Simplify:

a) $\frac{\sqrt{8}\times\sqrt{6}}{\sqrt{12}}$ b) $\frac{\sqrt{2}\times\sqrt{27}}{\sqrt{6}}$ c) $\frac{\sqrt{5}\times\sqrt{40}}{\sqrt{8}}$

d) $\frac{\sqrt{15}\times\sqrt{20}}{\sqrt{6}}$ e) $\frac{\sqrt{15}}{\sqrt{20}\times\sqrt{12}}$ f) $\frac{\sqrt{21}}{\sqrt{28}\times\sqrt{48}}$.

9 Express each of these surds with a rational denominator:

a) $\frac{1}{\sqrt{2}}$ b) $\frac{1}{\sqrt{3}}$ c) $\frac{1}{\sqrt{5}}$

d) $\frac{2}{\sqrt{2}}$ e) $\frac{6}{\sqrt{3}}$ f) $\frac{10}{\sqrt{5}}$

g) $\frac{3}{\sqrt{5}}$ h) $\frac{20}{\sqrt{2}}$ i) $\frac{3}{2\sqrt{5}}$

j) $\frac{4}{5\sqrt{2}}$ k) $\frac{3}{5\sqrt{5}}$ l) $\frac{\sqrt{3}}{2\sqrt{6}}$.

10 Express each denominator in its simplest form and hence rationalise the denominator of each fraction:

a) $\frac{1}{\sqrt{20}}$ b) $\frac{1}{\sqrt{50}}$ c) $\frac{10}{\sqrt{12}}$

d) $\frac{4}{\sqrt{8}}$ e) $\frac{\sqrt{5}}{\sqrt{20}}$ f) $\frac{\sqrt{6}}{\sqrt{75}}$.

11 Express each of the following in its simplest form with a rational denominator:

a) $\frac{\sqrt{4}}{\sqrt{3}}$ b) $\frac{\sqrt{5}}{\sqrt{2}}$ c) $\sqrt{\frac{9}{10}}$

d) $\frac{\sqrt{15}}{\sqrt{10}}$ e) $\sqrt{\frac{1}{7}}$ f) $\sqrt{\frac{3}{5}}$.

12 Express each of the following in its simplest form with a rational denominator:

a) $3\sqrt{2}+\frac{2}{\sqrt{2}}$ b) $\sqrt{5}+\frac{1}{\sqrt{5}}$ c) $\frac{1}{\sqrt{3}}+\frac{1}{\sqrt{12}}$

d) $\sqrt{3}-\frac{1}{\sqrt{3}}$ e) $\frac{10+8\sqrt{2}}{\sqrt{2}}$ f) $\sqrt{27}+\frac{6}{\sqrt{3}}$

g) $\sqrt{8}+\frac{3}{\sqrt{2}}$ h) $\frac{\sqrt{75}}{5}-\frac{7}{\sqrt{75}}$.

Reminder

When a compound surd, e.g. $2 + \sqrt{3}$, appears as the denominator of a fraction, the fraction can still be expressed with a rational denominator, by multiplying top and bottom by its conjugate surd, $2 - \sqrt{3}$.

Example Express $\dfrac{4}{3 + \sqrt{5}}$ with a rational denominator.

Solution

$$\frac{4}{3+\sqrt{5}} = \frac{4}{3+\sqrt{5}} \times \frac{3-\sqrt{5}}{3-\sqrt{5}} = \frac{4(3-\sqrt{5})}{(3)^2-(\sqrt{5})^2} = \frac{4(3-\sqrt{5})}{9-5} = 3-\sqrt{5}$$

Conjugate surds have the property that their sum and product are both rational:

e.g. $(2 + \sqrt{3}) + (2 - \sqrt{3}) = 4$

and $(2 + \sqrt{3}) \times (2 - \sqrt{3}) = (2)^2 - (\sqrt{3})^2 = 1$.

13B Express each of the following with a rational denominator:

a) $\dfrac{1}{1+\sqrt{2}}$ b) $\dfrac{2}{\sqrt{2}-1}$ c) $\dfrac{3}{3-\sqrt{3}}$

d) $\dfrac{5}{\sqrt{7}-2}$ e) $\dfrac{2}{\sqrt{3}+4}$ f) $\dfrac{3}{\sqrt{11}-5}$

g) $\dfrac{7}{\sqrt{13}-3}$ h) $\dfrac{6}{7+\sqrt{7}}$ i) $\dfrac{3}{\sqrt{5}+\sqrt{2}}$

j) $\dfrac{8}{\sqrt{7}-\sqrt{5}}$ k) $\dfrac{\sqrt{2}}{\sqrt{5}-\sqrt{3}}$ l) $\dfrac{7}{11-2\sqrt{2}}$.

14H Express each of the following with a rational denominator:

a) $\dfrac{\sqrt{7}-\sqrt{3}}{\sqrt{7}+\sqrt{3}}$ b) $\dfrac{1+\dfrac{1}{\sqrt{3}}}{1-\dfrac{1}{\sqrt{3}}}$ c) $\dfrac{1-\dfrac{1}{\sqrt{2}}}{1+\dfrac{1}{\sqrt{2}}}$

d) $\dfrac{2+\dfrac{1}{\sqrt{5}}}{2-\dfrac{1}{\sqrt{5}}}$ e) $\dfrac{\sqrt{3}-\dfrac{1}{\sqrt{3}}}{2}$ f) $\dfrac{\sqrt{5}+\dfrac{1}{\sqrt{5}}}{3}$

g) $\dfrac{x-y}{1+xy}$, where $x = \dfrac{1}{\sqrt{3}}$ and $y = \dfrac{1}{\sqrt{3}}$.

79 The Laws of Indices

Reminder

x^2 means the product of two factors, both of which are x.
x^p means the product of p factors, each of which is x.
The first law of indices: $a^p \times a^q = a^{p+q}$.
i.e. when we **multiply** powers of a, we **add the indices**.

Example Simplify a) $a^5 \times a^3$ b) $b^7 \times b^7$ c) $3c^4 \times 5c^3$ d) $d^2 \times d^3 \times d$.

Solution
a) $a^5 \times a^3 = a^{5+3} = a^8$
b) $b^7 \times b^7 = b^{7+7} = b^{14}$
c) $3c^4 \times 5c^3 = (3 \times 5) \times (c^{4+3}) = 15c^7$
d) $d^2 \times d^3 \times d = d^{2+3+1} = d^6$

Exercise 79

1 Express in index form:

a) $3 \times 3 \times 3$
b) $2 \times 2 \times 2 \times 2$
c) $5 \times 5 \times 5$
d) $7 \times 7 \times 7 \times 7 \times 7$
e) $10 \times 10 \times 10$
f) $a \times a \times a \times a$
g) $x \times x \times x$
h) $a \times a \times b \times a \times b$
i) $m \times n \times m \times n \times n$
j) $3 \times a \times a \times 4 \times b$
k) $a \times b \times c \times a$
l) $3 \times p \times p \times 2 \times p$
m) $2 \times f \times g \times g \times f$
n) $x \times x \times y \times y \times y \times z$
o) $m \times 2 \times n \times 3 \times m \times n$.

2 Express as a product of factors (without using indices):

a) p^5
b) x^3
c) y^4
d) m^2n^3
e) $3m^3$
f) $2x^2y^3$.

3 Simplify:

a) $3^3 \times 3^2$
b) 5×5^2
c) $2^3 \times 2^3$
d) $5^6 \times 5^3$
e) $7^8 \times 7^3$
f) $10^3 \times 10^6$
g) $a^2 \times a^3$
h) $a^3 \times a$
i) $p^3 \times p^2$
j) $p^4 \times p^5$
k) $m^6 \times m^2$
l) $y^2 \times y^5$
m) $2y^2 \times 3y^6$
n) $3x \times 4x^3$
o) $2a^5 \times 3a^4$
p) $10^2 \times 10 \times 10^3$
q) $x^4 \times x^5 \times x^6$
r) $x^2 \times y \times y^4$
s) $3a^4 \times 5a^3$
t) $6p^2 \times 2p^5$
u) $2x \times 3x \times 4x$
v) $q^2 \times p^2 \times q^4$
w) $a \times b^2 \times b^3 \times a^6$
x) $x^2y \times xy^4$.

Reminder

The second law of indices: $a^p \div a^q = a^{p-q}$, when $p > q$.
i.e. when we **divide** powers of a, we **subtract the indices**.
At this stage we have an additional version for $q > p$: $a^p \div a^q = \frac{1}{a^{q-p}}$.

Example Simplify a) $a^5 \div a^3$ b) $a^3 \div a^5$ c) $6b^4 \div 2b^2$.

Solution

a) $a^5 \div a^3 = a^{5-3} = a^2$ $\left(\text{since } a^5 \div a^3 = \frac{a^5}{a^3} = \frac{a.a.\not{a}.\not{a}.\not{a}}{\not{a}.\not{a}.\not{a}} = a^2\right)$

b) $a^3 \div a^5 = \frac{1}{a^{5-3}} = \frac{1}{a^2}$ $\left(\text{since } a^3 \div a^5 = \frac{a^3}{a^5} = \frac{\not{a}.\not{a}.\not{a}}{a.a.\not{a}.\not{a}.\not{a}} = \frac{1}{a^2}\right)$

c) $6b^4 \div 2b^2 = (6 \div 2)(b^{4-2}) = 3b^2$

4 Express these numerators and denominators as products of factors and simplify (as if you didn't know the second law of indices):

a) $\frac{3^3}{3^2}$ b) $\frac{2^5}{2^3}$ c) $\frac{6^4}{6^3}$

d) $\frac{5^7}{5^4}$ e) $\frac{x^5}{x^2}$ f) $\frac{y^6}{y^4}$.

5 Apply the second law of indices to simplify:

a) $2^9 \div 2^5$ b) $3^8 \div 3^3$ c) $10^6 \div 10^4$ d) $10^{19} \div 10^{13}$

e) $\frac{x^3}{x}$ f) $\frac{a^5}{a^4}$ g) $\frac{p^6}{p^2}$ h) $\frac{q^7}{q^4}$

i) $x^5 \div x^2$ j) $x^9 \div x^4$ k) $a^6 \div a$ l) $p^7 \div p^2$

m) $12m^6 \div 3m^5$ n) $9y^8 \div 3y^5$ o) $8z^6 \div 4z$ p) $4t^3 \div 2t$

q) $\frac{6p^4}{2p^2}$ r) $\frac{12x^6}{3x^2}$ s) $\frac{2a^2b^3}{ab^2}$ t) $\frac{10p^3q^7}{2p^2q^3}$.

6 Simplify:

a) $a^6 \div a^8$ b) $b^4 \div b^5$ c) $c^3 \div c^6$

d) $d^5 \div d^7$ e) $2x^5 \div x^8$ f) $3y^3 \div y^4$

g) $z^4 \div 2z^2$ h) $k^4 \div 2k^6$ i) $4p^7 \div 2p^8$

j) $a^2b^3 \div a^3b^2$ k) $x^4y^3 \div x^3y^4$ l) $p^2q^5 \div p^4q$

m) $r^2s^3 \div r^3s^2$ n) $2pq \div 3p^2q^3$ o) $2x^2z^3 \div 4xz^5$

p) $(6x^4 \times 4x^3) \div 8x^8$.

Reminder

The third law of indices: $(a^p)^q = a^{pq}$
i.e. when we **raise** a power of a to another power, we **multiply the indices**.

Example Simplify a) $(a^3)^3$ b) $(b^2)^4$.

Solution a) $(a^3)^3 = a^{3\times3} = a^9$
b) $(b^2)^4 = b^{2\times4} = b^8$

7 Simplify each of the following by writing out in full as if you didn't know the third law of indices:

a) $(2^2)^3$ b) $(3^2)^3$ c) $(5^4)^2$
d) $(2^3)^4$ e) $(5^2)^3$ f) $(6^2)^4$.

8 Use the third law of indices to simplify:

a) $(x^3)^2$ b) $(y^2)^4$ c) $(z^5)^2$ d) $(x^3)^4$
e) $(p^4)^3$ f) $(q^5)^3$ g) $(r^5)^4$ h) $(s^2)^6$
i) $(t^3)^n$ j) $(u^8)^m$ k) $(v^x)^2$ l) $(w^p)^5$
m) $(x^a)^5$ n) $(y^a)^b$ o) $(m^x)^y$ p) $(n^p)^q$
q) $(w^{2a})^5$ r) $(x^{3s})^4$ s) $(y^{3a})^{2b}$ t) $(z^{2y})^{3x}$.

Reminder

The fourth law of indices: $(ab)^n = a^n \times b^n$
(This law is used as often 'right to left' as it is 'left to right'.)

Example a) Expand $(xy^2)^3$.
b) Express $4x^2$ as a perfect square.

Solution a) $(xy^2)^3 = (x)^3 \times (y^2)^3$ (by the fourth law of indices)
$= x^3 \times y^{2\times3}$ (by the third law of indices)
$= x^3y^6$

b) $4x^2 = 2^2 \times x^2 = (2 \times x)^2 = (2x)^2$

9 By writing out in full (as if you didn't know the fourth law of indices), simplify:

a) $(ab)^2$ b) $(pq)^3$ c) $(xy)^4$.

10 Using the fourth law of indices, simplify:

a) $(ab)^3$ b) $(pq)^4$ c) $(x^2y)^2$
d) $(p^2q^3)^3$ e) $(2x)^4$ f) $(5x)^3$
g) $(2a)^5$ h) $(3p^2)^3$ i) $(km)^x$
j) $(c^2d)^p$ k) $(fg^3)^{2x}$ l) $(abc)^7$.

11 Use the appropriate law(s) of indices to simplify:

a) $3^5 \times 3^2$ b) $3^7 \div 3^4$ c) $\frac{2^8}{2^5}$
d) $(3^2)^5$ e) $(2^3)^4$ f) $2^3 \times 2^4$
g) $10^6 \times 10^7$ h) $10^6 \div 10^3$ i) $(5^6)^3$
j) $(a^m)^2$ k) $a^m \times a^n$ l) $p^4 \div p$
m) $a^4 \times a$ n) $t^6 \times t^4$ o) $(u^2)^3 \times u^2$
p) $(v^2)^4 \div v^5$ q) $(w^3)^3 \div w^6$ r) $(x^3)^4 \div (x^2)^3$
s) $(3^3)^4 \div 3^6$ t) $(8^2)^4 \div 8^6$ u) $(5^2)^4 \times 5^3$
v) $3x^2 \times 4x^3$ w) $3y^4 \div 4y^3$ x) $(2z^3)^2 \div 4z^6$
y) $10^3 \times 10^5 \div 10^6$ z) $\frac{a^4 \times a^2}{a^6}$ aa) $\frac{p^2 \times p^5}{p^4}$
bb) $\frac{2b^3 \times 3b^2}{5b^5}$ cc) $\frac{6a^4 \times 3b^2}{2ab}$ dd) $(p^2q)^3 \times (pq^2)^2$
ee) $\frac{(2x)^3 \times (3x^2)^2}{(6x^5)^2}$ ff) $\frac{(3y^3)^2 \times (2z^2)^3}{(12y^4z^5)^2}$.

80 Zero and Negative Indices

Reminder

Suppose a^0 does exist and obeys the laws of indices, then $a^0 = a^{n-n} = \frac{a^n}{a^n} = 1$.

And suppose a^{-n} does exist and obeys the laws of indices, then $a^{-n} = a^{0-n} = \frac{a^0}{a^n} = \frac{1}{a^n}$.

These two facts must now also be memorised: $a^0 = 1$ and $a^{-n} = \frac{1}{a^n}$.

Note:
- You should now see why the notation 10^{-1}, 10^{-2}, … was used with standard form.
- We no longer need two forms of the second law of indices, because, for example,

$\frac{k^3}{k^5} = k^{3-5} = k^{-2} = \frac{1}{k^2}$ which is the same as $\frac{k^3}{k^5} = \frac{1}{k^{5-3}} = \frac{1}{k^2}$.

Reminder continued

- Learn to use the laws of indices with negative indices,

  e.g. $\frac{x^{-7}}{x^{-4}} = x^{-7-(-4)} = x^{-3}$ is much easier than $\frac{x^{-7}}{x^{-4}} = \frac{1}{x^7} \div \frac{1}{x^4} = \frac{1}{x^7} \times \frac{x^4}{1} = \frac{x^4}{x^7} = \frac{1}{x^3}$!

  or $(a^{-2})^{-3} = a^{(-2)(-3)} = a^6$ is much easier than $(a^{-2})^{-3} = \left(\frac{1}{a^2}\right)^{-3} = \frac{1}{\left(\frac{1}{a^2}\right)^3} = \frac{1}{\left(\frac{1}{a^6}\right)} = a^6$!

  Do not confuse

  $2x^0$ $(= 2 \times 1 = 2)$ with $(2x)^0 = 1$ or

  $2x^{-3}$ $\left(= 2 \times \frac{1}{x^3} = \frac{2}{x^3}\right)$ with $(2x)^{-3} = \frac{1}{(2x)^3} = \frac{1}{8x^3}$.

Exercise 80

1 Express each of the following with a positive index:

a) 2^{-8} b) 10^{-3} c) 2^{-4} d) 10^{-6}
e) 10^{-1} f) 3^{-2} g) 8^{-4} h) 3^{-4}
i) a^{-3} j) $2b^{-1}$ k) $3k^{-3}$ l) $5t^{-2}$.

2 Express each of the following with a negative index:

a) $\frac{1}{5}$ b) $\frac{1}{7^5}$ c) $\frac{1}{9^3}$ d) $\frac{1}{6^4}$
e) $\frac{1}{t^7}$ f) $\frac{1}{z^4}$.

3 Use the rules for zero and negative indices to give equivalent expressions for:

a) 10^{-2} b) $\frac{1}{10^3}$ c) 10^0 d) $\frac{1}{10}$
e) $\frac{1}{2}$ f) 2^{-2} g) $\frac{1}{2^4}$ h) 2^{-3}
i) $\frac{1}{4^3}$ j) 5^{-6} k) 7^{-3} l) $\frac{1}{5}$
m) $\frac{1}{7^{-3}}$ n) $\frac{1}{10^5}$ o) $\frac{1}{3^3}$ p) $\frac{1}{a^{-5}}$
q) 5^{-3} r) 10^{-8} s) $2x^{-2}$ t) $(2x)^{-2}$
u) $3z^{-1}$ v) $5y^{-4}$ w) $(3x)^{-4}$ x) $7 \times (3x^2)^0$.

4 Simplify, giving only positive indices in your answers:

a) $3^3 \times 3^{-4}$ b) $2^{-6} \times 2^4$ c) $2^3 \div 2^5$
d) $3^6 \div 3^4$ e) $5^6 \div 5^8$ f) $(6^3)^{-2}$
g) $(2^{-1})^3$ h) $5^2 \times 5^{-4}$ i) $\frac{t^5}{t^3}$
j) $\frac{u}{u^4}$ k) $\frac{v^3}{v^4}$ l) $w^3 \times w^{-5}$

m) $x^4 \div x^{-6}$ n) $y^5 \times y^{-8}$ o) $z^3 \div z^{-7}$

p) $\dfrac{a^4}{a^4}$ q) $2x^4 \div x^{-3}$ r) $7y^{-2} \div y^{-1}$.

5 Simplify:

a) $3x^3 \times 2x^{-2}$ b) $(4ab)^2 \div (2a^2b^{-1})^3$ c) $4a^3 \times 2b^{-2} \times 3a^2b$

d) $6a^3 \times 2b^2 \div 8a^2b^{-1}$ e) $2x^3 \times (3y^{-1})^2 \div 4xy^3$ f) $(2x^{-1}y^2)^{-3} \div (3x^3y^{-2})^2$.

6 Simplify, expressing each answer with positive indices:

a) $a^5 \times a^3 \times a$ b) $2x^2 \times x^3$ c) $2y^3 \times 3y^4$

d) $x^5 \div x^2$ e) $x^2 \div x^{-3}$ f) $4x^3 \div 2x^5$

g) $b^2 \times b^5 \times b^{-4}$ h) $(x^{-2})^3$ i) $(y^{-4})^{-2}$

j) $6x^3 \div 3x^{-1}$ k) $\dfrac{8y^3}{4y^{-2}}$ l) x^2y^{-3}

m) $x^{-3} \div 2x^2$ n) $(x^0)^2$ o) $(3^{-6} \div 3^5) \times 3^8$

p) $\dfrac{x^3 \times y^2}{x^5y}$.

7B Simplify:

a) $\dfrac{3x^2 \times 2x^{-4}}{8x^3}$ b) $\dfrac{2y^2 \times 3x^3}{4x^2y^4}$ c) $\dfrac{2ac^2 \times a^2b}{4a^3b}$

d) $\dfrac{(3a^2b)^3}{9a^5b^2}$ e) $\dfrac{(2xy^{-2})^3}{2x^2y^{-4}}$ f) $(a^{-1})^{-1} \div (a^2)^{-2}$

g) $(x^5 \div x^5) + (x^5 \div x^{-5})$ h) $(x^2y^3)^0 + (x^2y^3)^{-2}$ i) $\dfrac{(p^3)^4 \times p^{-5}}{p} \div \dfrac{p^2}{(p^{-2})^3}$.

81 Fractional Indices

Reminder

$x^{\frac{p}{q}} = \sqrt[q]{x^p} = (\sqrt[q]{x})^p$

This definition is seldom required in practice.

It is usually easier to evaluate such a surd using the laws of indices.

It is useful to remember, however, that $x^{\frac{1}{2}} = \sqrt{x}$.

Example Evaluate a) $4^{\frac{3}{2}}$ b) $8^{\frac{2}{3}}$ c) $27^{-\frac{4}{3}}$.

Solution

a) $4^{\frac{3}{2}} = (2^2)^{\frac{3}{2}} = 2^{2\times\frac{3}{2}} = 2^3 = 8$

b) $8^{\frac{2}{3}} = (2^3)^{\frac{2}{3}} = 2^{3\times\frac{2}{3}} = 2^2 = 4$

c) $27^{-\frac{4}{3}} = (3^3)^{-\frac{4}{3}} = 3^{3\times-\frac{4}{3}} = 3^{-4} = \dfrac{1}{3^4} = \dfrac{1}{81}$

Exercise 81

1B Evaluate:

a) $9^{\frac{3}{2}}$ b) $25^{\frac{1}{2}}$ c) $64^{\frac{2}{3}}$ d) $8^{-\frac{2}{3}}$
e) $16^{-\frac{3}{4}}$ f) $36^{\frac{3}{2}}$ g) $4^{-\frac{5}{2}}$ h) $81^{\frac{3}{4}}$
i) $125^{-\frac{2}{3}}$ j) $100^{-\frac{5}{2}}$.

2B Simplify:

a) $a^{\frac{2}{3}} \times a^{\frac{5}{3}}$ b) $b^{\frac{3}{2}} \times b^{\frac{5}{2}}$ c) $c^{\frac{4}{3}} \times c^{\frac{2}{3}}$ d) $d^{\frac{7}{3}} \times d^{-\frac{1}{3}}$
e) $\left(y^{-\frac{1}{4}}\right)^2$ f) $\left(f^{\frac{1}{2}} \times f^{\frac{5}{2}}\right)^2$ g) $2g^{\frac{5}{3}} \times 3g^{-\frac{2}{3}}$ h) $3h^2 \times 4h^{-\frac{1}{2}}$
i) $i^{\frac{4}{3}} \div i^{\frac{2}{3}}$ j) $j^{\frac{7}{5}} \div j^{\frac{3}{5}}$ k) $k^{\frac{5}{8}} \div k^{-\frac{1}{8}}$ l) $l^2 \div l^{\frac{1}{4}}$
m) $m^{\frac{1}{2}} \div m^{-\frac{1}{2}}$ n) $n^{\frac{2}{3}} \div n^{-\frac{1}{3}}$ o) $6x^{\frac{5}{3}} \div 3x^{\frac{2}{3}}$ p) $12p^{-\frac{3}{4}} \div 4p^{\frac{3}{4}}$.

3B Simplify:

a) $a^{\frac{1}{2}} \times a^{\frac{1}{3}}$ b) $b^{\frac{2}{3}} \times b^{\frac{1}{6}}$ c) $c^{\frac{3}{4}} \times c^{\frac{2}{3}}$ d) $d^{\frac{5}{8}} \times d^{\frac{5}{6}}$
e) $y^{\frac{2}{3}} \div y^{\frac{1}{4}}$ f) $f^{\frac{3}{2}} \div f^{\frac{2}{3}}$ g) $g^{\frac{7}{3}} \div g^{-\frac{1}{2}}$ h) $h^{\frac{3}{7}} \div h^{-\frac{3}{2}}$
i) $2i^{\frac{1}{2}} \times 3i^{-\frac{3}{2}}$ j) $3j^{\frac{5}{2}} \times 4j^{\frac{1}{3}}$ k) $8k^{-\frac{1}{2}} \div 4k^{-\frac{1}{3}}$ l) $5l^{-\frac{5}{2}} \div 15l^{-\frac{3}{2}}$.

4B Evaluate the following where $a = 16$ and $b = 27$:

a) $2a^{\frac{1}{4}}$ b) $3b^{\frac{1}{3}}$ c) $4a^{-\frac{3}{2}}$ d) $9b^{-\frac{2}{3}}$
e) $a^{\frac{1}{2}} \times b^{\frac{2}{3}}$ f) $3a^{\frac{3}{2}} \times 2b^{-\frac{1}{3}}$ g) $(2a)^{\frac{2}{5}} \times (3b)^{-\frac{1}{4}}$ h) $\left[a^{\frac{1}{2}} - b^{\frac{1}{3}}\right]^{-2}$.

5B Solve:

a) $3^y = 81$ b) $2^x = \frac{1}{8}$ c) $4^z = 32$ d) $27^k = \frac{1}{81}$.

6H Expand the brackets:

a) $x(x + 1)$ b) $x^2(x + 2)$ c) $x^3(x^{-2} + 2x^{-1})$
d) $2x^4(3x^{-2} - x^{-1})$ e) $y^{-1}(y - 1)$ f) $y^{-2}(2y + y^3)$
g) $3z^{-3}(z^4 - 2z)$ h) $5z^2(2z^2 - 3z^{-2})$ i) $p^{\frac{1}{2}}\left(p^{\frac{3}{2}} - p\right)$
j) $q^{\frac{3}{2}}\left(2q^{-\frac{1}{2}} - 3q^{\frac{1}{2}}\right)$ k) $3r^{-1}\left(r^{\frac{2}{3}} + 2r^{-\frac{1}{3}}\right)$ l) $4s^{-\frac{3}{4}}\left(3s^{-2} - 5s^{-\frac{1}{3}}\right)$.

7H Simplify:

a) $(2a^{-2}b^3)^2$ b) $\left(3a^{\frac{1}{2}}b^{\frac{1}{3}}\right)^2$ c) $\left(4a^{-1}b^{\frac{2}{3}}\right)^3$ d) $\left(5x^{\frac{4}{3}}y^{-\frac{1}{3}}\right)^3$
e) $(9p^{-2}q^4)^{-\frac{1}{2}}$ f) $(16r^{-1}s^2)^{-\frac{3}{2}}$ g) $\left(x^{\frac{2}{5}}y^{-\frac{4}{3}}\right)^{-\frac{1}{2}}$ h) $\left(4m^{\frac{2}{5}}n^{-4}\right)^{\frac{5}{2}}$.

8H Show that $2x^{-\frac{3}{2}} - 3x^{-\frac{2}{3}}$ can be written in the form $\frac{2 - 3x^a}{x^b}$.

82 Revision of Further Algebra

Exercise 82

1 Reduce these vulgar fractions to their lowest terms:

a) $\frac{4}{8}$ b) $\frac{16}{48}$ c) $\frac{36}{54}$ d) $\frac{38}{95}$.

2 Simplify:

a) $\frac{x^2y}{xy^3}$ b) $\frac{20z^4}{5z}$ c) $\frac{(p^2 + q^2)}{(p^2 + q^2)^3}$ d) $\frac{5t}{10t + 20}$.

3 Simplify:

a) $\frac{4x - 12}{9 - 3x}$ b) $\frac{3y + 6}{y^2 - 4}$ c) $\frac{x^2 + 8x + 16}{x^2 - 16}$ d) $\frac{x^2 + x - 6}{x^2 - x - 12}$.

4 Express as a single fraction:

a) $\frac{3}{4} + \frac{4}{5}$ b) $2 + \frac{3}{t}$ c) $\frac{x + 1}{x} - \frac{1}{x - 1}$ d) $\frac{x}{x+1} + \frac{x-1}{x+2}$.

5 Simplify

a) $\frac{5}{9} \times \frac{21}{10}$ b) $\frac{x^2 + x}{x - 2} \times \frac{2x - 4}{x}$

c) $\frac{4}{5} \div \frac{18}{25}$ d) $\frac{x^3 + 3x^2}{x - 1} \div \frac{x^2 + 3x}{3x - 3}$.

6 Change the subject of the formula:

a) $v = u + ft$ to f b) $I = \frac{PTR}{100}$ to R

c) $a = \frac{1}{2}(p + q)r$ to p d) $k = \frac{1}{2}mv^2$ to v.

7 Simplify:

a) $\sqrt{32} + \sqrt{72}$ b) $\sqrt{50} + \sqrt{27} + \sqrt{3} - \sqrt{18}$

c) $3\sqrt{10} \times 4\sqrt{6}$ d) $(2 + 3\sqrt{5})^2$.

8 Express with a rational denominator:

a) $\frac{2}{\sqrt{7}}$ b) $\frac{\sqrt{3}}{\sqrt{5}}$ c) $\frac{1 + \sqrt{3}}{\sqrt{2}}$ d) $\frac{2}{5 - \sqrt{3}}$.

9 Simplify:

a) $\frac{x^2 \times x^3}{x^4}$ b) $\frac{y^{-4} \times y^3}{y^{-2}}$ c) $(2x)^3(3y)^{-2}$ d) $p^{\frac{3}{4}} \times p^{-\frac{1}{2}}$.

10 Evaluate:

a) $100^{\frac{1}{2}}$ b) $9^{-\frac{1}{2}}$ c) $4^{\frac{3}{2}} + 5^0$ d) $(3^2 + 4^2)^{\frac{1}{2}}$.

83 Test on Further Algebra

Allow 45 minutes for this test

1 Simplify:

a) $\sqrt{12} + \sqrt{75}$

b) $\sqrt{72} - \sqrt{48} + \sqrt{18} - \sqrt{12}$

c) $3\sqrt{15} \times 5\sqrt{21}$

d) $(2\sqrt{2} - 3\sqrt{3})^2$.

2 Express with a rational denominator:

a) $\dfrac{3}{\sqrt{5}}$

b) $\dfrac{\sqrt{2}}{\sqrt{3}}$

c) $\dfrac{5 - \sqrt{2}}{\sqrt{3}}$

d) $\dfrac{11}{7 - \sqrt{5}}$.

3 Simplify:

a) $p^{\frac{2}{3}} \times p^{\frac{5}{3}}$

b) $q^{\frac{8}{5}} \div q^{\frac{2}{5}}$

c) $\dfrac{r^{\frac{3}{4}} \times r^{\frac{5}{4}}}{r^{-1}}$

d) $\dfrac{(2p^2q^3)^3 \times (3pq)}{(64p^9q^6)^{\frac{1}{3}}}$.

4 Evaluate:

a) $25^{\frac{1}{2}}$

b) $16^{\frac{3}{2}}$

c) $27^{-\frac{2}{3}}$

d) $(5^2 + 12^2)^{\frac{1}{2}}$.

5 Change the subject of the formula:

a) $A = 2\pi rh$ to h

b) $x = a + wt$ to w

c) $p = a\left(t + \dfrac{2}{u}\right)$ to u

d) $F = \dfrac{\text{GME}}{R^2}$ to R

e) $v = w\sqrt{a^2 - x^2}$ to x.

6 Simplify:

a) $\dfrac{(x^2 + y^2)^2}{(x^2 + y^2)}$

b) $\dfrac{3xy^2}{x^2y}$

c) $\dfrac{5t^2u}{10tu^2}$

d) $\dfrac{5x + 10y}{3x + 6y}$.

7 Simplify:

a) $\dfrac{x^2 - 2x}{3x - 6}$

b) $\dfrac{5y + 15}{y^3 + 3y}$

c) $\dfrac{x^2 - 3x + 2}{4 - x^2}$

d) $\dfrac{x^2 - x - 12}{2x^2 + 7x + 3}$.

8 Express as a single fraction:

a) $2 + \dfrac{5}{x}$

b) $1 - \dfrac{1}{x^2}$

c) $\dfrac{y + 2}{y} - \dfrac{y}{y + 2}$

d) $\dfrac{z}{z - 1} + \dfrac{z + 1}{z + 2}$.

9 Simplify:

a) $\dfrac{x^2 - 1}{x^2 + 3x + 2} \times \dfrac{x^2 - 4}{x^2 - 3x + 2}$

b) $\dfrac{x^2 - 9}{x^2 + 5x + 6} \times \dfrac{x^2 + 6x + 8}{x^2 - 16}$

c) $\dfrac{x^3 - 5x^2}{x^2 - 6x + 5} \times \dfrac{x^2 + x - 2}{x^2 + 2x}$

d) $\dfrac{x^2 - 25}{x^2 + x} \div \dfrac{x^2 + 5x}{x^2 - 1}$.

QUADRATIC FUNCTIONS

84 Finding the Equation from the Graph

Reminder

The graph of a quadratic function is called a parabola. The plural is parabolae, but parabolas is nowadays more commonly used.

Recall that in coordinate geometry, we considered two forms of the straight line; those which passed through the origin and those which did not.

The former had an equation of the form $y = mx$ and the latter $y = mx + c$.

Similarly we consider two types of parabola; those which have their vertex at the origin, and those which do not.

The former have an equation of the form $y = kx^2$, the latter $y = k(x + \text{a})^2 + b$.

Example Find the equations of the quadratic functions whose graphs are shown.

a)

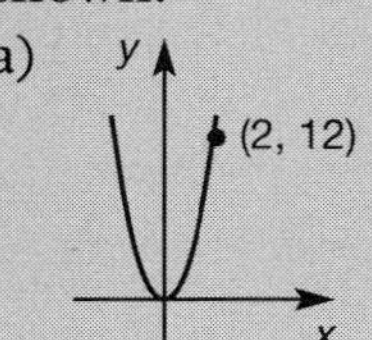

b)

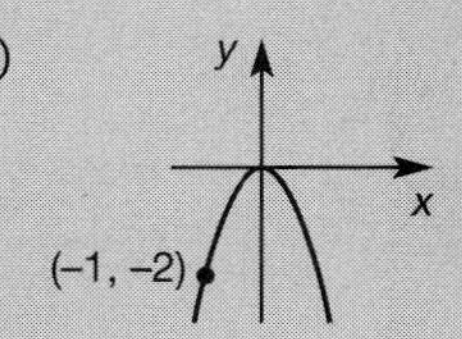

Solution

a) A parabola with a vertex at the origin has an equation of the form $y = kx^2$.

$$\left.\begin{matrix} x = 2 \\ y = 12 \end{matrix}\right\} \Rightarrow \quad 12 = k \times 2^2 \quad \Rightarrow \quad k = \frac{12}{4} = 3 \quad \Rightarrow \quad y = 3x^2.$$

b) Similarly, we can begin with $y = kx^2$.

$$\left.\begin{matrix} x = -1 \\ y = -2 \end{matrix}\right\} \Rightarrow \quad -2 = k(-1)^2 \quad \Rightarrow \quad k = \frac{-2}{1} = -2 \quad \Rightarrow \quad y = -2x^2.$$

Exercise 84

1 Find the quadratic functions whose graphs are shown.

a) 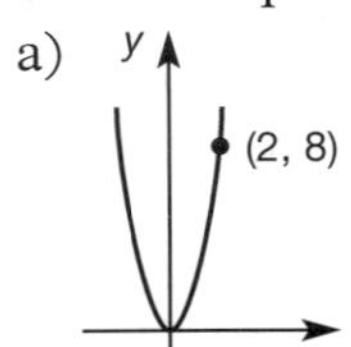

b) 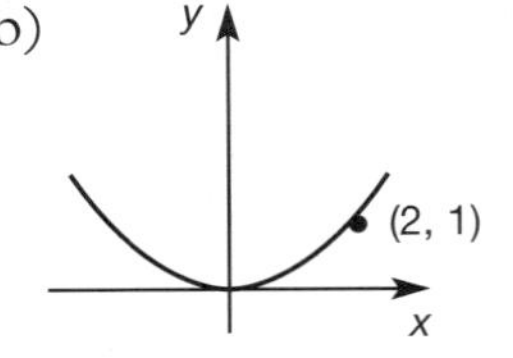

c) 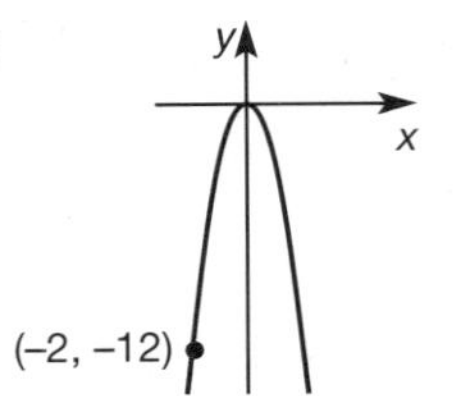

d) 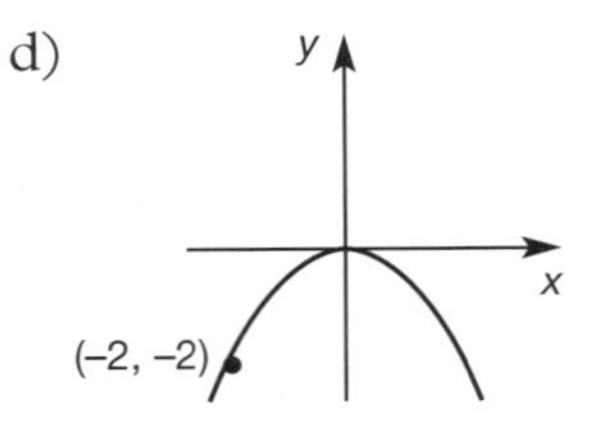

2 Each of these graphs has an equation of the form $y = kx^2$. Find the value of k in each case.

a) 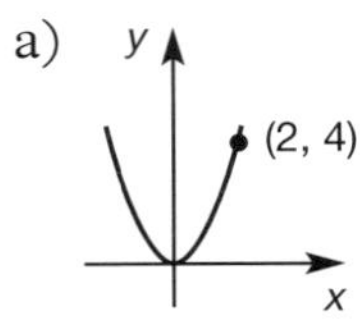

b) 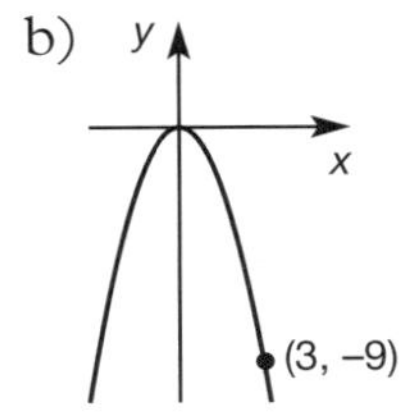

c) 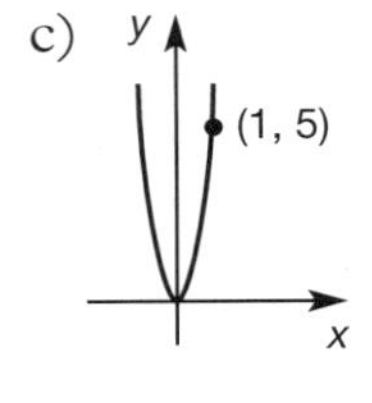

d) 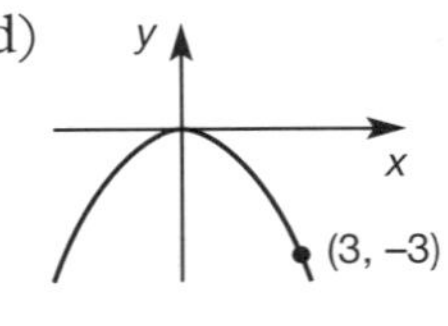

3 a) Use a graphic calculator, computer software or quick sketches to draw the family of parabolas with equation $y = kx^2$, for $k = 1, 2, \frac{1}{2}, 3, \frac{1}{3}, ..$ and $k = -1, -2, -\frac{1}{2}, -3, -\frac{1}{3},$ ... and comment on the effect of the value of k.

b) Similarly, for revision and comparison, investigate the family of straight lines through the origin with equation $y = kx$ for $k = 1, 2, \frac{1}{2}, 3, \frac{1}{3}, ..$ and $k = -1, -2, -\frac{1}{2}, -3, -\frac{1}{3}, ..$ and comment on the effect of the value of k.

Reminder

The quadratic function $(x + a)^2 + b$ has a minimum value of b when $x = -a$.

This is because when $x = -a$, $(x + a)^2 = 0$. Otherwise $(x + a)^2 > 0$.

Example Find the equations of the quadratic functions shown.

a) 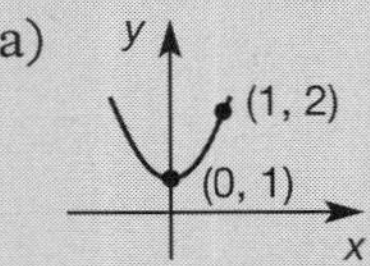

b) 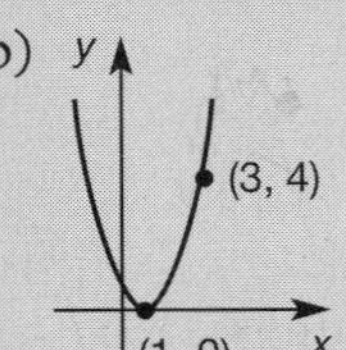

Reminder continued

Solution

a) minimum turning point at (0, 1)

$\Rightarrow \quad y = k(x - 0)^2 + 1$

i.e. $\quad y = kx^2 + 1$

when $x = 1, y = 2 \quad$ so $2 = k \times 1^2 + 1$

$\Rightarrow \quad k = 1 \quad$ hence $\quad y = x^2 + 1$

b) minimum turning point at (1, 0)

$\Rightarrow \quad y = k(x - 1)^2 + 0$

i.e. $\quad y = k(x - 1)^2$

when $x = 3, y = 4 \quad$ so $4 = k(3 - 1)^2$

$\Rightarrow \quad k = 1 \quad$ hence $\quad y = (x - 1)^2$.

4 Find the equations of the quadratic functions whose graphs are shown:

a) 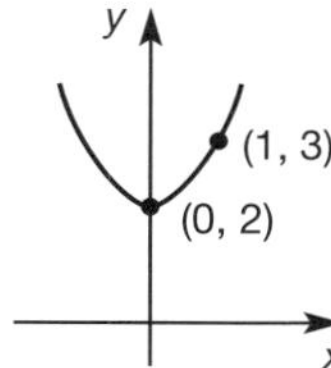

b) 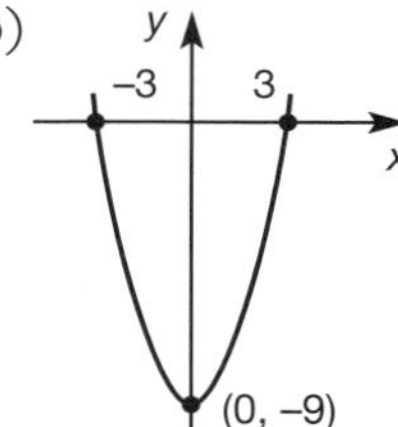

c) 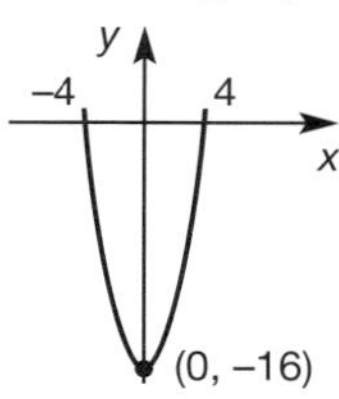

d) 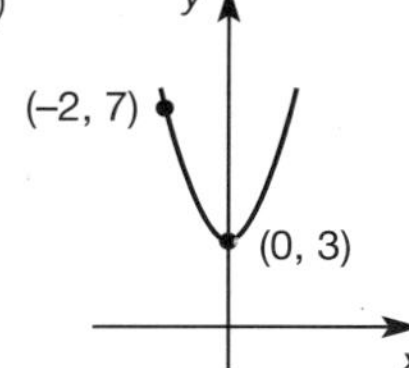

5 a) Use a graphic calculator, computer software or quick sketches to draw the family of parabolas with equation $y = x^2 + k$, for $k = 0, 1, 2, 3, 4, ..$ and $k = -1, -2, -3, -4, ...$ and comment on the effect of the value of k.

b) Similarly, for revision and comparison, investigate the family of straight lines with equation $y = x + k$ for $k = 0, 1, 2, 3, 4, ..$ and $k = -1, -2, -3, -4, ...$ and comment on the effect of the value of k.

6 Find the equations of the quadratic functions whose graphs are shown:

a) 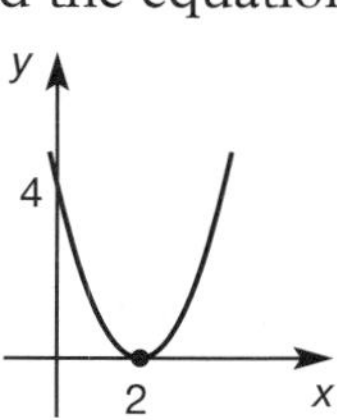

b) 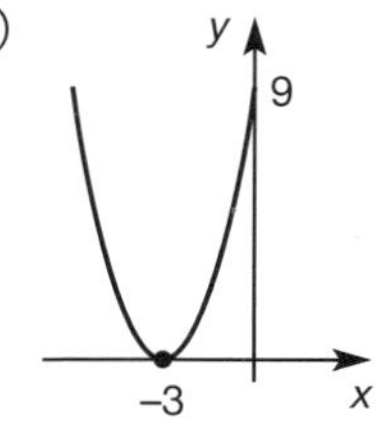

c) 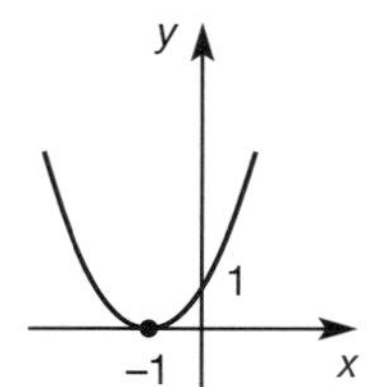

d) 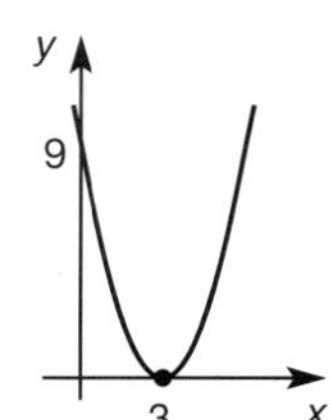

7 a) Use a graphic calculator, computer software or quick sketches to draw the family of parabolas with equation $y = (x - a)^2$, for $a = 0, 1, 2, 3, 4, ..$ and $a = -1, -2, -3, -4, ...$ and comment on the effect of the value of a.

b) Similarly, for revision and comparison, investigate the family of straight lines with equation $y = x - a$ for $a = 0, 1, 2, 3, 4, ...$ and $a = -1, -2, -3, -4, ..$ .and comment on the effect of the value of a.

8 Find the equations of the quadratic functions whose graphs are shown:

a)
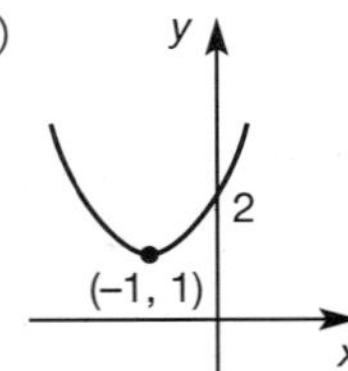

b)
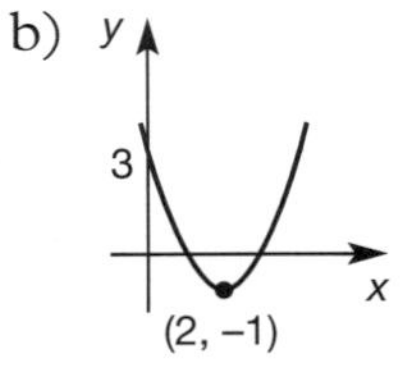

c)
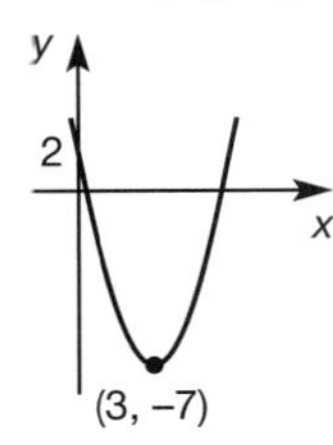

d)
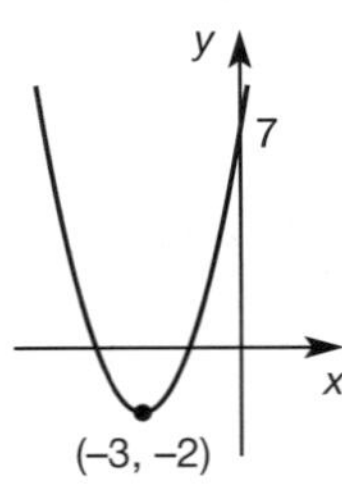

9 Find the equations of these quadratic functions whose graphs all pass through the origin:

a)
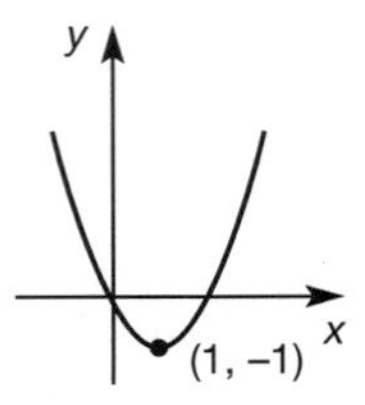

b)
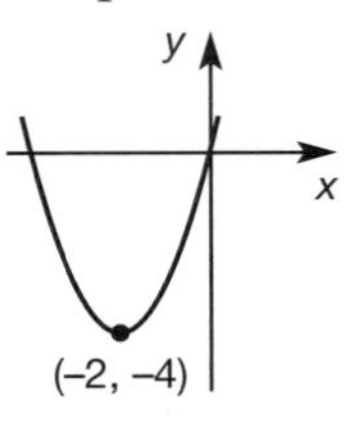

c)
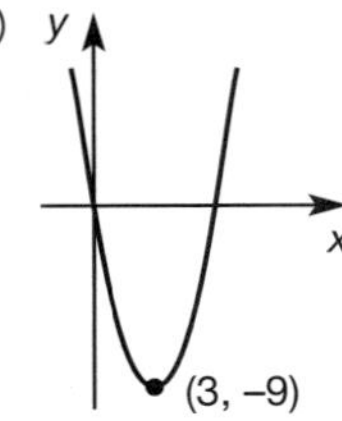

d)
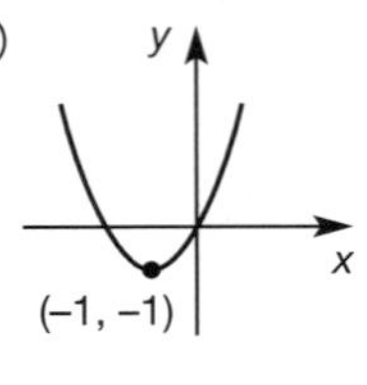

Reminder

The quadratic function $b - (x + a)^2$ has a maximum value of b when $x = -a$.

10H Find the equations of the quadratic functions with these graphs:

a)
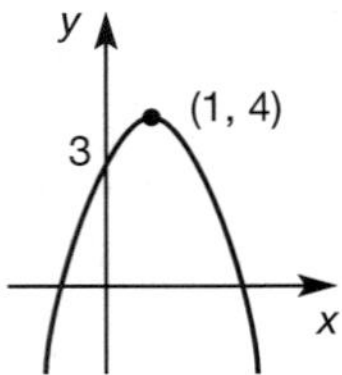

b)
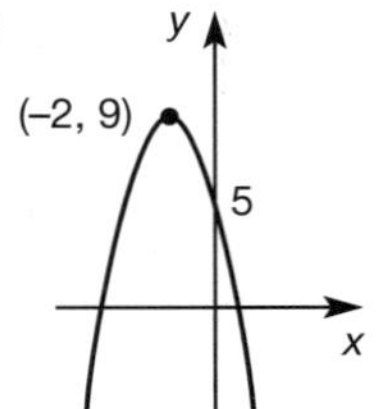

c)
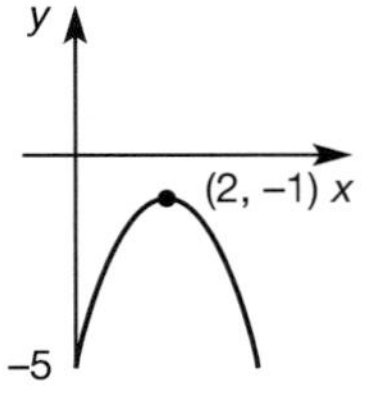

11H Find the equations of the quadratic functions with these graphs, where k might not be equal to 1:

a)
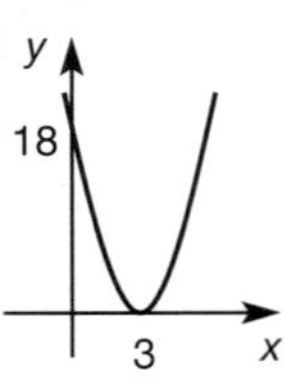

b)
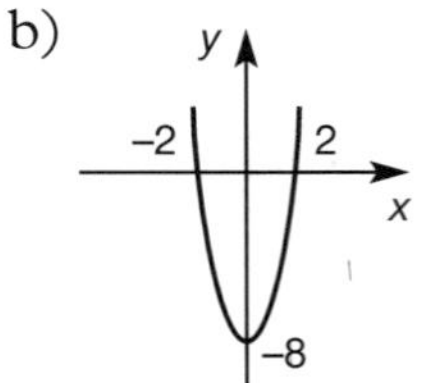

c)
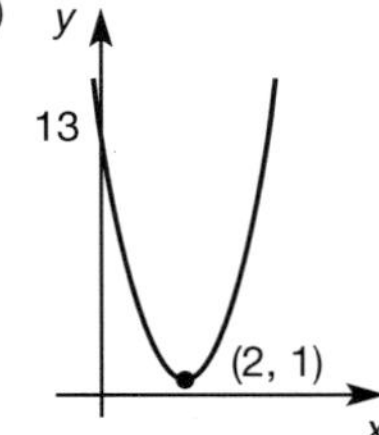

d)
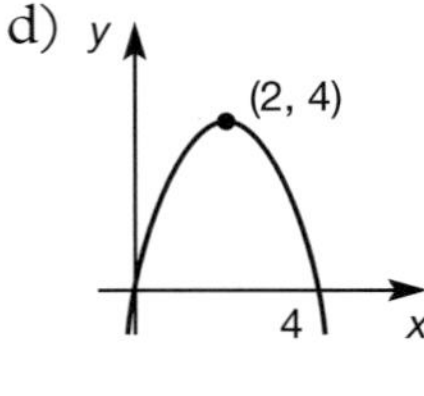

85 Finding Properties of the Graph from the Equation

Reminder

The graph of $y = k(x + a)^2 + b$ has a turning point at $(-a, b)$.
If $k = +1$, then it is a minimum turning point.
If $k = -1$, then it is a maximum turning point.

This also applies to $y = kx^2 + b$, i.e. when $a = 0$.

The axis of symmetry always passes through the turning point.
Therefore it has equation $x = -a$.

Example

For each of the following quadratic graphs, find the coordinates of the turning point, its nature, and the equation of the axis of symmetry:

a) $y = (x + 3)^2 - 8$
b) $y = 7 - (x - 2)^2$
c) $y = -4x^2$.

Solution

a) compare $y = (x + 3)^2 - 8$ with $y = k(x + a)^2 + b$,
clearly $k = 1$, and there is a minimum turning point at $(-3, -8)$
the axis of symmetry is $x = -3$ (which passes through $(-3, -8)$)

b) compare $y = 7 - (x - 2)^2$ with $y = k(x + a)^2 + b$,
clearly $k = -1$, and there is a maximum turning point at $(2, 7)$
the axis of symmetry is $x = 2$

c) compare $y = -4x^2$ with $y = k(x + a)^2 + b$
clearly $k = -4$ and $a = b = 0$
so there is a maximum turning point at $(0, 0)$
the axis of symmetry is $x = 0$ (the y-axis).

Exercise 85

1 Copy and complete:

quadratic function	coordinates of turning point	nature	equation of axis of symmetry
$y = 2x^2$			
$y = -3x^2$			
$y = (x + 5)^2$			
$y = (x - 2)^2$			
$y = (x + 4)^2 + 3$			
$y = (x - 3)^2 + 4$			
$y = 3 - (x + 6)^2$			
$y = 8 - (x - 7)^2$			
$y = (x - 5)^2 + 6$			
$y = 6 - (x + 5)^2$			
$y = 6x^2$			

2 This diagram shows the graph of $y = (x + 3)^2 - 11$.

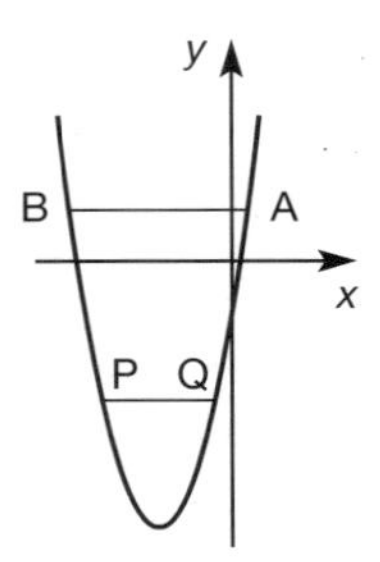

a) State the coordinates of the turning point.
b) State the equation of the axis of symmetry.
c) State the coordinates of the point where the graph crosses the y-axis.
d) If P $(-5, t)$ lies on the graph, find the value of t.
e) If PQ is parallel to the x-axis, find (using the symmetry) the coordinates of Q.
f) If A $(1, k)$ lies on the graph, find the value of k.
g) If BA is parallel to the x-axis, find the coordinates of B.

3 This diagram shows the graph of $y = 4 - (x - 1)^2$.

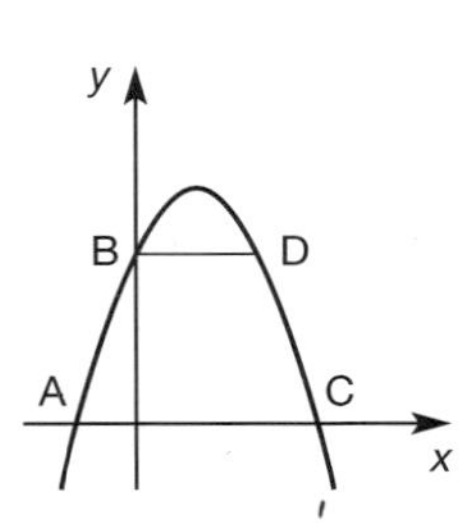

a) State the coordinates of the maximum turning point.
b) State the equation of the axis of symmetry.
c) State the coordinates of the point B.
d) Given that BD is parallel to AC, find the coordinates of D.
e) Given that A is $(-1, 0)$, state the coordinates of C.

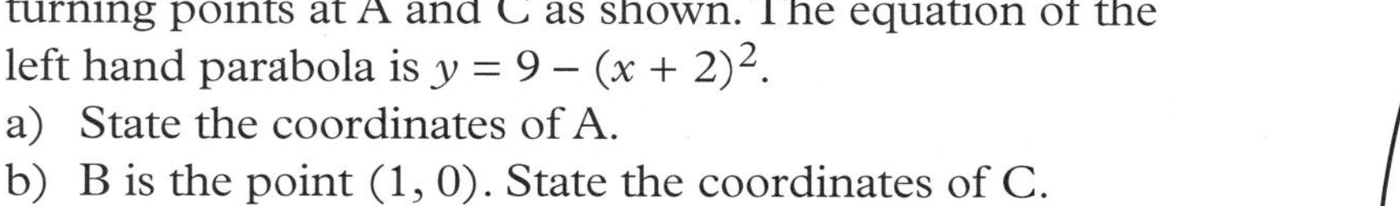

4 McFarlane's pie shop has a large M as a logo.
It consists of two congruent parabolas, with maximum turning points at A and C as shown. The equation of the left hand parabola is $y = 9 - (x + 2)^2$.

a) State the coordinates of A.
b) B is the point (1, 0). State the coordinates of C.
c) Find the equation of the right hand parabola.

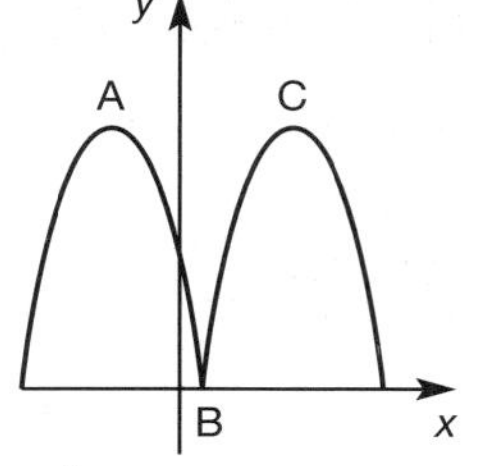

5 Part of a monumental sculpture has two adjacent channels for carrying water from a fountain.
The cross-sections of these channels are congruent parabolas with minimum turning points F and G.
The equation of the right hand channel is $y = (x - 1)^2 - 4$.

a) State the coordinates of G.
b) H is the point (–1, 0). State the coordinates of F.
c) Find the equation of the left hand channel.

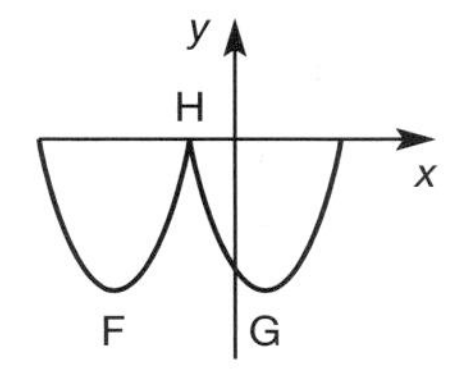

86 Graphical Solution of Quadratic Equations

Reminder

The quadratic function $x^2 + x - 12$ has value zero when $x = 3$ or when $x = -4$.
$[3^2 + 3 - 12 = 9 + 3 - 12 = 0$, and $(-4)^2 + (-4) - 12 = 16 - 4 - 12 = 0]$
Hence 3 and – 4 are the roots of the quadratic equation $x^2 + x - 12 = 0$.
One way of finding the roots of a quadratic equation is to draw the graph of the appropriate quadratic function and to read off where it crosses the x-axis.

Example Solve the quadratic equation $4x^2 + 4x - 35 = 0$ graphically.

Solution Find the values of the function $y = 4x^2 + 4x - 35$ for integral values of x between –4 and 4.

x	–4	–3	–2	–1	0	1	2	3	4
$4x^2$	64	36	16	4	0	4	16	36	64
$4x$	–16	–12	–8	–4	0	4	8	12	16
–35	–35	–35	–35	–35	–35	–35	–35	–35	–35
$(y =)\ 4x^2 + 4x - 35$	13	–11	–27	–35	–35	–27	–11	13	45

Reading from the graph (page 198), the roots are $-3\frac{1}{2}$ and $2\frac{1}{2}$.

It should be checked that $4\left(-3\frac{1}{2}\right)^2 + 4\left(-3\frac{1}{2}\right) - 35 = 0$ and $4\left(2\frac{1}{2}\right)^2 + 4\left(2\frac{1}{2}\right) - 35 = 0$.

Reminder continued

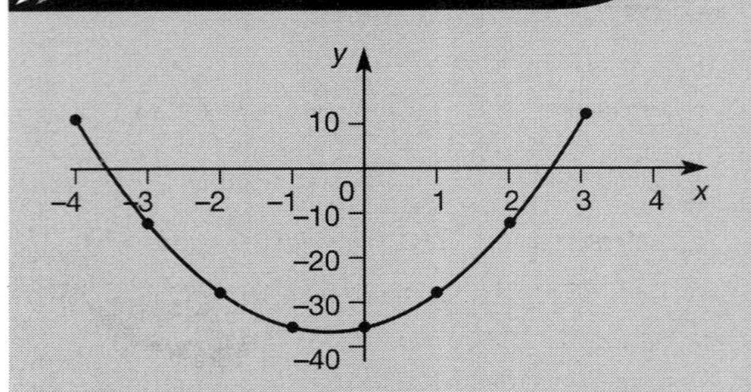

The coordinates of the minimum turning point $\left(-\frac{1}{2}, -36\right)$ can also be read off the graph. (Alternatively, the axis of symmetry is halfway between -1 and 0, i.e. $x = -\frac{1}{2}$; and $4\left(-\frac{1}{2}\right)^2 + 4\left(-\frac{1}{2}\right) - 35 = 1 - 2 - 35 = -36$.)

When you produce this graph on a graphic calculator, you can zoom in on the roots, but you should also carry out the evaluations at $x = -3\frac{1}{2}, -\frac{1}{2}$ and $2\frac{1}{2}$.

Exercise 86

1 a) Complete this table of values for the quadratic function $x^2 - 3x - 10$:

x	-4	-3	-2	-1	0	1	2	3	4	5	6
x^2											
$-3x$											
-10											
$x^2 - 3x - 10$											

b) Draw a smooth curve through the appropriate points to find the roots of $x^2 - 3x - 10 = 0$.

c) Note that the roots were obvious from the table of values, but now that you have obtained the graph, use it to find the coordinates of the turning point.

2 Use a graphic calculator, appropriate software, or plots from tables of values to obtain the graphs of the following quadratic functions. Hence find the roots of these quadratic equations, stating also the coordinates of the turning point on each graph:

a) $x^2 - 4x + 3 = 0$ b) $x^2 - 2x - 3 = 0$ c) $x^2 + 2x - 8 = 0$
d) $x^2 + 4x + 3 = 0$ e) $x^2 + x - 6 = 0$ f) $x^2 - 25 = 0$
g) $x^2 - x - 6 = 0$ h) $x^2 + 3x - 4 = 0$ i) $x^2 + 3x + 2 = 0$.

3B Solve graphically these quadratic equations:

a) $4x^2 - 1 = 0$ b) $8x^2 - 14x - 15 = 0$ c) $6x^2 + x - 12 = 0$
d) $63x^2 + 23x - 56 = 0$.

87 Solution of Quadratic Equations by Factorising

Reminder

The last question in the previous exercise illustrates the shortcomings of graphical methods for solving equations. Algebraic methods of solution provide greater precision. Factorising the quadratic function is the usual method, provided that the quadratic function can be factorised.
The solution depends on the fact that if the product of two factors is zero, then either factor being zero would satisfy the equation, e.g.
if $(x + 2)(x - 3) = 0$, then either $(x + 2) = 0$ or $(x - 3) = 0$ i.e. $x = -2$ or 3.

[Be very careful. This is only true when the right hand side is *zero*, e.g.
if $(x + 2)(x - 3) = 1$, then it is *false* to conclude that $x + 2 = 1$ or $x - 3 = 1$.
Check this for yourself.]

Example Solve $15x^2 + 4x - 3 = 0$ by factorising.

Solution

$$15x^2 + 4x - 3 = 0 \quad \Rightarrow \quad (3x - 1)(5x + 3) = 0$$
$$\Rightarrow \quad 3x - 1 = 0 \quad \text{or} \quad 5x + 3 = 0$$
$$\Rightarrow \quad x = \frac{1}{3} \quad \text{or} \quad x = -\frac{3}{5}$$
$$\text{i.e.} \quad x = \frac{1}{3}, -\frac{3}{5}$$

Exercise 87

1 Write down the roots of these quadratic equations:

a) $(x - 1)(x - 3) = 0$
b) $(y + 2)(y - 1) = 0$
c) $(z - 3)(z + 4) = 0$
d) $(p - 2)(p - 3) = 0$
e) $(q - 5)(q - 7) = 0$
f) $(r + 2)(r + 3) = 0$
g) $(s - 4)(s + 5) = 0$
h) $(t - 2)(t + 2) = 0$
i) $(u - 3)(u + 3) = 0$
j) $(v - 2)(v - 2) = 0$
k) $(w - 3)(w - 3) = 0$
l) $(x + 4)(x - 7) = 0$
m) $(y - 5)(y + 5) = 0$
n) $(z - 7)(z - 7) = 0$
o) $(a - 3)(a + 8) = 0$.

2 Solve these quadratic equations by factorising:

a) $x^2 - 4 = 0$
b) $x^2 - 4x = 0$
c) $x^2 - 9 = 0$
d) $x^2 - 9x = 0$
e) $x^2 - 6x = 0$
f) $x^2 - 6x + 9 = 0$
g) $x^2 - 36 = 0$
h) $x^2 - 10x + 25 = 0$
i) $x^2 + 10x + 25 = 0$
j) $x^2 + 10x + 16 = 0$
k) $x^2 - 3x + 2 = 0$
l) $x^2 - 9x + 20 = 0$
m) $x^2 + x - 6 = 0$
n) $x^2 - x - 12 = 0$
o) $x^2 + 4x - 5 = 0$.

3 Find the x-coordinates of the points where the parabola with the given equation crosses the x-axis:

a) $y = x^2 + 7x$ b) $y = x^2 - x - 2$ c) $y = x^2 - 5x + 6$
d) $y = x^2 - 25$ e) $y = x^2 - 8x - 15$ f) $y = x^2 - 4x - 21$
g) $y = x^2 + 8x$ h) $y = x^2 - 7x - 10$ i) $y = x^2 + 3x - 28$
j) $y = x^2 - 49$ k) $y = x^2$ l) $y = x^2 + 2$.

Reminder

Not every quadratic equation comes in the standard form $ax^2 + bx + c = 0$. Sometimes you have to re-arrange it into this form in order to factorise.

Example Solve $x(x + 1) = 12$.

Solution
Expand the brackets: $x^2 + x = 12$
take every term to the same side: $x^2 + x - 12 = 0$
factorise: $(x - 3)(x + 4) = 0$
solve: $x - 3 = 0$ or $x + 4 = 0 \Rightarrow x = 3, -4$

4B Solve these quadratic equations by re-arranging them in standard form and factorising:

a) $x(x - 1) = 6$ b) $x^2 + 2 = 3x$ c) $x(x - 5) = 14$
d) $(x - 4)^2 = 4 - x$ e) $(x + 1)^2 = x + 7$ f) $(x + 3)(x + 5) = 2(3x + 25)$.

5B Solve these quadratic equations by factorising:

a) $2x^2 - x - 10 = 0$ b) $2x^2 - 7x - 15 = 0$ c) $2x^2 - x - 21 = 0$
d) $3x^2 - 11x - 4 = 0$ e) $2x^2 - 11x + 15 = 0$ f) $5x^2 + 24x - 5 = 0$
g) $3x^2 + 2x - 8 = 0$ h) $3x^2 + 17x + 20 = 0$ i) $5x^2 - 13x + 6 = 0$
j) $6x^2 - x - 5 = 0$ k) $4x^2 - 8x + 3 = 0$ l) $4x^2 - 8x - 5 = 0$
m) $6x^2 - x - 1 = 0$ n) $15x^2 + 14x - 8 = 0$ o) $12x^2 + 7x - 12 = 0$.

88 The Quadratic Formula

Reminder

Not every quadratic equation involves a quadratic function which can be factorised. In fact not every quadratic equation has (real) roots.
The *quadratic formula* can be used to solve any quadratic equation whether factorisation is possible or not.

$ax^2 + bx + c = 0$ has the two solutions (roots) $x = \dfrac{-b \pm \sqrt{b^2 - 4ac}}{2a}$.

Reminder continued

Example Solve these equations using the quadratic formula:

a) $3x^2 - 5x - 7 = 0$

b) $3x^2 - 5x + 7 = 0$.

Solution

a) Compare $3x^2 - 5x - 7 = 0$
with $ax^2 + bx + c = 0$;
then $a = 3 \quad b = -5 \quad c = -7$

$$\text{so } x = \frac{-b \pm \sqrt{b^2 - 4ac}}{2a} = \frac{-(-5) \pm \sqrt{(-5)^2 - 4(3)(-7)}}{2 \times 3}$$

$$= \frac{5 \pm \sqrt{25 + 84}}{6} = \frac{5 \pm \sqrt{109}}{6}$$

i.e. the exact solutions of this equation are $\frac{1}{6}(5 + \sqrt{109})$ and $\frac{1}{6}(5 - \sqrt{109})$

Note that these roots are conjugate surds.
It is more common to be asked to evaluate the roots to some specific degree of accuracy, e.g. to 3 decimal places, in which case $x = 2{\cdot}573, -0{\cdot}907$.

b) Compare $3x^2 - 5x + 7 = 0$
with $ax^2 + bx + c = 0$;
then $a = 3 \quad b = -5 \quad c = 7$

$$\text{so } x = \frac{-b \pm \sqrt{b^2 - 4ac}}{2a} = \frac{-(-5) \pm \sqrt{(-5)^2 - 4(3)(7)}}{2 \times 3}$$

$$= \frac{5 \pm \sqrt{25 - 84}}{6} = \frac{5 \pm \sqrt{-59}}{6}$$

but the square root of a negative quantity does not exist, so this equation has no (real) solutions.

Exercise 88

1 Solve these quadratic equations correct to 3 decimal places, where possible, by using the quadratic formula:

a) $2x^2 - 5x + 3 = 0$
b) $3x^2 + 7x + 4 = 0$
c) $x^2 + 2x - 5 = 0$
d) $x^2 - 3x + 1 = 0$
e) $2x^2 + 7x - 1 = 0$
f) $4x^2 + 5x - 2 = 0$
g) $x^2 + 7x - 6 = 0$
h) $x^2 - 7x + 6 = 0$
i) $3x^2 - 4x + 7 = 0$
j) $5x^2 - x - 3 = 0$
k) $4x^2 + 3x - 5 = 0$
l) $3x^2 + 4x + 5 = 0$
m) $x^2 - 4x - 5 = 0$
n) $2x^2 + 4x + 3 = 0$
o) $3x^2 - 8x + 1 = 0$.

89 Quadratics in Context

Reminder

Example The diagram shows the graph of $y = (x - 2)^2 - 16$.

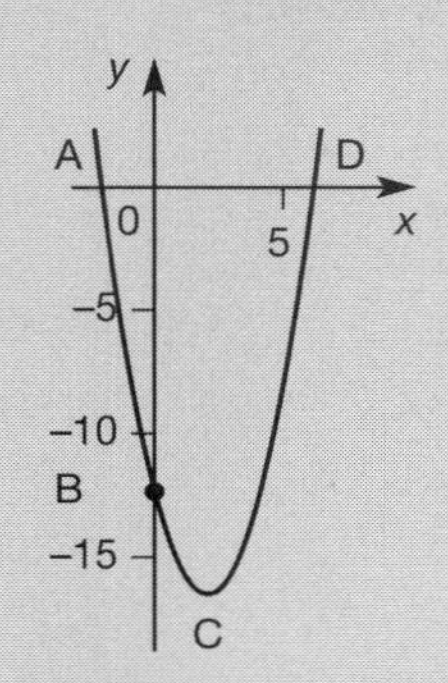

Find the coordinates of

a) C
b) B
c) A and D.

Solution

a) As before, $(x - 2)^2 - 16$ has a minimum value of -16 when $x = 2$
$\Rightarrow$ C is $(2, -16)$
b) At B, $x = 0$ so $y = (0 - 2)^2 - 16 = -12$ $\Rightarrow$ B is $(0, -12)$
c) At A and D, $y = 0$ i.e. $(x - 2)^2 - 16 = 0$
It is important to recognise this as a quadratic equation, and then solve it.

$$\begin{aligned} & (x-2)^2 - 16 = 0 \\ \Rightarrow\ & x^2 - 4x + 4 - 16 = 0 \\ \Rightarrow\ & x^2 - 4x - 12 = 0 \\ \Rightarrow\ & (x+2)(x-6) = 0 \\ \Rightarrow\ & x = -2, 6 \end{aligned}$$

hence A is $(-2, 0)$ and D $(6, 0)$

Example

a) Factorise $16 + 6x - x^2$.
b) Write down the roots of $16 + 6x - x^2 = 0$.
c) The graph of $y = 16 + 6x - x^2$ crosses the x-axis at A and B. Find the length AB.
d) Find the coordinates of the turning point and determine its nature.

Solution

a) Look for factors of 16 with a difference of 6,
hence $16 + 6x - x^2 = (8 - x)(2 + x)$
b) $16 + 6x - x^2 = 0 \Rightarrow (8 - x)(2 + x) = 0 \Rightarrow x = -2, 8$
c) A and B are $(-2, 0)$ and $(8, 0)$, so AB = 10
d) The axis of symmetry is halfway between the roots, and so has equation $x = 3$;
$x = 3 \Rightarrow y = 16 + 18 - 9 = 25$, so the turning point is $(3, 25)$
Since the turning point lies above A and B, it is a maximum turning point

Exercise 89

1 a) Factorise $x^2 - 4x - 5$.
b) Solve $x^2 - 4x - 5 = 0$.
c) Find where the graph of $y = x^2 - 4x - 5$ cuts the x-axis.
d) Find the coordinates of the turning point and determine its nature.

2 This diagram shows the graph of $y = 4 - (x - 1)^2$.
Find the coordinates of

a) C
b) B
c) A and D.

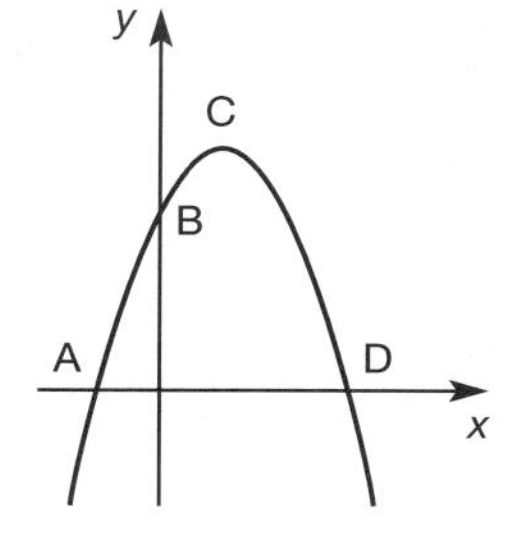

3 This diagram shows the graph of $y = 8 - 2x - x^2$.
PQ and KL are parallel to the x-axis.
The equation of KL is $y = -7$.
Find the coordinates of
a) Q
b) P
c) K and L.

4 A satellite dish is laid on a workbench.
The cross-section of the inner surface is a parabola with equation $y = 0.13\,x^2$ relative to the axes shown.
If the diameter of the circular rim of the dish is 2 metres, calculate the depth of the dish.

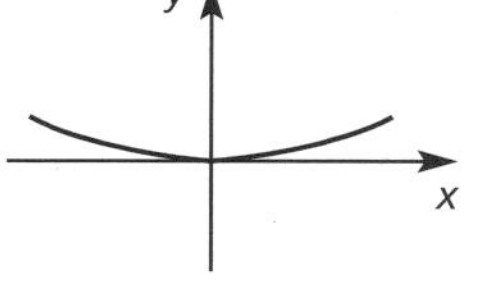

5 A 0·22 calibre bullet is in the shape of a cylinder with a paraboloid nose.
The bullet is placed on a table resting on its circular base.
With axes taken as shown, the equation of the parabolic cross-section of the nose is $y = 0{\cdot}363 - 10(x - 0{\cdot}11)^2$.
(Sizes are in inches.)

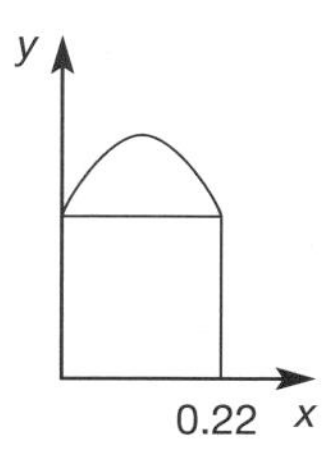

a) Calculate the overall height of the bullet.
b) Calculate the height of the cylindrical part of the bullet.
c) Show that the height of the nose is half the height of the cylinder.

6B A fireman is at the top of a turntable ladder directing a jet of water on to a blazing building. The trajectory of the water jet is a parabola with equation $y = 20 - (x - 20)^2$ relative to axes with the fireman at the origin. (Sizes are in metres)

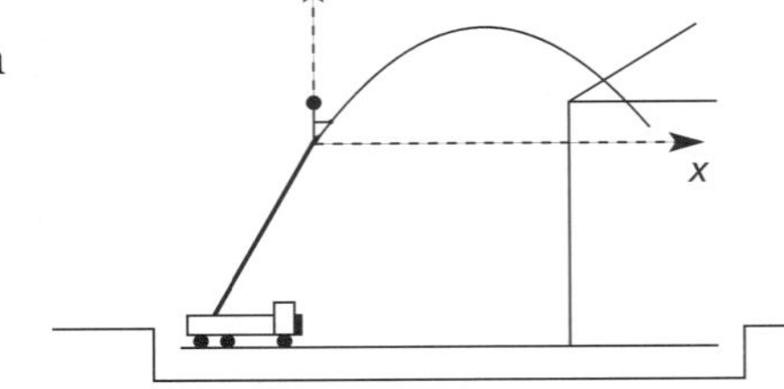

a) How far does the jet of water travel horizontally towards the fire before it stops rising and starts to descend?

b) The fireman is 30 m above the road. How high does the water jet reach above the road?

7B A large ship is tied up in port. A cross-section of the hull midway along the ship is in the shape of a parabola with equation $y = \frac{1}{5}(x - 20)^2$ relative to the coordinate axes shown. (Sizes are in metres)

Find

a) the coordinates of the keel in this cross-section.

b) the depth of the keel below the top of the hull.

c) If the line PQ represents a car deck 20m above the keel, find the width of this deck

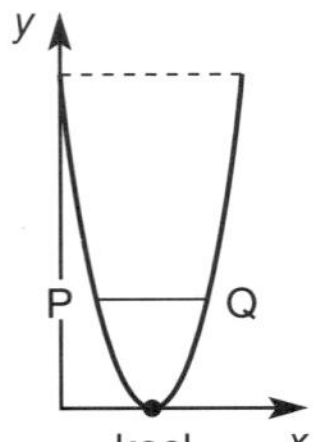

8B In a convent school in Barcelona designed by Antonio Gaudi (1852–1926), the corridor roofs are supported by cast concrete arches. Each archway consists of two congruent parabolas as shown, with the floor level half way up the lower parabola.

a) Write down the equation of

(i) the lower parabola

(ii) the upper parabola

(iii) the floor

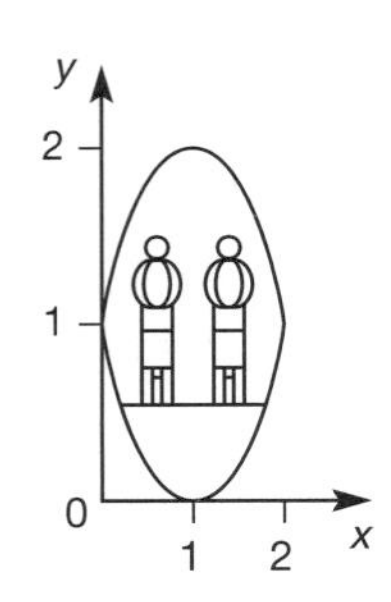

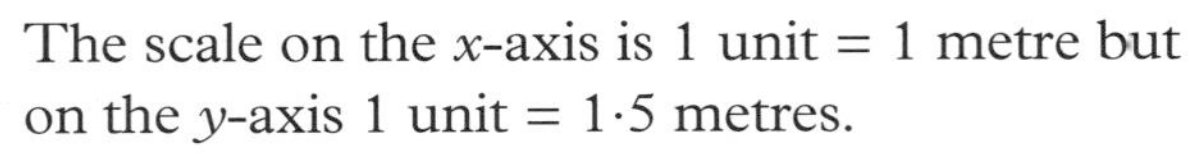
The scale on the x-axis is 1 unit = 1 metre but on the y-axis 1 unit = 1·5 metres.

b) Calculate:

(i) the height of the top of the arch above the floor

(ii) the width of the floor (as a surd).

90 Revision of Quadratic Functions

Exercise 90

1 Each diagram shows the graph of $y = x^2$ and two other parabolas.
Write down the equation of the other two parabolas in each case.

a) (p and q)

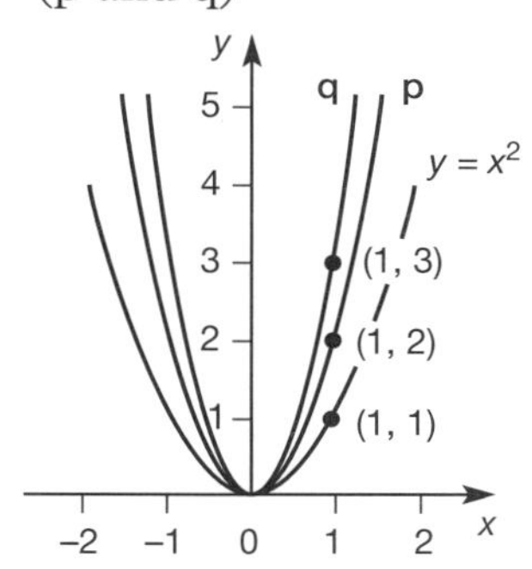

b) (r and s)

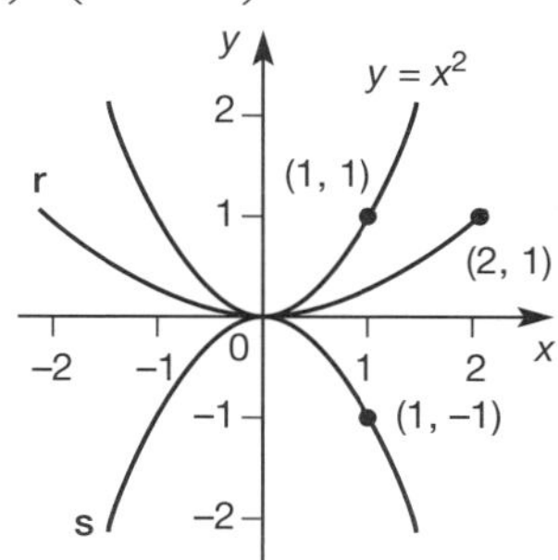

c) (u and t)

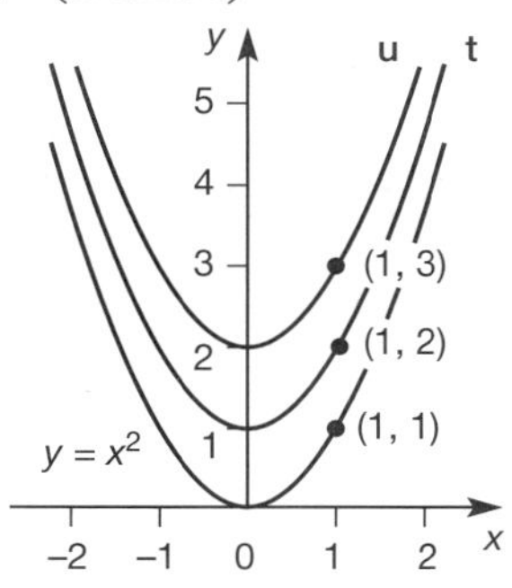

d) (v and w)

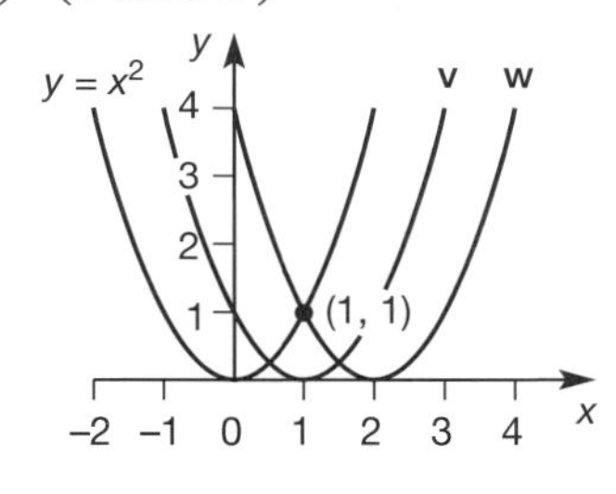

e) (g and h)

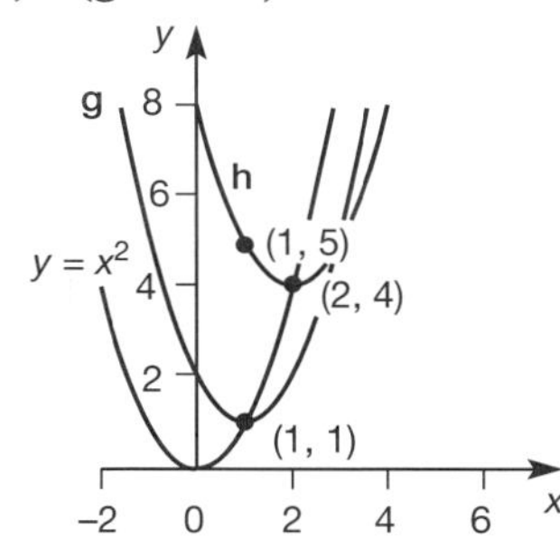

f) (k and m)

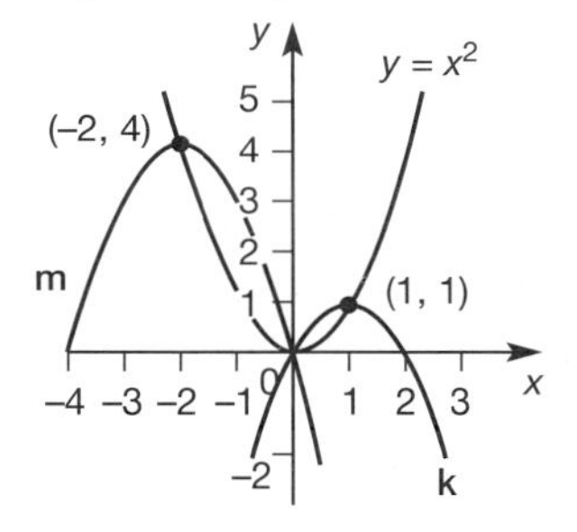

2 Copy and complete:

quadratic function	coordinates of turning point	nature	equation of axis of symmetry
$(x-2)^2 + 3$			
$4 - (x-1)^2$			
$-3x^2$			
$(x-5)^2$			

3 Solve graphically:

a) $x^2 + 2x - 3 = 0$

b) $4x(x-2) = 21$

c) $3x^2 + 2x - 1 = 0$

d) $5x^2 - 8x + 2 = 0.$

4 Solve by factorising:

a) $x^2 - 11x + 30 = 0$

b) $x^2 - 4x - 21 = 0$

c) $x(x+2) = 15$

d) $6x^2 + x - 2 = 0.$

5 Solve, correct to 3 decimal places, by the quadratic formula:

a) $x^2 - 24x + 143 = 0$ b) $6x^2 - 11x - 35 = 0$

c) $4x^2 + 5x - 2 = 0$ d) $3x^2 - 11x + 7 = 0$.

6B Sketch the graphs of:

a) $y = (x - 2)^2 + 1$ b) $y = (x + 3)^2 - 2$

c) $y = 5x^2$ d) $y = -2x^2$

e) $y = 4 - (x - 1)^2$.

91 Test on Quadratic Functions

Allow 45 minutes for this test

1 Find the equation of each quadratic graph shown:

a)
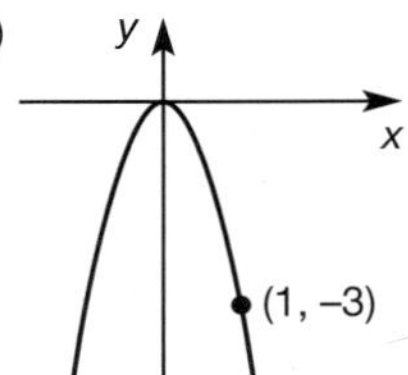

b)
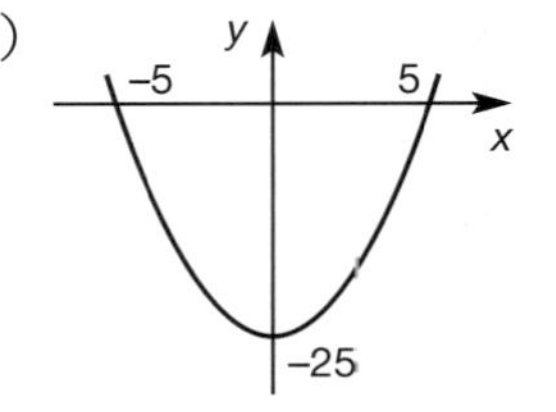

c)
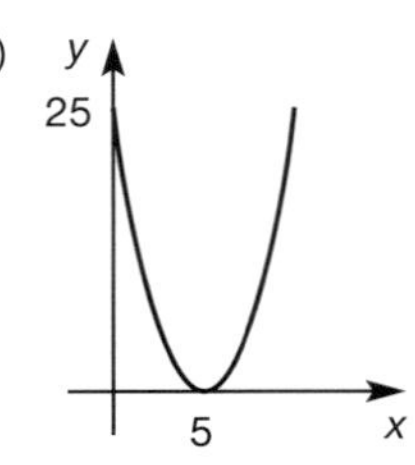

d)
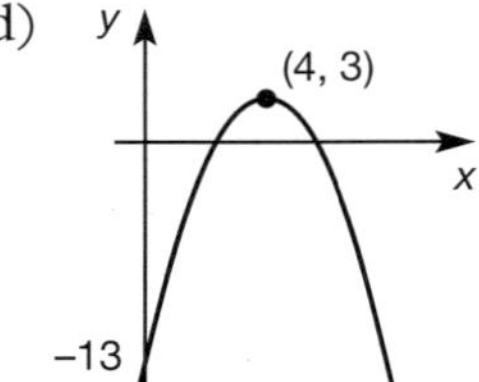

e)
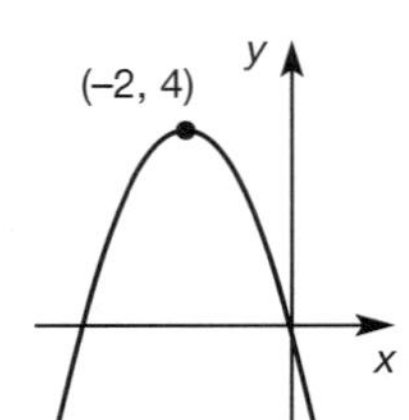

f)
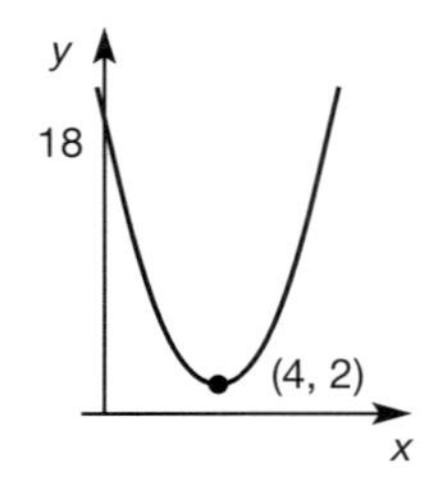

2 For each of the following parabolas write down:

(i) the coordinates of the turning point

(ii) the nature of the turning point

(iii) the equation of the axis of symmetry.

a) $y = (x - 2)^2 + 5$ b) $y = 7 - (x - 4)^2$ c) $y = 9 - (x - 1)^2$.

3 Solve graphically:

a) $5 - x^2 = 0$ b) $2x^2 + 3x = 0$ c) $3x^2 - 5x + 1 = 0$.

4 Solve by factorising:

a) $x^2 - 12x = 0$ b) $6x^2 - 17x + 12 = 0$ c) $2(x^2 + 6) = 11x$.

5 Solve, correct to 2 decimal places:

a) $x^2 - 4x + 2 = 0$ b) $x^2 - 5x + 1 = 0$ c) $2x^2 + 6x + 3 = 0$.

6B Sketch the parabola which has equation:

a) $y = \frac{1}{2}x^2$ b) $y = (x - 3)^2 + 2$.

FURTHER TRIGONOMETRY

92 Related Graphs 1 [$y = kf(x)$]

Reminder

Earlier we saw the graphs of $\sin x°$, $\cos x°$ and $\tan x°$ for the interval $0 \leqslant x \leqslant 360$:

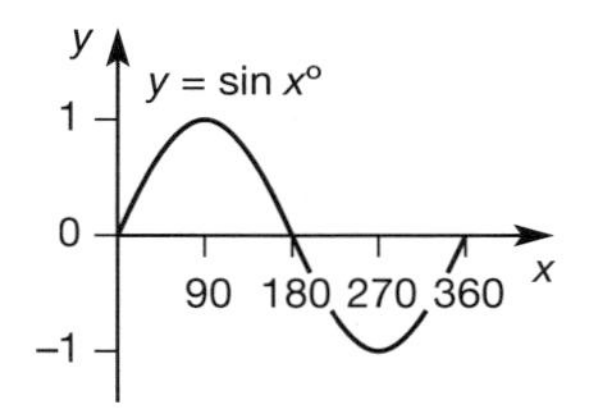

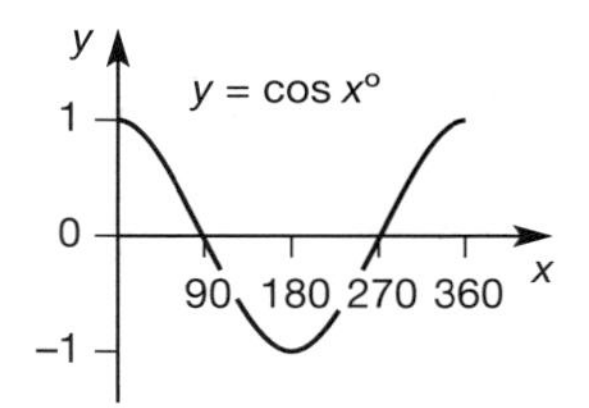

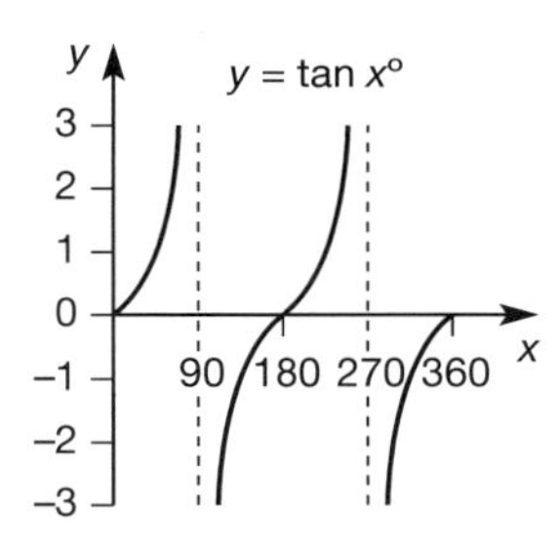

We need to make use of these graphs again in this section.

In Exercise 84 question 3, we noted that when the line with equation $y = x$ and the parabola with equation $y = x^2$ were subjected to a 'stretch' parallel to the y-axis with a scale factor of say 3, they became the line with equation $y = 3x$ and the parabola with equation $y = 3x^2$.

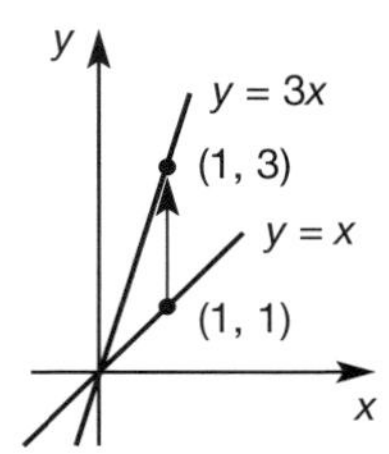

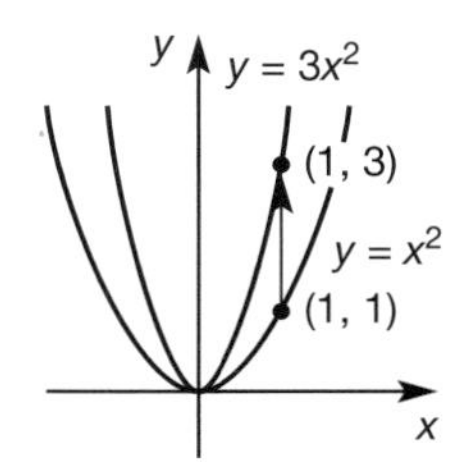

Similarly, under this same 'stretch', the graph of $y = \sin x°$ would become $y = 3\sin x°$:

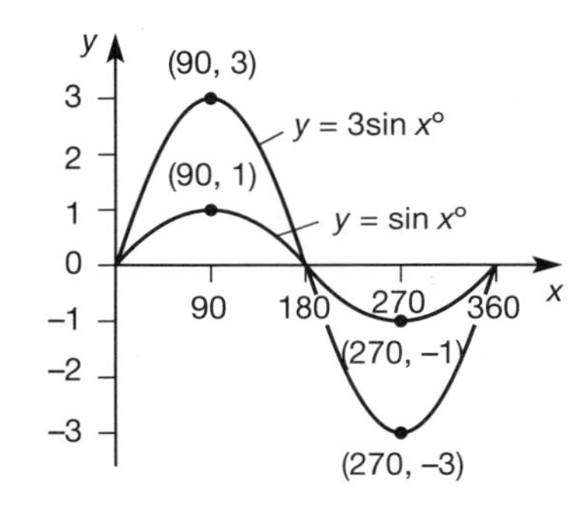

Reminder continued

Example Sketch, on the same diagram, and in the range $0 \leqslant x \leqslant 360$, the graphs which have equations

a) $y = \cos x°$

b) $y = 2\cos x°$.

Solution

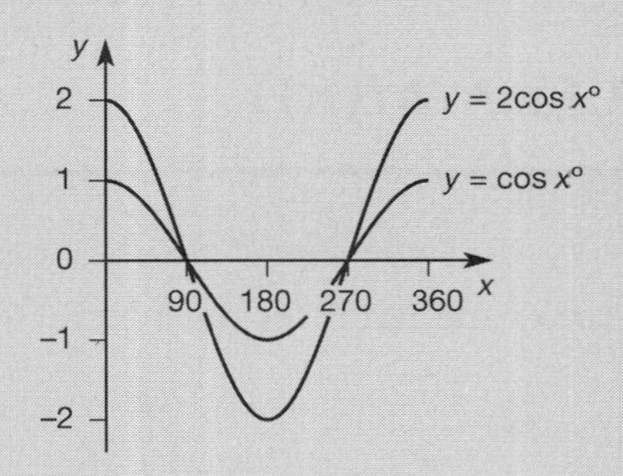

Notice how $y = \cos x°$ and $y = 2\cos x°$ have the same zeros, since $2 \times 0 = 0$.

Exercise 92

1 a) Sketch, on the same diagram, the lines with equations:
(i) $y = x$ (ii) $y = 2x$ indicating the points where $x = 1$.

b) Sketch, on the same diagram, the parabolas with equations:
(i) $y = x^2$ (ii) $y = 2x^2$ indicating the points where $x = 1$.

c) Sketch, on the same diagram, and in the range $0 \leqslant x \leqslant 360$, the trig. functions with equations:
(i) $y = \sin x°$ (ii) $y = 2\sin x°$ indicating all the zeros and turning points.

2 Sketch, on the same diagram, and in the range $0 \leqslant x \leqslant 360$, the graphs with equations:
(i) $y = \cos x°$ (ii) $y = 3\cos x°$ indicating all the zeros and turning points.

3 Sketch, on the same diagram, and in the range $0 \leqslant x \leqslant 360$, the graphs with equations:
(i) $y = \tan x°$ (ii) $y = 2\tan x°$.

4 Sketch, on the same diagram, and in the range $0 \leqslant x \leqslant 360$, the graphs with equations:
(i) $y = \sin x°$ (ii) $y = -\sin x°$ indicating all the zeros and turning points.

5 Sketch, on the same diagram, and in the range $0 \leqslant x \leqslant 360$, the graphs with equations:
(i) $y = \cos x°$ (ii) $y = -2\cos x°$ indicating all the zeros and turning points.

6 State the values of a and b in each of these sketch graphs:

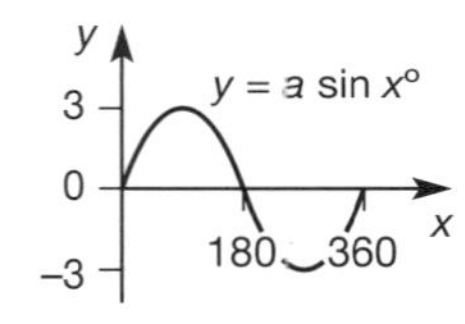

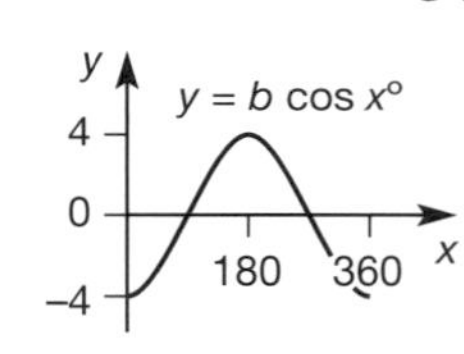

93 Related Graphs 2 [$y = f(x) + k$]

Reminder

In Exercise 84 question 5 we noted that under a translation of 2 units upwards, the line with equation $y = x$ became the line with equation $y = x + 2$, and the parabola with equation $y = x^2$ became the parabola with equation $y = x^2 + 2$.

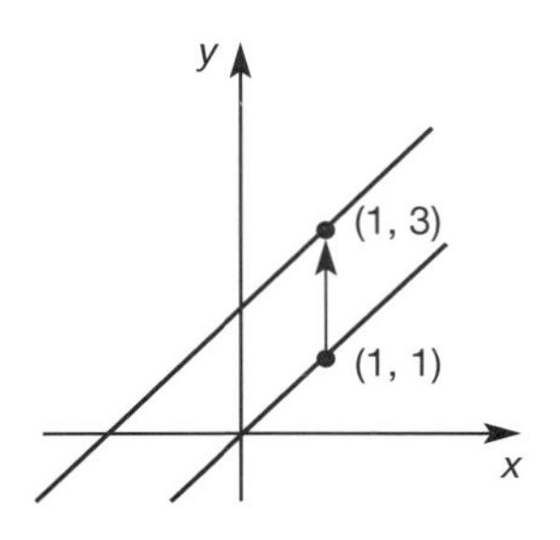

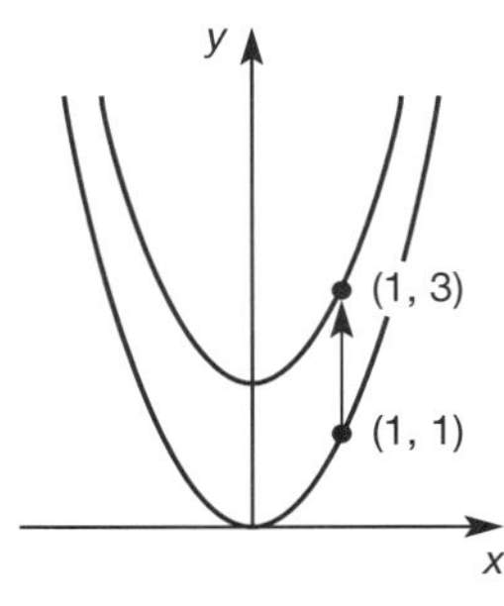

Similarly, under this same translation, the graph of $y = \sin x°$ would become $y = \sin x° + 2$:

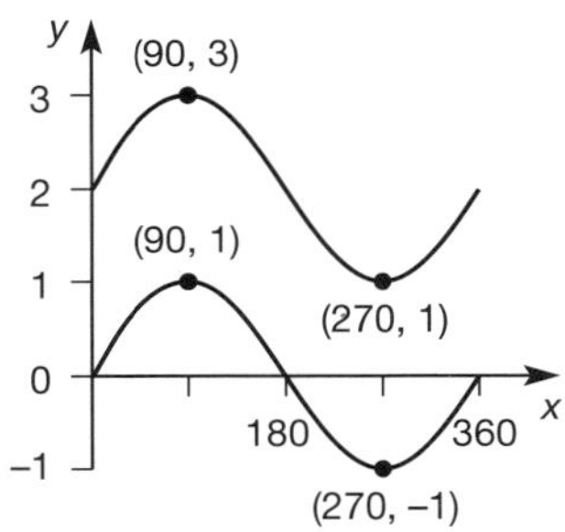

Example Sketch on the same diagram, and in the range $0 \leqslant x \leqslant 360$, the graphs with equations:

a) $y = \cos x°$

b) $y = 1 + \cos x°$.

Solution This involves a translation of 1 unit up the y-axis.

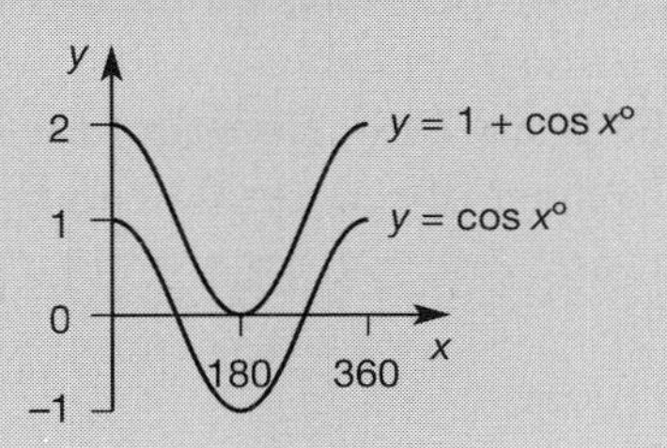

Exercise 93

1 a) Sketch, on the same diagram, the lines with equations:
(i) $y = x$ (ii) $y = x + 3$ indicating the points where $x = 1$.

b) Sketch, on the same diagram, the parabolas with equations:
(i) $y = x^2$ (ii) $y = x^2 + 3$ indicating the points where $x = 1$.

c) Sketch, on the same diagram, and in the range $0 \leqslant x \leqslant 360$, the trig. functions with equations:
(i) $y = \cos x°$ (ii) $y = \cos x° + 3$ indicating all the turning points.

2 Sketch, on the same diagram, and in the range $0 \leqslant x \leqslant 360$, the graphs with equations:
(i) $y = \sin x°$ (ii) $y = \sin x° + 1$ indicating all the zeros and turning points.

3 Sketch, on the same diagram, and in the range $0 \leqslant x \leqslant 360$, the graphs with equations:
(i) $y = \tan x°$ (ii) $y = \tan x° + 1$.

4 Sketch, on the same diagram, and in the range $0 \leqslant x \leqslant 360$, the graphs with equations:
(i) $y = \sin x°$ (ii) $y = \sin x° - 1$ indicating all the zeros and turning points.

5 Sketch, on the same diagram, and in the range $0 \leqslant x \leqslant 360$, the graphs with equations:
(i) $y = \cos x°$ (ii) $y = \cos x° - 2$ indicating all the turning points.

6 State the values of a and b in each of these sketch graphs:

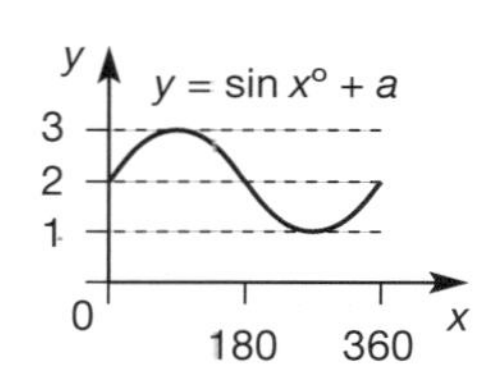

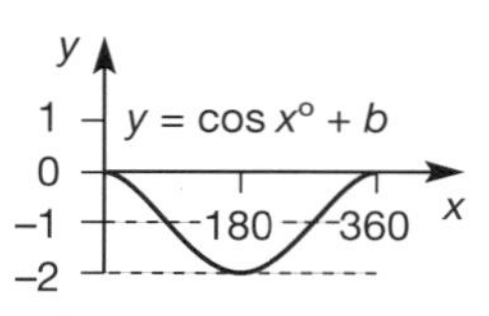

94 Related Graphs 3 [periodicity and $y = f(kx)$]

Reminder

In trigonometry, angles are measured by a line (with one end fixed at the origin) which rotates from the positive x-axis in an anticlockwise direction. This line will have the same finishing position after rotating through 45°, 405°, or 765° (i.e. 45°, 360° + 45°, or 2×360° + 45°).
Hence sin 45° = sin 405° = sin 765°. Similarly for their cosines and tangents. The graphs we saw at the start of section 92 can therefore be repeated indefinitely to both right and left, e.g. for $y = \sin x°$:

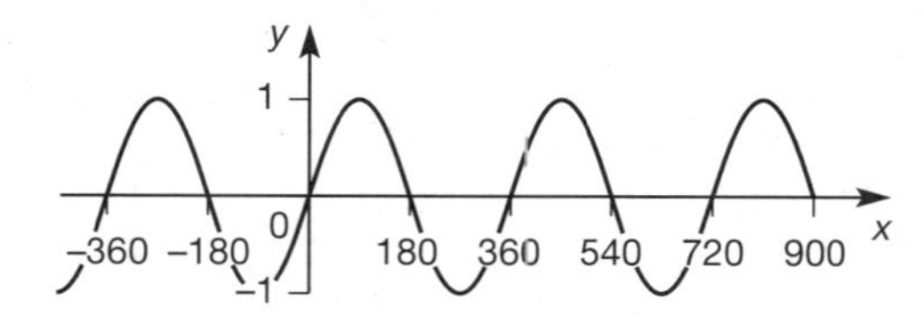

Reminder continued

Since this graph repeats itself every 360°, we say that it has a period of 360°.
Note that 360 is the shortest 'wavelength' which repeats.
Imagine the y-axis moving to where the line $x = 90$ currently is, and you obtain a cosine graph. Thus the cosine graph also has a period of 360°.
The graph of $y = \tan x°$ repeats every 180°, so it has a period of 180°.
We shall mention *periodicity* later.

There are two ways of considering the connection between the graphs of $y = x$ and $y = 3x$. To obtain $y = 3x$ from $y = x$ we previously used a stretching scale factor 3, parallel to the y-axis. Equally good is a reduction by a scale factor of $\frac{1}{3}$ parallel to the x-axis.

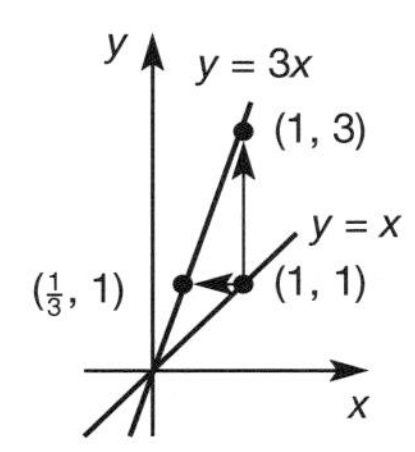

Thus replacing x by $3x$ creates a reduction scale factor of $\frac{1}{3}$, parallel to the x-axis, and so in the case of a trig. function, reduces the period to $\frac{1}{3}$.
e.g. consider $y = \sin(3x)$:
if $x = 60$ then $y = \sin(3 \times 60)° = \sin 180° = 0$;
if $x = 120$ then $y = \sin(3 \times 120)° = \sin 360° = 0$
so that the period is now 120°

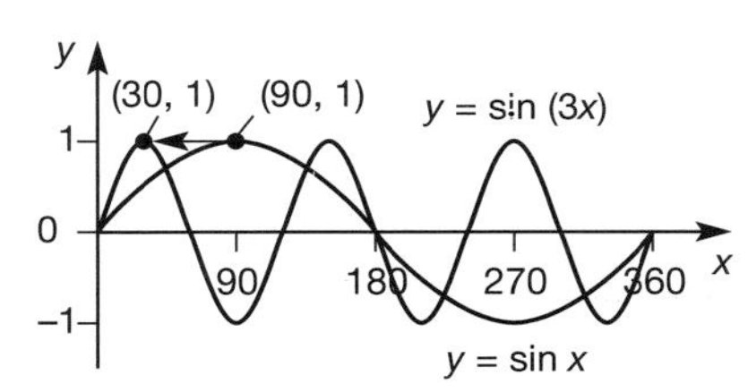

Example

a) State the period of each of these functions:
(i) $y = \cos x°$ (ii) $y = \cos(3x)°$ (iii) $y = -\cos x°$

b) Sketch the graphs of these functions (in the range $0 \leqslant x \leqslant 360$):
(i) $y = \cos 3x°$ (ii) $y = 2\cos 3x°$ (iii) $y = -\cos 3x°$.

Solution

a) (i) 360° (This is a fact that you must remember.)
(ii) 120° (Replacing x by $3x$ reduces the period to $\frac{1}{3}$.)
(iii) 360° (The minus sign has no effect on the period.)

b) (i) we need 3 cycles between 0 and 360
(ii) stretch (×2) the graph to (i)
(iii) the minus sign produces a reflection in the x-axis

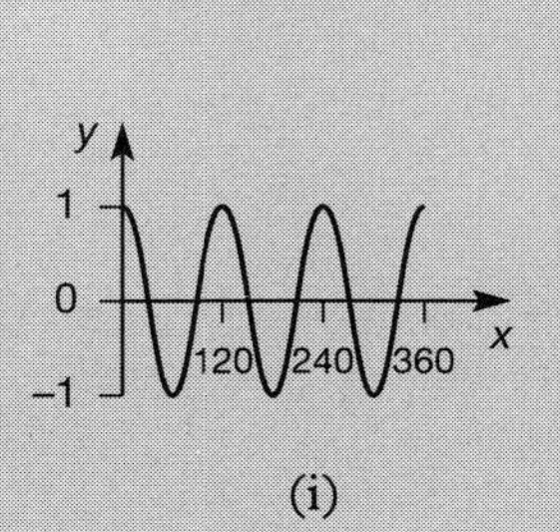

(i)

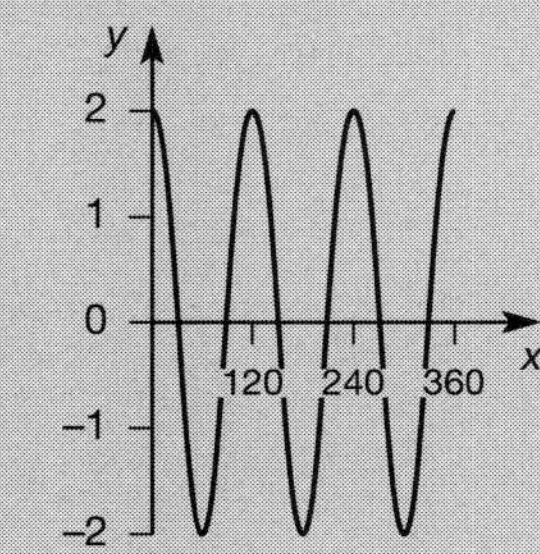

(ii)

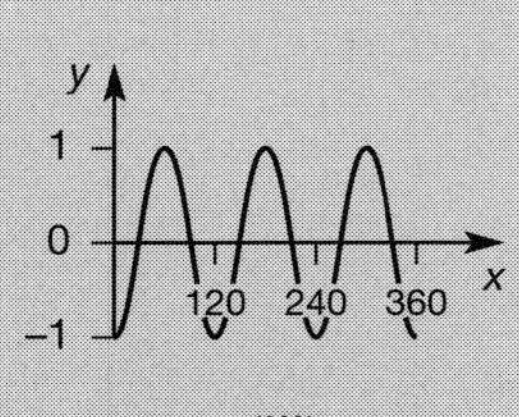

(iii)

Exercise 94

1 State the period of each function:

a) $y = \sin x°$ b) $y = \sin 2x°$ c) $y = \sin 5x°$

d) $y = \cos x°$ e) $y = \cos 4x°$ f) $y = -\cos 2x°$

g) $y = \tan x°$ h) $y = \tan 3x°$ i) $y = \tan (-x)°$.

2 Sketch the graphs of these functions in the interval $0 \leqslant x \leqslant 360$ indicating the coordinates of the turning points:

a) $y = \sin 2x°$ b) $y = 3 \sin 2x°$ c) $y = -\sin 2x°$.

3 Sketch the graphs of these functions in the interval $0 \leqslant x \leqslant 360$ indicating the coordinates of the turning points:

a) $y = \cos 2x°$ b) $y = 3 \cos 2x°$ c) $y = -\cos 2x°$.

4 Sketch the graphs of these functions in the interval $0 \leqslant x \leqslant 360$ indicating the intersections with the coordinate axes:

a) $y = \tan 2x°$ b) $y = 2 \tan 2x°$ c) $y = -\tan 2x°$.

5 Sketch the graphs of these functions in the interval $0 \leqslant x \leqslant 360$:

a) $y = \sin \frac{1}{2}x°$ b) $y = \cos \frac{1}{3}x°$ c) $y = \tan \frac{1}{4}x°$.

6 Find the values of a and b for these graphs:

$(y = \sin (ax)°)$

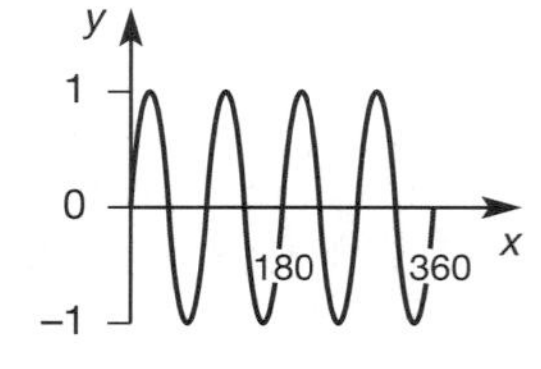

$(y = \cos (bx)°)$

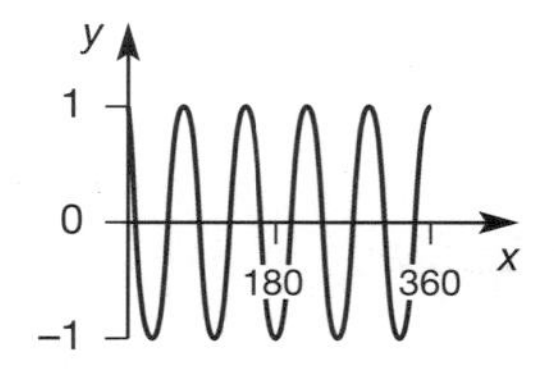

7 Find the values of p, q, r, s, u, v and w for these graphs:

$y = p \sin (qx)°$

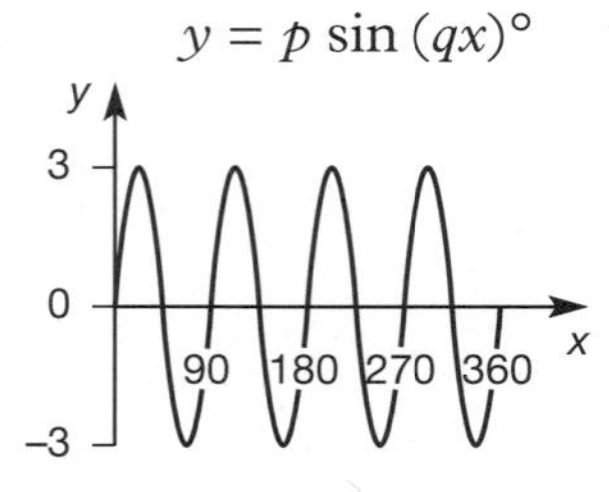

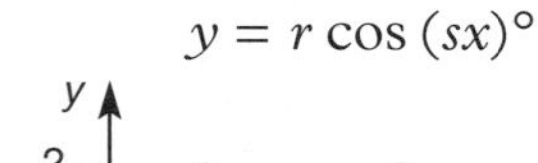

$y = r \cos (sx)°$

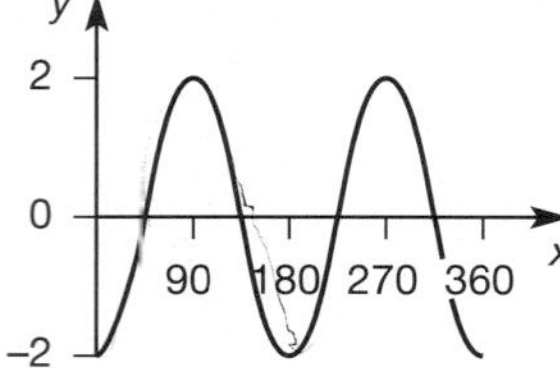

$y = 4 \tan (ux)°$

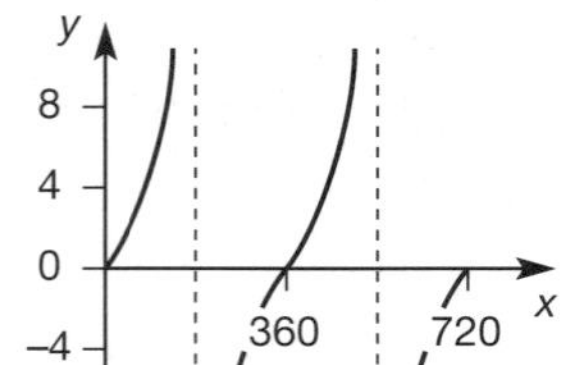

$y = v \sin (wx)°$

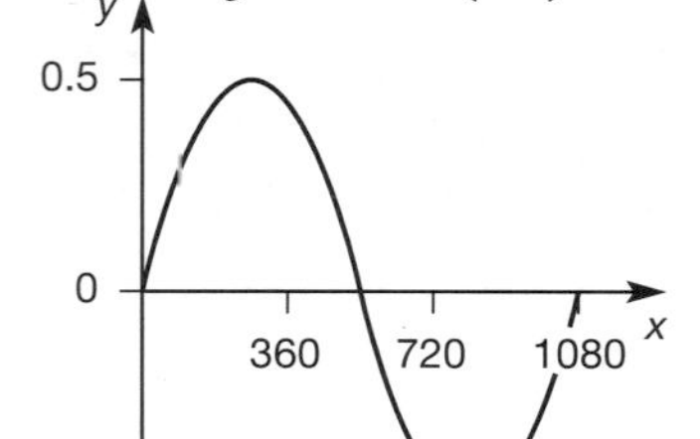

95 Related Graphs 4 [$y = f(x - k)$]

Reminder

In Exercise 84 question 7, we noted that replacing x by $(x - 2)$ produced a translation of 2 units to the right, for example:

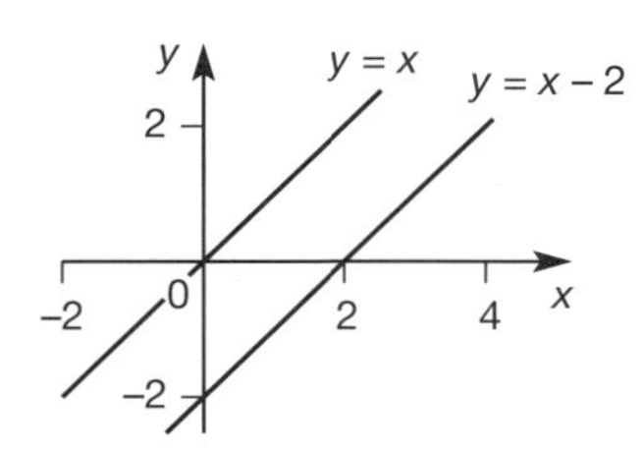

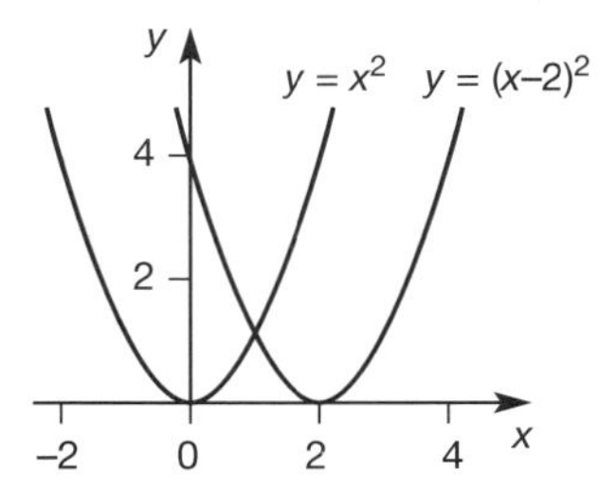

Similarly, replacing x by $(x - 30)$ moves a trig. graph 30 units to the right:

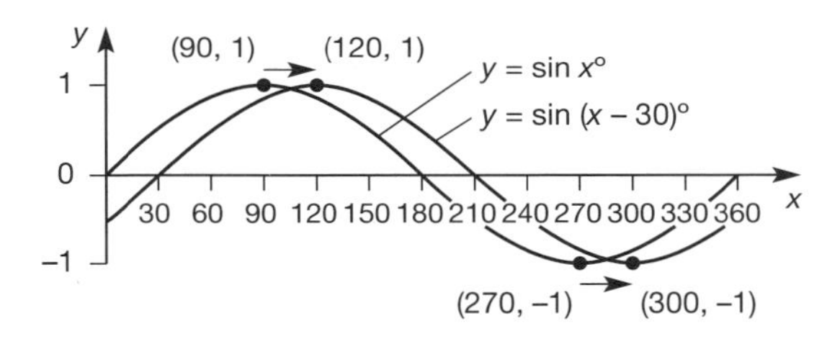

Example Sketch the graph of $y = \sin(x + 30)°$ for $0 \leqslant x \leqslant 360$.

Solution Replacing the x in $\sin x°$ by $(x + 30)$ moves the graph of $\sin x°$ 30° to the left;

so $(0, 0) \longrightarrow (-30, 0)$, $(90, 1) \longrightarrow (60, 1)$, $(180, 0) \longrightarrow (150, 0)$, $(270, -1), \longrightarrow (240, -1)$, $(360, 0) \longrightarrow (330, 0)$.

hence

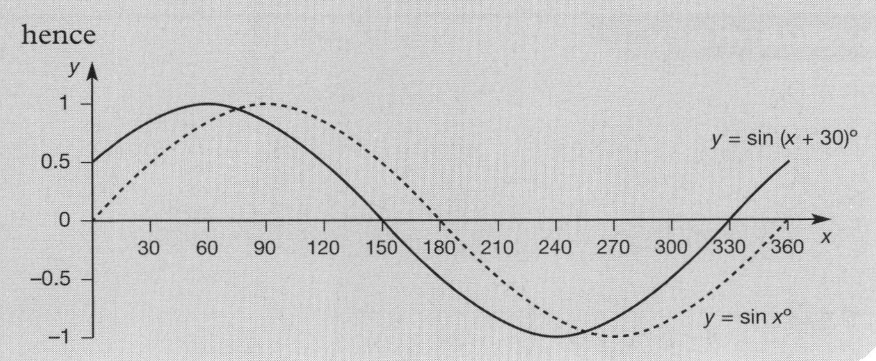

Exercise 95

1B Sketch the graphs of these functions in the range $0 \leqslant x \leqslant 360$ indicating turning points and intersections with axes.

a) $y = \sin(x - 60)°$
b) $y = \sin(x + 60)°$
c) $y = \cos(x - 90)°$
d) $y = \cos(x - 120)°$
e) $y = \cos(x + 60)°$
f) $y = \tan(x - 45)°$
g) $y = \tan(x + 90)°$
h) $y = \sin(x - 180)°$.

2B Find values for p, q, r and s $(-360 \leqslant p, q, r, s \leqslant 360)$ by studying the following graphs:

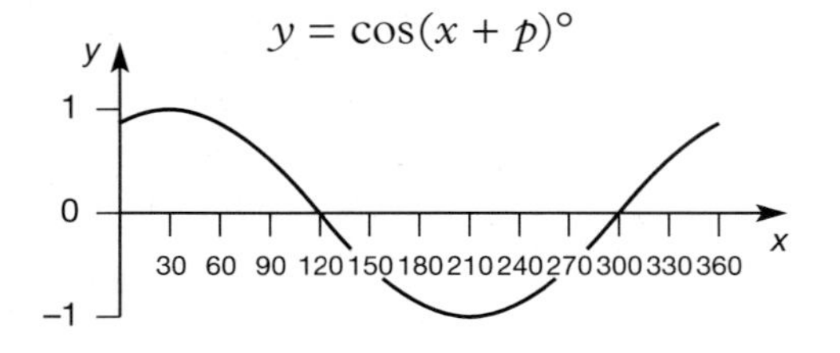

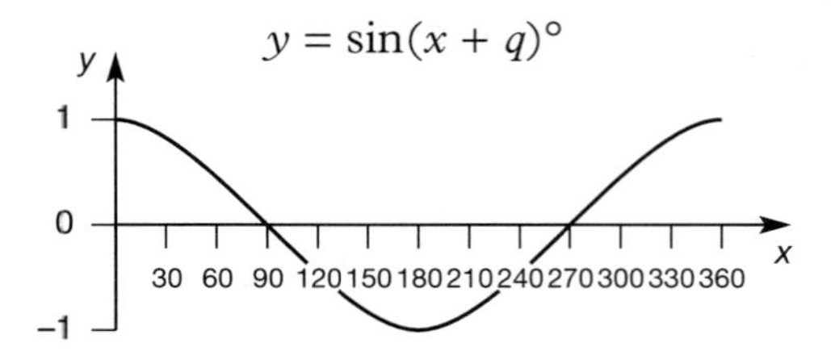

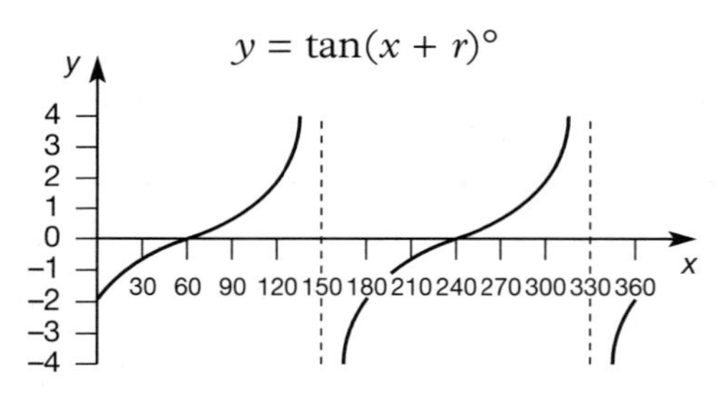

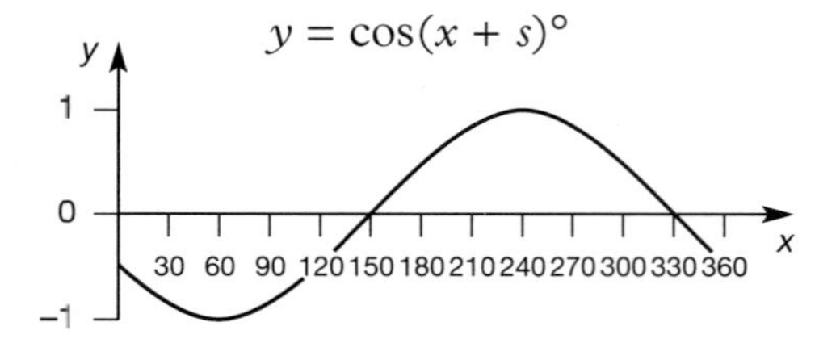

3B Sketch the graphs of these functions in the range $0 \leqslant x \leqslant 360$ indicating turning points and intersections with axes:

a) $2\cos(x + 30)°$
b) $3\sin(x - 60)°$
c) $-\tan(x - 45)°$
d) $1 + \sin 2x°$
e) $1 - \cos 2x°$
f) $2 + 3\sin(x - 20)°$.

Summary

$y = kf(x)$

$y = f(x) + k$

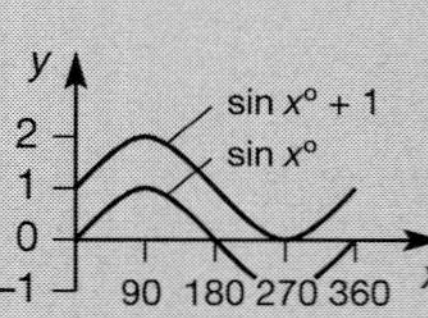

$y = f(kx)$

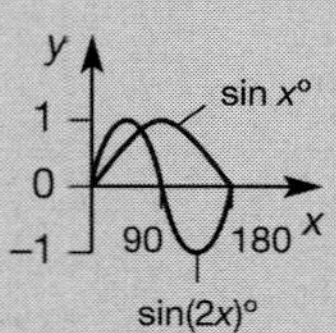

$y = f(x - k)$

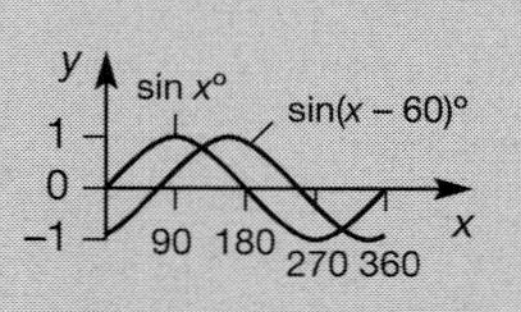

96 Trigonometric Equations

Reminder

You should review the reminders in section 43 before proceeding. In particular the use of the 'all, sin, tan, cos' principle, and the rules for the other quadrants: $180° - A$, $180° + A$, $360° - A$.

[Remember that we are not interested in $90° \pm A$ or $270° \pm A$.]

Example Solve a) $5\cos p° - 4 = 0$ b) $5\sin q° + 2 = 0$ for $0 \leqslant x \leqslant 360$.

Solution a) $5\cos p° - 4 = 0 \Rightarrow 5\cos p° = 4 \Rightarrow \cos p° = \frac{4}{5} = 0{\cdot}8$

the angle whose cosine is 0·8 is 36·9°; (from a calculator)
cosine is positive in the first and fourth quadrants,
so we need x and $360 - x$
i.e. $p = 36{\cdot}9$ or $360 - 36{\cdot}9 \Rightarrow p = 36{\cdot}9, 323{\cdot}1$

b) $5\sin q° + 2 = 0 \Rightarrow 5\sin q° = -2 \Rightarrow \sin q° = -\frac{2}{5} = -0{\cdot}4$

the angle whose sine is 0·4 is 23·6; (ignore the minus initially)
sine is negative in the third and fourth quadrants,
so we need $180 + x$, $360 - x$
i.e. $q = 180 + 23{\cdot}6$ or $360 - 23{\cdot}6 \Rightarrow q = 203{\cdot}6, 336{\cdot}4$

Understanding of the solution of $5\cos p° - 4 = 0$ may be further enhanced by considering a graphical approach. The solutions of this equation, written in the form $5\cos p° = 4$, are the x-coordinates of the points of intersection of the graphs with equations $y = 5\cos p°$ and $y = 4$.

The accuracy of the roots can be improved by zooming in. In practice this would be a time consuming method for the solution of a simple trig equation, but remember that diagrams often provide clarification in mathematics.

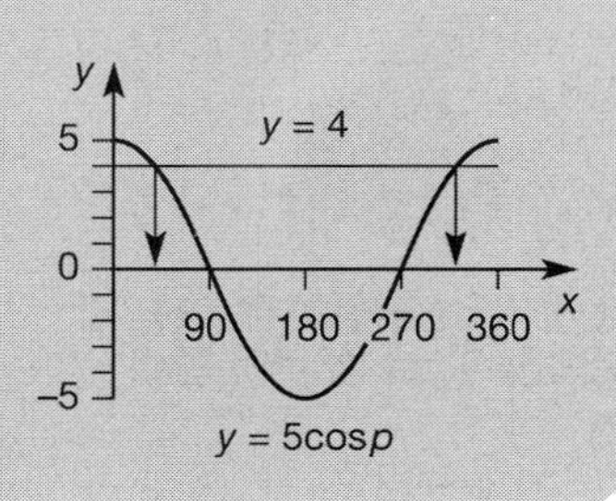

Exercise 96

1 For each of the following equations find two solutions for x, $(0 \leqslant x \leqslant 360)$:

a) $\sin x° = \sin 30°$
b) $\cos x° = \cos 60°$
c) $\tan x° = \tan 45°$
d) $\sin x° = \sin 72°$
e) $\cos x° = \cos 85°$
f) $\tan x° = \tan 21°$
g) $\cos x° = \cos 300°$
h) $\sin x° = \sin 172°$
i) $\tan x° = \tan 200°$
j) $\sin x° = -\sin 20°$
k) $\cos x° = -\cos 50°$
l) $\tan x° = -\tan 80°$
m) $\sin x° = \sin 210°$
n) $\cos x° = \cos 250°$
o) $\tan x° = \tan 100°$
p) $\sin x° = -\sin 200°$
q) $\cos x° = -\cos 128°$
r) $\tan x° = -\tan 230°$
s) $\tan x° = \tan 234°$
t) $\sin x° = \sin 188°$.

2 Solve these equations, correct to 1 decimal place where necessary, for angles between 0° and 360° inclusive:

a) $\tan x° = 1$
b) $\tan x° = -1$
c) $\cos x° = \frac{1}{2}$
d) $\cos x° = -\frac{1}{2}$
e) $\sin x° = \frac{\sqrt{3}}{2}$
f) $\tan x° = -\sqrt{3}$
g) $\sin x° = -\frac{1}{\sqrt{2}}$
h) $\tan y° = \frac{1}{2}$
i) $\cos q° = \frac{1}{4}$
j) $\sin q° = \frac{1}{2}$
k) $\sin k° = \frac{1}{3}$
l) $\cos r° = -\frac{\sqrt{3}}{2}$
m) $\sin s° = -\frac{1}{2}$
n) $\cos t° = -\frac{1}{\sqrt{2}}$
o) $\cos u° = -\frac{1}{4}$
p) $\sin v° = 0{\cdot}729$
q) $\sin w° = 0{\cdot}101$
r) $\cos q° = 0{\cdot}302$
s) $\tan r° = 2{\cdot}75$
t) $\sin s° = -0{\cdot}345$
u) $\cos t° = -0{\cdot}253$
v) $\tan u° = -1{\cdot}5$
w) $\sin v° = -\frac{5}{7}$
x) $\cos w° = \frac{7}{9}$.

3 Solve these equations for angles between 0° and 360° inclusive:

a) $2\sin x° + 1 = 0$
b) $3\cos y° - 2 = 0$
c) $4\tan z° - 5 = 0$
d) $5\sin p° + 3 = 0$
e) $6\cos q° - 5 = 0$
f) $7\tan r° - 2 = 0$
g) $8\sin u° + 7 = 0$
h) $9\cos x° - 5 = 0$
i) $10\tan t° + 11 = 0$
j) $1000\tan v° + 9 = 0$
k) $14\sin x° - 5 = 8$
l) $23\cos y° + 11 = 29$.

97 Trigonometric Identities

Reminder

An example of an identity in algebra is $x^2 - 1 = (x - 1)(x + 1)$.
This 'equation' cannot be solved for x. It is true for every value of x.
In order to stress this, it is sometimes written as $x^2 - 1 \equiv (x - 1)(x + 1)$, meaning that the two sides are not just equal but identically equal.

Reminder continued

Similarly, there are two identities in trigonometry, which must be known:

$\sin^2 A + \cos^2 A = 1$ and $\tan A = \dfrac{\sin A}{\cos A}$ Note: $\sin^2 A$ denotes $(\sin A)^2$

The former is also often used in either of these two equivalent forms:

$\sin^2 A = 1 - \cos^2 A$ or $\cos^2 A = 1 - \sin^2 A$

Example Express $4 + 5\cos^2 x$ in terms of $\sin^2 x$.

Solution Replace $\cos^2 x$ by $1 - \sin^2 x$, so
$4 + 5\cos^2 x = 4 + 5[1 - \sin^2 x] = 4 + 5 - 5\sin^2 x = 9 - 5\sin^2 x$.

Example Prove that $3 + 2\tan^2 x \cos^2 x = 5 - 2\cos^2 x$

Solution If this question had simply been expressed as 'Simplify $3 + 2\tan^2 x \cos^2 x$', there would have been some debate as to what the simplest form was. Asking for the proof of an identity clarifies how much manipulation you are expected to do. In the following exercise, always start with the left hand side (l.h.s.) and manipulate it until you produce the right hand side (r.h.s.).

Many examples which involve $\tan x$ can be solved by replacing $\tan x$ by $\dfrac{\sin x}{\cos x}$ and using the algebra of fractions.

Proof: l.h.s. $= 3 + 2\tan^2 x \cos^2 x$

$= 3 + 2\left(\dfrac{\sin^2 x}{\cos^2 x}\right)\cos^2 x$

$= 3 + 2\sin^2 x = 3 + 2(1 - \cos^2 x) = 3 + 2 - 2\cos^2 x$

$= 5 - 2\cos^2 x =$ r.h.s.

Exercise 97

1 Prove these identities by making use of $\sin^2 A + \cos^2 A = 1$:

a) $3\sin^2 x - 2 = 1 - 3\cos^2 x$

b) $5\cos^2 x - 3 = 2 - 5\sin^2 x$

c) $3\cos^2 x + 4 = 7 - 3\sin^2 x$

d) $4\sin^2 A + 1 = 5 - 4\cos^2 A$

e) $8 - 6\cos^2 A = 6\sin^2 A + 2$

f) $2 - 5\cos^2 \theta = 5\sin^2 \theta - 3$

g) $10 - 7\sin^2 \theta = 7\cos^2 \theta + 3$

h) $\cos^3 x + \sin^2 x \cos x = \cos x$

i) $\sin^3 x + \cos^2 x \sin x = \sin x$.

2 Prove these identities by making use of $\tan A = \dfrac{\sin A}{\cos A}$.

a) $\tan x \cos x = \sin x$

b) $\dfrac{1}{\tan x} = \dfrac{\cos x}{\sin x}$

c) $1 + \tan A = \dfrac{\sin A + \cos A}{\cos A}$

d) $1 + \dfrac{1}{\tan A} = \dfrac{\sin A + \cos A}{\sin A}$

e) $1 - \tan A = \dfrac{\cos A - \sin A}{\cos A}$

f) $1 - \dfrac{1}{\tan A} = \dfrac{\sin A - \cos A}{\sin A}$

g) $1 - \dfrac{1}{\tan^2 A} = \dfrac{\sin^2 A - \cos^2 A}{\sin^2 A}$

h) $\dfrac{\sin A + \cos A}{\tan A} = \cos A + \dfrac{\cos^2 A}{\sin A}$.

3 Establish the following identities $\left(\text{by making use of } \sin^2 A + \cos^2 A = 1 \text{ and } \tan A = \dfrac{\sin A}{\cos A}\right)$

a) $\sin A \cos A \tan A = 1 - \cos^2 A$

b) $\tan A + \dfrac{1}{\tan A} = \dfrac{1}{\sin A \cos A}$

c) $1 + \dfrac{1}{\tan^2 A} = \dfrac{1}{\sin^2 A}$

d) $\dfrac{2}{\tan^2 A} - 1 = \dfrac{2}{\sin^2 A} - 3$

e) $1 + \dfrac{3}{\tan^2 A} = \dfrac{3}{\sin^2 A} - 2$

f) $\sin^2 A \cos^2 A \tan^2 A = \sin^2 A - \cos^2 A + \cos^4 A$.

98 Revision of Further Trigonometry

Exercise 98

1 For the interval $0 \leqslant x \leqslant 360$, sketch these graphs:

a) $y = \sin x°$

b) $y = 2 \sin x°$

c) $y = \sin x° + 2$

d) $y = \sin 2x°$

e) $y = \sin(x - 45)°$

f) $y = 2 \sin(x + 45)°$.

2 For the interval $0 \leqslant x \leqslant 360$, sketch these graphs:

a) $y = \cos x°$

b) $y = 3 \cos x°$

c) $y = \cos x° + 3$

d) $y = \cos 3x°$

e) $y = \cos(x + 45)°$

f) $y = 3 \cos(x - 45)°$.

3 State the values of a, b, c, p, q and r, by referring to these two graphs:

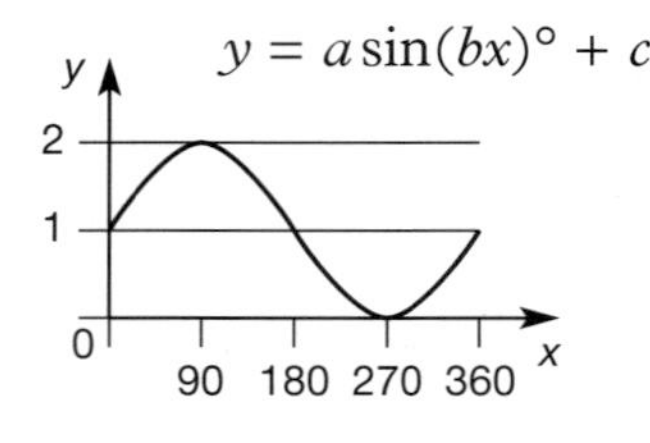

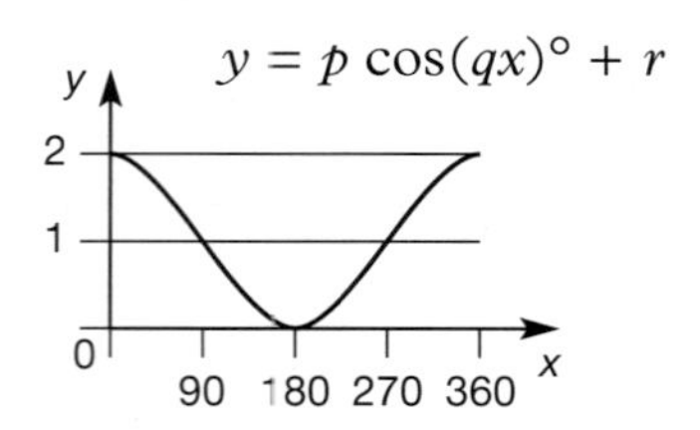

4 State the values of A and B by referring to these two graphs:

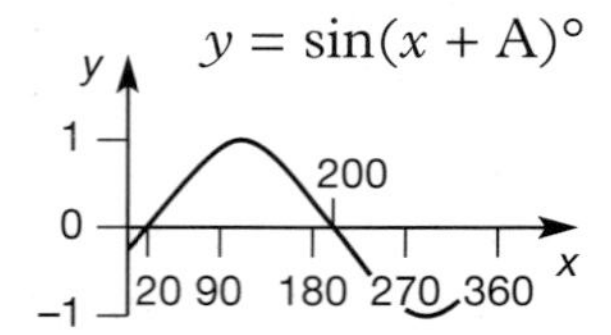

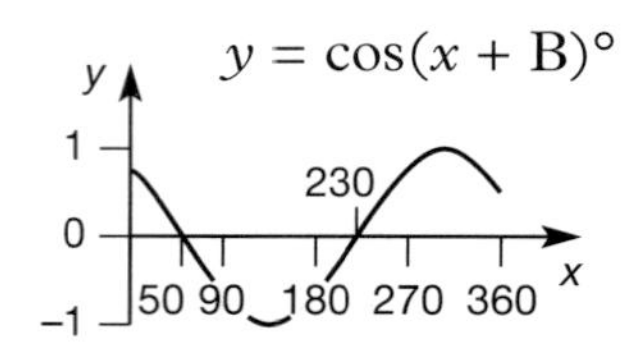

5 State the values of a, b, p, and q, by referring to these two graphs:

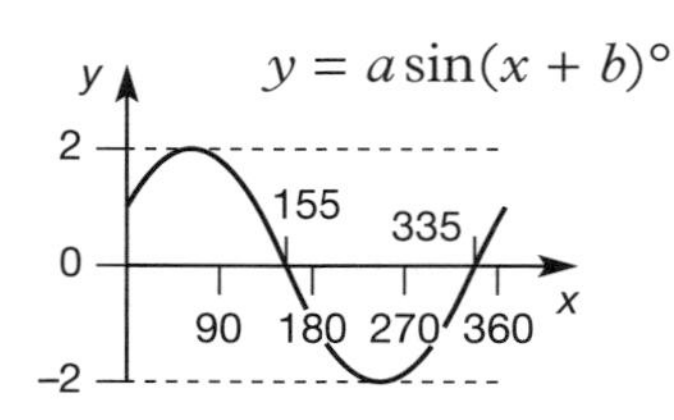

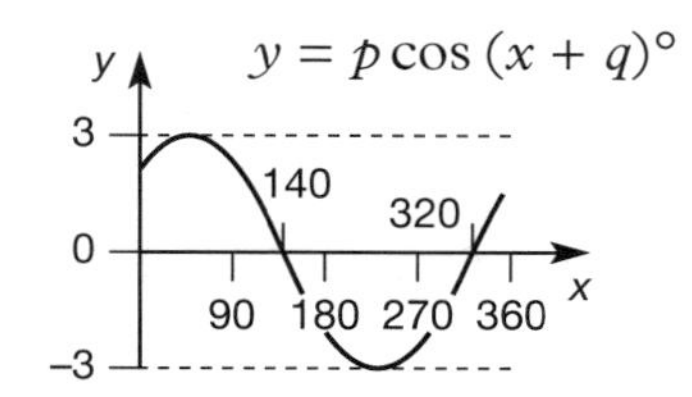

6 State the values of t, u, r, s and q, by referring to these two graphs:

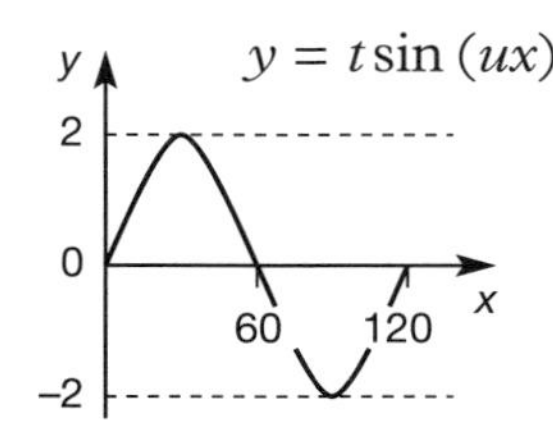

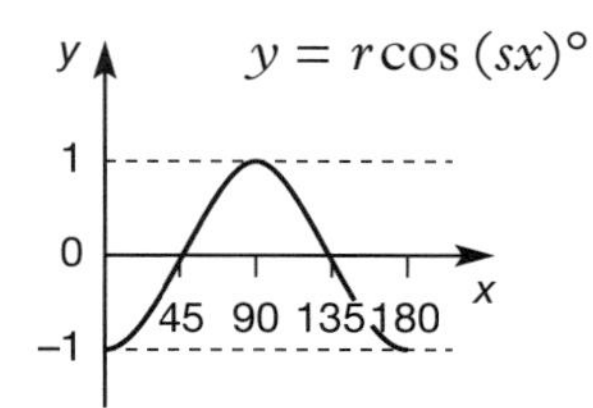

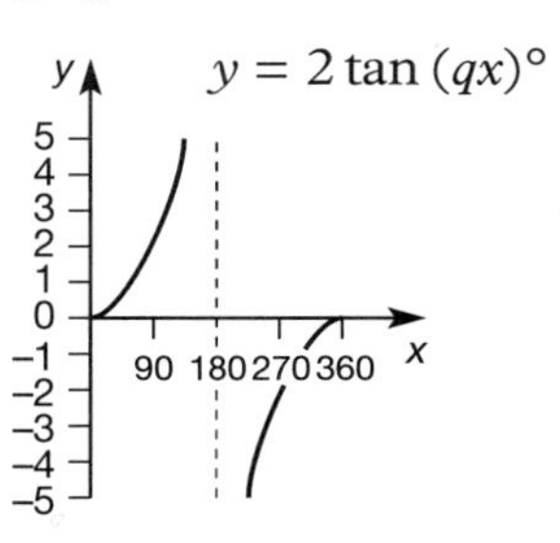

7 Solve, correct to 1 decimal place, for angles between 0 and 360 (inclusive):

a) $\cos x° = -1$ b) $\tan y° = 0$ c) $\sin z° = 1$

d) $\tan p° = \frac{1}{2}$ e) $\cos q° = -\frac{1}{4}$ f) $\sin r° = -\frac{2}{5}$

g) $3 \cos x° - 2 = 0$ h) $\tan y° + 4 = 0$ i) $5 \sin z° + 3 = 0$

j) $3 \sin u° - 5 = 0$ k) $3 \tan y° - 5 = 0$ l) $5 \cos x° = 0$.

8B a) Express $3 - 4\cos^2 x$ in terms of $\sin^2 x$.

b) Express $5\sin^2 y - 3$ in terms of $\cos^2 y$.

c) Express $1 + \tan^2 A$ in terms of $\cos^2 A$.

9B Establish the identity: a) $3 + \tan A = \dfrac{\sin A + 3\cos A}{\cos A}$.

b) $\tan A + \cos A = \dfrac{1 + \sin A - \sin^2 A}{\cos A}$.

10B The height, h metres (above street level) of the floor of a capsule on the 'Edinburgh Eye' is given by $h = 12 + 10\cos(3t)°$ where t is the number of seconds since it passed through the lowest position.

a) Calculate the height of a capsule 30 seconds after it passed through the lowest position.

b) Calculate how long it takes to travel from the position where the capsule is 17 m above street level and rising to the position where the capsule is 17 m above street level and falling.

11B The minute hand of the clock in my study is 6 cm long. The height, h cm, of the tip of this hand above the horizontal line through the centre of the circular clockface is given by $h = 6\sin(90 - 6t)°$ where t is the number of minutes past the hour.

a) Calculate the height after 5 minutes.

b) Find two times between 2 o'clock and 3 o'clock when the the tip of the minute hand is 3 cm below the centre of the clock face.

99 Test on Further Trigonometry

Allow 45 minutes for this test

1 State the period of:
a) $y = \cos 4x°$ b) $y = \sin(3x - 29)°$ c) $y = \tan 5x°$ d) $y = \cos\left(\frac{1}{3}x\right)°$.

2 For the interval $0 \leqslant x \leqslant 360$, sketch the graph of:
a) $y = 2\cos x°$ b) $y = \cos x° + 2$ c) $y = \cos(2x)°$.

3 For the interval $0 \leqslant x \leqslant 360$, sketch the graph of:
a) $y = \sin(x - 90)°$ b) $y = 3\sin(x + 40)°$.

4 State the values of a, b, p, q, s, t and u, by referring to these graphs:

$y = a\sin(bx)°$ $y = p\cos(qx)°$ $y = s\cos(tx)° + u$

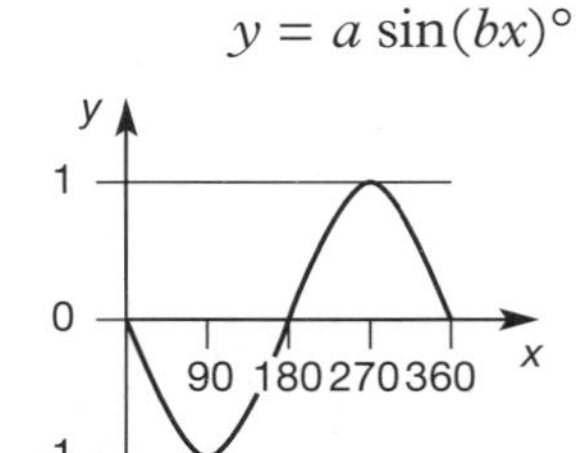

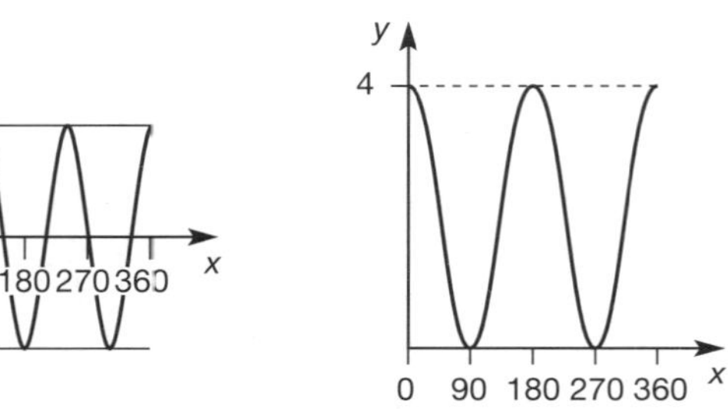

5 Solve, correct to 1 decimal place, for angles in the interval $0 \leqslant x \leqslant 360$:
a) $\sin x° = 0{\cdot}75$ b) $\cos x° + 0{\cdot}23 = 0$ c) $2\tan x° + 2{\cdot}8 = 0$
d) $3\sin x° - 2 = 0$ e) $\cos x° + 2 = 0$.

6B Simplify: a) $6\sin^2 x + 5\cos^2 x$
b) $3 + 4\cos^2 x + 5\sin^2 x$
c) $\sin A \cos^2 A \tan A$.

7B Establish the identity: $\cos A\,(\sin A + \tan A) = \sin A(1 + \cos A)$.

SOCIAL ARITHMETIC

100 Earnings

People in different jobs can be paid in different ways.

Example *Slashers* is a nationwide low-cost grocery chain.

a) Tom is a store manager who earns an annual salary of £27 504. Calculate his gross monthly income.

b) Emily is the supervisor of the delicatessen. She earns a basic annual salary of £15 000 with commission of $2\frac{1}{2}$% of her sales during the previous month.
Calculate her income for a month when the delicatessen had previous month's sales of £40 000.

c) Edith is a till operator who works 7 hours per day for 5 days each week. Her hourly rate is £6·30. Calculate her gross weekly wage.

d) Desmond is a shift worker who is paid 15% more per hour than Edith because of his unsociable hours. If he also works a 35-hour week, calculate his gross annual income (for 52 weeks).

e) Finlay is a maintenance man and is paid £8·25 per hour for a 40-hour week. If he works more than 8 hours in one day, or works on a Saturday morning, he is paid overtime at 'time and a half'. He is paid double time for overtime work on a Saturday afternoon or Sunday. Calculate his gross pay for a full week's work with 2 hours overtime on Tuesday night and from 8 a.m. till 1 p.m. on Saturday.

f) Kevin is a self-employed tiler who has contracted to re-tile the staff toilets for £450. (*Slashers* has already purchased the tiles.) If it takes Kevin 51 hours to do this job, is he earning more or less than Finlay's basic rate?

g) If Kevin had originally estimated that it would take 60 hours to complete this job, and *Slashers* give him a bonus of £50 for completing the work ahead of schedule, what percentage of the total is this bonus paid to Kevin?

Reminder continued

Solution

a) monthly income = annual income ÷ 12 = £27 504 ÷ 12 = £2292

b) basic monthly income = annual income ÷ 12 = £15 000 ÷ 12
= £1250; commission = $2\frac{1}{2}$% of £40 000 = 0·025 × £40 000
= £1000; gross income for the month = £1250 + £1000 = £2250

c) 5 × 7 hours = 35 hours ⇒ gross wage = 35 × £6·30 = £220·50

d) [Remember that an increase of 15% is the same as multiplying by 1·15.] Using part c), £220·50 × 1·15 = £253·58 ⇒
£253·58 × 52 = £13 186·16

e) hours worked: 40 (basic) + (2 × $1\frac{1}{2}$ (Tue)) + (4 × $1\frac{1}{2}$ (Sat a.m.))
+ (1 × 2 (Sat p.m.))
= 40 + 3 + 6 + 2 = 51 ⇒ gross pay = 51 × £8·25 = £420·75

f) £450 ÷ 51 = £8·82 > £8·25 i.e. more than Finlay

g) total paid = £450 + £50 = £500 of which £50 is 10%.

Exercise 100

1 Arthur negotiates a salary of £18 000 with a 5% annual rise. Calculate his gross monthly income in his third year of employment.

2 Teresa was on a salary of £16 500 last year. Her first gross monthly pay this year was £1430. What percentage rise did she receive?

3 Holly and Ivy left college at the same time and obtained different jobs. Holly was paid £14 000 each year for 5 years. Ivy started at £12 500 per annum in her first year with an annual increment of 5%. In total, which girl had earned more money in the first five years of employment, and how much more?

4 David has an annual pension of £4758, and his wife Amanda a weekly pension of £78·60.
Calculate a) David's weekly pension
b) Amanda's annual pension
c) their average total monthly pension.

5 Rob earns £235 per week (for 52 weeks) and Sam earns £1020 per month. Who earns more per year, and how much more?

6 Charlie sells earth-moving equipment. He has a basic salary of £29 000 p.a. plus 1% commission on annual sales. Calculate his salary in a year when he sold £800 000 worth of equipment.

7 Audrey sells furniture. Her basic salary is £15 000 p.a. plus 2% commission on annual sales over £10 000. Calculate her total salary if she sells £50 000 worth of furniture in one year.

8 Honest Harry sells second-hand cars. He has a basic salary of £10 000 p.a. plus commission of 5% of sales. What level of sales must he achieve in a year to double his basic salary?

9 Sandra works in a fashion store. Her annual salary is £12 000 plus commission. Her sales for last month amounted to £8500 and her pay was £1255. Calculate the percentage rate of her commission.

10 Grace and Helen work in a department store. Grace is paid £80 per week plus 4% commission on all her sales. Helen is paid £85 per week plus 3% commission on all her sales.
a) Calculate Grace's pay for a week in which she sold £800 worth of goods.
b) What value of goods would Helen have to sell to earn the same?
c) One week both girls sold the same value of goods and earned the same total wage. Calculate the value of the goods sold and the wage earned. [Use algebra.]

11 Jim is a joiner who is paid £8·35 per hour for a 40-hour week. Calculate his gross annual earnings (for 52 weeks).

A building trade union negotiates the following rates for overtime. These should be used for questions 12 to 14:

the first 8 hours each weekday	basic rate
any work in excess of 8 hours in one day	time and a half
any work done on a Saturday before noon	time and a half
any work done on a Saturday afternoon or on a Sunday	double time

12 Here is Frank's time sheet (Lunch is 12:30 – 1300)

Monday	08:00–12:30	13:00–16:30
Tuesday	08:00–12:30	13:00–16:30
Wednesday	08:00–12:30	13:00–17:30
Thursday	08:00–12:30	13:00–17:30
Friday	08:00–12:30	13:00–18:30
Saturday	08:00–12:30	13:00–15:00

Calculate his gross earnings for the week at £7·50 per hour.

13 In one week Mike worked his basic 40 hours plus 3 hours overtime on Tuesday night. Calculate his gross weekly wage if his hourly rate was £9·36.

14 There was a rush to get a job finished by a certain deadline and the squad of men involved worked a 9-hour day (08:00 till 17:30) all week (Monday to Saturday). If their hourly rate was £7·75, calculate the gross weekly wage for each man.

15 Thomas, Richard, and Harold agreed to do a landscaping job for £820. Thomas worked for 5 days on the job, Richard for 3, and Harold for 2.
a) How much should each receive?
b) If each day involved 8 hours of work, calculate the hourly rate they were charging.

16 A firm of painters repainted the interior of a small supermarket for £1777. The store paid a bonus of £200 for the job being completed ahead of schedule.
a) Express the bonus as a percentage of the original cost of the work.
b) The painters were each paid £7·96 per hour. To what did this rate rise because of the percentage increase received.

17 Three tradesmen completed a job. Arthur charged £8 per hour, Bob £7, and Calum £5. They received a bonus of £150, which they shared in proportion to their rates of pay. Calculate how much bonus each received.

101 Deductions

Reminder

When you earn money, the government charges you *income tax* and *national insurance contributions*. Your employer may also set aside superannuation; a fixed proportion of your earnings saved to give you a pension in addition to the state pension. There is a threshold for paying NIC: £91 per week or £395 per month; then you pay 11% of the rest of your gross income (in NIC). There is an annual threshold for paying tax: £4745 plus any superannuation you may pay (plus any other allowances to which you are entitled); then you pay 10% of the first £2020 of taxable income, 22% of the next £29 380, and 40% of any taxable income still left.

To be able to make accurate calculations, you have to know the current allowances and rates of tax. These change nearly every year, as do wage rates, but to obtain the answers at the back of this book, you should base your calculations on the following information:

National Insurance Contributions: 11% of gross monthly income above £395
or 11% of gross weekly income above £91
Reduced rate for contracted-out superannuation schemes : 9·4%

Income Tax: (Personal Allowance £4745)

Tax Bands		
starting rate	10%	0 – £2020
basic rate	22%	£2021– £31 400
higher rate	40%	over £31 400

You should always calculate the income tax due for the year. Thereafter this can be divided up into monthly or weekly amounts.

Reminder continued

Example

Tanya is a primary school teacher earning a salary of £21 750. She pays 8% of this salary in superannuation. Her monthly N I contributions are 9·4% of what she earns in excess of £395. In addition to her personal tax allowance she has professional allowances of £160. Calculate her net monthly take-home pay.

Solution

Allowances:	superannuation	8% of £21 750 =	1740
	personal allowance		4745
	professional allowances		160
	Total allowances		6645

Income Tax:	salary:	= 21 750		
	(– allowances	= –6 645)		
	so taxable income	= 15 105		tax
	(– starting band	= –2 020)	(@ 10%) ⇒	202·00
	so basic band	= 13 085	(@ 22%) ⇒	2878·70
	total tax due			3080·70

National Insurance: 9·4% of [(21 750 – (12 × 395)] = 1598·94

Deductions:	superannuation	1740
	income tax	3080·70
	N I contibutions	1598·94
	total deductions	6419·64

Net Pay:	salary	21 750·00
	deductions	–6 419·64
⇒	net annual income	15 330·36
⇒	net monthly income	£1 277·53

Example

Eric is a gardener earning £7·75 per hour for a 35-hour week. His weekly N I contributions are 11% of what he earns in excess of £91. Calculate his take home pay.

Solution

Earnings: 35 × £7·75 × 52 = £271·25 × 52 = 14105
(– personal allowance = –4745)
so taxable income = 9360

Income tax:	taxable income	= 9360		
	(– starting band	= –2020)	(@ 10%) ⇒	202
	so basic band	= 7340	(@ 22%) ⇒	1614.80
	and total annual tax due			1816·80
	hence weekly tax due		=	34·93
	N I contributions = 11% of (271·25 – 91)		=	19·82
	hence total weekly deductions		=	54·75
	and net weekly wage (271·25 – 54·75)		=	£216·50

Exercise 101

1 Using the allowances and rates of tax given calculate the monthly tax payable by each of these individuals:

Name	Annual Salary	Superannuation Rate
Anthea	£18 450	0
Barclay	£21 400	6%
Colin	£25 750	8%
Davina	£35 430	9%
Elsie	£42 000	5%.

2 Using the allowances and rates of tax given, calculate the weekly tax payable by each of these individuals:

Name	Fergus	Gregor	Harriet	Ina	Julie
Weekly Pay	£235	£279	£343	£400	£750

3 National Insurance deductions for monthly paid earners are calculated as 11% of the gross monthly income over £395, or 9·4% if in a contracted-out superannuation scheme.

a) Calculate the net monthly salary for each of these monthly paid earners:

Name	Annual Salary	Superannuation Rate
Kelly	£15 500	not superannuated
Lesley	£17 000	not superannuated
Malcolm	£18 400	6% (contracted-out)
Norman	£20 750	7% (contracted-out)
Olga	£31 000	8% (contracted-out).

National Insurance Contributions are also paid to the Inland Revenue by employers for each employee. They are calculated as 12·8% of the gross monthly income over £395, or 11·2% for contracted-out employees.

b) Calculate the employer's monthly NIC for each of the above employees.

4 Bert is an architectural technician earning £20 000 per annum and paying 6% of his salary into a superannuation fund.

a) What percentage of his gross salary is collected (in income tax and NIC) by the Inland Revenue?

b) Calculate his employer's NIC for him for the year (11·2% of gross income in excess of £395 per month).

5 This is a copy of bricklayer Colin Dunbar's weekly pay slip from his firm *Westering Homes*:

Name	Nat. Ins. No.	Works No.	Week No.	
Colin Dunbar	AB 12 34 56 C	X345	23	
hours worked basic	overtime time and a half	overtime double time	hourly rate	gross pay
35	4	1	£11·26	A
NIC	Income Tax	total Deductions		net pay
B	C	D		E

Calculate the entries for A, B, C, D, and E.

6 This is a copy of a monthly pay slip for a salesperson who earns commission of 5% on sales over £2000.

Name	Nat. Ins. No.	Ref No.		Month
Edith Fleming	DE 65 43 21F	T125		February
basic pay	sales	commission	bonus	gross pay
£1000	£6500	A	£25	B
NIC	Income Tax			net pay
C	D			E

Calculate the entries for A, B, C, D, and E.

7 This is a monthly pay slip for a civil engineer who pays 5% of her salary into a contracted-out superannuation fund.

Name	Nat. Ins. No.	Employee Ref.		Month
Gillian Hardy	GH 21 43 65 I	TS23		May
gross pay	superannuation	NIC	Income Tax	net pay
£2850	A	B	C	D

Calculate the entries for A, B, C, and D.

102 Borrowing

Reminder

When you take a loan from a bank or a finance company you generally repay an agreed fixed amount every month. This figure is usually obtained from a loan table. Larger loans often attract lower rates of interest. When you buy something expensive with a credit card there is usually more freedom to vary the monthly repayments, but the interest rate will probably be higher and variable. There is usually a minimum monthly repayment on any outstanding debt.

Example Vic and Winnie wish to borrow £10 000 from their bank towards installing a new kitchen. They wish to repay the loan over 5 years. Here is the bank's loan table:

APR	Loan	5 yrs	5 yrs	10 yrs	10 yrs	15 yrs	15 yrs	25 yrs	25 yrs
		no ins	+ ins	no ins	+ ins	no ins	+ ins	no ins	+ ins
7·7%	£75 000	1501·15	1808·35	888·39	1069·94	693·22	834·71	551·90	664·34
	£50 000	1000·77	1290·61	592·27	763·61	462·16	595·73	367·94	474·13
	£30 000	600·47	774·53	355·34	458·39	277·28	370·36	220·77	284·36
7·9%	£25 000	502·95	648·87	298·96	385·66	234·16	309·09	187·50	251·25
8·9%	£20 000	411·22	538·70	248·95	326·12	198·04	255·47	162·30	209·37
	£15 000	308·42	404·03	186·73	244·62	148·53	191·60	121·72	157·02
9·9%	£10 000	210.10	281·53	129·50	173·53	104·52	137·97	87·50	115·50
	£5000	105·05	140·77	64·75	86·77	52·26	68·98	43·75	57·75

Calculate the cost of this loan if they borrow:

a) without repayment protection
b) with repayment protection.

Solution

a) monthly repayment = £210·10
total for 5 years = 60 × £210·10
= £12 606
cost of loan = £2606

b) monthly repayment = £281·53
total for 5 years = 60 × £281·53
= £16 891·80
cost of loan = £6891·80

Reminder continued

Example Vic and Winnie decide to take the above £10 000 loan without repayment protection. They also wish to borrow £5000 towards carpets and furniture, again repaying over 5 years without repayment protection. Is it cheaper to take two separate loans or a single loan for £15 000? By how much and why?

Solution

separate loans (£10 000 + £5,000):
kitchen (from above) = £12 606
carpets etc: 60 × £105·05 = £6303
total = £18 909

single loan of £15 000:
60 × £308·42 = £18 505·20

hence single loan is £403·80 cheaper
(It is cheaper because the larger loan attracts a lower APR)

Exercise 102

1 Using the loan table given for Vic and Winnie's bank, state the monthly repayments for the following loans:
a) £75 000 over 25 years with repayment protection
b) £20 000 over 15 years without repayment protection
c) £10 000 over 15 years without repayment protection
d) £30 000 over 10 years with repayment protection
e) £25 000 over 5 years without repayment protection.

Drumadairy Mutual currently uses the following loan table. Use it for questions 2 to 5.

		2 yrs	2 yrs	4 yrs	4 yrs	5 yrs	5 yrs
Loan	APR	+ prot	no prot	+ prot	no prot	+ prot	no prot
£3000	19·7%	166·40	145·50	101·42	83·00	90·87	70·50
£5000	6·9%	252·94	223·37	142·02	119·20	122·16	98·37
£10 000	6·5%	503·65	445·00	281·60	236·66	241·74	195·00
£15 000	6·5%	755·47	667·50	422·42	355·00	362·62	292·50

2 Andrew wishes to borrow £3000 without repayment protection over two years.
Find: a) his monthly repayment
b) his total repayment
c) the cost of this loan.

3 Barbara wishes to borrow £5000 over 4 years with repayment protection.
Find: a) her monthly repayment
b) her total repayment
c) the cost of this loan.

4 Catriona wishes to borrow £5000 over 5 years with repayment protection.
Find: a) her monthly repayment
b) her total repayment
c) the cost of this loan.

5 Donald wishes to borrow £10 000 over 4 years.
Calculate how much more it costs with repayment protection than it does without.

6 The *Boreland & Bennochy* Building Society is offering loans at 17·7% APR, using the following loan table:

loan	60 months with prot.	60 months without prot.	48 months with prot.	48 months without prot.	36 months with prot.	36 months without prot.	24 months with prot.	24 months without prot.
£1000	30·18	24·54	34·26	28·55	40·31	35·36	55·06	49·16
£2000	60·37	49·08	68·52	57·10	80·62	70·72	110·12	98·32
£3000	90·55	73·62	102·78	65·86	120·93	106·08	165·18	147·48
£5000	144·79	122·70	165·59	142·75	198·90	176·80	272·84	245·80
£10 000	289·57	245·40	331·18	285·50	397·80	353·60	545·68	491·60

Gina takes out a loan of £3000 without protection and pays it back over 4 years. Calculate the cost of the loan.

7 Henry takes a 17·7% APR loan of £4000 from *Boreland & Bennochy* Building Society without protection, paying it back over 3 years. How much does this loan cost him?

8 Similarly Isaac takes a *Boreland & Bennochy* Building Society loan of £10 000 with protection over 5 years. How much does he spend on protection?

The *Pundit* Credit Card applies these conditions:
- any uncleared balance accrues interest at the rate of 1·65% per month
- the minimum payment is 3% of the balance due or £5, whichever is the greater
- payment protection costs 75p per month per £100 on the current statement balance
- there is a 1·5% handling charge (minimum £1·50) on all cash advances

9 Holly has used her new *Pundit* credit card once to buy a bike for £245. She does not have payment protection. Calculate the minimum payment due when her next statement arrives.

10 Kelly has used her new *Pundit* credit card once to buy a cooker for £318. She does not have payment protection. Calculate the minimum payment due when her next statement arrives.

11 Ian has made purchases on his *Pundit* credit card bringing his balance up to £145·73. Calculate the cost of payment protection for this balance.

12 Janice has made purchases on her *Pundit* credit card bringing her balance up to £247·34. Calculate the cost of payment protection for this balance.

13 Gordon is in a camera shop and wishes to buy a new digital camera costing £180. He can obtain a discount of 3% by paying cash. How much can he save by collecting £180 in cash from an autoteller using his *Pundit* credit card (read the rules above) in order to obtain the discount for cash, (given that he pays off the balance in his next statement).

14 Here is Evelyn's recent monthly statement from *Pundit*:

previous balance	121·36
payment received; thank you	121·36
balance	0·00
Grocer	35·23
Garage	41·26
Theatre	26·50
Supermarket	70·85
Filling Station	40·23
£40 cash advance (inc. handling charge)	A
payment protection	B
current balance	C
minimum payment	D

Calculate the entries for A, B, C and D.

15 [You will need to be able to use the x^y button on your calculator for this question.] If the monthly interest rate on a loan is r%, then the APR is given by

$$\left[100\left[\left(1+\frac{r}{100}\right)^{12}-1\right]\right]$$

a) Confirm that the Pundit monthly interest rate of 1·65% is equivalent to an APR of 21·7%.

b) Find the APR equivalent to a monthly interest rate of 1·21%.

16 There is also a formula for finding the monthly interest rate (r) which is equivalent to a given APR: $r = 100\left[\left(1+\frac{\text{APR}}{100}\right)^{\frac{1}{12}}-1\right]$.

a) Check that this formula gives a monthly interest rate of 1·65% when the APR is 21·7%.

b) Find the monthly interest rate equivalent to an APR of 19·7%.

103 Revision of Social Arithmetic

Exercise 103

1 a) Aaron earns £18 720 per annum. How much is this per month?
b) Beverley earns £225 per week. How much is this per year?
c) Clem earns £8·42 per hour for a 40-hour week. How much is this per year (52 weeks)?
d) Debbie earns £2450 per month. How much is this per year?

2 Find the new salary or rate of pay for each of these employees following the pay rise indicated:

a) Esther	£23 750 p.a.	receives $2\frac{1}{2}$%
b) Fred	£346 per week	receives $1\frac{1}{2}$%
c) Gabriella	£1925 per month	receives 3%
d) Harry	£6·73 per hour	receives 4%.

3 Ivor earns £235 per week (for 52 weeks per year) and Jimmy earns £1020 per month. Who earns more and by how much?

4 Calculate the annual income for each of the following salespersons:

a) Kathleen	£120 per week plus commission:	$3\frac{1}{2}$% of £5600
b) Larry	£960 per month plus commission:	4% of £17 800
c) Magda	£12 000 per annum plus commission:	5.5% of £120 000.

5 Neil who has a basic salary of £14 000 p.a. earns commission of 2% on all sales in excess of £5000.
a) Calculate his gross annual salary when his annual sales amount to £230 000.
b) How much is this per month?

6 Calculate the gross weekly income for each of the following:

Name	basic hours	overtime at time and a half	overtime at double time	hourly rate
Oliver	40	4	1	£7·75
Paul	35	1	0	£8·44
Rita	$37\frac{1}{2}$	0	0	£8·04

7 Sammy worked 35 hours at the basic rate, 6 hours overtime at time and a half, and 3 hours overtime at double time. The gross wage earned for the week was £461·50. Calculate the basic hourly rate of pay.

8 Calculate the National Insurance contributions due on a gross weekly wage of:
a) £89 b) £95 c) £123 d) £400.

9 Calculate the National Insurance contributions due on a gross monthly wage of:
a) £390 b) £565 c) £1205 d) £1835.

10 Calculate the monthly income tax deduction for the following people:
a) Theo, who earns £19 560 p.a.
b) Una, who earns £21 470 p.a.
c) Violet, who earns £2450 per month.
d) Walter, who earns £2635 per month.

11 *Lomond Securities* are offering loans at 19·9% and use this loan table:

loan	60 months with prot.	60 months without prot.	48 months with prot.	48 months without prot.	36 months with prot.	36 months without prot.	24 months with prot.	24 months without prot.
£1000	31·43	25·55	35·44	29·53	41·38	36·30	56·07	50·06
£2000	62·85	51·10	70·87	59·06	82·76	72·60	112·13	100·12
£3000	94·28	76·65	106·31	88·59	124·15	108·90	168·20	150·18
£5000	150·74	127·75	171·27	147·65	204·19	181·50	277·83	250·30
£10 000	301·49	255·50	342·55	295·30	408·38	363·00	555·67	500·60

Derek arranges a loan of £5000 re-payable over 4 years with payment protection.
Find: a) the monthly repayment
b) the total repaid
c) the cost of the loan
d) the cost of the repayment protection.

12 Here is Graham's recent monthly statement from his credit card supplier:

previous balance	563·42
payment received; thank you	400·00
balance	A
Interest (1·55% per month)	B
Furniture Store	123·65
Supermarket	85·47
Travel Agent	256·00
Filling Station	35·67
Supermarket	52·13
payment protection (75p per £100 of balance)	C
current balance	D
minimum payment (3% of balance due)	E

Calculate the entries for A, B, C, D, and E.

104 Test on Social Arithmetic

Allow 45 minutes for this test

1 Murray takes a loan of £2000 without repayment protection from *Lomond Securities* (see previous page) for 2 years.
Find: a) the monthly repayment
b) the total repaid
c) the cost of the loan.

2 Tim is paid £9·45 per hour for a basic $37\frac{1}{2}$ hour week. Find his gross pay for a week when he works 3 hours overtime at time and a half and a further 2 hours at double time.

3 Gareth is paid a basic annual salary of £11 550. He is paid commision of 6% of sales in excess of £10 000 each month. Calculate his gross pay for a month in which his sales amounted to £15 400.

4 Calculate the annual tax due on a taxable income of:
a) £2000 b) £4500 c) £15 000 d) £40 000.

5 Calculate the weekly National Insurance Contribution on a gross weekly wage of:
a) £325 b) £596.

6 Calculate the monthly National Insurance Contribution on a gross monthly wage of:
a) £1327 b) £2596.

7 Jonathan, who has a basic salary of £13 500 p.a. earns commission of 3% on all sales in excess of £6000.
a) Calculate his gross annual salary when his annual sales amount to £190 000.
b) Calculate his net monthly take-home pay.

8 Here is Philip's recent monthly statement from his credit card supplier: (Philip has opted not to have payment protection.)

previous balance	423·26
payment received; thank you	350·00
balance	A
Interest (1·75% per month)	B
Theatre	82·00
Supermarket	53·27
Garage	223·56
Filling Station	37·85
Supermarket	42·19
current balance	C
minimum payment (3% of balance due)	D

Calculate the entries for A, B, C, and D.

LOGIC DIAGRAMS

105 Tree Diagrams

A tree diagram is a useful way of listing all the possibilities when there is a choice of categories or outcomes.

Example Find how many different possible meals may be served when the menu consists of:

(Starters)	*(Main Courses)*	*(Sweets)*
lobster bisque (L)	sirloin steak (S)	bread and butter pudding (B)
grilled grapefruit (G)	baked halibut (H)	sticky toffee pudding (T)
	roast duck salad (R)	assorted ice cream (I)
	pasta of the day (P)	

Solution We label the choices with a single letter to minimise the writing.

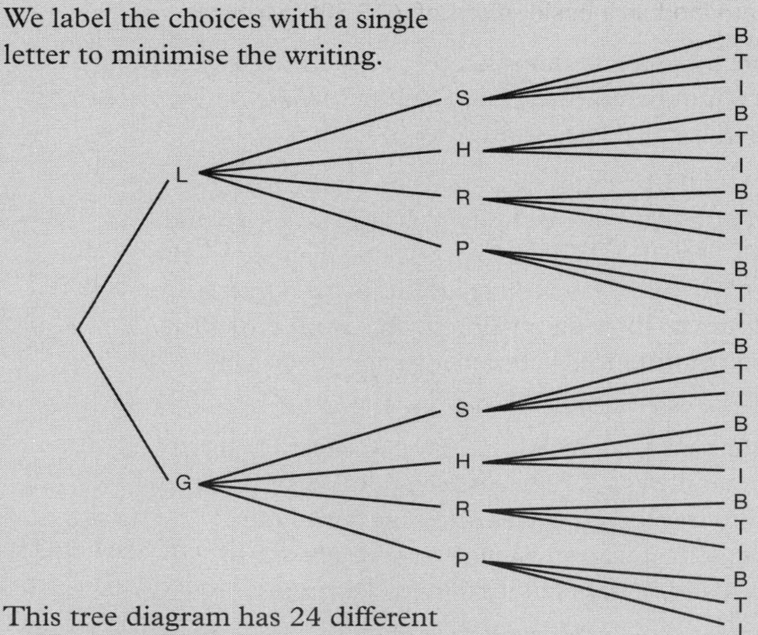

This tree diagram has 24 different ends, indicating 24 different choices of meal.
The top line of branches for example, indicates the choice of lobster bisque; steak; and bread and butter pudding.

Reminder continued

Notes

If this were an *à la carte menu* (i.e. each dish priced individually), we could write the price above the branches and find the total cost of each meal by adding the prices above the branches.
If all choices were equally likely (which they probably would not be in fact), then each choice of meal would have a probability of $\frac{1}{24}$.
Other probabilities could also be calculated, e.g. of having grapefruit and steak but not ice cream.

Exercise 105

1 Use a tree diagram to list all the possible choices of a two-course meal from the following menu:

(*Starters*)	(*Main Courses*)
soup (S)	haddock (H)
melon (M)	rump steak (R)
fruit juice (F)	pork fillets (P)
	tagliatelli (T).

2 In an Olympic 800 m race, three Scotsmen, Finlay, Gregor, and Hamish, took the gold, silver and bronze medals, but you are not told who won which medal. Construct a tree diagram to help you list all the possible arrangements.

3 There are three roads (numbered 1, 2, 3) from Abty to Bolum and two roads numbered (4, 5) from Bolum to Catlyn. Use a tree diagram to list all routes from Abty to Catlyn.

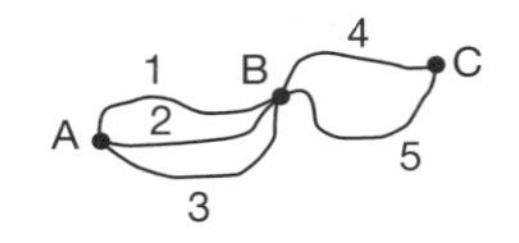

4 P, Q, R, S, T, U, V, and W are towns in Strathclyde. There are three roads from P to T (via Q, R, or S), and two roads from T to W (via U or V). The map shows the distances (in miles) between adjacent towns. Draw a tree diagram to show all possible routes from P to W, including the distances between adjacent towns, and hence find the shortest route.

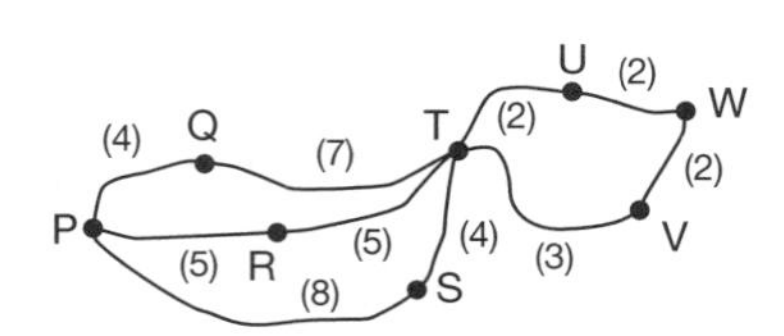

5 This is the same map as for question 4, except that the numbers on the roads represent the number of minutes required to travel each road during the rush hour. Construct a tree diagram similar to your answer to question 4 and hence identify the quickest route.

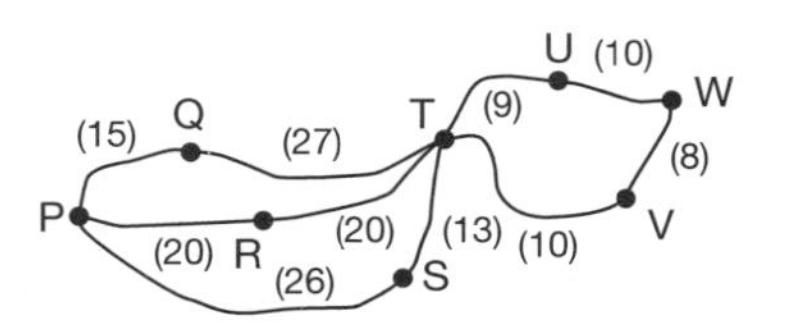

6 a) This spinner is in the shape of a regular pentagon. Draw a tree diagram to list all the possible outcomes of spinning this spinner and tossing a coin.

b) Hence write down the probability of obtaining a head and an even number in one trial.

7 A 10p and a 2p coin are tossed simultaneously, each with outcome heads (H) or tails (T).
   a) Construct a tree diagram to list all the possible outcomes.
   b) Write down the probability of obtaining two heads.
   c) Explain why it is more likely to have one head and one tail than two heads.

8 This map shows five villages P, Q, R, S, and T and the distances (in km) between them. The bakery is in the village P and the van driver has to deliver to villages Q, R, S, and T. He wishes to do so without passing through any village more than once.

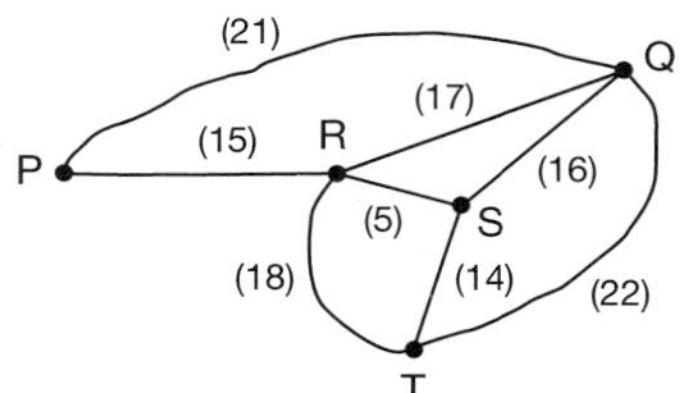

   a) Draw a tree diagram to show that there are 12 ways in which this can be done (before returning to P) and identify the shortest route.
   b) From the end point of each of the above routes, add (by inspection) the shortest distance back to P, and hence calculate the length of the shortest round trip, identifying two ways that this might be achieved.

106 Network Diagrams

Reminder

A network diagram consists of *vertices* (or *nodes*) which may or may not be connected by *edges* (or *arcs*). It is often required to find a path (a succession of vertices) which contains either all the vertices or all the edges. Sometimes we need a closed path (a path which starts and ends at the same vertex). The *degree* of a vertex is the number of edges meeting there.

Example This is a map showing nine towns, and the main roads joining them.

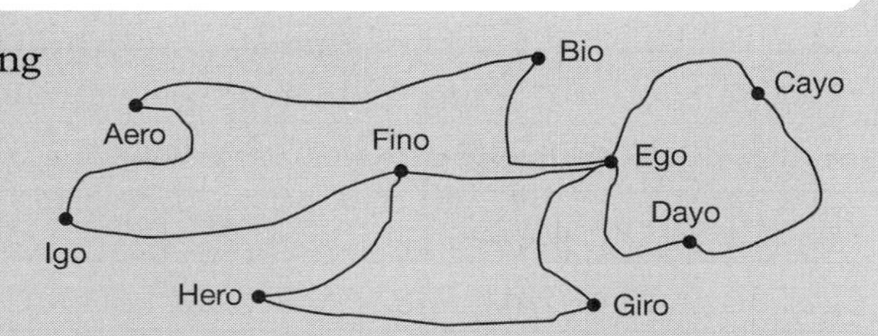

a) Make a network diagram to represent the map.
b) There is one church in every town, and a tourist, staying in a hotel in Aero, wishes to visit each in a single trip.
   (i) Suggest an acceptable route (starting and finishing in Aero).
   (ii) Does this route cover every road on the map?
   (iii) Does your route involve passing through any town already visited?
c) Following bad weather, a roads inspector must inspect every main road.
   (i) Suggest a route which covers every road, but none more than once.

Reminder continued

(ii) What do you notice about the towns through which he passes only once?

(iii) Is there an acceptable route which starts and finishes in the same town?

Solution a)

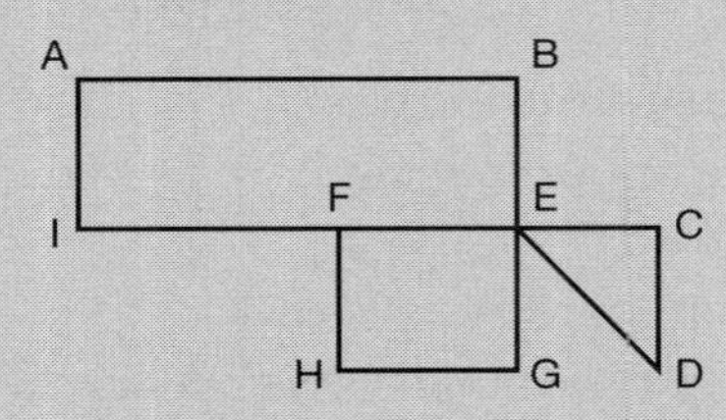

or

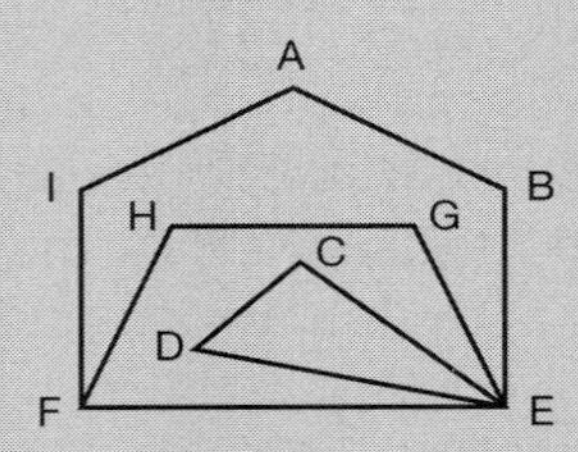

b) (i) ABECDEGHFIA This could also be reversed. The route round 'triangle' ECD could also be reversed, so giving 4 possibilities.

(ii) No, EF has not been covered.

(iii) Yes, E.

c) (i) To answer this question it is sensible to examine the degree of every vertex:

vertex	A	B	C	D	E	F	G	H	I
degree	2	2	2	2	5	3	2	2	2

Clearly, any route covering all roads must start and end at vertices with an odd degree. Thus start at E, end at F, (and travel) FHGEFIABEDCE, or the reverse.
(The engineer could reverse his route round ECD and EFGH, so giving 8 possibilities.)

(ii) They have degree 2.

(iii) No, not in this case.

Exercise 106

1 a) For each network diagram, find a pathway which passes through each vertex. (You do not need to finish where you started.)

(i)

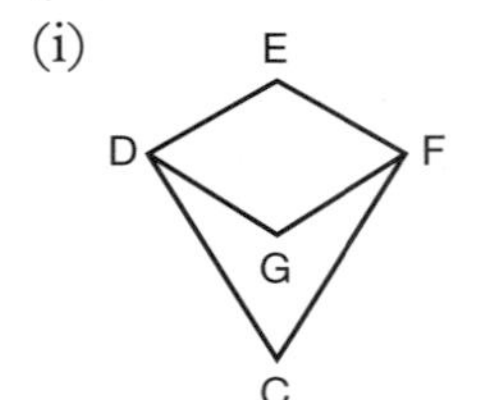

(ii) I

H

J

K

(iii)

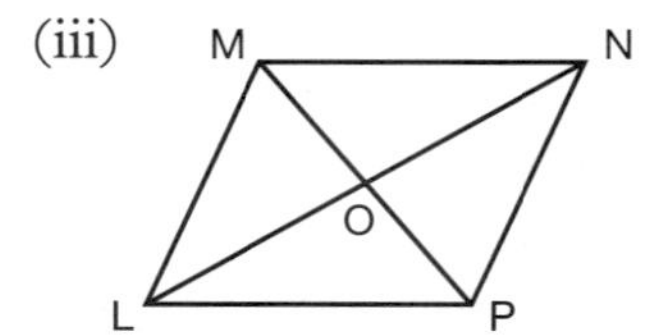

b) State the degree of each vertex in these networks.

c) (i) In diagram (i), why is it not possible to trace the network starting and finishing at the same point unless you go over an arc more than once?
(ii) Which arc could be added to this diagram so that you could trace the network without lifting the pencil off the page, starting and finishing at the same point? (Any point?)

d) For diagrams (i) and (ii), find a pathway which uses each arc exactly once.

e) Why is this not possible for diagram (iii)?

2 Which map corresponds to which network diagram?

a) 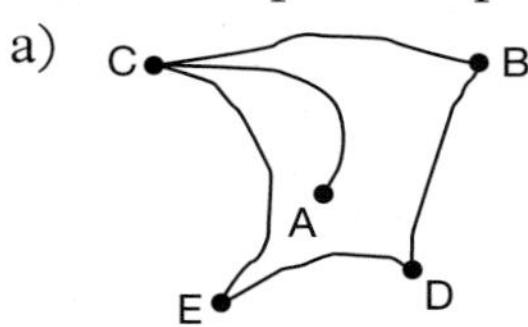

b) 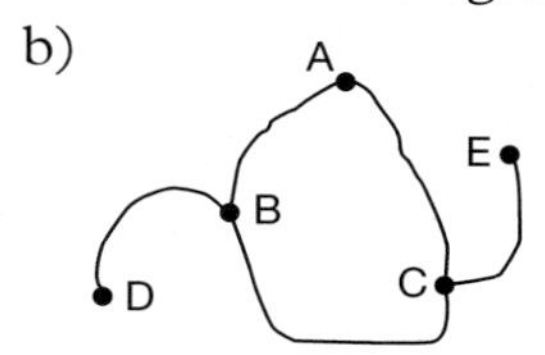

c) 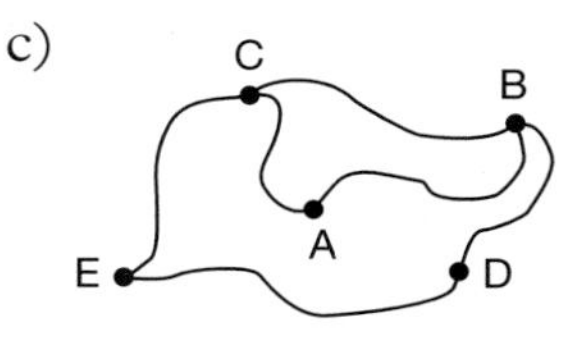

(i) A, B, C, D, E

(ii) 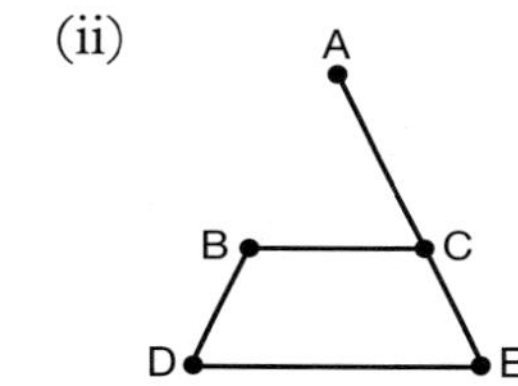

(iii) 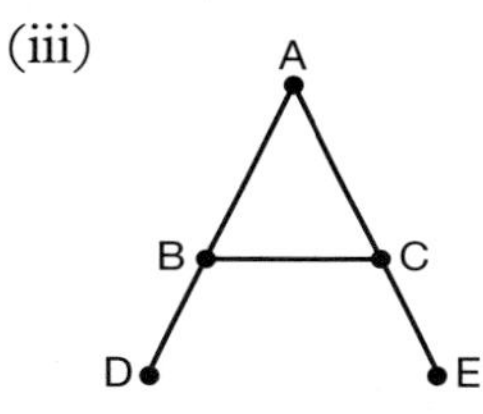

(iv) 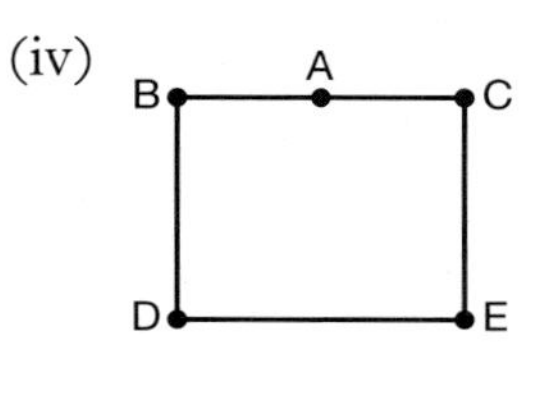

3 a) State the degree of each vertex in these network diagrams:

(i) A, E, D, B, C

(ii) 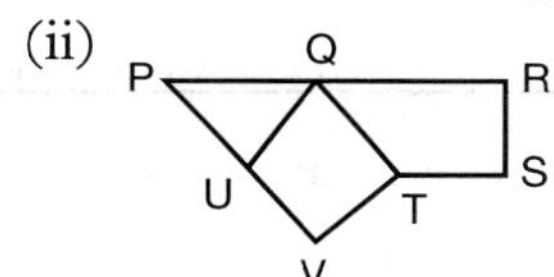

(iii) 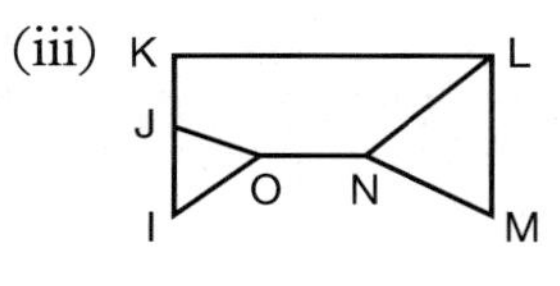

b) In networks (i) and (ii) there is a path which uses each edge exactly once. Which points in each diagram must be the start and finish of these paths?

c) Hence write down a path for each of diagrams (i) and (ii) which uses each edge exactly once.

d) Explain why there is no such path for network diagram (iii).

4 (i) 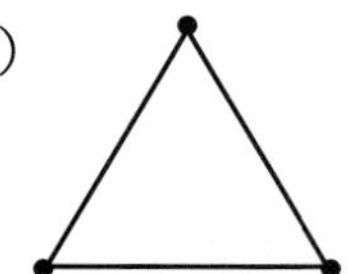

(ii)

(iii) 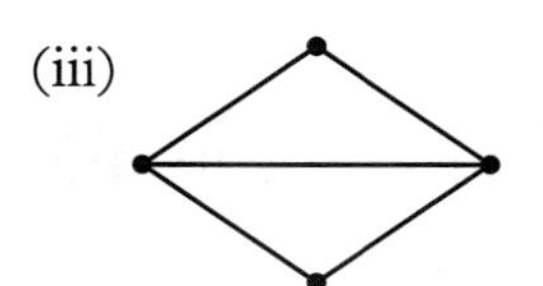

(iv) 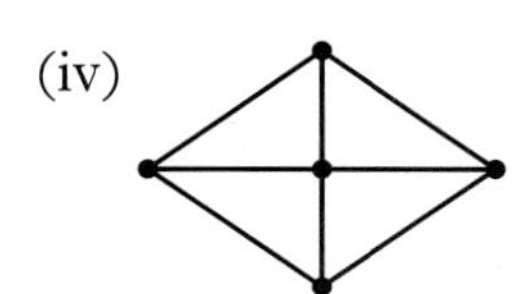

Diagram (i) is a 2,2,2 network, because it has three vertices of degree 2.
Diagram (ii) is a 2,2,2,2 network, because it has four vertices of degree 2.
Diagram (iii) is a 2,2,3,3 network, because it has two vertices of degree 2 and two of degree 3.

a) State the corresponding description of diagram (iv).

b) Make a sketch of a 2,2,2,3,3 network.

5 Here is part of the McClerty family tree:

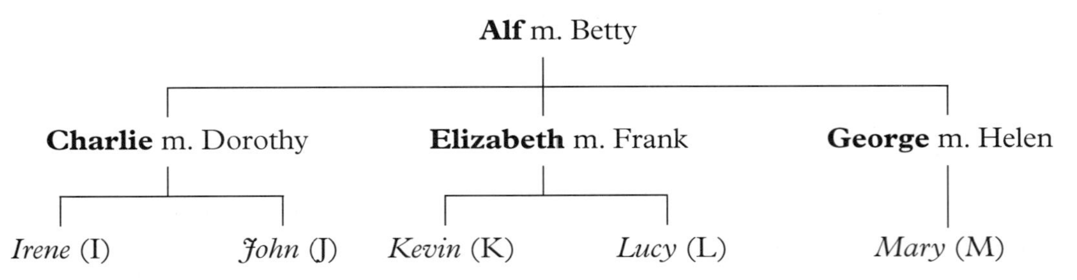

a) Construct a network diagram with vertices I, J, K, L, and M and arcs which denote 'is a cousin of'.

b) How many arcs are there in this diagram?

c) Find the degree of each vertex.

d) Can you spot any connection between the number of arcs and the sum of the degrees of all the vertices?

6 Here is another family tree:

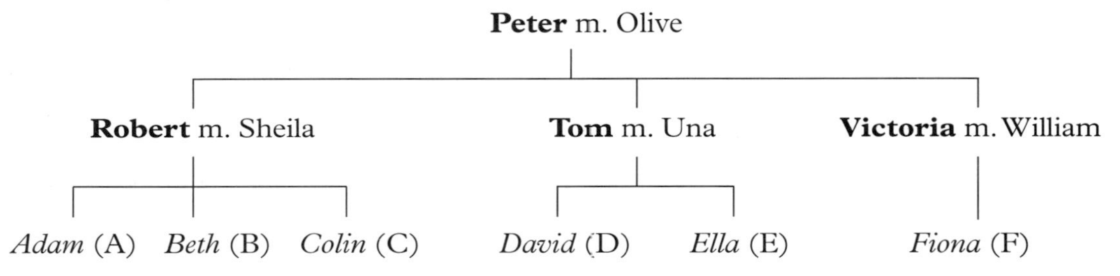

a) Construct a network diagram with vertices A, B, C, D, E, and F and arcs which denote 'is a cousin of'.

b) Find the degree of each vertex.

c) If T is the total of the degrees of all the vertices, and E is the number of edges, state the connection between T and E.

7 This network diagram represents the streets of a village. The intersections of the streets are labelled. Trevor lost his wallet when he was out on his bike, and wishes to cycle round the village looking for it using a route that will cover each street exactly once.

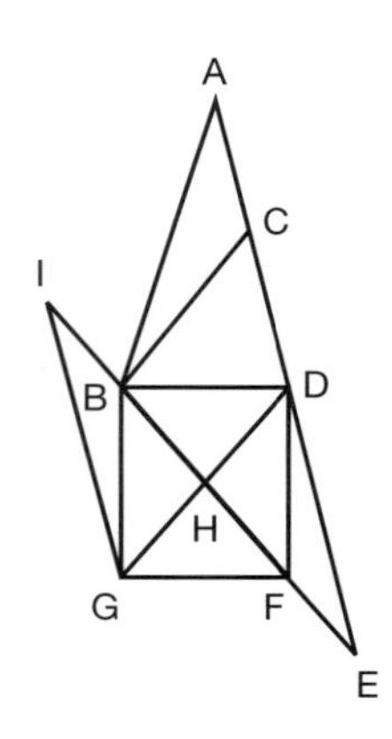

a) At which street intersections could such a route start and finish?

b) On this route, how many times will Trevor pass through junction:

(i) A (ii) B (iii) H?

c) Give an example of an acceptable route which contains the complete street from E to I as a part of it.

8 This network diagram shows some distances (in miles) between Aberdeen, Dundee, Edinburgh, Perth, Stirling and Glasgow.

Aberdeen					
	Dundee				
		Edinburgh			
			Glasgow		
		42		Perth	
					Stirling

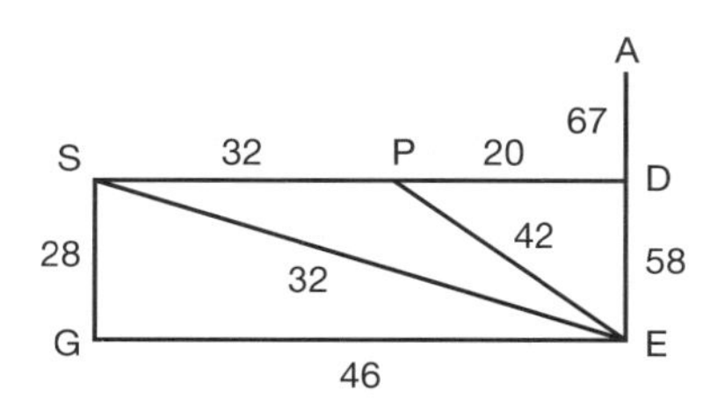

a) Use the network diagram to complete the mileage chart.

b) Of these six towns, which two are: (i) closest (ii) furthest apart?

c) (i) How much longer is it to drive from Dundee to Edinburgh via Perth than going directly?

(ii) Why might going via Perth be quicker?

d) Here is another network diagram which represents the same information.

(i) Find the degree of each vertex in both network diagrams.

(ii) Identify which points in the original network diagram the vertices I, J, K, L, M, and N now represent.

9 This mileage chart shows some distances (in miles) between five places in the U.K.

Cardiff					
252	Hull				
	430	Inverness			
256	165	305	Kendal		
155	188		267	London	

a) Construct a network diagram to show the eight roads referred to in the mileage chart.

b) Hence find the shortest distance from:

(i) Cardiff to Inverness

(ii) Inverness to London.

10 Mr and Mrs Munro decide to go hillwalking. The tasks requiring to be done before they leave are these:

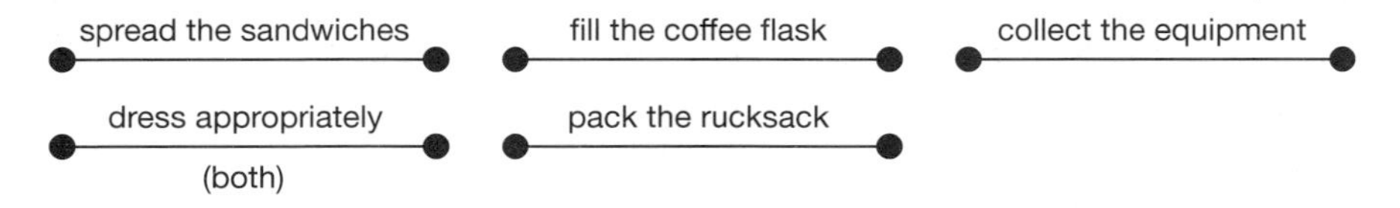

Each task takes 5 minutes to carry out. Arrange them in the following network diagram to show one way in which all the tasks can be complete, so that they can leave 15 minutes after deciding to go.

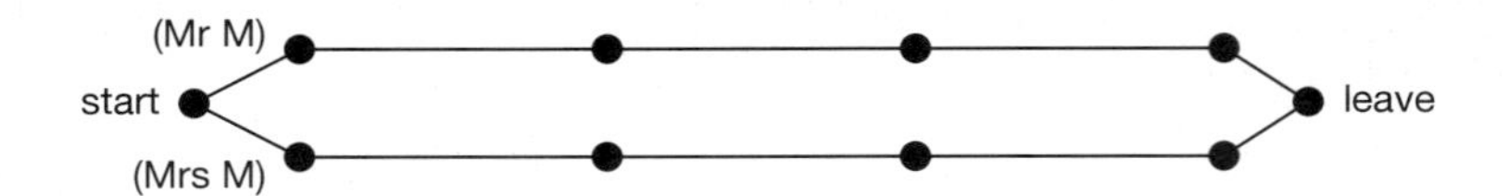

(This is sometimes called an *activity diagram*, and the idea can be extended to more complex operations.)

107 Flowcharts

Reminder

A flowchart is a useful way of organising a set of instructions. It is a convention that each flowchart has a title, a start box and a stop box. Each instruction appears within a rectangular outline. A diamond shaped outline is used for a decision, which can only be taken in answer to a question with answer 'yes' or 'no'.

Example This flowchart shows how to calculate the monthly pay for a salesperson, given their annual salary and monthly sales, with commission paid at 3% of sales in excess of £3000 (up to £10 000) and 4% in excess of £10 000. Use it to find the gross monthly pay for a salesperson on £12 000 p.a. plus monthly sales of £8000.

To calculate gross monthly pay:

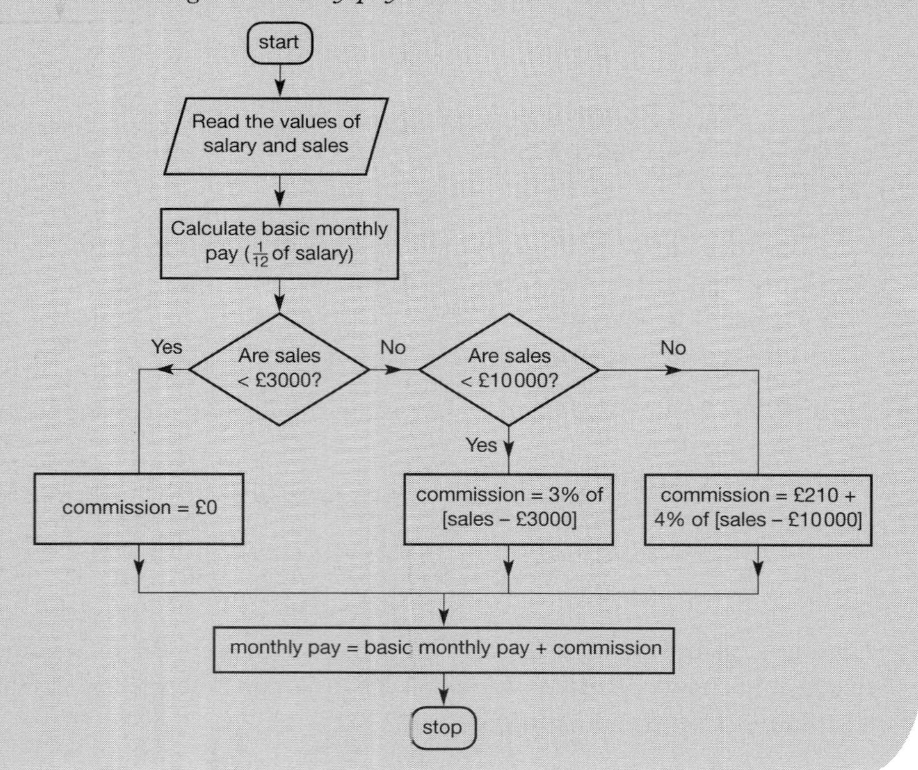

Reminder continued

Solution basic monthly pay = $\frac{1}{12}$ of £12 000 = £1000
commission = 3% of (£8000 – £3000) = 3% of £5000 = £150
monthly pay = £1000 + £150 = £1150.

Exercise 107

1 Use this flowchart to find the value of:

a) f(2)
b) f(10)
c) f(1·6)

To evaluate $f(x)$
where $f(x) = 3x^2 - 2x + 5$:

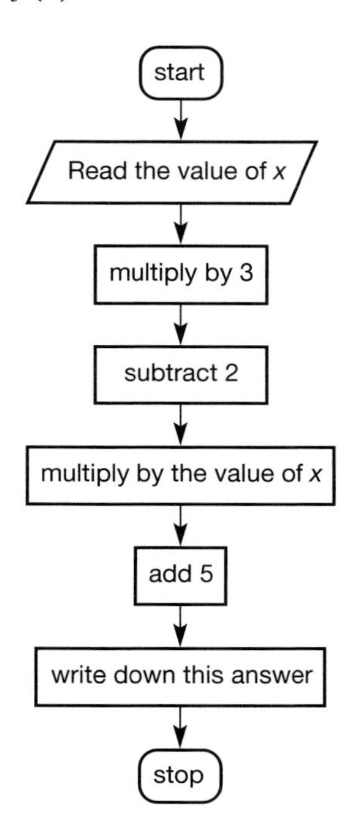

2 Use this flowchart to find the value of

a) f(1)
b) f(0)
c) f(2)
d) f(5)

To evaluate $f(x)$
where $f(x) = x^3 - 2x^2 + 3x - 4$:

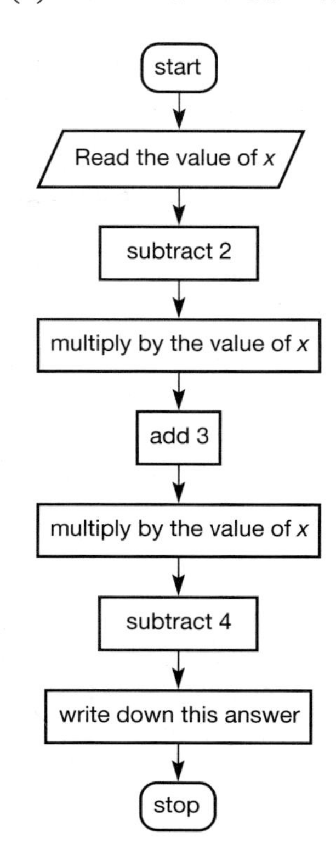

3 Use this flowchart to calculate the monthly wage of a salesman with a basic annual salary of £15 000 and commission of 5% on sales over £5000, when his monthly sales amount to:

a) £4000 b) £9000 c) £17 500.

To calculate a monthly wage:

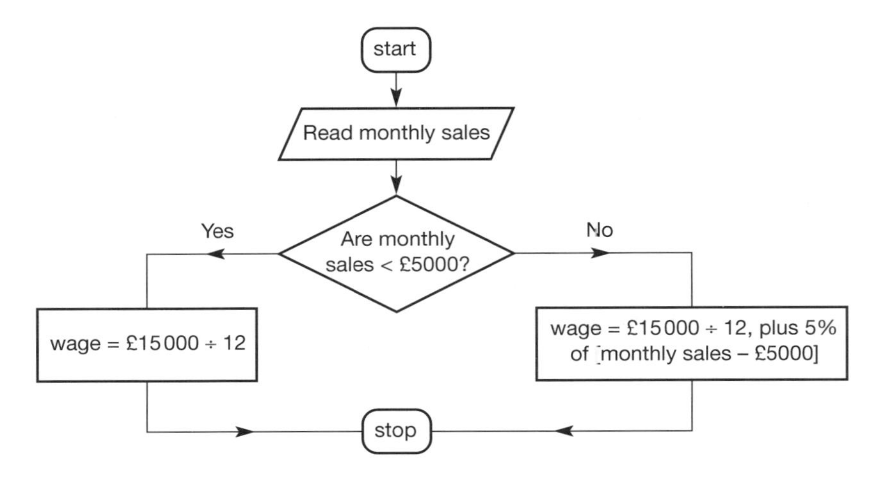

4 Use this flowchart to calculate the sale price of goods originally priced at:

a) £55 b) £180 c) £20 d) £65.

To calculate a sale price:

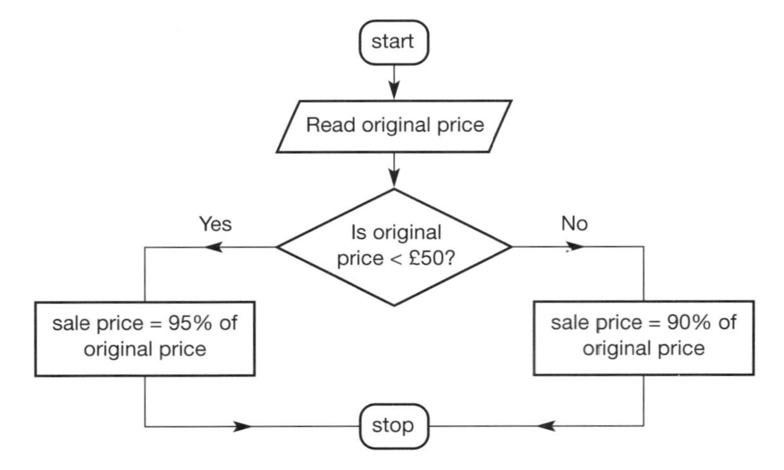

5 Use this flowchart to calculate the income tax to be paid where the taxable income is:
a) £12 000 b) £2000 c) £8000 d) £32 500.

To calculate annual tax due:

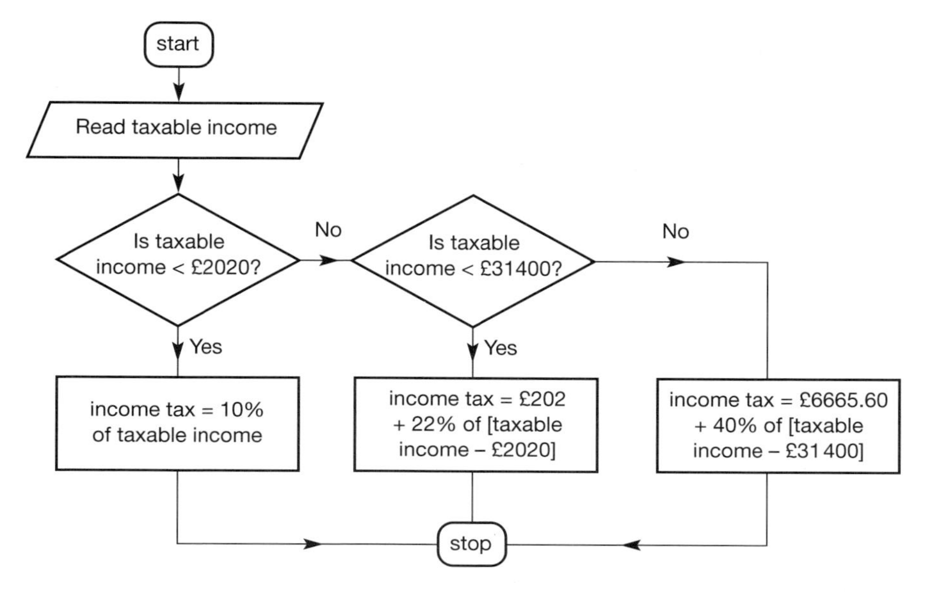

6 A certain component costs 5p (each) if 100 or less are bought, 4p (each) if between 100 and 200 are bought, and 3p (each) if 200 or over are bought. Use the flowchart to calculate the cost of these numbers of components:
a) 25 b) 320 c) 127 d) 200.

To calculate the cost of a batch of components:

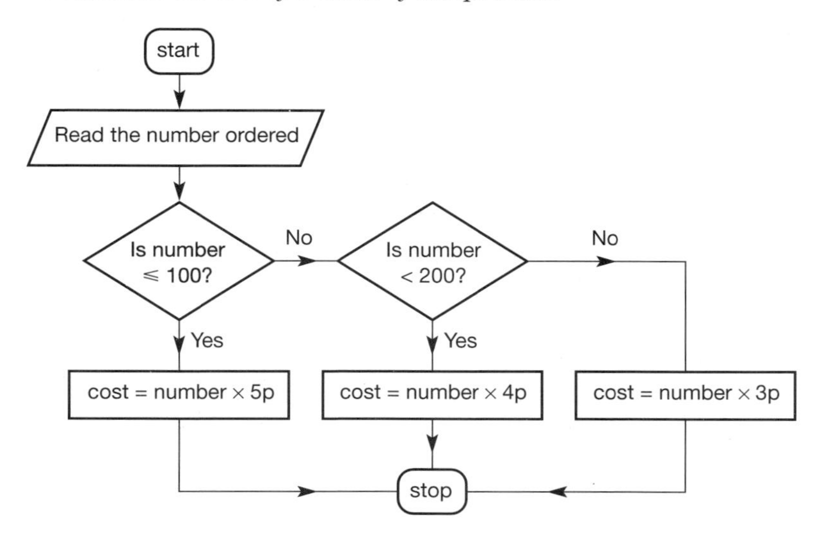

7 Use this flowchart to calculate the cost of posting a letter of mass:

a) 50 g b) 90 g c) 140 g

d) 250 g e) 500 g f) 180 g.

To calculate the cost of posting a first class letter:

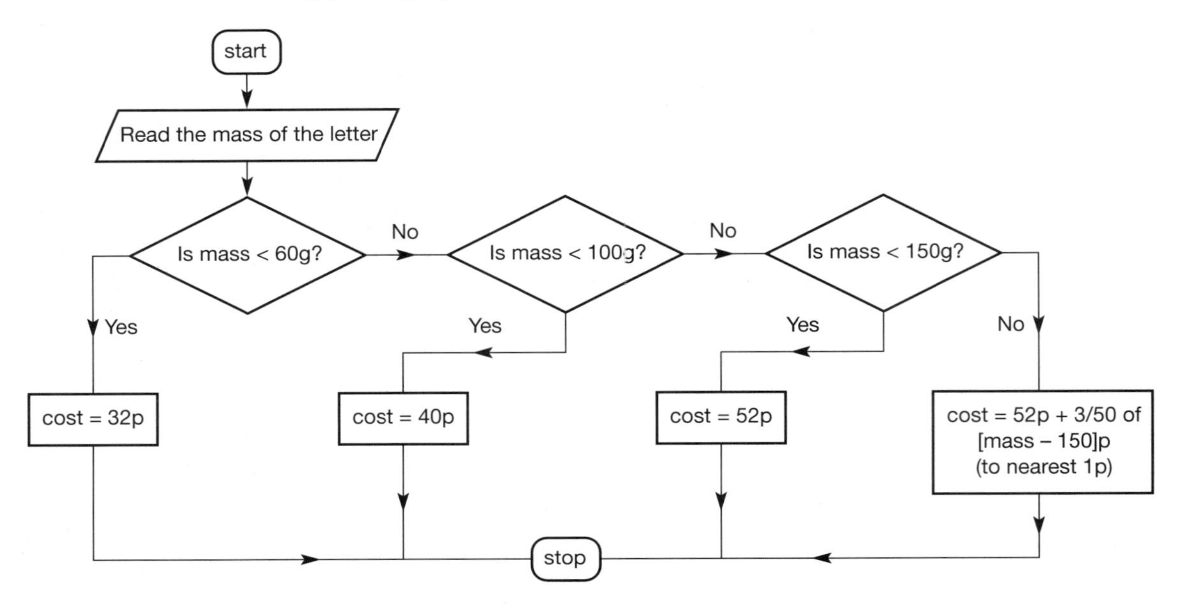

108 Spreadsheets

Reminder

Spreadsheets are very useful for analysing data, especially where the same kind of calculation is repeated several times. They can also be used to create diagrams and graphs. It is to be hoped that you already have some experience of using spreadsheets in other subjects, or on your own home computer. You do not have access to a computer during the Intermediate 2 exam, so this exercise concentrates on the kind of question you are likely to meet in the exam, e.g. interpreting a spreadsheet or inserting a formula in a cell.

Example A motorist buys a new car and records the total distance travelled every time he buys petrol. He zeroes the trip reading and records that as well. From this data he calculates the rate of consumption of petrol (in litres per 100 km) for the previous tankful and since his purchase of the car. He enters data in columns A, B, C and D.

Reminder continued

	A	B	C	D	E	F	G	H
1	date	odometer	trip	petrol	consumption	check	total petrol	overall
2		(km)	(km)	(litres)	(litres/100km)		used up	cons.
3	01/07/04	0	0	40				
4	14/07/04	498	498	42	8.03	498	40	8.03
5	28/07/04	997	499	39	8.42	499	82	8.22
6	10/08/04	1477	480	41	8.13	480	121	8.19
7	25/08/04	1972	495	40	8.28	495	162	8.22

a) What formula was entered in cell E4 ?
b) How was the rest of column E computed?
c) What formula was entered in cell F4 to check the trip meter against the odometer?
d) How was the rest of column F computed?
e) What formulae were entered in cells G4 and G5 and how was the rest of column G computed?
f) What formula was entered in cell H4 ?
g) If you clicked on cell H7, what formula would appear in the edit bar?
h) If you clicked on cell B5, which cell would be highlighted after you
   (i) pressed the TAB key
   (ii) pressed the RETURN key
   (iii) ticked the edit bar?

Solution

a) = D3*100/C4
b) highlight cell E4 and those cells below it and fill down
c) = B4 – B3
d) highlight cell F4 and those cells below it and fill down
e) = D3 and = G4 + D4 and fill down from cell G5
f) = G4*100/B4
g) = G7*100/B7
h) (i) C5 (ii) B6 (iii) B5

Exercise 108

1 A teacher gives her class three tests over the term and records their marks in cells B4 to D17 in the spreadsheet below.

	A	B	C	D	E	F	G
1		test A	test B	test C	total	percentage	
2	possible mark	56	58	66			
3	pupil						
4	Andrew	51	49	60			
5	Betty	47	50	59			
6	Carol	39	47	51			
7	David	35	48	49			
8	Ewan	55	52	63			
9	Fiona	50	51	59			
10	George	43	45	49			
11	Helen	32	37	40			
12	Ian	29	35	39			
13	Janice	52	53	64			
14	Kelly	48	51	61			
15	Lewis	39	42	55			
16	Mandy	23	35	41			
17	Neil	55	52	59			
18	Average						

a) What formula would she enter in cell B18 to calculate the class average for test A?

b) What formula would she enter in cell E2 to calculate the total possible mark for the three tests?

c) Describe the quickest way of finding the total mark for each pupil.

d) Describe the quickest way of finding the entries for cells C18, D18 and E18.

e) If cell F4 is to contain Andrew's percentage mark as a whole number but without the percentage sign, explain how this can be achieved.

f) If cell G4 is to display the same as cell F4 but with the percentage sign included, explain how this can be achieved.

2 A sum of money has been borrowed and is being paid back at the rate of £1250 per month. Interest is charged at 1% per month. The data for the year is shown in the spreadsheet.
Column B : 1st: : The loan outstanding on the 1st of the month after payment of £1250;
Column C : 31st : the loan outstanding at the end of the month including interest for that month.

	A	B	C	D
1	month	1st	31st	
2	January	£65 240.32	£65 892.72	
3	February	£64 642.72	£65 289.15	
4	March	£64 039.15	£64 679.54	
5	April	£63 429.54	£64 063.84	
6	May	£62 813.84	£63 441.98	
7	June	£62 191.98	£62 813.90	
8	July	£61 563.90	£62 179.53	
9	August	£60 929.53	£61 538.83	
10	September	£60 288.83	£60 891.72	
11	October	£59 641.72	£60 238.14	
12	November	£58 988.14	£59 578.02	
13	December	£58 328.02	£58 911.30	

a) What formula would have been entered in cell C2 ?
b) What formula would have been entered in cell B3 ?
c) How would the remaining entries be obtained?
d) Indicate how you could use column D to calculate the total interest paid over the year, with the answer appearing in cell D14.

3 Members of Banorky Badminton Club pay £2·50 each evening that they attend. Carol keeps the books and enters all details on a spreadsheet, which she also uses to keep a check on how much cash she should have in her 'badminton purse'.

	A	B	C	D	F	G	H	I	K
1	**Income**				**Expenditure**				cash
2	date			total	date			total	in hand
3	1–8	cash in hand	£45.50	£45.50	25–8	shuttlecocks	£36.00	£36.00	£9.50
4	2–9	subscriptions	£20.00	£65.50	2–9	hire of hall	£15.00	£51.00	£14.50
5				£65.50		juice & biscuits	£4.50	£55.50	£10.00
6	9–9	subscriptions	£22.50	£88.00	9–9	hire of hall	£15.00	£70.50	£17.50
7				£88.00		juice & biscuits	£4.00	£74.50	£13.50
8	16–9	subscriptions	£17.50	£105.50	16–9	hire of hall	£15.00	£89.50	£16.00
9				£105.50		juice & biscuits	£5.00	£94.50	£11.00
10	23–9	subscriptions	£15.00	£120.50	23–9	hire of hall	£15.00	£109.50	£11.00
11				£120.50		juice & biscuits	£4.00	£113.50	£7.00
12	30–9	subscriptions	£25.50	£145.50	30–9	hire of hall	£15.00	£128.50	£17.00
13				£145.50		juice & biscuits	£3.50	£132.00	£13.50
14	7–10	subscriptions	£22.50		7–9	hire of hall	£15.00		
15						juice & biscuits	£3.50		

a) How does Carol ensure that all the entries include the '£' sign.

b) In order to complete column D by filling down, what formulae should be entered in cells D3 and D4 ?

c) Write down the formulae for cells I3 and I4.

d) Column K is also completed by filling down.

  (i) State the formula for cell K3.

  (ii) If you then click on cell K11, what formula appears in the edit bar?

e) Deduce the amounts which should appear in cells D14, D15, I14, I15, K14 and K15.

4 *Fred's Super Furniture Store* has taken 20% off all its stock for the summer sale. Fred has a spreadsheet with all his stock details.

	A	B	C	D	E	F	G
1	**item**	**cost price**	**selling price**	**reduction**	**sale price**	**profit**	**% age profit**
2	double bed	£250.00	£325.00	£65.00	£260.00	£10.00	4%
3	wardrobe	£210.00	£275.00	£55.00	£220.00	£10.00	5%
4	sideboard	£300.00	£385.00				
5	oil painting	£165.00	£225.00				

a) What formula did Fred enter in cell: (i) D2 (ii) E2 (iii) F2 (iv) G2 ?
b) How did Fred ensure that all the entries in column G included the '%' sign and were all whole numbers?
c) Calculate the entries for cells D4 to G5.

5 In order to solve the equation $x^2 - 12x + 35 = 0$, Edgar wrote $x^2 - 12x + 35$ as $(x - 12)x + 35$. Then he chose values of x from 1 to 10 and computed corresponding values of $(x - 12)$, $x(x - 12)$ and $(x - 12)x + 35$, from which he solved the equation.

	A	B	C	D
1	x	$x - 12$	$x(x - 12)$	$x(x - 12) + 35$
2	1	−11	−11	24
3	2	−10	−20	15
4	3	−9	−27	8
5	4	−8	−32	3
6	5	−7	−35	0
7	6	−6	−36	−1
8	7	−5	−35	0
9	8	−4	−32	3
10	9	−3	−27	8
11	10	−2	−20	15

a) Explain how to obtain column A by filling down.
b) State the formula entered in cell: (i) B2 (ii) C2 (iii) D2.
c) What will appear in the edit bar if you click on cell D9 ?
d) State the roots of this quadratic equation.
e) Write down the factors of $x^2 - 12x + 35$.

6 Flushed with his success in solving the previous equation, Edgar set about solving the equation $12x^2 - 53x + 56 = 0$. He entered values of x in column A and computed values of $12x^2 - 53x + 56$ in column B.

	A	B	C	D	E
1	x	12 * x ^ 2 – 53 x + 56		x	12 * x ^ 2 – 53 x + 56
2	0	56		1.0	15.00
3	1	15		1.1	12.22
4	2	–2		1.2	9.68
5	3	5		1.3	7.38
6	4	36		1.4	5.32
7	5	91		1.5	3.50
8	6	170		1.6	1.92
9	7	273		1.7	0.58
10	8	400		1.8	–0.52
11	9	551		1.9	–1.38
12	10	726		2.0	–2.00
13				2.1	–2.38
14				2.2	–2.52
15				2.3	–2.42
16				2.4	–2.08
17				2.5	–1.50
18				2.6	–0.68
19				2.7	0.38
20				2.8	1.68
21				2.9	3.22
22				3.0	5.00

a) What formula did Edgar enter in cell B2 ?

b) Edgar could see no zeros in column B, like he did in column D last time, but he deduced that since the value of $12x^2 - 53x + 56$ went from 15 to –2 between $x = 1$ and $x = 2$, there must be a root of the equation between $x = 1$ and $x = 2$. Where else could he deduce that there must be another root?

Edgar therefore chose values of x from 1 to 3, as shown in column D.

c) He copied cell B3 and pasted it in to cell E2. What did the formula become?

d) Estimate the roots of this quadratic equation correct to 2 decimal places.

109 Revision of Logic Diagrams

Exercise 109

1 This map shows four places A, B, C, and D and the distances between them in kilometres. Construct a tree diagram to find all the possible sensible routes from A to D, and hence find the shortest route.

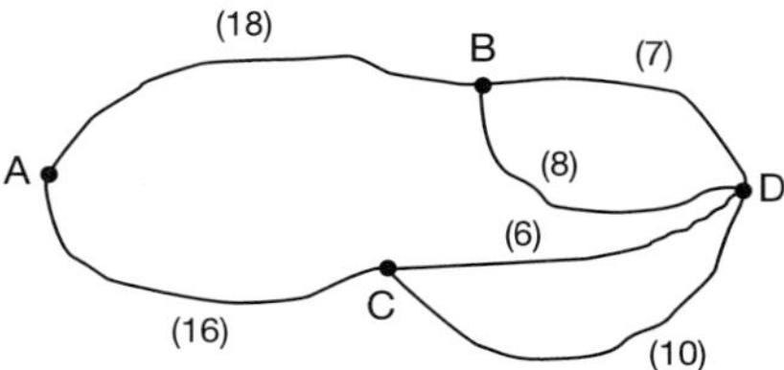

2 (i) A F B E D C (ii) G H J I (iii) K L O N M

a) Find the degree of each vertex in these three network diagrams.
b) Which diagram cannot be drawn by a continuous pencil line without lifting the pencil from the page or going over an edge more than once, and why?
c) For each of the other two diagrams, state a suitable pathway for drawing the network by a continuous pencil line without lifting the pencil from the page or going over any edge more than once.

3 Use this flowchart to calculate the discounted amount using a mail order catalogue when the original balance is:

a) £15·75 b) £37·60 c) £63.

To calculate the discounted amount using a mail order catalogue:

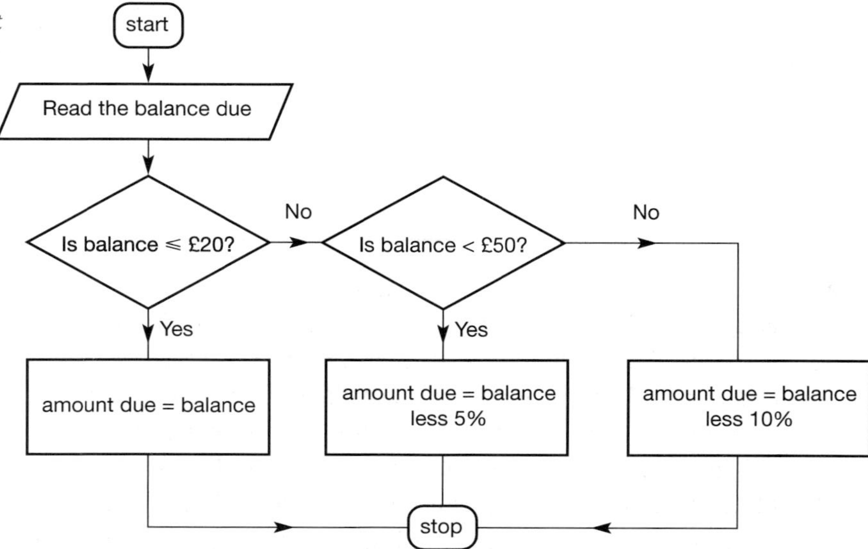

4 Maggie has devised a spreadsheet to check her electricity bills. Each quarterly bill is the sum of the standing charge, £12, and the cost of the units @ 6·45p per unit. She enters her meter readings in columns A and B.

	A	B	C	D	E	F	G
1	previous	current					
2	reading	reading	units used	cost of units	gross total	VAT @ 5%	total cost
3	96 534	98 194	1660	£107.07	£119.07	£5.95	£125.02
4	98 194	99 530	1336	£86.17	£98.17	£4.91	£103.08

a) State the formula required in each of the following cells:
(i) C3 (ii) D3 (iii) E3 (iv) F3 (v) G3.

b) If Maggie highlighted cell A3, which cell would be highlighted after she:
(i) pressed the RETURN key
(ii) pressed the TAB key
(iii) clicked on the √ in the edit bar ?

c) Suppose Maggie 'was in' cell B3 and intended to go to cell B4, but ended up in cell C3 by pressing the wrong key. Other than simply clicking on cell B4, how else could she return to cell B4?

d) Maggie entered the formula '=B3' in cell A4. Describe the different effects of:
(i) filling down columns A to G
(ii) filling down column A followed by filling down columns C to G.

110 Test on Logic Diagrams

Allow 45 minutes for this test

1 This map shows four towns P, Q, R, and S and the distances between them in kilometres. The local newspaper is printed in town P and has to be delivered to Q, R, and S.

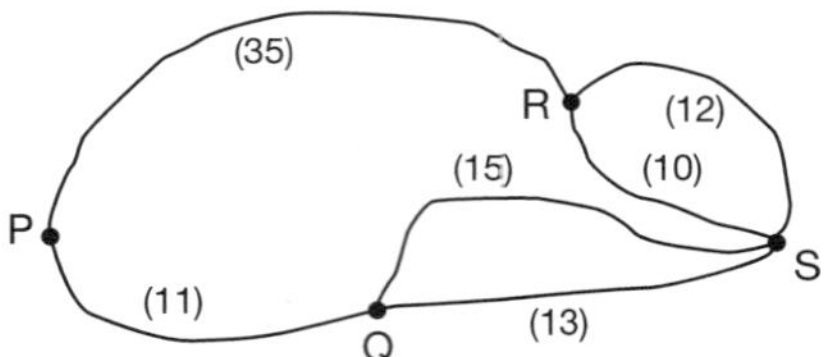

a) Construct a tree diagram to find the eight obvious routes for the van driver to consider, and calculate the length of each of these journeys, starting and finishing at P.
b) Why are there only four different possible distances?
c) Find a ninth route which is even shorter than all of these eight in the tree diagram.

2 a) Find the degree of each vertex in this network diagram.

b) How can you use your answer to part a) to find the number of arcs in the diagram?
Check by counting them.

c) Find a pathway which can be used to draw the network by a continuous pencil line without lifting the pencil from the page or going over any edge more than once, starting and finishing at the same vertex.

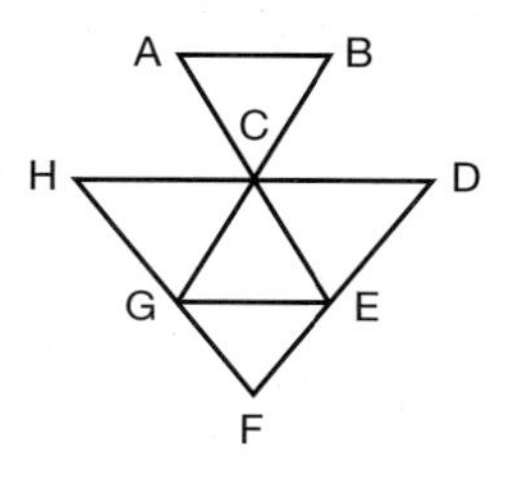

d) If the edge AH is added to the diagram, is it still possible to find such a pathway? [Find an example, or justify that it cannot be done.]

3 Use this flowchart to calculate the net cost of a package holiday trip with *Bairn Travel* when the number of pupils in the group is:

a) 8 at a basic cost of £185

b) 17 at a basic cost of £225

c) 25 at a basic cost of £510.

To calculate group holiday costs:

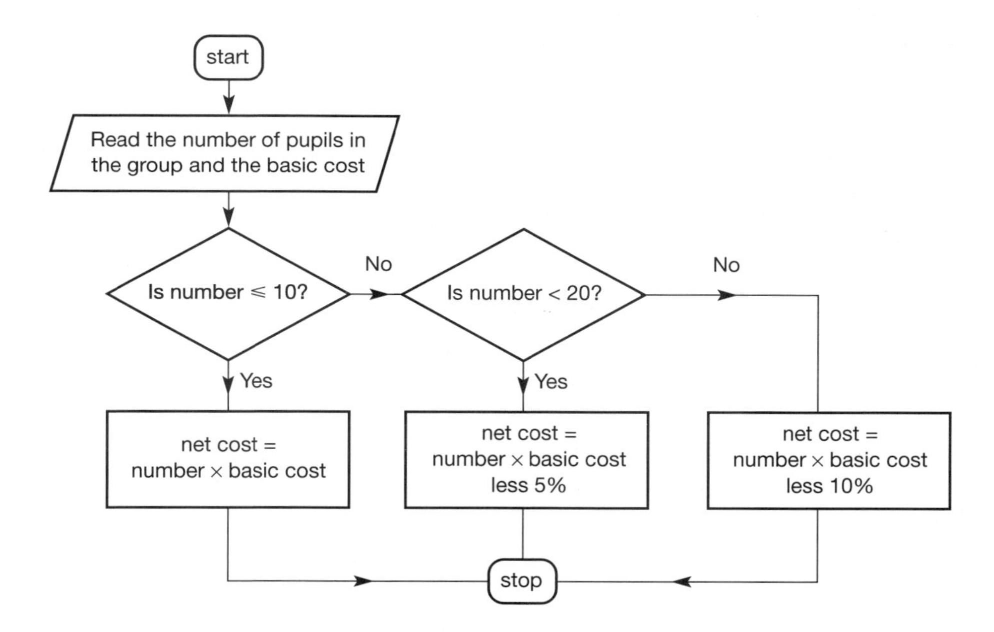

4 Some football match statistics were recorded on a spreadsheet:

	A	B	C	D	E
1		throw-ins taken	corners conceded	free kicks conceded	goals scored
2	Andy Young	12	2	0	0
3	Willie McNaught	2	1	0	0
4	Andy Leigh	13	3	0	1
5		9	6		

a) In the default arrangement for entries in a spreadsheet, what sort of alignment do you expect to see for:
   (i) text entries (ii) numerical entries?

b) (i) State the formula in cell B5 required to calculate the average for this column.
   (ii) State the formula in cell C5 required to calculate the sum of this column.

c) Who was most likely to be the centre-half (the central defender)?

FORMULAE

111 Revision of Formulae in Symbols

Reminder

This is not a new topic. You have used many formulae in the past. We are only going to get some more practice in using and evaluating them.

Example Calculate the volume of this soup can:

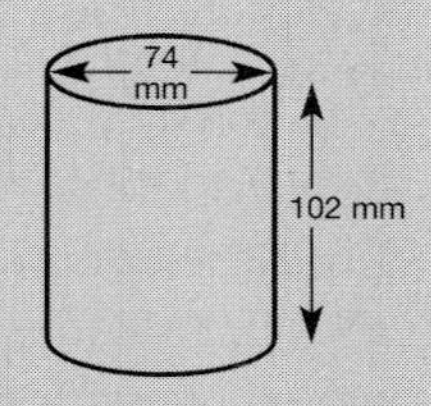

Solution The formula for the volume, V mm^3, of a cylinder with base radius r mm and height h mm is given by $V = \pi r^2 h$.
We need to replace the symbols (or variables) in this formula by the values appropriate to this calculation. For the given soup can, $r = \frac{1}{2} \times (74) = 37$ and $h = 102$
Hence $V = \pi \times 37^2 \times 102 = 438\ 685{\cdot}715$ (by calculator)
$= 439\ 000$ mm^3 (to 3 significant figures)

Exercise 111

1 The formula for the perimeter, P mm, of a rectangle of length l mm and breadth b mm is $P = 2(l + b)$.
Calculate the perimeter when:
a) $l = 4$ and $b = 7$
b) $l = 26$ and $b = 31$
c) $l = 2{\cdot}5$ and $b = 7{\cdot}8$.

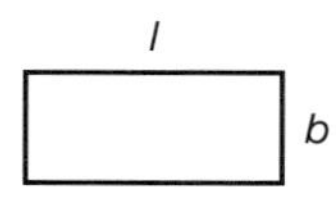

2 The formula for the perimeter, P mm, of a triangle with sides of length a mm, b mm and c mm is $P = a + b + c$.
Calculate the perimeter when:
a) $a = 2$, $b = 3$ and $c = 4$
b) $a = 11$, $b = 17$ and $c = 25$
c) $a = 2{\cdot}6$, $b = 3{\cdot}7$ and $c = 5{\cdot}1$.

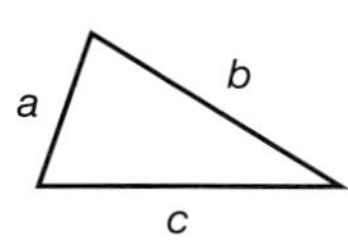

3 The formula for the area, A mm^2, of this triangle with base b mm and height h mm is $A = \frac{1}{2}bh$.

Calculate the area when:

a) $b = 50$ and $h = 20$

b) $b = 125$ and $h = 75$

c) $b = 47$ and $h = 13$.

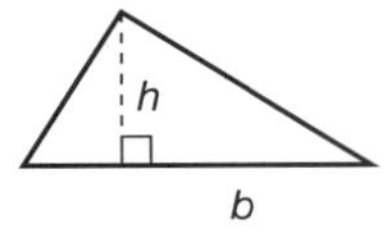

4 a) The formula for the circumference, C mm, of a circle of radius r mm is $C = 2\pi r$.

Calculate the circumference when:

i) $r = 2$ ii) $r = 3{\cdot}2$ iii) $r = 6{\cdot}5$.

b) The formula for the area, A mm^2, of this circle is $A = \pi r^2$.

Calculate the area when:

i) $r = 3$ ii) $r = 7$ iii) $r = 11$.

5 The volume, V mm^3, of a cube of side λ mm is given by $V = \lambda^3$.

Calculate the volume when:

a) $\lambda = 1$ mm b) $\lambda = 2$ mm c) $\lambda = 5{\cdot}2$ mm.

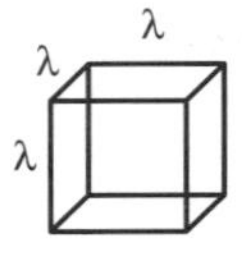

6 The equation of the line passing through the point (0, b) with gradient a is $y = ax + b$.

Find the equation of the line passing through:

a) (0, 1) with gradient 2

b) (0, 2) with gradient −1

c) (0, −1) with gradient 3.

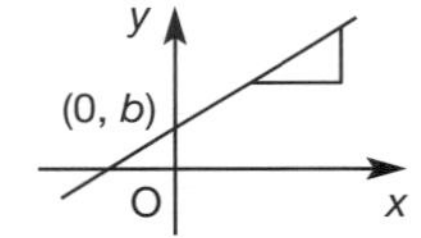

7 The volume, V mm^3, of the loaf shape shown, which has length λ mm and cross-sectional area A mm^2, is given by $V = A\lambda$.

Calculate the volume when:

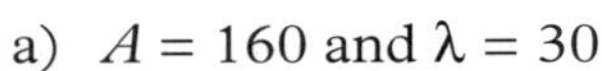

a) $A = 160$ and $\lambda = 30$

b) $A = 185$ and $\lambda = 42$

c) $A = 192$ and $\lambda = 37$.

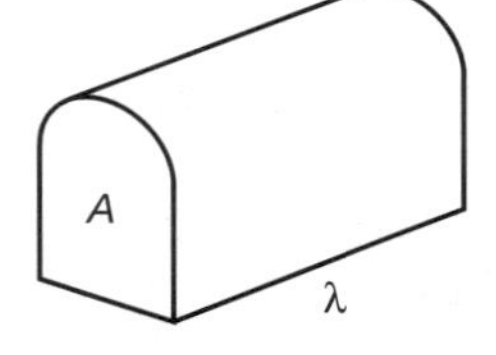

8 The formula for the volume, V mm^3, of a cuboid of length λ mm, breadth b mm and height h mm is $V = lbh$.

Calculate the volume when:

a) $\lambda = 14$, $b = 5$ and $h = 6$

b) $\lambda = 10$, $b = 5$ and $h = 5$

c) $\lambda = 12{\cdot}3$, $b = 7{\cdot}4$ and $h = 2{\cdot}1$.

9 If £P is invested for T years at R% simple interest, then the interest due, £I, is given by the formula:

$$I = \frac{PTR}{100}.$$

Calculate the interest due on:

a) £200 invested for 2 years at 5% p.a.

b) £1000 invested for 5 years at 6% p.a.

c) £500 invested for 6 months at 4% p.a.

10 The volume, V cm^3, of a cone with base radius r cm and height h cm is given by $V = \frac{1}{3}\pi r^2 h$

Expressing your answers to the nearest litre, (1 litre = 1000 cm^3) calculate the volume of a cone with:

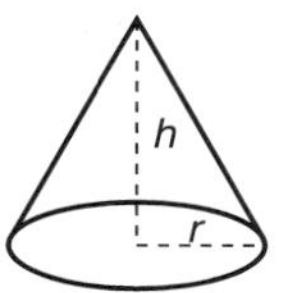

a) $r = 50$ and $h = 60$
b) $r = 75$ and $h = 180$
c) $r = 125$ and $h = 200$.

11 The volume, V mm^3, of a sphere of radius r mm is given by $V = \frac{4}{3}\pi r^3$.
Calculate, correct to 3 significant figures, the volume of a sphere of radius:

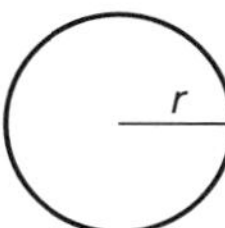

a) 7 mm   b) 18 mm   c) 102 mm.

12B The standard deviation (s) of a data set can be calculated from the formula:

$$s = \sqrt{\frac{\sum x^2 - \frac{(\sum x)^2}{n}}{n-1}}.$$

Calculate the standard deviation of a data set where:
a) $\sum x^2 = 216786$ $\sum x = 2628$ $n = 32$
b) $\sum x^2 = 176061$ $\sum x = 2135$ $n = 26$
c) $\sum x^2 = 21434$ $\sum x = 770$ $n = 28$.

112 Formulae in Words

Reminder

Many formulae are just as easily remembered or explained in words.

Example Calculate the mean of the data set 11 17 23 28 34 46.

Solution The formula for the mean is $mean = \frac{\textit{the total of all the numbers}}{\textit{the number of numbers}}$ $\left(\text{or } \bar{x} = \frac{\sum x}{n}\right)$.

The total of all these numbers is 159.
There are 6 numbers so the mean is: $\frac{159}{6} = 26{\cdot}5$.

Exercise 112

1 In a network diagram, (*the sum of the degrees of the vertices*) = 2 × (*the number of edges*).
Find the sum of the degrees of the vertices when the number of edges is:
a) 4   b) 5   c) 11.

2 When a trial has equally likely outcomes, the formula for the probability of an event is:

$$P(\text{an event}) = \frac{\textit{number of outcomes favourable to the event}}{\textit{total number of possible outcomes.}}$$

One card is drawn at random from a standard pack of 52 cards.
Find the probability of drawing:
a) a red ten
b) a black ace or a red queen
c) a face card (i.e. Jack, Queen, King or Ace).

3 The traditional way of calculating the length of time required to cook a turkey was by the formula: *cooking time = 20 minutes + 20 minutes per lb.*
Calculate the cooking time required for a turkey of mass:
a) 10 lb b) 12 lb c) 15 lb.

4 A child's dose of medicine can be calculated from the formula:

$$\textit{child's dose} = \frac{\textit{mass of child in kg}}{70} \times \textit{adult dose.}$$

Rounding your answers to the nearest 5 mg, calculate the child's dose for a child of mass:
a) 25 kg where the adult dose is 250 mg
b) 20 kg where the adult dose is 500 mg
c) 27 kg where the adult dose is 125 mg.

5 The equation of a line is $y = (\text{gradient} \times x) + (\text{intercept})$
Write down the equation of the line with:
a) gradient 3 and intercept 2
b) gradient 1 and intercept –5
c) gradient –2 and intercept –3.

6 The area (A mm^2) of a triangle with lengths of sides measured in mm is given by
A = *(length of one side)* × *(length of another side)* × *(sine of the angle between these sides)* ÷ 2.
Calculate the area of these triangles:

a)
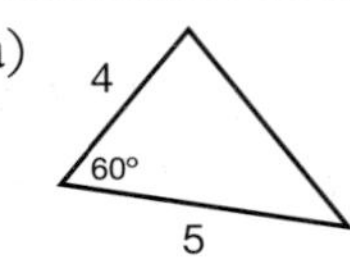

b)
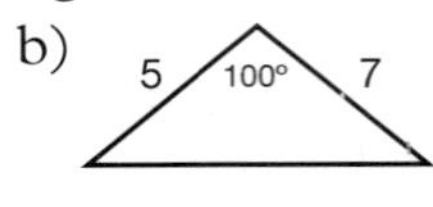

c)
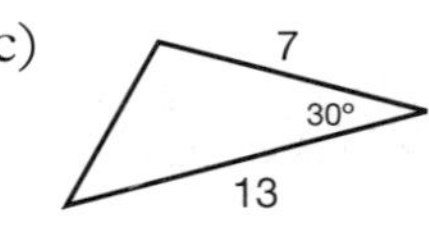

7 When planning a hill walk, walkers must allow more time for climbing as their progress then is slower than when walking on the level. In 1892, the Scottish climber Naismith devised the rule:

$$\textit{time required} = \frac{\textit{distance in miles}}{\textit{3 miles per hour}} + \textit{30 mins for every 1000 feet of climbing.}$$

Apply Naismith's rule to estimate the time required for a hill walk of:
a) length 18 miles climbing 1000 feet
b) length 15 miles climbing 3000 feet
c) length 20 miles climbing 2500 feet.

8 The metric version of Naismith's Rule is:

$$\textit{time required} = \frac{\textit{distance in km}}{\textit{5 km per hour}} + \textit{30 mins for every 300 m of climbing.}$$

Apply Naismith's rule to estimate the time required for a hill walk of:

a) length 20 km climbing 600 m
b) length 15 km climbing 750 m
c) length 23 km climbing 1000 m.

9 There are three North lines on an ordnance survey map: true north, grid north and magnetic north. The direction of magnetic north is different at different points on the earth's surface, as well as changing over time. In Scotland, in 2004 the magnetic variation, i.e. the deviation of magnetic north from grid north, was 4° west and was decreasing by $\frac{1}{2}^{\circ}$ every 5 years.

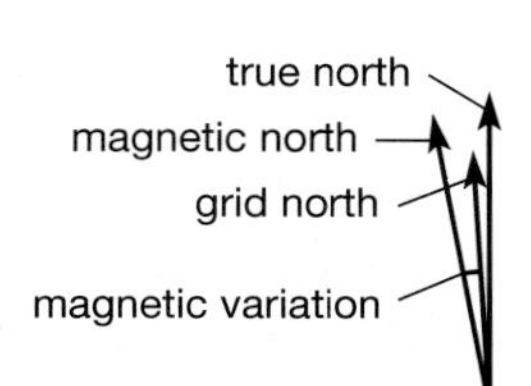

Hence the formula: $\textit{magnetic variation} = \left[4 - \frac{\textit{number of years since 2004}}{10}\right]^{\circ}.$

Calculate the magnetic variation in:

a) 2009 b) 2014 c) 2054.

10 Producers of home-made wine can calculate the percentage of alcohol by volume of their wines by measuring the specific gravity of the must before fermentation starts, and after it has finished. The formula required is

$$\textit{percentage of alcohol by volume} = \frac{\textit{(initial specific gravity)} - \textit{(final specific gravity)}}{0.00736}.$$

Calculate the percentage of alcohol by volume for a wine where, during fermentation, the specific gravity drops from:

a) 1·115 to 1·003 b) 1·109 to 1·002 c) 1·107 to 0·998.

113 Formulae in Symbols

Reminder

All the formulae used in Exercise 111 should have been familiar to you. Those used in this exercise will not be. You do not need to know anything about the subject matter when there is a formula given. Simply substitute the values of the variables into the formula and work out what is required.

Reminder continued

Example The speed, v ms^{-1}, of a particle performing simple harmonic motion of amplitude a m is given by the formula, $v = \omega\sqrt{a^2 - x^2}$. where ω is a constant (for the given motion) and x is the displacement from the centre of the oscillation.
Calculate the speed of the particle when the displacement is 3 m, the amplitude 5 m and the value of ω is 4.

Solution We do not need to know anything about simple harmonic motion to answer this question. All we need to know is to use $\omega = 4$, $a = 5$ and $x = 3$.

So $v = \omega\sqrt{a^2 - x^2} \Rightarrow v = 4\sqrt{5^2 - 3^2} = 4\sqrt{25 - 9} = 4 \times \sqrt{16}$
$= 4 \times 4 = 16$

i.e. the speed is 16 ms^{-1}.

Exercise 113

1 A block of wood is placed on a horizontal hinged desk lid and the lid slowly raised. The block begins to slide when the desk lid is inclined at an angle of $a°$ to the horizontal. This indicates that the coefficient of friction between the block and the desk lid is μ, where $\mu = \tan a°$.
Calculate μ when a is:

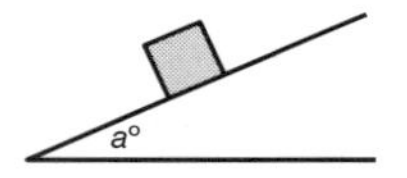

a) 20 b) 37 c) 41.

2 The impulse, I units, of a force which changes the speed of a body of mass m kg from u m/s to v m/s is given by $I = m(v - u)$.
Calculate the impulse of a force acting on a body of mass:
a) 3 kg and changing its speed from 4 m/s to 8 m/s
b) 2 kg and changing its speed from 1 m/s to 7 m/s
c) 5 kg and changing its speed from 0 m/s to 9 m/s.

3 Statisticians often calculate the coefficient of correlation (r) between two matched sets of data. There is also a coefficient of alienation (k) between these sets of data, defined by:

$$k = \sqrt{1 - r^2}.$$

Calculate the coefficient of alienation when the coefficient of correlation is:
a) 0 b) $\frac{1}{2}$ c) 1.

4 A ball which is dropped from a height of H units on to a fixed surface and rebounds to a height of h units is said to have a coefficient of restitution, e, given by: $e = \sqrt{\frac{h}{H}}$.

Calculate the coefficient of restitution when:

a) a golf ball is dropped 8 feet on to a concrete slab and rebounds $4\frac{1}{2}$ feet
b) a tennis ball is dropped 2 m on to grass and rebounds 0·5 m
c) a ball bearing is dropped 1 m on to an anvil and rebounds 0·36 m.

5 When a random sample of size n of a population is measured for some statistic, the mean (m) and the standard deviation (s) of the sample are usually calculated. The credibility of the value of m as an estimate of the mean of the underlying population is measured by a quantity, SE_m, called the standard error of the mean, which is calculated from the formula:

$$SE_m = \frac{s}{\sqrt{n}}.$$

Calculate the standard error of the mean for a sample of size:

a) 25 and standard deviation 5
b) 100 and standard deviation 3·7
c) 400 and standard deviation 10·2.

6 A point on the circumference of a disc of radius r m spinning at ω revolutions per minute has a linear speed of v m/s where v is given by the formula

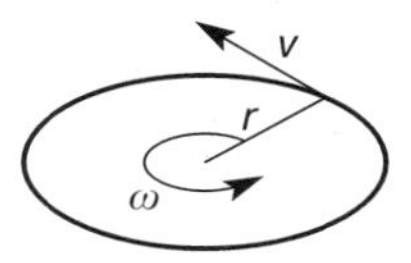

$$v = \frac{\pi}{30} r\omega.$$

Calculate the linear speed of a point on the circumference of a disc of radius:

a) 150 mm spinning at 45 r.p.m. (a vinyl single) (watch the units)
b) 87 mm spinning at $33\frac{1}{3}$ r.p.m. (a vinyl L.P.) (watch the units)
c) 0.2 m spinning at 100 r.p.m.

7 The distance s m fallen by a body in the first t seconds after being dropped is given by $s = \frac{1}{2}gt^2$ where $g = 9{\cdot}81$ ms^{-2} (the acceleration due to gravity).

Calculate the distance fallen in the first:

a) 1 s
b) 1·5 s
c) 2 s.

8 If a mass of m kg rests on a smooth horizontal surface, and a force of P newtons acts on it for t seconds, then the distance, s m, it will cover from rest is given by:

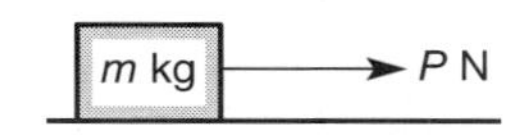

$s = \frac{Pt^2}{2m}$, and the speed, v ms^{-1}, it will attain is given by $v = \frac{Pt}{m}$.

Calculate the distance moved and the speed attained by a mass of:

a) 2 kg acted on by a force of 3 N for 5 s
b) 3 kg acted on by a force of 5 N for 4 s
c) 5 kg acted on by a force of 10 N for 6 s.

9 The period, T seconds (i.e. the time taken for one revolution), of a conical pendulum of height h m as shown is given by

$T = 2\pi\sqrt{\dfrac{h}{g}}$, where $g = 9{\cdot}81\ \text{ms}^{-2}$ (acceleration due to gravity).

Find the period of a conical pendulum of height:

a) 1 m b) 0·75 m c) 0·5 m.

10B The coordinates of the point of intersection of the straight lines with equations

$$\begin{aligned} ax + by &= c \\ px + qy &= r \end{aligned} \quad \text{are given by } x = \frac{cq - br}{aq - bp},\quad y = \frac{ar - cp}{aq - bp}.$$

Find the coordinates of the point of intersection of the lines with equations:

a) $\begin{aligned} x - y &= 1 \\ x + y &= 3 \end{aligned}$ b) $\begin{aligned} 4x - 2y &= 1 \\ 3x + y &= 7 \end{aligned}$ c) $\begin{aligned} x - 2y &= -2 \\ 9x - 3y &= -28 \end{aligned}$

11B The quadratic formula, which gives the roots of the quadratic equation $ax^2 + bx + c = 0$, is:

$$x_1 = \frac{-b + \sqrt{b^2 - 4ac}}{2a} \qquad x_2 = \frac{-b - \sqrt{b^2 - 4ac}}{2a}.$$

Find the roots of the equation:

a) $x^2 - 12x + 35 = 0$ b) $x^2 - x - 12 = 0$ c) $8x^2 - 22x + 15 = 0$.

12B When a body is projected with velocity u m/s at an elevation of $a°$, the horizontal range, R m, is given by:

$R = \dfrac{u^2 \sin(2a°)}{g}$ and the greatest height attained, H m, by

$H = \dfrac{(u \sin a°)^2}{2g}$, where $g = 9{\cdot}81\ \text{m/s}^2$ as before.

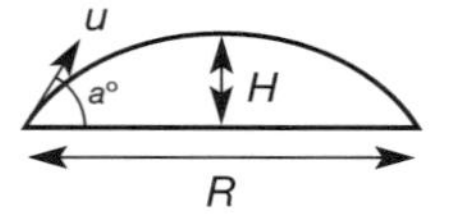

Calculate the range and greatest height for a body projected at:

a) 10 m/s at an elevation of 40°
b) 20 m/s at an elevation of 50°
c) 15 m/s at an elevation of 30°.

114 Transposing Formulae

Reminder

So far in this chapter you have been asked to calculate the value of the subject of the formula, i.e. the variable on its own on the left hand side of the equation. Occasionally you have to calculate the value of one of the other variables, probably on the right hand side. This is not new, you have already met this sort of problem, for example, in using the Sine Rule.

Reminder continued

Example For a network diagram, we have met the formula $T = 2E$, where T is the total number of the degrees of all the vertices and E is the number of edges.
Calculate the number of edges when the sum of the vertex degrees is 18.

Solution

Method 1 (Solve an equation)

$T = 2E$

$\Rightarrow\ 18 = 2E$

$\Rightarrow\ E = 18 \div 2 = 9$

Method 2 (Change the subject)

$T = 2E$

$\Rightarrow\ E = \frac{1}{2}T$

$\Rightarrow\ E = \frac{1}{2}(18) = 9$

Exercise 114

1 A rectangular flower bed has breadth 5 m and length λ m.
The formula for the perimeter is therefore $P = 2(\lambda + 5)$.
Calculate the length of the flower bed when the perimeter is:

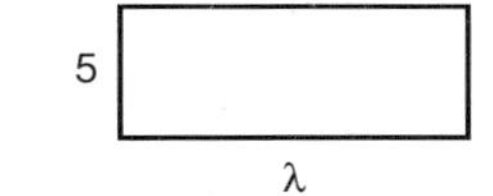

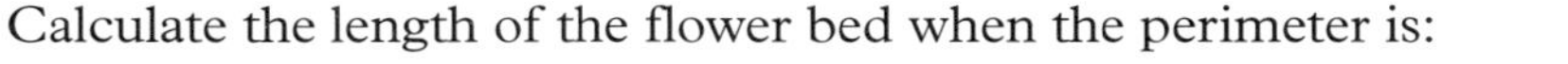

a) 20 m   b) 40 m   c) 75 m.

2 The perimeter, P m, of an isosceles triangle with equal sides of length a m and a third side of length b m is given by $P = 2a + b$.
Calculate the length of side b when the perimeter is:

a) 20 m and the length of side a is 5 m
b) 25 m and the length of side a is 10 m
c) 30 m and the length of side a is 9 m.

3 The volume (V m^3) of a pipe of cross-sectional area A m^2 and length λ m is given by:

$V = A\lambda$.

Calculate the length of pipe which has a cross-sectional area of:

a) 0·50 m^2 and a volume of 10 m^3
b) 0·25 m^2 and a volume of 4 m^3
c) 0·80 m^2 and a volume of 24 m^3.

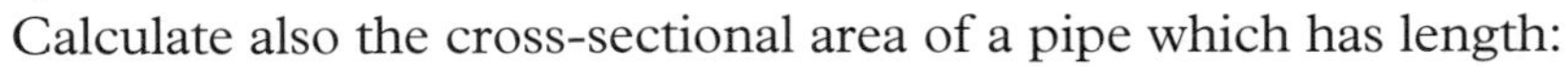

Calculate also the cross-sectional area of a pipe which has length:

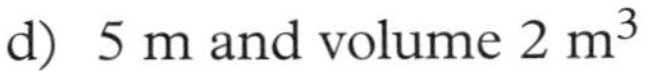

d) 5 m and volume 2 m^3
e) 8 m and volume 3 m^3
f) 4 m and volume 2·5 m^3.

4 Calculate the radius of a circle which has circumference:

a) 5·027 m   b) 8·168 m   c) 4712 mm.

5 Calculate the radius of a circle which has area:

a) 0·1257 m^2 b) 0·3848 m^2 c) 4·524 m^2.

6B For the formula $p = qt + r$, find the value of:

a) p when $q = 2$, $t = 3$ and $r = 4$
b) q when $p = 33$, $t = 4$ and $r = 5$
c) t when $p = 43$, $q = 5$ and $r = 3$
d) r when $p = 70$, $q = 6$ and $t = 11$.

7B The variables w, x, y and z are related by the formula $w^2 = x^2 + 2yz$. Find the value of:

a) w when $x = 2$, $y = 6$ and $z = 5$
b) x when $w = 7$, $y = 2$ and $z = 10$
c) y when $w = 4$, $x = 2$ and $z = 1$
d) z when $w = 5$, $x = 1$ and $y = 6$.

8H Calculate the radius of a sphere which has volume:

a) $4{\cdot}1888\ m^3$
b) $65{\cdot}45\ m^3$
c) $2{\cdot}1447\ m^3$.

115 Revision of Formulae

Exercise 115

1 The surface area, A mm^2, of a sphere of radius r mm is given by the formula $A = 4\pi r^2$. Calculate the surface area of a sphere of radius:

a) 5 mm
b) 15 mm
c) 23 mm.

2 The surface area, A mm^2, of a closed cylinder (e.g. a can of beans) with base radius r mm and height h mm is given by $A = 2\pi r^2 + 2\pi rh$.
Calculate the surface area of a closed cylinder with base radius:

a) 40 mm and height 110 mm
b) 35 mm and height 90 mm
c) 102 mm and height 350 mm.

3 The average speed for a journey is calculated from the formula:

$$\textit{average speed} = \frac{\textit{total distance}}{\textit{total time}}.$$

Calculate the average speed for a journey of:

a) 4 km in 1 hour followed by 5 km in 1 hour
b) 8·5 km in $1\frac{1}{2}$ hours followed by 8 km in $1\frac{1}{4}$ hours
c) 11 km in $2\frac{1}{4}$ hours followed by $12\frac{3}{4}$ km in $2\frac{1}{2}$ hours.

4 Fuel consumption of motor vehicles is often measured in litres per 100 km, so the formula is:

$$\textit{consumption} = 100 \times \frac{\textit{number of litres used}}{\textit{number of km driven}}.$$

Calculate the fuel consumption for:

a) a Morris Minor which used 25 litres of petrol to travel 280 km
b) an Austin Mini which used 20 litres of petrol to travel 250 km
c) a Sunbeam Talbot which used 40 litres of petrol to travel 300 km.

5 a) When two resistances, R_1 and R_2 ohms, are wired up in parallel in an electrical circuit, the effective resistance, R ohms, is given by:

$$R = \frac{R_1 \times R_2}{R_1 + R_2}.$$

Calculate the effective resistance when the individual resistances are:

(i) 2 ohms and 3 ohms
(ii) 3 ohms and 5 ohms
(iii) 7 ohms and 11 ohms.

b) When three resistances; R_1, R_2 and R_3 ohms, are wired up in parallel in an electrical circuit, the effective resistance, R ohms, is given by

$$R = \frac{R_1 \times R_2 \times R_3}{(R_1 \times R_2) + (R_2 \times R_3) + (R_3 \times R_1)}.$$

Calculate the effective resistance when the individual resistances are:

(i) 2, 3 and 5 ohms (ii) 3, 5 and 7 ohms (iii) 6, 8 and 10 ohms.

6B Find the value of the unknown quantity asked for:

a) x where $P = 3x$ and $P = 33$
b) y where $Q = 5y^2$ and $Q = 20$
c) z where $R = 2(5 + z)$ and $R = 32$
d) r where $A = 1 + r^2$ and $A = 65$
e) p where $B = 4(p + q)$, $B = 48$ and $q = 7$
f) q where $C = k^2 + 4qr$, $C = 81$, $k = 5$ and $r = 2$.

116 Test on Formulae

Allow 45 minutes for this test

1 The area, A mm^2, of an annulus (e.g. a washer) with outer diameter D mm and inner diameter d mm is given by the formula:

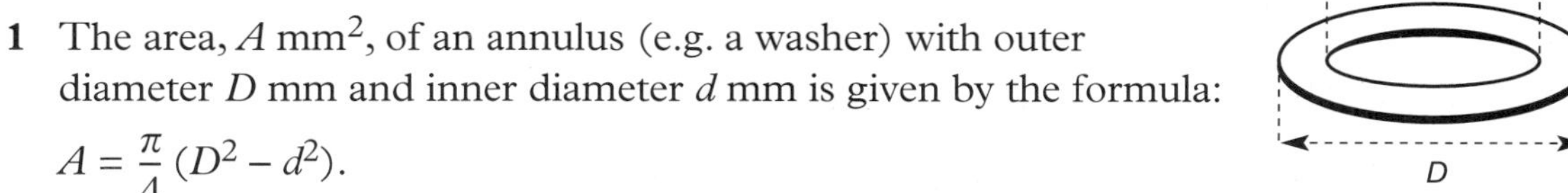

$$A = \frac{\pi}{4}(D^2 - d^2).$$

Calculate the area of an annulus with outer and inner diameters of:

a) 30 mm and 20 mm b) 30 mm and 10 mm c) 20 mm and 5 mm.

2 The volume, V m^3, of the swimming pool shown (with lengths in metres) is given by the formula:

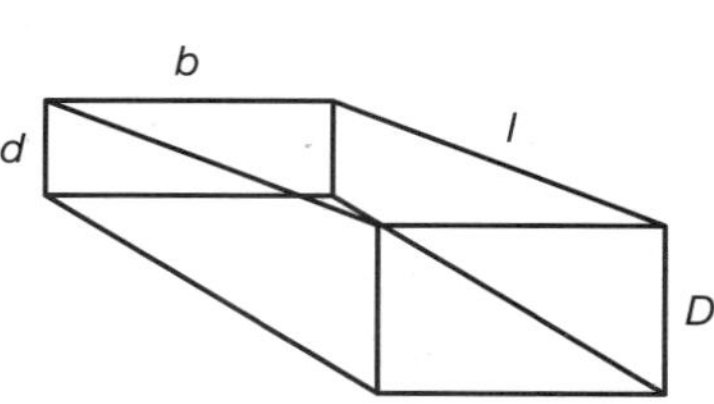

$$V = \frac{1}{2}(d + D)\,lb.$$

Calculate the volume of a pool of this shape when:

a) $d = 1$, $D = 3$, $b = 8$ and $l = 50$
b) $d = 1{\cdot}5$, $D = 2{\cdot}5$, $b = 8$ and $l = 60$
c) $d = 1$, $D = 2$, $b = 6$ and $l = 40$.

3 The volume of a prism can be found from the formula:

volume = (*area of cross-section*) × (*length*).

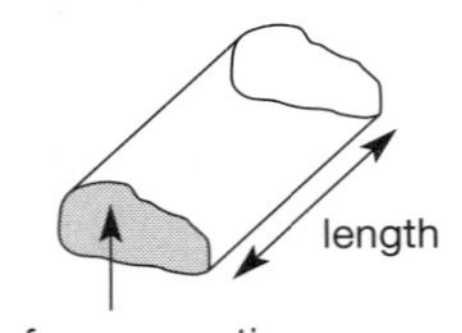

Calculate the volume of a prism having a cross-sectional area of:
a) 50 mm^2 and a length 200 mm
b) 75 mm^2 and a length 150 mm
c) 150 mm^2 and a length 100 mm.

4 The coordinates of the midpoint of a line can be obtained by the formula:

$$(x, y) = \left(\frac{\textit{sum of the two x-coordinates}}{2}, \frac{\textit{sum of the two y-coordinates}}{2}\right).$$

Find the coordinates of the midpoint of the line joining:
a) (3, 4) and (7, 8) b) (2, –1) and (9, 5) c) (–2, –3) and (6, –4).

5 Parallel rays of light pass through a convex lens and meet at the focus (F) which is at a distance (f), called the focal length, from the centre of the lens.

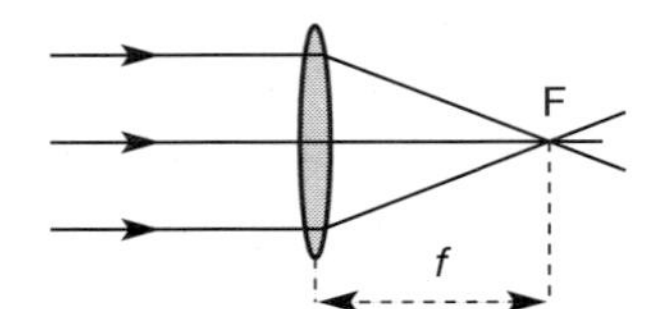

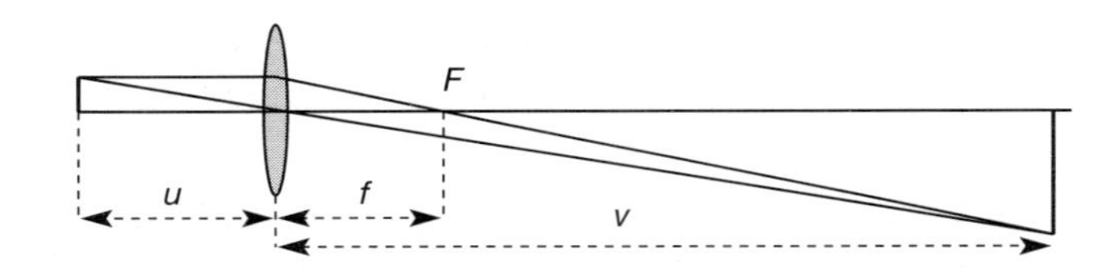

An object at a distance u from the lens can produce an image at a distance v from the lens, as shown above.
The relationship between the three lengths u, v and f can be expressed in the form:

form: $f = \dfrac{u \times v}{u + v}$.

Calculate the focal length of a lens for which:
a) an object distance of 50 mm produces an image distance of 75 mm
b) an object distance of 20 mm produces an image distance of 30 mm
c) an object distance of 40 mm produces an image distance of 70 mm.

6 The area of wallpaper required for cuboid-shaped rooms with length l m, breadth b m and height h m is given by the formula: $A = 2(l + b)h$.
a) Calculate the area of wallpaper required for a room measuring 4 m × 3 m with height 2·5 m.
b) Calculate the length of a room which requires an area of 37m^2 of wallpaper and has a breadth of 2·7 m and height 2·8 m.

FURTHER STATISTICS

117 Grouped Data

Reminder

Discrete data is data which can be counted e.g. number of goals scored, number of teeth missing. **Continuous** data arises from measurement. e.g. heights of pupils, time to complete a task. The data sets which appeared in Unit 2 used either discrete data or continuous data which had been conveniently rounded. When the variate is continuous, it is more common to use grouped data, e.g. times for running 100 m : 12·0 – 12·1; 12·1 – 12·2; 12·2 – 12·3 (seconds); carefully defining '12·1 – 12·2' as $12{\cdot}1 \leq t < 12{\cdot}2$, because 12·2 cannot appear twice. Discrete data is also grouped when the range is inconveniently large.

Example

Potential contestants for a T.V. quiz show were given a straightforward mental arithmetical problem to solve and told to type their answer on a computer. The time taken by each to complete the task was measured by the computer, and the following grouped frequency table was obtained:

time (t seconds)	6–12	12–18	18–24	24–30	30–36	36–42	42–48	48–54	54–60
frequency	7	11	20	28	31	16	9	6	2

Estimate the mean and the mode of this frequency distribution and make a box plot.

Solution

(Note that the question asks for an *estimate* of these statistics. It is not possible to give exact answers for these calculations because the process of grouping data results in some loss of accuracy. We assume, for convenience, that the frequency allotted to each class interval is evenly spread throughout that interval. The mean is calculated as in Section 68, but using the midpoint of each class interval as the variate value. It is usual to use the cumulative frequency column for the calculation of the median and quartiles.)

Reminder continued

class interval	midpoint x	frequency f	xf	cumulative frequency
6–12	9	7	63	7
12–18	15	11	165	18
18–24	21	20	420	38
24–30	27	28	756	66
30–36	33	31	1023	97
36–42	39	16	624	113
42–48	45	9	405	122
48–54	51	6	306	128
54–60	57	2	114	130
Totals		130	3876	

$$\text{mean} = \frac{\Sigma xf}{\Sigma f} = \frac{3876}{130} = 29{\cdot}8$$

The modal class is the interval with the highest frequency, i.e. 30–36. The class containing the median is the class containing the 65th and 66th scores, i.e. 24–30.

To obtain a value for the mode rather than just a class interval, we can construct a bar graph and interpolate using the cells on either side of the modal class, as shown. Only the three relevant cells need be reproduced. Follow the arrow down from F to read the mode off the scale, or use similar triangles: AB:CD = 3:15 = 1:5, so EF:FG = 1:5 or EF:EG = 1:6 so EF is one sixth of EG or one sixth of 6, i.e. 1. Hence the mode is 30 + 1 = 31.
[You would not need to include these labels A, B,…, G.]

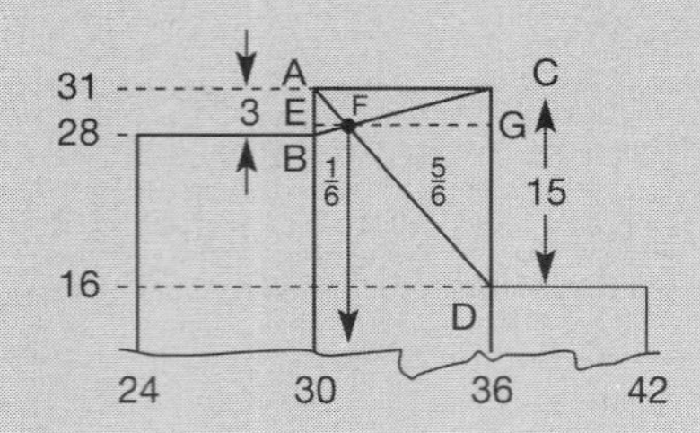

Reminder continued

To obtain values for the median and the quartiles, rather than just class intervals, we can construct an ogive (cumulative frequency curve) and read off the 65th and 66th scores (and the 33rd and 98th scores for the quartiles). Then we can construct a box plot.

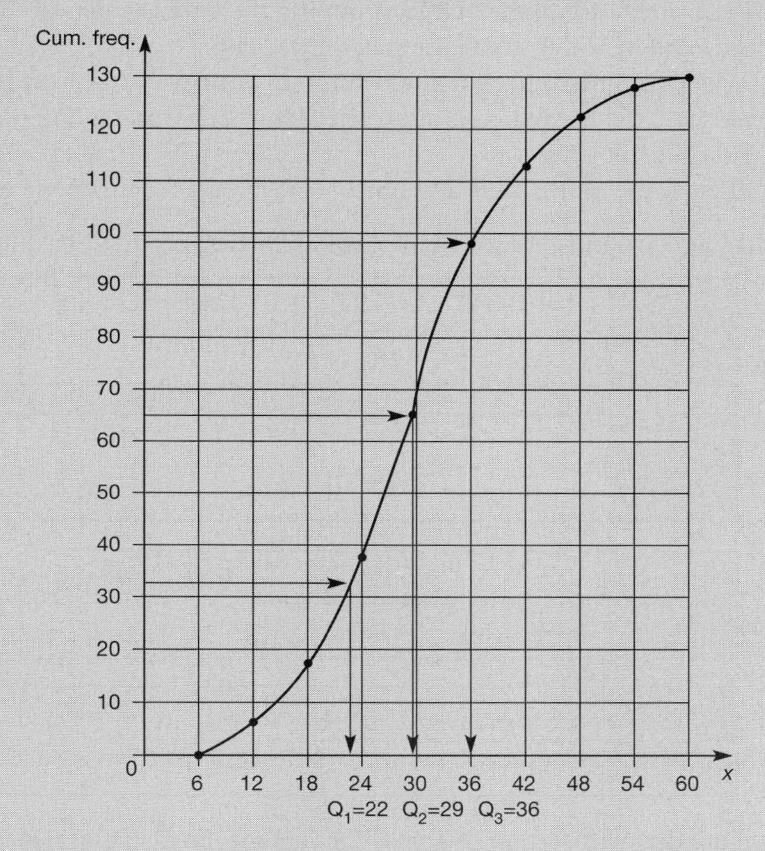

We need to know the range of the data before we can complete a box plot. Just as the other statistics were estimates for our grouped data, so are the outer values of the distribution. The best we can do is use the midpoints of the outer class intervals. Hence:

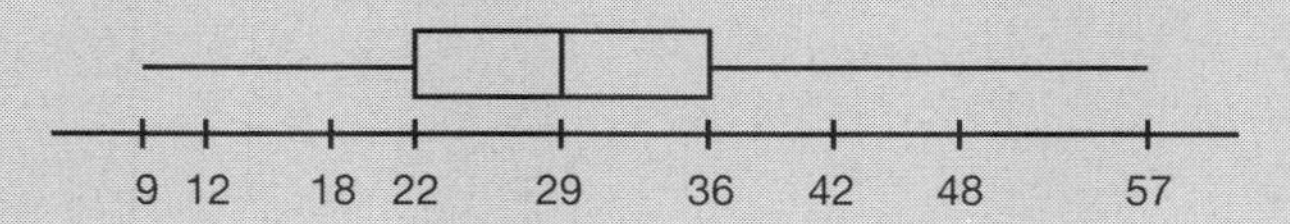

Exercise 117

1 The golf handicaps of the 20 players on a school staff outing were:

3	3	6	7	9	12	12	14	16	16
18	18	18	18	20	24	24	26	28	28.

a) Construct a frequency table, grouping the data in intervals 3 – 5, 6 – 8, 9 – 11…. etc., and hence estimate the mean handicap.
b) Add a cumulative frequency column to your previous working and use it to estimate the median and quartiles. Hence construct a box plot, stating the semi-interquartile range.
c) Identify the modal class and obtain an estimate for the mode using the method shown.

2 Calculate a mean score for each of the following grouped frequency tables:

a)

x	10–14	15–19	21–24	25–29	30–34
f	31	28	13	2	1

b)

x	1–3	4–6	7–9	10–12	13–15
f	5	8	18	30	9

c)

x	15–25	25–35	35–45	45–55
f	8	11	15	6

d)

x	21–23	23–25	25–27	27–29	29–31
f	11	8	9	15	7

3 Estimate the median and quartiles for each of the distributions in question 2.

4 Estimate the mode in each of the distributions in question 2.

5 A group of 36 pupils each counted his or her number of aunts, uncles, and cousins, and their totals were:

5	16	10	16	0	19	20	15	12	6	13	23
13	14	10	19	17	24	26	7	27	10	20	20
30	19	28	8	3	19	25	9	21	16	17	16

a) Construct a frequency table, grouping the data in intervals 0 – 4, 5 – 9, 10 – 14, …. etc., and hence estimate the mean number of such relations.
b) Add a cumulative frequency column to your working and use it to estimate the median, quartiles, and semi-interquartile range.
c) Identify the modal class and obtain an estimate for the mode.
d) Find the probability of a pupil chosen at random from this group belonging to the modal class.

6 A random sample of television viewers were asked how many hours per week they spent viewing. The results are as shown:

number of hours	2–8	9–15	16–22	23–29	30–36
frequency	7	8	17	13	5

a) State the modal class and obtain an estimate for the mode.
b) Obtain an estimate for the mean of the sample.
c) Obtain estimates for the median and quartiles.
d) Find the probability that a person in the sample chosen at random watched more than 22 hours of television per week.

7 This frequency table records the prelim marks for an Intermediate 2 class.

mark (%)	30–39	40–49	50–59	60–69	70–79	80–89
frequency	2	6	8	10	2	2

a) Obtain an estimate for the mode.
b) Obtain an estimate for the mean percentage mark of the class.
c) Obtain estimates for the median and quartiles.
d) What percentage of the class scored less than half marks?

8 The verbal reasoning quotients of 100 army recruits were distributed as follows:

VRQ	55–69	70–84	85–99	100–114	115–129	130–144
frequency	2	15	34	34	14	1

a) Obtain an estimate for the mean VRQ of this batch of recruits.
b) Obtain estimates for the median and quartiles of this distribution.
c) Obtain an estimate for the mode of this distribution.
d) The standard deviation of the above distribution is 15. Compare this with the semi-interquartile range.

9 The practice nurse at a doctor's surgery took the diastolic blood pressure (measured in mm of mercury) of many patients during August, and recorded her results as follows:

blood pressure	68–71	72–75	76–79	80–83	84–87	88–91	92–95
frequency	3	10	31	75	47	28	6

a) Estimate the mean, median, and mode of this distribution.
b) Construct a box plot to illustrate the distribution.
c) If one of these patients is chosen at random, find the probability that the patient has a diastolic blood pressure of 92 or more.

10 The typing rates of a group of beginners on the keyboard (in words per minute) were:

79	47	62	72	64	62	83	77
75	71	65	83	71	77	79	71
81	50	74	79	74	72	46	62
78	70	83	75	68	68	62	73
49	65	75	57	80	65	75	78

a) Construct a grouped frequency table, using intervals 45–49, 50–54, 55–59, …. and hence estimate the mean number of words per minute.
b) Estimate the median, quartiles, and semi-interquartile range.
c) Identify the modal class and estimate the mode.
d) Find the probability of a pupil chosen at random from this group having a typing speed of at least 75 words per minute.

118 Your Statistical Assignment

There is no specified exercise of examples for you to do in this section. Instead, as outcome 4 of the unit test, you have to carry out a short statistical project to show that you can apply the statistical processes that you have learned. You need to find some numbers to crunch to illustrate these skills. Basically you need to obtain data from two fairly similar sources and compare them. The SQA suggests, for example, that you collect two samples of leaves from two different types of rhododendron, measure the lengths of the leaves and then analyse your data.

The essential aspects of your assignment are as follows:

a) Your tabulation of data must include:
- a statement of what it is you are setting out to compare and for *each* of your samples of data:
  - a grouped frequency table (using the same class intervals for both) showing clearly the:
    - midpoints of the class intervals
    - frequencies
    - cumulative frequencies.

b) The analysis of your data must include calculation/estimation of:
- the mean
- the mode
- the median
- the quartiles
- the semi-interquartile range.

c) Your summary and conclusions must show:
- a table of results for the items listed above
- both data sets in a single frequency table showing two frequency columns
- a comparison of the two sample means, the two sample medians, and the two sample modes, together with:

- some comment regarding one of the comparisons (in the context of your assignment)
- another comment which compares all three measures and summarises the situation
- a comparison of the variability of the two samples with a comment on this variability (in the context of your assignment).

Perhaps the best way to give guidance on what is required for this task is to provide an (extended) worked example on which you might model your own assignment:

Specimen Statistical Assignment: (To compare the lengths of chapters in the novels of two authors.)

I chose Frederick Forsyth and Jack Higgins as my two novelists, and randomly selected three novels by each author. The type fonts and page sizes were comparable throughout, so I counted the number of pages required for each chapter of each novel. The Forsyth novels were *The Dogs of War*, *The Odessa File* and *The Day of the Jackal.* The Higgins novels were *The Eagle has Landed*, *Year of the Tiger* and *A Season in Hell.* The raw data is given in the following table.

(Chapter)	FF1	FF2	FF3	JH1	JH2	JH3
(Preface)	18					
(1)	16	22	20	15	18	16
(2)	16	29	42	25	17	25
(3)	19	13	18	20	14	13
(4)	21	18	25	28	15	21
(5)	22	13	20	24	11	20
(6)	27	14	35	24	13	20
(7)	22	19	19	22	15	20
(8)	22	19	31	15	17	19
(9)	15	9	27	16	11	18
(10)	21	12	13	14	13	19
(11)	22	19	17	16	13	19
(12)	17	18	20	11	10	21
(13)	24	17	19	14	9	22
(14)	20	18	27	20	10	20
(15)	16	17	30	14	9	13
(16)	17	14	23	23	12	14
(17)	18	16	22	18	6	
(18)	17	15	19	18	13	
(19)	19		18	20		
(20)	20		25	17		
(21)	21		12			

(FF) class interval	freq. (f)	mid-point (x)	(xf)	cum freq.
6-10	1	8	8	1
11–15	9	13	117	10
16–20	30	18	540	40
21–25	13	23	299	53
26–30	5	28	140	58
31–35	2	33	66	60
36–40	0	38	0	60
41–45	1	43	43	61
	61		1213	

(JH) class interval	freq. (f)	mid-point (x)	(xf)	cum freq.
6-10	5	8	40	5
11–15	19	13	247	24
16–20	20	18	360	44
21–25	9	23	207	53
26–30	1	28	28	54
	54		882	

mean $= \bar{x} = \dfrac{\Sigma xf}{\Sigma f} = \dfrac{1213}{61} = 19{\cdot}9$

median : 31st
quartiles : 15th/16th 46th/47th

mean $= \bar{x} = \dfrac{\Sigma xf}{\Sigma f} = \dfrac{882}{54} = 16{\cdot}3$

median : 27th/28th
quartiles : 14th, 41st

These can be found easily by drawing the ogive for each distribution. Alternatively we can use arithmetic:

(FF)
30 chapters (the 11th to the 40th) belong to the class interval 16–20, so distributing them evenly:

pages	16	17	18	19	20
no. of chapters	6	6	6	6	6
places	11–16	17–22	23–28	29–34	35–40

hence median is 19
This analysis also gives the bonus of finding the lower quartile: lower quartile is 16

13 chapters (the 41st to the 53rd) belong to the class interval 21–25 so distributing them evenly:

pages	21	22	23	24	25
no. of chapters	3	2	3	2	3
places	41–43	44–45	46–48	49–50	51–53

hence upper quartile is 23, and so S.I.Q.R. $= \frac{1}{2}$ (23–16) $= 3{\cdot}5$

(JH)

20 chapters (the 25th to the 44th) belong to the class interval 16–20 so distributing them evenly:

pages	16	17	18	19	20
no. of chapters	4	4	4	4	4
places	25–28	29–32	33–36	37–40	41–44

hence median is 16

This analysis also gives the bonus of finding the upper quartile: upper quartile is 20

19 chapters (the 6th to the 24th) belong to the class interval 11–15 so distributing them evenly:

pages	11	12	13	14	15
no. of chapters	4	4	3	4	4
places	6–9	10–13	14–16	17–20	21–24

hence lower quartile is 13, and so S.I.Q.R. = $\frac{1}{2}$ (20–13) = 3·5

To find the modes we either use similar triangles as shown here or make an accurate bar graph and follow the arrows down to read the mode from the scale on the axis.

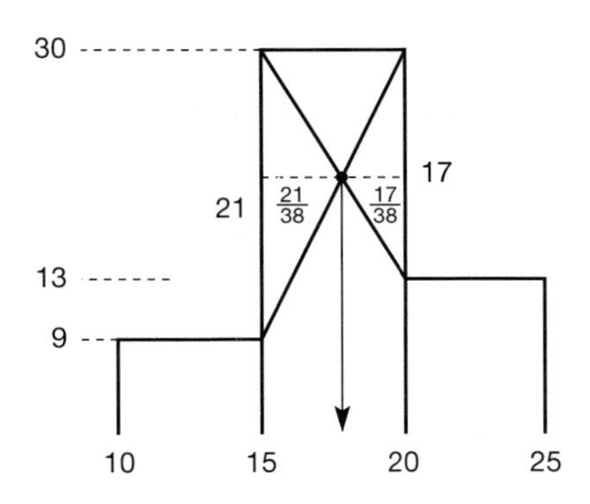

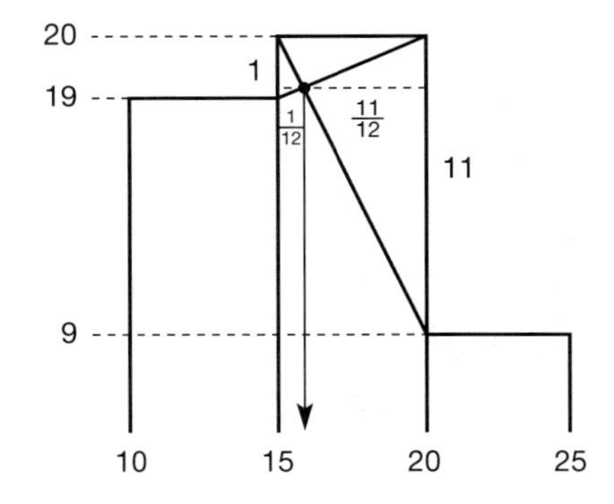

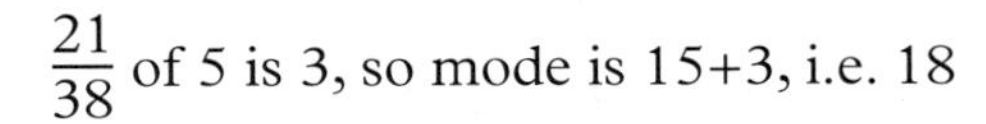

$\frac{21}{38}$ of 5 is 3, so mode is 15+3, i.e. 18

$\frac{1}{12}$ of 5 = $\frac{5}{12}$, so mode = $15 + \frac{5}{12} \approx 15$

Comparing these two sets of results:

Frequency tables:

class interval	FF	JH
6–10	1	5
11–15	9	19
16–20	30	20
21–25	13	9
26–30	5	1
31–35	2	0
36–40	0	0
41–45	1	0

Summary:

	FF	JH
mean	19·9	16·3
mode	18	15
minimum	9	6
lower quartile	16	13
median	19	16
upper quartile	23	20
maximum	42	28
semi-interquartile range	3·5	3·5

Box plots:

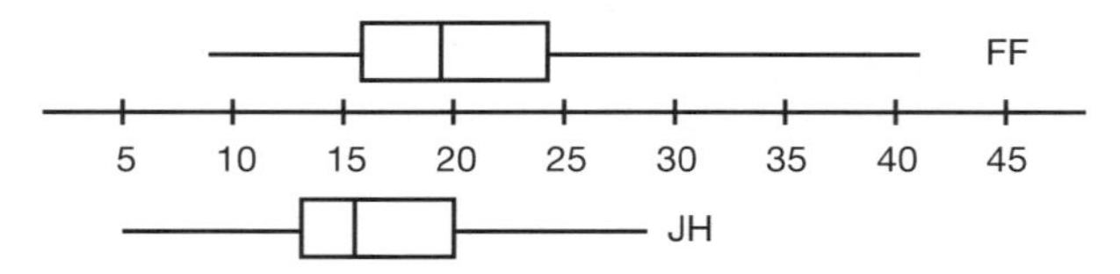

Further comments:

Frederick Forsyth writes more pages in each chapter than Jack Higgins does. His mean number of pages per chapter is 22% higher, his mode 20% higher and his median 18·75% higher. So 20% more pages per chapter is an acceptable comparison.

Only one quarter of FF's chapters but half of JH's chapters have less than 17 pages. Half of FF's chapters have more than 19 pages but only one quarter of JH's chapters have more than 20 pages.

The table of raw data also suggests that FF writes more chapters in each book than JH, hence his books are generally longer, but a sample of only three books per author is too small to reach any definite conclusion in this matter.

The semi-interquartile range is exactly the same for both samples, indicating that both authors 'stray' equally from their average number of pages per chapter when writing.

In both cases the mean and the median are very close, which suggests a fairly symmetrical distribution. This is possible because there are very few chapters with a single figure number of pages, which could have caused the distributions to be squeezed at the lower end. The frequency tables indicate that FF had one very long chapter which makes the box plot look misleadingly lop-sided.

In conclusion, it is interesting to discover these numerical differences in the styles of the novelists. How entertaining the books are does not, however, lend itself to such analysis.

ANSWERS

Exercise 1, page 1

1 a) 0·27 b) 0·89 c) 0·06 d) 0·355.
2 a) 30% b) 40% c) 75% d) 37·5%.
3 a) 28% b) 60% c) 8% d) 26·5%.
4 a) $\frac{37}{100}$ b) $\frac{9}{25}$ c) $\frac{21}{25}$ d) $\frac{1}{40}$.
5

22%	$\frac{11}{50}$	0·22
$37\frac{1}{2}$%	$\frac{3}{8}$	0·375
17%	$\frac{17}{100}$	0·17
90%	$\frac{9}{10}$	0·9
34%	$\frac{17}{50}$	0·34
0.5%	$\frac{1}{200}$	0·005

6 a) 66·667% b) 44·444% c) 83·333% d) 42·857%.

Exercise 2, page 3

1 a) £36 b) £26 c) 60 km d) 17·5 kg.
2 £43·60. **3** 50p. **4** £3175. **5** £11·25.
6 308 tonnes. **7** $2\frac{1}{2}$g. **8** £300. **9** £57·05.
10 £1998·50.
11 a) £40 b) £67·50 c) £180 d) £171.
12 a) £6 b) £6 c) £42 d) £67·50.

Exercise 3, page 4

1 a) 20% b) 40% c) 75% d) 40%.
2 45%. **3** 2·8% protein, 3·2% iron. **4** 55%.
5 a) 85% b) $12\frac{1}{2}$% c) $8\frac{1}{2}$% d) $26\frac{2}{3}$% e) 40% f) 16%.
6 20. **7** 60. **8** 200. **9** 32. **10** 5%.
11 1%. **12** 2 years. **13** 3 years. **14** £80.

Exercise 4, page 6

1 7. **2** 2346. **3** 159. **4** 4·4. **5** 322.
6 $49\frac{1}{2}$. **7** 240. **8** £231. **9** $29\frac{3}{4}$ h.
10 £51 230.

Exercise 5, page 7

1 20%. **2** 60%. **3** $37\frac{1}{2}$ %.
4 $12\frac{1}{2}$% profit. **5** $33\frac{1}{3}$% profit. **6** $8\frac{1}{3}$%.
7 $4\frac{1}{6}$% profit. **8** $23\frac{1}{3}$ % profit. **9** £9·60.
10 £195.

Exercise 6, page 9

1 a) £110·23 b) 41·65 km c) 69·12 kg d) 17·39 tonnes.
2 a) 75% b) 80% c) 78% d) 80% e) 57·5%.
3 £916·11. **4** £81·64. **5** 44%. **6** 20%.
7 a) 81·9 b) 9·912 c) 120·99 d) 509·34.
8 a) 57·5% b) 81·25% c) 8·75% d) 51·25%.
9 a) £59·80 b) 184·50 c) 184·32 d) £634·50.
10 a) £765·38 b) £1302·73 c) £1836·08 d) £2995·07.
11 a) £12 b) £78·75 c) £159·60 d) £124·50.
12 a) 7·5% b) 3·5%.
13 a) 3 years 3 months b) 2 years 9 months.
14 £6·50.

Exercise 7, page 12

1 a) 34·7 b) 147·23 c) 0.002 d) 27·0360.
2 a) 23·5 b) 53 c) 0·0016 d) 5150.
3 a) 57·14% b) 57·1%.
4 a) 55·56% b) 55·6%.
5 a) 34·4% b) 34%.
6 a) 83·33% b) 83·3%.
7 a) 0·50 b) 0·5.
8 a) 0·2 b) 0·19.
9 a) 0·8 b) 0·79.
10 a) 0·672 b) 0·67.

Exercise 8, page 13

1 a) 31·4 b) 31·36.
2 a) 2·480 b) 2·4797.
3 a) 41·1 b) 91·37 c) 6·2 d) 7·160.
4 30·8%. **5** a) 140·1 b) 140.
6 a) 214·57 b) 215.
7 a) 5·27 b) 3·48 c) 15·28 d) 16·00.
8 a) 44·54 b) 44·54494 c) 44·545.
9 a) 0·636 b) 0·308 c) 0·417 d) 0·263.
10 a) 154·13 b) 20·15 c) 15·23 d) 7·049.
11 22·45%.
12 a) £9·54 b) £14·68 c) £13·63 d) £5·71.

Exercise 9, page 15

1 a) $2{\cdot}35\times10^2$ b) $4{\cdot}367\times10^3$ c) $5{\cdot}326\times10$ d) $2{\cdot}5\times10^{-2}$ e) 7×10^{-3} f) $7{\cdot}53\times10^2$ g) $9{\cdot}465\times10^3$ h) $2{\cdot}7\times10^{-3}$ i) $5{\cdot}3\times10^{-2}$ j) $6{\cdot}85\times10^{-1}$ k) 9×10^{-4} l) $8{\cdot}3\times10^{-1}$.
2 a) 260 b) 3200 c) 461 d) 3265 e) 11 000 f) 278 g) 57 300 h) 24 000 i) 0·23 j) 0·0361 k) 0·0014 l) 0·0763.

3 a) 37·1 b) 34·2 c) 0·0862 d) 0·818.
4 17·8%. **5** 46·7%. **6** £597·37.
7 817. **8** 7·01%.
9 a) 11·4% b) £161·50 (or £161·53 if you rounded your previous answer).

Exercise 10, page 16

1 £6·45. **2** £48·48. **3** £5543·59.
4 £3350·38. **5** £216·65. **6** £30·23.
7 £154·00. **8** £3183·83. **9** £62·75.
10 £2·38 (£117·36 – £114·98).

Exercise 11, page 18

1 8·8 kg.
(To make sense, the answers for questions 2 to 9 can only be whole numbers, so have been rounded accordingly.)
2 75. **3** 261 566. **4** 51.
5 1539. **6** 3947. **7** 2579.
8 9 complete years. **9** 23 years. **10** 10%.

Exercise 12, page 19

1 53·72. **2** 566·0 tonnes. **3** 53·6%.
4 10 pupils. **5** £964. **6** £6·99.
7 apparently improving (73·9% to 76·5%).
8 96. **9** £560. **10** £32. **11** £5000.
12 800. **13** 70 000. **14** 63p.
15 a) £38·29 b) £45·37.

Exercise 13 (Test), page 20

1 58812. **2** 1038.
3 apparently improving (79·3% to 82·6%).
4 yes (either because 3·4% $<$ 4% or because 52 $<$ 61).
5 £211·34. **6** £67·66. **7** £3244·80.
8 4 years.

Exercise 14, page 23

1 a) 4 km^2 b) 33 mm^2 c) 21 cm^2
d) 16π (= 50·27) cm^2.
2 a) 68 cm^2 b) $49 - 4\pi$ (= 36·4) cm^2
c) 52 m^2 d) 51·47 m^2.
3 a) 20 units^2 b) 984 units^2 c) 15·6 cm^2
d) 75π (= 235·6) units^2.
4 a) 54 cm^2 b) 118 m^2 c) 336 mm^2
d) 200π (= 628 cm^2).
5 a) 7 cm b) 13 cm c) 19 km d) 29 mm.
6 23 cm.

Exercise 15, page 25

1 a) 8 cm^3 b) 231 m^3 c) 225 cm^3
d) 517 cm^3 e) 51 cm^3 f) 341 units^3
g) 63π (= 198) units^3
h) 6912π (= 21 715) units^3 i) 779 units^3
j) 12 990 units^3.
2 a) 192π (= 603) cm^3 b) 45π (= 141) cm^3
c) 271 cm^3 d) 287 cm^3.
3 a) 210 units^3 b) 82·5 units^3.
4 a) 10 units b) 7 units c) 8 cm d) 7 units
e) 5 units f) 4 cm.
5 13 cm. **6** 2·1 cm.

Exercise 16, page 26

1 a) 523·6 cm^3 b) 1436·8 cm^3
c) 40 478·8 cm^3 d) 747 504·5 cm^3.
2 a) 2673π (= 8397·5) cm^3 b) 10 094·4 cm^3
c) 5553·6 cm^3 d) 7056π (= 22 167·1) cm^3.
3 31·4 cm. **4** 532 cm^3. **5** 1111.
6 8·64 cm.

Exercise 17, page 27

1 a) 769·7 cm^3 b) 986·5 cm^3 c) 4040·9 cm^3
d) 1583·4 cm^3.
2 a) 3332·5 cm^3 b) 157·7 cm^3
c) 6284·6 cm^3 d) 40 511·9 cm^3
e) 422·2 cm^3.
3 12·9 cm. **4** 1761 cm^3. **5** 2.25 cm.
6 4 cm.

Exercise 18, page 28

1 a) 55 cm^3 b) 73 cm^3 c) 445 cm^3
d) 187 cm^3 e) 265 cm^3 f) 3823 cm^3.
2 a) 704 units^3 b) 275 units^3 c) 3 m^3
d) 9 cm^3 e) 555 cm^3.
3 3363 cm^3.
4 a) 327 cm^3 b) 4·3 cm.
5 a) 625 m^3 b) 62·5 litres.
6 2·625 m^3.

Exercise 19 (Test), page 30

1 a) 20 cm^3 b) 60 cm^3 c) 4300 cm^3
d) 180 cm^3 e) 65 cm^3 f) 590 cm^3.
2 a) 16 cm^3 b) 370 cm^3.
3 810 cm^3.

Exercise 20, page 32

1 a) $\frac{2}{3}$ b) $\frac{3}{4}$ c) 1 d) 2 e) $\frac{3}{2}$.
2 a) $-\frac{3}{4}$ b) $-\frac{3}{2}$ c) -1 d) $-\frac{1}{4}$ e) -3.
3 a) $\frac{4}{3}$ b) -2 c) $\frac{1}{3}$ d) $-\frac{2}{5}$ e) 0
f) not defined.

Exercise 21, page 33

1 a) $\frac{5}{2}$ b) 2 c) $\frac{3}{7}$ d) $\frac{5}{3}$ e) $\frac{1}{2}$ f) $\frac{3}{2}$
g) 1 h) $\frac{2}{3}$ i) $\frac{7}{2}$.

2 a) -3 b) $-\frac{2}{7}$ c) -2 d) -5 e) $-\frac{1}{3}$ f) $-\frac{2}{3}$
g) -1 h) $-\frac{3}{2}$ i) $-\frac{4}{3}$.

3 a) $-\frac{3}{4}$ b) -1 c) -2 d) $\frac{1}{7}$ e) -3 f) $-\frac{11}{5}$
g) $-\frac{7}{6}$ h) -4 i) $-\frac{1}{2}$.

4 a) A $(-3, -4)$, B $(2, 5)$, C $(5, 2)$
b) $m_{AB} = \frac{9}{5}$ $m_{AC} = \frac{3}{4}$ $m_{BC} = -1$.

5 a) P $(-2, 5)$, Q $(6, 3)$, R $(2, -4)$, S $(-7, 2)$
b) $m_{PQ} = -\frac{1}{4}$ $m_{QR} = \frac{7}{4}$ $m_{RS} = -\frac{2}{3}$ $m_{SP} = \frac{3}{5}$
c) $m_{SQ} = \frac{1}{13}$ $m_{PR} = -\frac{9}{4}$.

6 a) sketch
b) EF and GH have gradient $\frac{5}{4}$, EF and HG $-\frac{3}{2}$.

7 a) sketch b) each gradient is $\frac{4}{5}$
c) they lie on a straight line.

8 a) sketch
b) $m_{PQ} = \frac{7}{10}$ $m_{QR} = \frac{7}{11}$, so PQ is not parallel to QR, hence not in a line.

9 2.

10 4.

11 a) (i) sketch (ii) $\frac{7}{3}, -\frac{7}{3}, 0$ (iii) isosceles
(iv) $\angle$SVT (v) they are negatives of each other.
b) (i) sketch (ii) $-\frac{5}{9}, \frac{5}{9}$, not defined
(iii) isosceles (iv) $\angle$OLM
(v) they are negatives of each other
c) they are negatives of each other.

12 a) 0 b) 0 c) ∞ (or not defined) d) ∞.

13 a) sketch b) 0 c) (i) 0 (ii) $y = 4$
d) (i) $y = 5$ (ii) $y = -2$ e) $y = 0$.

14 a) sketch b) ∞ c) (i) ∞ (ii) $x = 5$
d) (i) $x = 4$ (ii) $x = -2$ e) $x = 0$.

15 a) sketch b) (i) $y = 4$ (ii) $x = 4$.

16 a) sketch
b) $m_{OW} = \frac{1}{2}$ $m_{OX} = -2$ $m_{OY} = -1$ $m_{OZ} = 6$
c) (i) $x = 1$ (ii) $y = 2$
d) $-2 \leqslant h \leqslant 4$ e) $k < -2$ or $k > 6$.

Exercise 22, page 36

1 a) sketch b) 1, 2, 3, 4, 5
c) $y = x$, $y = 2x$, $y = 3x$, $y = 4x$, $y = 5x$.

2 a) sketch b) $1, \frac{1}{2}, \frac{1}{3}, \frac{1}{4}, \frac{1}{5}$
c) $y = x$, $y = \frac{1}{2}x$, $y = \frac{1}{3}x$, $y = \frac{1}{4}x$, $y = \frac{1}{5}x$.

3 a) sketch b) $-\frac{1}{5}, -\frac{1}{3}, -1, -3, -5$
c) $y = -\frac{1}{5}x$, $y = -\frac{1}{3}x$, $y = -x$, $y = -3x$, $y = -5x$.

4 a) sketch b) $1, -1$ c) $y = x$, $y = -x$.

5 a) 0 0·25 0·50 0·75 1·00 1·25
d) $y = \frac{1}{4}x$
e)

0	0·50	1	1·50	2
0	1	2	3	4
0	1·50	3	4·50	6
0	2	4	6	8
0	3	6	9	12
0	4	8	12	
0	6	12		

g) they all pass through the origin.
h) the larger the gradient, the closer the line is to the y-axis.

6 a) $0 < m < 1$ b) $m > 1$
c) UT: $m < -1$ VW: $-1 < m < 0$.

7 a) sketch b) $\frac{2}{3}, \frac{5}{2}, -\frac{4}{3}, -\frac{1}{4}$
c) $y = \frac{2}{3}x$, $y = \frac{5}{2}x$, $y = -\frac{4}{3}x$, $y = -\frac{1}{4}x$
d) they have gradients 3 and -3.

8 a) sketch b) $y = 4$, $x = 5$, $x = -5$, $y = -4$
c) $y = \frac{4}{5}x$, $y = -\frac{4}{5}x$ d) 80 units^2.

9 a) sketch b) $2, \frac{4}{11}, 2, \frac{4}{11}$
c) a parallelogram
d) the origin is the midpoint of PR and QS.
e) (i) $y = \frac{5}{7}x$ (ii) $y = -\frac{1}{4}x$.

10 a) sketch b) $\frac{7}{4}, -1, \frac{2}{9}$
c) $y = x$, $x = 0$, $y = -\frac{2}{5}x$.

11 a) sketch b) $\frac{4}{7}, \frac{5}{4}, 6, 6, \frac{4}{7}, -\frac{1}{3}$
c) a parallelogram
d) $y = \frac{4}{7}x$, $y = \frac{5}{4}x$, $y = 6x$.

12 a) sketch b) 1, $y = 2x$, $x = 0$, $y = 0$
c) 50 units^2.

Exercise 23, page 39

1 a) $y = 4x + 2$ b) $y = x + 1$ c) $y = 3x - 1$
d) $y = -x + 3$ e) $y = \frac{1}{2}x - 2$.

2 a) $y = 2x + 3$ b) $y = -2x - 1$ c) $y = -\frac{1}{2}x + \frac{1}{2}$
d) $y = \frac{2}{3}x + 4$ e) $y = \frac{1}{4}x - \frac{1}{3}$.

3 a) 2, 3 b) 3, –2 c) –4, 5 d) $\frac{1}{2}$, –1 e) $-\frac{2}{3}, -\frac{3}{4}$.
4 a) 2, (0, $\frac{5}{2}$) b) 1, (0, 2) c) –2, (0, –3)
d) 2, (0, $\frac{1}{3}$) e) $-\frac{4}{5}$, (0, 2).
5 both gradients are a) 2 b) 5 c) $-\frac{2}{3}$ d) $\frac{5}{2}$.
6 b) and c) only.
7 a) $y = 2x$ b) $y = x$ c) $y = \frac{1}{2}x$ d) $y = -\frac{2}{3}x$
e) $y = -\frac{2}{5}x$.
8 a) $y = 3x + 1$ b) $y = \frac{3}{4}x + 1$ c) $y = -2x + 1$
d) $y = \frac{3}{2}x + 1$ e) $y = \frac{1}{5}x + 1$.
9 a) yes b) no c) yes d) no e) yes.
10 P, T only.
11 a) sketch b) they all pass through (0, 1)
c) if the coefficient is negative, the line slopes up to the left; the greater the coefficient is (numerically) the closer the line is to the y-axis.
12 a) sketch b) they are parallel
c) increasing the constant slides the line up the page.
13 a) sketch b) they are parallel.
14 a) sketch b) they are parallel c) yes.
15 a) $\frac{1}{2}$ b) $\frac{1}{3}$ c) 2 d) $-\frac{1}{2}$ e) 3 f) $-\frac{1}{4}$.
16 a)

x	0	10	20	30	40
y	30	32	34	36	38

b) sketch c) $y = \frac{1}{5}x + 30$.
17 a) sketch b) $y = 30x + 30$
c) $p = 30$, $q = 30$ d) 330 minutes.
18 a)

x	0	50	100	150	200
y	10	12·50	15	17·50	20

b) sketch c) $y = 0{\cdot}05x + 10$ d) £16.
19 a) £12 b) 3p c) $y = 0{\cdot}03x + 12$.
20 a) B (50, 0) b) –1 c) $y = 50 - x$.
21 a) there should only be a line of points, where $x = 0, 1, 2, \ldots$ (you can't buy $3\frac{1}{2}$ books)
b) Q (50, 0), –8 c) $y = 400 - 8x$.
22 a) sketch b) $y = -\frac{1}{2}x + 7$ c) yes.
23 (1, 0), (0, 1) (–1, –2).

Exercise 24, page 44

1 a) $\frac{3}{4}$ b) –4 c) 1 d) –2 e) 0.
2 a) A (1, 2), B (5, 4), C (–3, 2), D (1, –2), E (–3, –1), F (2, 2), G (–2, –1), H (3, –2)
b) $\frac{1}{2}$, –1, $\frac{3}{5}$, $-\frac{1}{5}$.
3 P (–3, –4), Q (2, 3); OP and OQ have different gradients, so POQ is not a straight line.
4 yes (each gradient is $\frac{5}{7}$).
5 a) $y = -\frac{2}{5}x - 2$ b) B and C.
6 a) sketch b) $y = 2x + 3$ c) P and Q.
7 $m_{PQ} = \frac{m - n}{m + n}$. **8** 1. **9** 6.
10 A (3, 1), B (1, 3), C (–1, 2), D (–3, –1) $\frac{1}{3}$, 3, –2, $\frac{1}{3}$.
11 $y = \frac{8}{5}x$.
12 a) (6, 10) b) (–3, –5) c) (9, 15).
13 (i) they both pass through the origin
(ii) they are equally inclined to the axes.
14 $2k$, $3k + 3$, $k + 3$.
15 a) $y = 2x + 3$ b) $y = x - 1$ c) $y = -2x + 4$
d) $y = \frac{1}{2}x - 2$.
16 a) 3, –4 b) –2, 5 c) $\frac{1}{4}, \frac{1}{3}$ d) $-\frac{1}{2}$, 2
e) $\frac{1}{2}, \frac{3}{2}$ f) $-\frac{2}{3}$, 0 g) $\frac{3}{4}$, 3 h) $-\frac{2}{5}, \frac{7}{5}$.
17 a) $y = x + 2$ b) $y = 2x - 1$ c) $y = -\frac{1}{2}x + 7$
d) $y = -x - 5$ e) $y = -\frac{2}{3}x + 3$.
18 a) $2x - y + 5 = 0$ b) $3x + 2y - 4 = 0$
c) $x - 2y + 8 = 0$ d) $x + 3y - 9 = 0$.
19 $\frac{1}{2}$, $y = \frac{1}{2}x - 2$. **20** $y = 3x - 3$.
21 a) (0, 2) b) $y = \frac{1}{3}x + 2$, $y = -\frac{1}{3}x + 2$.
22 a) D (–1, 1)
b) P (–3, –3), Q (0, –3), R (3, 3), S (0, 3)
c) $y = 2x + 3$, $y = 2x - 3$.
23 proof: $m_1 = m_2 \Rightarrow -\frac{a}{b} = -\frac{b}{a} \Rightarrow a^2 = b^2$.
24 $y = \frac{2}{5}x + t$.
25 a) $\frac{1}{3}$ b) $y = \frac{1}{3}x + 2k$ c) $y = -\frac{2}{3}x + 2k$.

Exercise 25 (Test), page 47

1 a) $\frac{2}{3}$ b) –2 c) $\frac{4}{5}$.
2 a) A (1, 2), B (3, 5), C (2, 1), D (3, –3), E (–4, –2), F (–1, 3) b) $\frac{3}{2}$, –4, $\frac{5}{3}$.
3 proof: gradients are all $\frac{2}{5}$.
4 a) $x = 4$, $y = 5$ b) $y = \frac{1}{4}x$, $y = \frac{5}{4}x$, $y = 5x$.
5 C (2, 1), P (2, 3), Q (4, 3), $y = \frac{3}{2}x$, $y = \frac{3}{4}x$.
6 $y = 4x - 3$.
7 2, $\left(0, \frac{5}{3}\right)$. **8** $y = 3x + 6$.
9 H (0, 3), $y = \frac{5}{3}x + 3$

Exercise 26, page 50

1 a) $2x - 8$ b) $3p + 15$ c) $4a - 12$ d) $5t - 10$
e) $24 - 4a$ f) $10 + 5t$ g) $12 - 3x$ h) $5k + 10$
i) $21 + 7z$ j) $8m - 8$ k) $12 - 6z$ l) $12 + 3p$
m) $6 - 2w$ n) $3h + 12$ o) $4t + 12$ p) $16 - 8b$.

2 a) $6 - 3t$ b) $5k + 5$ c) $18 - 6z$ d) $7w + 21$ e) $8 - 12y$ f) $10k + 5$ g) $6 - 9t$ h) $12t + 6$ i) $2x + 6y$ j) $3x - 6y$ k) $4p + 8q$ l) $35 - 15p$ m) $6x + 8y$ n) $6x - 9y$ o) $8p + 12q$ p) $15r - 20s$.

3 a) $x^2 + x$ b) $y^2 - y$ c) $t^2 + 2t$ d) $4p - p^2$ e) $x^2 + 2x$ f) $2t^2 - t$ g) $4w - w^2$ h) $3z - 2z^2$ i) $ax + ay$ j) $2bt - 3b$ k) $c - 4ck$ l) $2dx + d^2$ m) $2x^2 + 2x$ n) $3y - 6y^2$ o) $8z^2 + 12az$ p) $10wx - 15w^2$.

4 a) $-x - y$ b) $q - p$ c) $t - 1$ d) $-3 - r$ e) $-a - b$ f) $a - b$ g) $b - a$ h) x i) $r - 6$ j) $3 - c$ k) $-e - f$ l) $2h - g$ m) $-x - 2y$ n) $y - 2x$ o) $-3p - q$ p) $3 - 2p$.

5 a) $-2x - 2$ b) $3x - 9$ c) $8 - 2p$ d) $-5t - 15$ e) $-4x - 12$ f) $6 - 3x$ g) $20 - 4x$ h) $5x - 20$ i) $-12a - 15b$ j) $8q - 24p$ k) $-4z - 6y$ l) $6n - 9m$ m) $3ab - 6a^2$ n) $4xy - 8x^2$. o) $-10t^2 - 20tu$ p) $4v^2 - 3uv$.

6 a) $2x - 5$ b) $4p - 6$ c) $q + 12$ d) $6r + 10$ e) $x^2 + x$ f) $2y^2 + 2y$ g) $3z$ h) $3k^2 - k$ i) $3x - 5$ j) $22 + 4x$ k) $4x + 6$ l) $6 - y$ m) $11 - 3x$ n) $5 - 2t$ o) $6m - m^2$ p) $6q + 6$.

7 a) $x - 1$ b) $2y + 27$ c) $z + 14$ d) $p - 4$ e) $a^2 + b^2$ f) $a^2 + b^2$ g) $4c + 2d$ h) $f + g$ i) $x^2 - 2$ j) $2y^2 + y + 4$ k) $2z^2 - 7z - 9$ l) $2x^2 - xy + 3y^2$.

8 a) $7a + 2b$ b) $7x + 17y$ c) $x - 6$ d) $2ac + 2bc$ e) $4pr - 2qr$ f) $5x^2 + 6$ g) $5p + 3q - 3pq$ h) $3r$.

9 a) $4x + 6y + 2z$ b) $3x - 6y + 9z$ c) $14p + 21q - 7r$ d) $18f + 27g + 9h$ e) $15 + 4a - 8b$ f) $12p + 18q - 19$ g) $17 + 15x - 20y$ h) $20 + 21t - 28s$ i) $4 - 3x - 6y$ j) $4 - 6x + 8y$ k) $7 - 8a - 12b$ l) $29 - 27r + 36s$.

10 a) $12x + 5y + 7z - 4$ b) $p + 19q - 27r + 11$ c) $19a + 9b - 17c + 19$ d) $55 - x + 31y - 12z$.

11 a) $80\ \text{m}^2$ b) $24\ \text{m}^2$ c) $(80 + 24 =)\ 104\ \text{m}^2$ d) $(8 \times 13 =)\ 104\ \text{m}^2$ e) $8(10 + 3) = 8 \times 10 + 8 \times 3$.

12 a) $9a\ \text{m}^2$ b) $18\ \text{m}^2$ c) $(9a + 18)\ \text{m}^2$ d) $9(a + 2)\ \text{m}^2$ e) $9(a + 2) = 9 \times a + 9 \times 2$.

13 the methods used in parts c) and d) may be interchanged
a) $3p\ \text{ft}^2$ b) $3q\ \text{ft}^2$ c) $(3p + 3q)\ \text{ft}^2$ d) $3(p + q)\ \text{ft}^2$ e) $3(p + q) = 3 \times p + 3 \times q$.

14 the methods used in parts d) and e) may be interchanged
a) $7u\ \text{ft}^2$ b) $7v\ \text{ft}^2$ c) $7w\ \text{ft}^2$ d) $(7u + 7v + 7w)\ \text{ft}^2$ e) $7(u + v + w)\ \text{ft}^2$ f) $7(u + v + w) = 7 \times u + 7 \times v + 7 \times w$.

15 a) 1 b) 3 c) 5 d) -2 e) -2 f) 7 g) 6 h) $\frac{1}{2}$.

16 a) $pr + 3r$ b) $x^2 - x$ c) $qs - 5s$ d) $x^2 + xy$ e) $t^2 + 2t$ f) $t^3 + t$ g) $x^2 + 5x$ h) $y^2 - yz$ i) $-u^2 - 3u$ j) $2v - v^2$ k) $-2rs - s$ l) $2q - 3q^2$ m) $a^2 + 3a$ n) $9b - b^2$ o) $f^2 + 2f$ p) $6g + g^2$.

Exercise 27, page 54

1 a) $xy + py + qx + pq$ b) $xy - py + qx - pq$ c) $xy + py - qx - pq$ d) $xy - py - qx + pq$ e) $ab + bx + ay + xy$ f) $ab - bx + ay - xy$ g) $ab + bx - ay - xy$ h) $ab - bx - ay + xy$ i) $ax + a + xy + y$ j) $pq - 2q + pr - 2r$ k) $3r + rt - 3s - st$ l) $kq - nq - 4k + 4n$.

2 a) $x^2 + 3x + 2$ b) $t^2 - t - 12$ c) $y^2 + y - 6$ d) $z^2 - 4z + 4$ e) $p^2 - 2p - 15$ f) $q^2 + 2q - 8$ g) $r^2 + 9r + 14$ h) $s^2 - 9s + 20$ i) $x^2 - 7x + 12$ j) $y^2 + 5y + 6$ k) $z^2 + 9z - 10$ l) $w^2 + 10w + 21$.

3 a) $6x^2 + 11x + 4$ b) $6x^2 + 7x - 3$ c) $20y^2 + 11y - 3$ d) $4z^2 - 16z + 15$ e) $6p^2 - p - 12$ f) $20p^2 - 7p - 6$ g) $6q^2 - 25q + 21$ h) $6r^2 + 13r + 6$ i) $6a^2 - 23a + 20$ j) $6b^2 - 7b - 20$ k) $15c^2 - 19c + 6$ l) $30d^2 + 37d + 10$.

4 a) $x^2 + 3xy + 2y^2$ b) $x^2 - xy - 6y^2$ c) $2y^2 + 7yz - 4z^2$ d) $4p^2 - q^2$ e) $6s^2 + 5st - 6t$ f) $8u^2 + 22uv + 15v^2$ g) $12y^2 - 31yz + 20z^2$ h) $21a^2 + 5ab - 6b^2$ i) $20t^2 + 13tu - 15u^2$ j) $6b^2 - 19bc + 14c^2$ k) $25f^2 + 5fg - 12g^2$ l) $42p^2 + 79pq + 35q^2$.

5 a) $x^2 + 2x + 1$ b) $t^2 + 4t + 4$ c) $z^2 - 2z + 1$ d) $y^2 - 6y + 9$ e) $p^2 - 4p + 4$ f) $q^2 + 6q + 9$ g) $r^2 - 2rs + s^2$ h) $a^2 - 2ab + b^2$ i) $4x^2 + 4x + 1$ j) $1 - 6x + 9x^2$ k) $4 - 12y + 9y^2$ l) $4x^2 + 12xy + 9y^2$.

6 a) $12\ \text{m}^2$ b) $8\ \text{m}^2$ c) $6\ \text{m}^2$ d) $4\ \text{m}^2$ e) $(12+8+6+4=)\ 30\ \text{m}^2$ f) $[(3+2)(4+2) =]\ 30\ \text{m}^2$ g) $(3 + 2)(4 + 2) = 3 \times 4 + 2 \times 4 + 3 \times 2 + 2 \times 2$.

7 a) $pr\ \text{m}^2$ b) $qr\ \text{m}^2$ c) $ps\ \text{m}^2$ d) $qs\ \text{m}^2$ e) $pr + qr + ps + qs$ f) area $= (p + q)(r + s)$ etc.

8 a) $x^2 + 3x + 5$ b) $x^2 - 3x - 2$ c) $3p^2 + p - 12$ d) q^2 e) $k^2 - k + 5$ f) $m^2 - 4mn + 2n^2$ g) $x^2 + 2xy - 2y^2$

h) $xy - 12y^2$ i) $1 - 4x - 3x^2$ j) $3 + x - x^2$
k) $x^2 + 9x + 23$ l) $8 + 5x - 2x^2$
m) $x^2 - 3x + 10$ n) $4y - 4y^2$ o) 4.
p) $6p^2 - 11pq - 6q^2$.

9 a) proof
b) i) $7x - 10$ ii) $8x + 8$ iii) $6x + 15$
iv) $8x - 10$.

10 a) $x^4 + 7x^2 + 12$ b) $y^4 + y^2 - 2$ c) $z^4 - z^2 - 6$
d) $w^4 - 3w^2 + 2$ e) $6x^4 + 7x^2y^2 - 3y^4$
f) $6p^4 + 13p^2q^2 + 6q^4$ g) $12a^2 - ab^3 - b^6$
h) $2x^2 - xy^4 - y^8$ i) $a^4 - 4b^4$ j) $9p^4 - q^4$
k) $x^4 - x^3 - x + 1$ l) $x^4 - ax^3 - a^3x + a^4$.

Exercise 28, page 57

1 a) $2(x + y)$ b) $3(a + b)$ c) $3(a + 2b)$
d) $3(a - 2b)$ e) $2(p + 2q)$ f) $5(p - 2q)$
g) $4(r - 2)$ h) $9(s - 2)$ i) $3(2t + 1)$
j) $4(2u - 1)$ k) $3(3v + 1)$ l) $7(w + 2)$
m) $3(k - 4)$ n) $4(1 - 4m)$ o) $7(1 + 3n)$
p) $8(4t - 1)$.

2 a) $3(2x + 3)$ b) $2(2x - 3)$ c) $2(4x + 5)$
d) $3(4x - 3)$ e) $2(2 + 5y)$ f) $3(2 - 3y)$
g) $3(3 + 5y)$ h) $5(2 - 3y)$ i) $7(x - y)$
j) $2(4p + q)$ k) $3(k - 2m)$ l) $5(5t + y)$
m) $5(5p - 2q)$ n) $11(r + 2s)$ o) $5(3t + 7u)$
p) $10(3a - 4b)$.

3 a) $x(x + 1)$ b) $y(y - z)$ c) $a(x + y)$
d) $p(b - q)$ e) $t(t + 2)$ f) $t^2 (t + 2)$
g) $y(2 + y)$ h) $m(3 + m)$ i) $n(4 - n)$
j) $2r(2 - r)$ k) $3r(2r + 1)$ l) $2t(1 + t)$
m) $2x(x - 5)$ n) $2ar(r + 1)$
o) $3(x^2 + 2x + 3)$ p) $x(x^2 + 3x + 4)$.

4 a) $\pi(r - s)$ b) $2\pi(r - s)$ c) $\pi(r^2 + s^2)$
d) $2\pi r(h + r)$ e) $7(x + 1)$ f) $(a - b)(x + 1)$
g) $(x + 2)(p + q)$ h) $(a + b)(x + y)$
i) $(x^2 + 1)(x - 3)$ j) $(t^2 + 2)(t - 1)$
k) $(u + 2)(u - 3)$ l) $(v^2 + 1)(v - 1)$
m) $(x + 1)(x - 1)$ n) $(y - 2)(y + 2)$
o) $(z + 3)(t + u)$ p) $(a + 2b)c(1 - c)$.

5 a) $3(x + 2y - 3z)$ b) $3(k - 2l + m)$
c) $5(p + 2q - 3)$ d) $3(2a - 3b - 4)$
e) $x(x^2 + x + 1)$ f) $k(k^3 + 2k + 3)$
g) $p(2p^2 + 3p + 5)$ h) $q^2 (4q^2 + 3q + 2)$
i) $2a^2 (a^2 - 2a + 3)$ j) $3b^2 (b^2 + 2b - 1)$
k) $5t(t^2 + 2t + 3)$ l) $6t^2 (2 + 3t + t^2)$.

Exercise 29, page 58

1 a) $(f - g)(f + g)$ b) $(y - z)(y + z)$
c) $(m - n)(m + n)$ d) $(u - v)(u + v)$
e) $(a - 4)(a + 4)$ f) $(b - 3)(b + 3)$
g) $(t - 5)(t + 5)$ h) $(u - 6)(u + 6)$
i) $(p - 2)(p + 2)$ j) $(q - 3)(q + 3)$
k) $(r - 4)(r + 4)$ l) $(s - 5)(s + 5)$
m) $(6 - t)(6 + t)$ n) $(7 - u)(7 + u)$
o) $(4 - v)(4 + v)$ p) $(3 - w)(3 + w)$.

2 a) $(1 - 2b)(1 + 2b)$ b) $(6 - 5a)(6 + 5a)$
c) $(3c - 1)(3c + 1)$ d) $(5d - 2)(5d + 2)$
e) $(9 - 4s)(9 + 4s)$ f) $(8 - 3t)(8 + 3t)$
g) $(11u - 1)(11u + 1)$ h) $(12v - 5)(12v + 5)$
i) $(2p - 3q)(2p + 3q)$ j) $(4a - 3b)(4a + 3b)$
k) $(5x - 4y)((5x + 4y)$ l) $(2t - 5u)(2t + 5u)$
m) $(10u - 7v)(10u + 7v)$ n) $(9r - 8s)(9r + 8s)$
o) $(6p - 7q)(6p + 7q)$ p) $(5a - 8b)(5a + 8b)$.

3 a) $2(p - q)(p + q)$ b) $3(x - y)(x + y)$
c) $5(u - v)(u + v)$ d) $8(a - b)(a + b)$
e) $5(r - 2)(r + 2)$ f) $6(s - 3)(s + 3)$
g) $7(t - 2)(t + 2)$ h) $8(u - 5)(u + 5)$
i) $5(v - 2w)(v + 2w)$ j) $3(a - 3b)(a + 3b)$
k) $4(4c - d)(4c + d)$ l) $6(2f - g)(2f + g)$
m) $3(2a - 3bc)(2a + 3bc)$
n) $2(5p - 6qr)(5p + 6qr)$
o) $5t(4m - 3n)(4m + 3n)$
p) $2(1 - 7x)(1 + 7x)$.

4 a) 40 b) 80 c) 120 d) 72 e) 88
f) 180 g) 200 h) 120 i) 180 j) 240
k) 280 l) 998 000 m) 81 n) 25 o) 121
p) 49.

5 a) 3 b) 20 c) 16 d) 8.

6 a) 200π b) 40π c) 105π d) 133π.

7 a) $(a + b - c)(a + b + c)$
b) $(a - b - c)(a - b + c)$
c) $(a - b - c)(a + b + c)$
d) $(a - b + c)(a + b - c)$
e) $(p + q - r)(p + q + r)$
f) $(5 - v + w)(5 + v - w)$
g) $(r + s - 2t)(r + s + 2t)$
h) $(3a - 2b + 2c)(3a + 2b - 2c)$
i) $(a + c)(a + 2b - c)$ j) $5(2z - 1)$
k) $(9 - 2a)(7 + 2a)$ l) $4x(y - z)$.

8 a) $(x - 1)(x + 1)(x^2 + 1)$
b) $(y - 2)(y + 2)(y^2 + 4)$
c) $(x - y)(x + y)(x^2 + y^2)$
d) $(p - qr)(p + qr)(p^2 + q^2r^2)$
e) $(2z - 1)(2z + 1)(4z^2 + 1)$
f) $(u - 3v)(u + 3v)(u^2 + 9v^2)$.

9 a) $3(t - 1)(t + 1)(t^2 + 1)$
b) $2(u - 2)(u + 2)(u^2 + 4)$
c) $a(p - q)(p + q)(p^2 + q^2)$
d) $5(2v - 1)(2v + 1)(4v^2 + 1)$
e) $3(p - 3q)(p + 3q)(p^2 + 9q^2)$
f) $2(3u - 2v)(3u + 2v)(9u^2 + 4v^2)$.

Exercise 30, page 62

1 a) $(x+1)(x+2)$ b) $(y+3)(y+6)$
c) $(z+2)(z+5)$ d) $(w+3)(w+7)$
e) $(x+4)(x+5)$ f) $(y+2)(y+3)$
g) $(z+4)(z+7)$ h) $(w+5)(w+6)$
i) $(x-1)(x-2)$ j) $(y-3)(y-5)$
k) $(z-1)(z-3)$ l) $(w-2)(w-5)$
m) $(x-1)(x-5)$ n) $(y-2)(y-7)$
o) $(z-6)(z-3)$ p) $(w-5)(w-7)$.

2 a) $(x+2)(x-1)$ b) $(y+1)(y-2)$
c) $(z-2)(z+4)$ d) $(w+2)(w-4)$
e) $(u+5)(u-4)$ f) $(v-3)(v+6)$
g) $(r+1)(r-5)$ h) $(s+2)(s-7)$
i) $(t-4)(t+6)$ j) $(k-4)(k+7)$
k) $(m-2)(m+8)$ l) $(n+5)(n-9)$
m) $(a-2)(a+3)$ n) $(b+5)(b-6)$
o) $(c+2)(c-5)$ p) $(d+7)(d-8)$.

3 a) $(x+3)(x+5)$ b) $(y-2)(y+7)$
c) $(z-3)(z-11)$ d) $(w-7)(w-9)$
e) $(u+2)(u+11)$ f) $(v-5)(v+7)$
g) $(r+4)(r-9)$ h) $(s-3)(s-10)$
i) $(t-6)(t+7)$ j) $(k-8)(k-9)$
k) $(m-5)(m+7)$ l) $(n-3)(n+9)$
m) $(a+1)(a+3)$ n) $(b-7)(b-10)$
o) $(c-9)(c+11)$ p) $(d+3)(d-8)$.

4 a) $3(x-3)(x-4)$ b) $5(y-2)(y+7)$
c) $7(z-1)(z-11)$ d) $8(w+2)(w+9)$.

5 a) $(x+3)(x-4)$ b) $(x+1)(x+2)$
c) $(x-3)(x-8)$ d) $(x-2)(x+5)$
e) $(x+1)(x-2)$ f) $(x-3)(x+5)$
g) $(x+2)(x-4)$ h) $(x-1)(x-7)$.

6 a) $(\mathrm{a}+2\mathrm{b})(\mathrm{a}+3\mathrm{b})$ b) $(p+3q)(p-4q)$
c) $(t-5u)(t-7u)$ d) $(u-3v)(u+5v)$
e) $(k+t)(k-2t)$ f) $(x-4y)(x+7y)$
g) $(y+2z)^2$ h) $(f-3g)(f+3g)$.

7 a) $(x+a)(x+b)$ b) $(x-m)(x+n)$
c) $(x-p)(x-q)$ d) $(x+5a)(x-2b)$.

Exercise 31, page 64

1 a) $(2x+1)(x+2)$ b) $(3x+1)(x+3)$
c) $(5x+3)(x+1)$ d) $(7x+5)(x+1)$
e) $(5x-1)(x-5)$ f) $(3x-7)(x-1)$
g) $(3x-1)(x-3)$ h) $(5x-1)(x-3)$
i) $(2x+1)(x-2)$ j) $(3x-2)(x+1)$
k) $(2x-5)(x+1)$ l) $(3x-1)(x+2)$
m) $(5x-2)(x+1)$ n) $(7x+1)(x-3)$
o) $(5x-1)(x+2)$ p) $(3x-7)(x+1)$.

2 a) $(3x+1)(x+8)$ b) $(6y-1)(y-1)$
c) $(7y+2)(y+3)$ d) $(3z-7)(z-3)$
e) $(3z+1)(3z+5)$ f) $(5z-2)(z-3)$
g) $(5\mathrm{c}-1)(\mathrm{c}-4)$ h) $(5\mathrm{a}-2)(6\mathrm{a}-1)$
i) $(2x-5)(x+2)$ j) $(4y+3)(y-1)$
k) $(5y-1)(y+4)$ l) $(2z+7)(z-4)$
m) $(3a+4)(a-2)$ n) $(4b-1)(b+3)$
o) $(4x+5)(x-1)$ p) $(3x+2)(x-4)$.

3 a) $(2x+3)(3x+2)$ b) $(2x-5)(3x-4)$
c) $(2y+3)(5y+4)$ d) $(5y-3)(3y-5)$
e) $(2x-7)(2x+3)$ f) $(2x-5)(3x+4)$
g) $(4y-3)(2y+3)$ h) $(2y-3)(6y+5)$
i) $(2z-3)(2z+9)$ j) $(3z+7)(3z-4)$
k) $(3z+5)(5z+4)$ l) $(2z-5)(7z-3)$
m) $(3c+4)(7c-1)$ n) $(6d+5)(d+3)$
o) $(3b+2)(6b+5)$ p) $(2d-5)(10d+3)$.

4 a) $3(x+1)(x-2)$ b) $3(2x+1)(x-1)$
c) $2(2x+3)(3x-1)$ d) $4(2x-1)(3x-2)$
e) $5(3y-2)(2y-3)$ f) $2(2y+3)(3y+4)$
g) $3(3y-2)(2y+5)$ h) $7(4y-3)(2y+3)$.

5 a) $(3x-4)(7x+2)$ b) $(4x-3)(5x+4)$
c) $(2x-5)(3x-7)$ d) $(3x+2)(4x+5)$
e) $(3x+7)(2x+1)$ f) $(3x+2)(2x-5)$.

6 a) $(3x+2y)(2x-3y)$ b) $2(4x+3y)(x+2y)$
c) $(3x-4y)(4x-3y)$ d) $(3x-y)(5x+2y)$
e) $(3p+2q)(4p+3q)$ f) $(4p+5q)(5p-3q)$
g) $(6p-5q)(7p-2q)$ h) $(2p+3q)(6p-7q)$
i) $(2k-3l)(3k+4l)$.

Exercise 32, page 66

1 no factors. **2** $ax(x+1)$.
3 $(2x-1)(2x+1)$. **4** $(x+4)(x-5)$.
5 $2(x-3)(x+3)$. **6** $(x-3)(x+5)$.
7 $2(x-2)(x+3)$. **8** no factors.
9 $(2x+1)(x-5)$. **10** $2(x^2+9)$.
11 $3(x-1)(x+1)$. **12** $(x+2)(x-7)$.
13 $(3x+1)(3x+2)$. **14** $(3x-4)(2x+3)$.
15 $3(2x+3)(x-5)$. **16** no factors.
17 $ax(a-y)$. **18** $(7x-6y)(7x+6y)$.
19 $4(x-3)(x+3)$. **20** $(2x-3)^2$.
21 $(2x+3)(3x-7)$. **22** $(x-7)(x-9)$.
23 $a(x+\mathrm{a}y+a^2z)$. **24** $(7-6t)(7+6t)$.
25 $8(5x-4y)(5x+4y)$. **26** $5(x+1)^2$.
27 no factors. **28** $2(2x^2+y+4z)$.
29 $(5x-2)(7x+3)$. **30** $2(2x-3)(5x-9)$.

Exercise 33, page 66

1 a) $4x+8$ b) $7y-7$ c) $8x^2+24$
d) $20-5x$ e) $3x+3y+6z$ f) $28-8p+12q$
g) $3\mathrm{b}-\mathrm{c}-2\mathrm{a}$ h) $5r-15q-10p$.

2 a) $2x+7$ b) $3y+10$ c) $5z-7$ d) $17-7s$
e) $7+3x$ f) $14-5y$ g) $7z-15$ h) $12-t$.

3 a) x^2-x-2 b) y^2-y-12 c) $2z^2-7z-15$
d) $5k^2-7k-6$ e) $6k^2-5k-21$
f) $6z^2-25z+24$ g) $12p^2-pq-q^2$
h) $12a^2-11ab-15b^2$.

4 a) $x^3 + 2x^2 + 2x + 1$ b) $2x^3 + 5x^2 - 4x - 3$
c) $6x^3 + 13x^2 + 14x + 12$
d) $6x^3 - 16x^2 + 23x - 10$.
5 a) $3(x + 2y)$ b) $9(t + 9)$ c) $16(1 + 4w)$
d) $7(2u + 7v)$ e) $x(4x + 7)$ f) $3y(2y + 3)$
g) $2z(2z + 3)$ h) $4pq(2p + 3q)$.
6 a) $(x - 9)(x + 9)$ b) $(y - 7)(y + 7)$
c) $(7z - 8)(7z + 8)$ d) $(8p - 5q)(8p + 5q)$
e) $(t^2 - 9u)(t^2 + 9u)$ f) $(10f - 13g)(10f + 13g)$
g) $(11p - 6q)(11p + 6q)$
h) $(20a - 11b)(20a + 11b)$.
7 a) $2(x - 1)(x + 1)$ b) $3(y - 3)(y + 3)$
c) $5(5 - z)(5 + z)$ d) $4(7 - a)(7 + a)$
e) $7(p - q)(p + q)$ f) $a(x - y)(x + y)$
g) $3(10m - 9n)(10m + 9n)$
h) $7(5r - 4s)(5r + 4s)$.
8 a) $(x + 5)(x - 7)$ b) $(y - 3)(y + 9)$
c) $(z - 4)(z - 9)$ d) $(a + 2)(a + 7)$
e) $(b - 2)(b - 11)$ f) $(c + 3)(c + 12)$
g) $(d - 4)(d - 8)$ h) $(f - 6)(f + 7)$.
9 a) $(2x + 1)(3x - 1)$ b) $(5y - 2)(y + 1)$
c) $(3z - 2)(z - 1)$ d) $(2t + 5)(5t - 2)$
e) $(5u - 3)(6u + 5)$ f) $(3v - 1)(7v - 3)$
g) $(3w + 2)(5w + 4)$ h) $(4k - 3)(2k + 3)$.
10 a) no factors b) $5(x - 1)(x + 1)$
c) $(x - 5)(x + 7)$ d) $2(y - 3)(y - 12)$
e) $3(x - 1)(x - 2)$ f) $7(p - 7q)(p + 7q)$
g) no factors h) $2(3x - 4)(2x + 1)$.
11 a) $(7x - 11)(7x + 18)$ b) $(11x - 28)(11x + 39)$.

Exercise 34 (Test), page 67

1 a) $3x + 6$ b) $8y + 4z$ c) $3x + 3y - 6z$
d) $x^2 + 2xy$.
2 a) $2x + 5$ b) $5 - 3y$ c) $6z - 4$ d) $5p + 8q$.
3 a) $x^2 + 8x + 15$ b) $y^2 - y - 20$
c) $6z^2 - 7z + 2$ d) $8t^2 + 14t - 15$.
4 a) $2(x + 3y)$ b) $x(x + 3)$ c) $3(x^2 + 17)$
d) $2rs(3r + 2s)$.
5 a) $(10 - x)(10 + x)$ b) $(p - 2q)(p + 2q)$
c) $(5a - 4b)(5a + 4b)$ d) $(7u - 2v)(7u + 2v)$.
6 a) $2(x - 2)(x + 2)$ b) $5(y - 3)(y + 3)$
c) $3(1 - 5z)(1 + 5z)$ d) $5(2p - 5q)(2p + 5q)$.
7 a) $(x + 1)(x + 3)$ b) $(y - 3)(y - 12)$
c) $(z + 4)(z - 8)$ d) $(t - 2)(t + 7)$.
8 a) $(2x + 1)(3x - 1)$ b) $(5y - 2)(3y + 4)$
c) $(3z - 4)(5z + 3)$ d) $(5u - 2)(6u - 5)$.
9 a) $x^3 + 3x^2 + 5x + 3$ b) $3x^3 - 5x^2 + 3x - 1$
c) $2x^3 + 5x^2 + x - 3$ d) $6x^3 - 13x^2 + 18x - 8$.

Exercise 35, page 70

1 a) 314 mm b) 105 mm c) 79 mm
d) 236 mm e) 524 mm.
2 a) 37 mm b) 183 mm c) 55 mm
d) 275 mm e) 147 mm f) 403 mm.
3 a) 90° b) 188 mm. **4** a) 60° b) 84 mm.
5 6·283 m. **6** a) 15·708 m. b) 20.944 m.
7 3·142 m. **8** 49 inches.
9 a) 1·257 m b) 60° c) 610 mm.
10 a) 26° b) 73° c) 104° d) 156°
e) 234°.
11 a) 34° b) 90° c) 104° d) 154°
e) 191°.
12 401°. **13** a) 1·5 m b) 239 mm c) 727 mm.
14 370 mm.

Exercise 36, page 73

1 a) 22 619 mm² b) 3770 mm² c) 5655 mm²
d) 28 274 mm² e) 41 469 mm².
2 a) 3351 mm² b) 5027 mm² c) 6702 mm²
d) 8378 mm² e) 11729 mm² f) 13404 mm².
3 a) 90° b) 22 698 mm².
4 a) 60° b) 523 599 mm².
5 2 513 274 mm². **6** 2457 mm².
7 no, only room for 4·03 people.
8 a) 942 478 mm² b) 471 239 mm².
9 a) 70 686 mm² b) 255°.
10 a) 55° b) 92° c) 165° d) 229°.
11 a) 31 500 mm² b) 100 mm. **12** 395 mm.

Exercise 37, page 76

1 60°.
2 $p = 90°$ $q = 59°$ $r = 22°$ $s = 67°$ $t = 15°$.
3 5 m, 6 m, 7 m, 8 m, 20 m.
4 5 m, 61 m, 65 m, 41 m, 37 m.
5 15 m, 35 m, 60 m, 91 m, 56 m. **6** 50°.
7 124°. **8** 29°. **9** 65°. **10** 10°.
11 40°. **12** 40°.
13 proof, concluding with $R\hat{S}T = R\hat{S}Q + Q\hat{S}T = 65° + 25° = 90°$.
14 36·9°, 35·0°, 26·6°, 30·0°, 50·5°.
15 4·0 m, 4·6 m, 6·7 m, 5·7 m, 16 m.
16 15 m, 31 m, 49 m, 17 m, 19 m.

Exercise 38, page 79

1 40°, 30°, 55°, 25°, 68°.
2 101 mm, 73 mm, 85 mm, 145 mm.
3 15 mm, 48 mm, 18 mm, 60 mm. **4** 16 mm.
5 30°, 60°, 90°. **6** 36°, 54°, 90°. **7** 25°.
8 20°. **9** 40°. **10** 31°. **11** 26°.
12 $a = 22{\cdot}6°$ $b = 36{\cdot}0°$ $c = 32{\cdot}6°$ $d = 42{\cdot}8°$
$e = 55{\cdot}3°$.
13 BC = 7·1 m AD = 11 m AE = 12 m.
14 XZ = 20·2 m PQ = 21·9 m AB = 15·0 m.

Exercise 39, page 83

1 (3, 4). **2** (20, 24), radius 25. **3** 50 mm.
4 17 m. **5** 48 mm.
6 proof: $P\hat{S}Q = 90° = O\hat{R}Q$. **7** 112°. **8** 54°.
9 2·4 m. **10** a) 25 m. b) 35·5°.

Exercise 40, page 84

1 a) 10·5 mm b) 78·5 mm^2.
2 a) 50·0° b) 130·6 mm^2.
3 a) 140·0° b) 144·2 mm.
4 1095·9 mm^2. **5** 17·0 mm.
6 214 mm. **7** 349 m^2. **8** 9·7 mm.
9 a) 42° b) 21·6 units.
10 a) (i) 29° (ii) 29° b) 53·0 m.
11 a) $x = y$ b) 36·9° c) 4 units
d) AC = 2 × OM.
12 24mm. **13** 88 mm. **14** 144 mm.
15 198 mm. **16** 17 mm. **17** 85 mm.
18 a) 4 feet b) 43 feet c) 110·4 ft^2.
19 a) 100 mm b) 112 mm.

Exercise 41 (Test), page 87

1 46·5 mm. **2** 1911·5 mm^2. **3** 63°.
4 52°. **5** 49°. **6** 144°. **7** 15°.
8 a) 24 mm b) 7·2°.
9 a) 85 mm b) 51·2 mm. **10** 25°.

Exercise 42, page 90

1 a) $\frac{8}{17}$ b) $\frac{15}{17}$ c) $\frac{8}{15}$. **2** a) $\frac{21}{29}$ b) $\frac{20}{29}$ c) $\frac{21}{20}$.
3 a) $\frac{9}{41}$ b) $\frac{40}{41}$ c) $\frac{9}{40}$. **4** a) $\frac{7}{25}$ b) $\frac{24}{25}$ c) $\frac{7}{24}$.
5 a) $\frac{a}{c}$ b) $\frac{b}{c}$ c) $\frac{a}{b}$. **6** a) $\frac{e}{f}$ b) $\frac{d}{f}$ c) $\frac{e}{d}$.
7 a) $\frac{g}{k}$ b) $\frac{h}{k}$ c) $\frac{g}{h}$. **8** a) $\frac{n}{m}$ b) $\frac{l}{m}$ c) $\frac{n}{l}$.
9 $a = 38·7°$ $b = 53·1°$ $c = 36·9°$.
10 $p = 9·97$ $q = 6·88$ $r = 12·6$.
11 $x = 7·60$ $y = 13·7$ $z = 6·62$.
12 a) 9·06 km b) 4·23 km.
13 a) 3·759 m b) 1·368 m.
14 10·7°.
15 a) (i) 60° (ii) 30°.
b) No (e.g. $C\hat{T}F$ = 29·2° not 30°).
16 4·673 m.
17 AC = 5·464 m, BC = 9·464 m, area = 13·3 m^2.
18 a) 0 b) 1 c) 0 d) 0 e) 1 f) ∞.
19 a) 90 b) 90 c) 0 d) 0.

Exercise 43, page 94

1 a) 0·743 b) –0·669 c) –1·111 d) –0·669
e) –0·743 f) 0·900 g) –0·857 h) 0·515
i) –1·664.
2 a) 133° b) 97° c) 168·5°.
3 a) 157° b) 132° c) 96°.
4 a) 163° b) 103° c) 128°.
5 a) 35·3°, 144·7° b) 64·3°, 115·7°
c) 18·0°, 162·0°.
6 a) 123° b) 165·2° c) 137·6°.
7 a) 288° b) 108°, 252°.
8 a) 97° b) 263°, 277°.
9 a) 202° b) 158°, 338°.
10 a) 32°, 148° b) 212°, 328°.
11 a) 47°, 313° b) 133°, 227°.
12 a) 65°, 245° b) 115°, 295°.

Exercise 44, page 98

1 a) 55 mm^2 b) 36 mm^2 c) 14 mm^2
d) 49·5 mm^2.
2 a) 37·3 mm^2 b) 42·3 mm^2 c) 15·6 mm^2
d) 44·1 mm^2.
3 a) $b\sin\hat{C}$ b) $\frac{1}{2}ab\sin\hat{C}$.
4 a) 42·8 mm^2 b) 131·5 mm^2 c) 96·0 mm^2
d) 135·8 mm^2.
5 6·232 m^2. **6** 13 081·5 m^2. **7** 23 895·9 mm^2.
8 30°, 150°
3 4 30° 3 150°
9 82·2° or 97·8°.
10 a) 7964 mm^2 b) 6836·2 mm^2 c) 1127·8 mm^2.
11 374·1 m^2.
12 a) 23 776 mm^2 b) 2·828 m^2 c) 3·126 m^2.
13 4906 mm^2. **14** 26·8° or 153·2°.

Exercise 45, page 100

1 a) 16 b) 12 c) 16.
2 a) AD = $c\sin\hat{B}$ b) $b = \frac{AD}{\sin\hat{C}}$ c) $b = \frac{c\sin\hat{B}}{\sin\hat{C}}$.
3 a) 5·142 mm b) 5·677 mm c) 10·113 mm.
4 a) 9·1 mm b) 2·557 m c) 11·9 mm
d) 5·422 m.
5 a) $\frac{p}{\sin\hat{P}} = \frac{q}{\sin\hat{Q}} = \frac{R}{\sin\hat{R}}$ b) $\frac{x}{\sin\hat{X}} = \frac{y}{\sin\hat{Y}} = \frac{z}{\sin\hat{Z}}$
c) $\frac{k}{\sin\hat{K}} = \frac{l}{\sin\hat{L}} = \frac{m}{\sin\hat{M}}$.
6 AC = 506·8 m BC = 389·9 m.
7 DF = 10·6 km EF = 8·95 km.
8 520·9 m.
9 a) 5·9 km b) 7·9 km c) 3·4 km.
10 RS = 1·302 m QR = 2·571 m.
11 MN = 10·97 m LN = 21·727 m.

Exercise 46, page 103

1 36·2°. **2** 25·7°. **3** 31·3°. **4** 34·1°.
5 $\hat{D}$ = 40·8°, $\hat{E}$ = 25·2°. **6** 83·0°. **7** 58·4°.

8 33·1°. **9** 306°. **10** 27·9°.

Exercise 47, page 104

1 a) 8·1 b) 5·8 c) 12·5.
2 a) $c^2 = x^2 + y^2$ b) $DC = b - x$
c) (i) $a^2 = y^2 + DC^2$
(ii) $a^2 = x^2 + y^2 + b^2 - 2bx$
(iii) $a^2 = c^2 + b^2 - 2bx$
d) $x = c\cos\hat{A}$ $a^2 = b^2 + c^2 - 2bc\cos\hat{A}$.
3 a) $p^2 = q^2 + r^2 - 2qr\cos\hat{P}$
b) $m^2 = k^2 + l^2 - 2kl\cos\hat{M}$.
4 a) $= w^2$ d) $= v^2$.
5 14·2 km. **6** 21·25 km. **7** 55·1 km.
8 4·64 km. **9** 12·8 m. **10** 6 km/h.
11 39·5 mm.

Exercise 48, page 106

1 a) $\cos\hat{P} = \dfrac{q^2 + r^2 - p^2}{2qr}$
b) $\cos\hat{Z} = \dfrac{x^2 + y^2 - z^2}{2xy}$.
2 a) $= \cos\hat{W}$. b) $= \cos\hat{V}$. **3** 74·4°.
4 112·4°. **5** $\hat{L} = 55{\cdot}8°$ $\hat{N} = 82{\cdot}8°$.
6 $\hat{P} = 29{\cdot}9°$ $\hat{Q} = 56{\cdot}3°$. **7** 120°. **8** 43·6°.

Exercise 49, page 107

1 16·6 km 340° **2** 324·5 km, 093°.
3 11·2 m. **4** 13 km/h.
5 $b = 6{\cdot}245$m $\hat{C} = 76{\cdot}1°$ $\hat{A} = 43{\cdot}9°$. **6** 4·5 m.

Exercise 50, page 108

1 6·428 m. **2** 2·911 m. **3** 39·8°.
4 7·660 m. **5** 29·856 m. **6** 26·4°.
7 9·271km. **8** 9·8 miles.
9 a) 30, 150 b) 60, 300 c) 45, 225
d) 72, 108 e) 85, 275 f) 21, 201
g) 8, 172 h) 30, 330 i) 20, 200
j) 200, 340 k) 130, 230 l) 100, 280
m) 210, 330 n) 110, 250 o) 100, 280.
10 a) (see Exercise 43 Reminder)
b) (i) 0 (ii) 0 (iii) 0 (iv) –1 (v) 1
(vi) 0 (vii) 1 (viii) 1 (ix) 0 (x) 0
(xi) 0 (xii) ∞
c) (i) 90 (ii) 180 (iii) 270
(iv) 0, 180, 360 (v) 0, 360
(vi) 0, 180, 360 (vii) 90, 270.
11 a) 5580 mm^2 b) 105 mm.
12 a) 20 m b) 148 m^2.
13 26·6°. **14** 56·8°.

Exercise 51 (Test), page 109

1 a) –0·829 b) –0·682 c) –0·869.
2 a) 15·7, 164·3 b) 195·7, 344·3
c) 61·3, 298·7 d) 118·7, 241·3
e) 58·0, 238·0 f) 122·0, 302·0.
3 13 374 mm^2. **4** a) 8·521 m b) 71·0°.
5 a) 7·402 m b) 50·5°.

Exercise 52, page 111

1 a) 125 150 175 200 225 250
b)

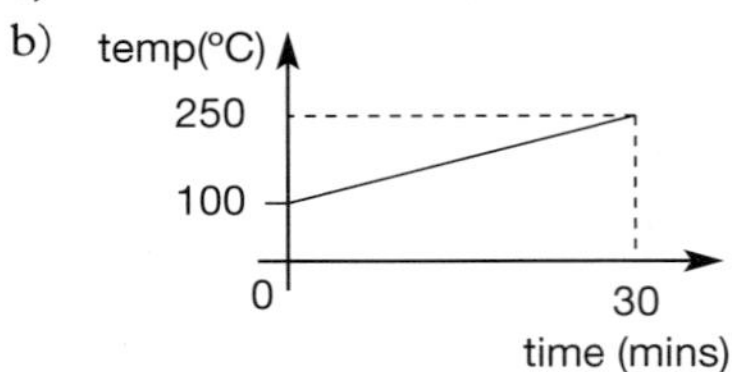

c) *yes*, 160°.
2 a)

cost (£)
130
100
70
40
0
1 2 3
time (hours)

b) £80 c) £80.
3 a)

cash left (£)
10
1
0
4
battery sets

b) 4 c) £1·25.
4 a)

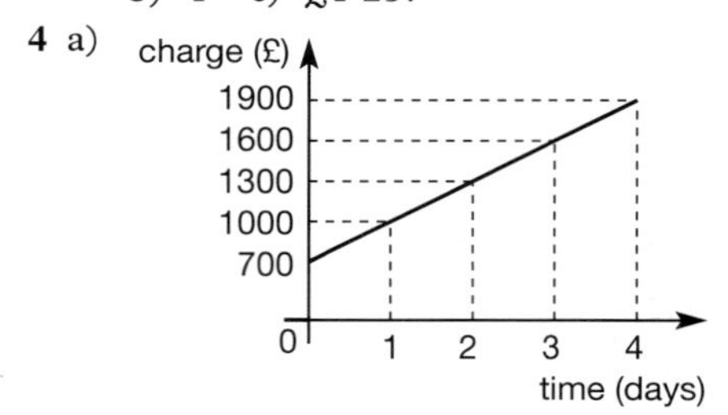

b) probably 4 days.
5 a)

nails left
250
201
152
103
54
5
0
7 14 21 28 35
time (mins)

b) 40 c) 40.

6 a)

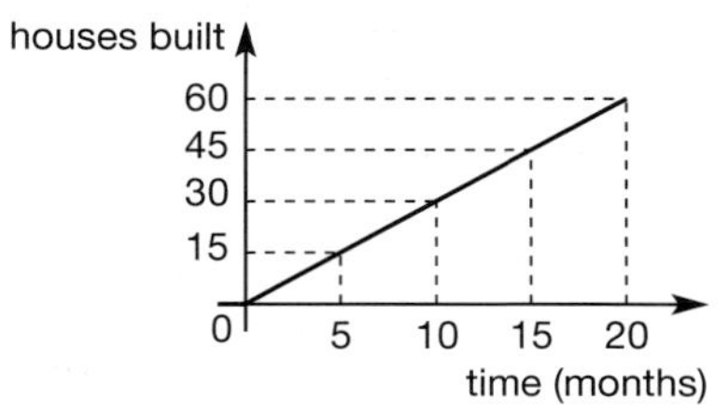

b)

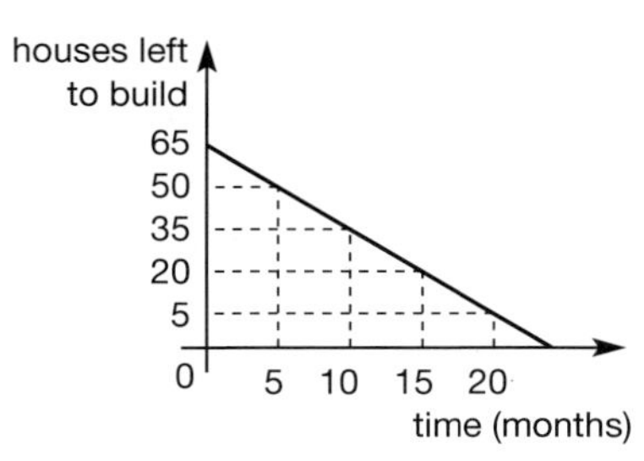

c) No, 10 months: 30 built, 35 to build
11 months: 33 built, 32 to build.

7 a)

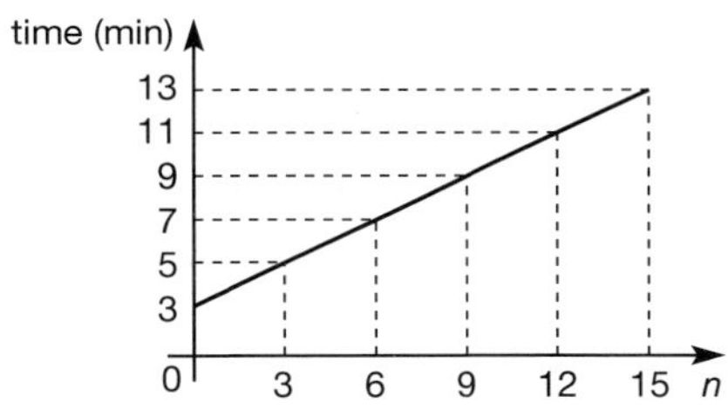

b) gradient $= \frac{2}{3}$ $\qquad t = \frac{2}{3}n + 3$

c)

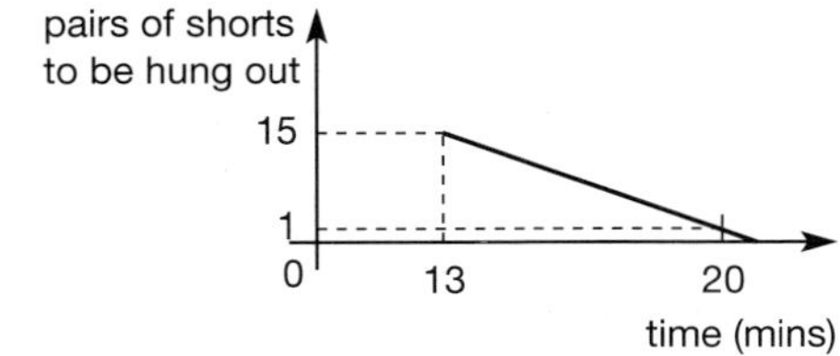

d) $20\frac{1}{2}$ mins.

Exercise 53, page 113

1 a)

cost (£)
294
182
154
0
(M)
(F)
2800
no. of miles

b) Mymotors is cheaper for less than 2800 miles.
Flashcars is cheaper for more than 2800 miles.
There is no difference for exactly 2800 miles.

2 a)

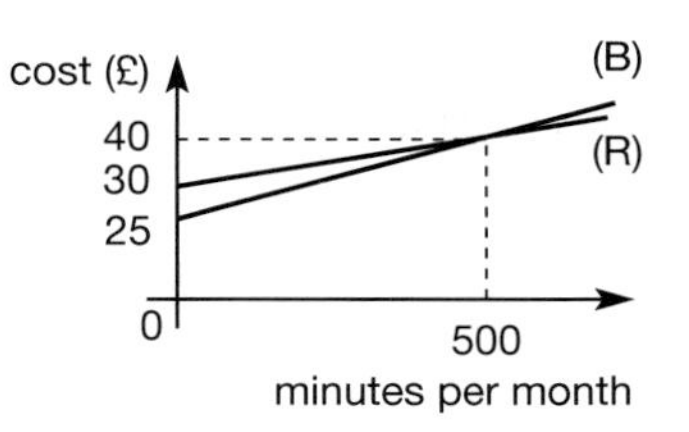

b) Budgtel is cheaper for less than 500 min.
Rapicall is cheaper for more than 500 min.
There is no difference for exactly 500 min.

3 a)

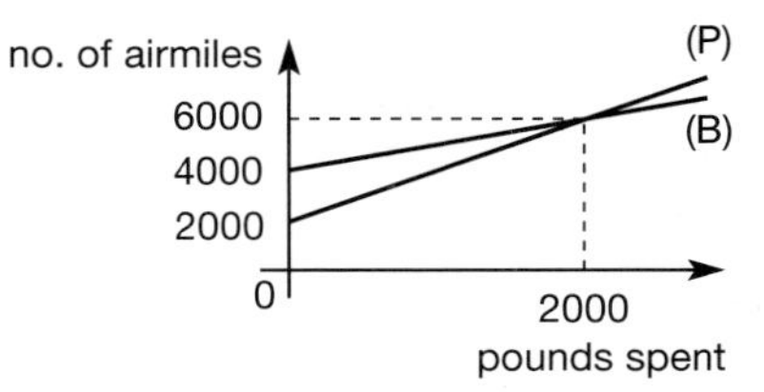

b) Primrose, £3500.

4 a)

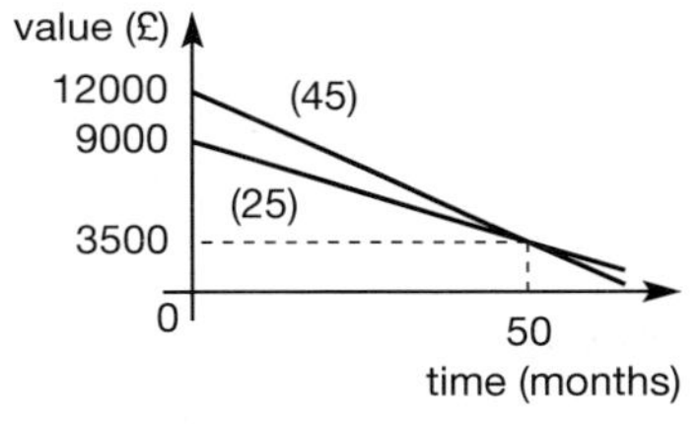

b) Trade in a Rover 45 before 50 months.
Trade in a Rover 25 after 50 months.

5 a)

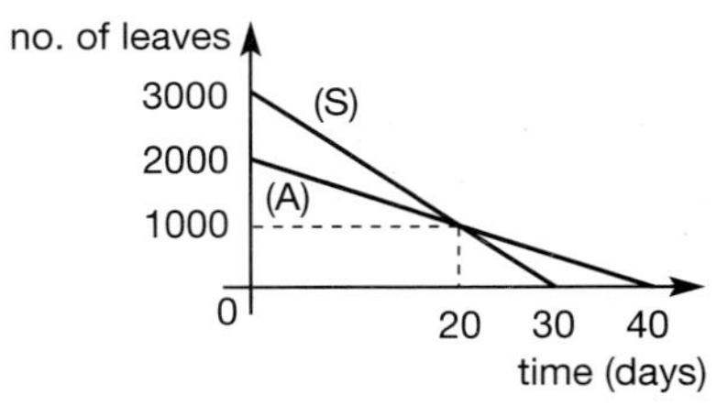

b) Yes, after 20 days.

6 a)

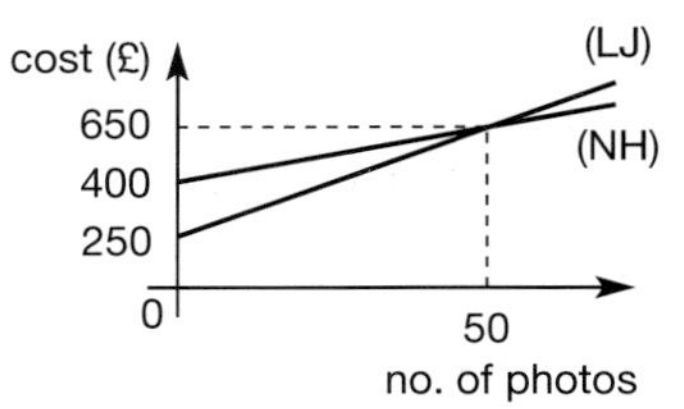

b) L J is cheaper for less than 50 photos.
N H is cheaper for more than 50 photos.
There is no difference for exactly 50 photos.

Exercise 54, page 115

1 a) S: $y = 30 + 0{\cdot}02x$ F: $y = 20 + 0{\cdot}03x$

b) cost (£), no. of therms, S, F, 1000

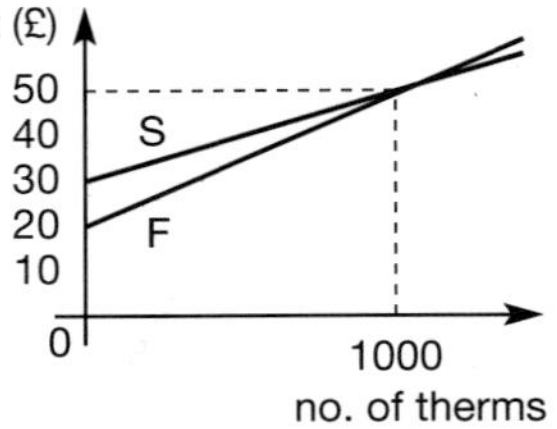

c) Fifegas is cheaper for less than 1000 therms. Scotgas is cheaper for more than 1000 therms. There is no difference for exactly 1000 therms.

2 a) CP: $y = 1{\cdot}5 + 0{\cdot}25x$ L: $y = 0{\cdot}5 + 0{\cdot}45x$

b)

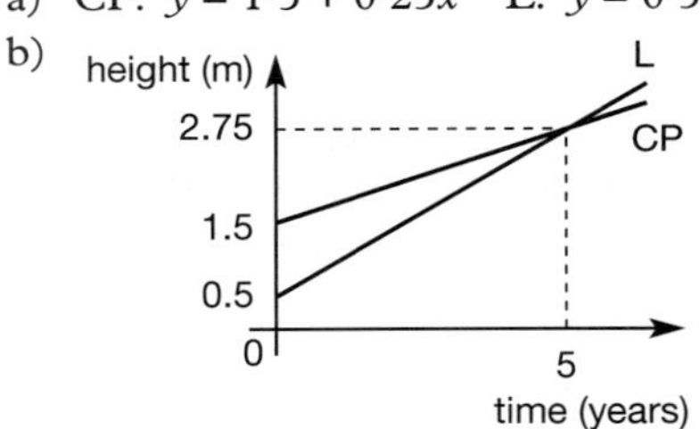

c) after 5 years.

3

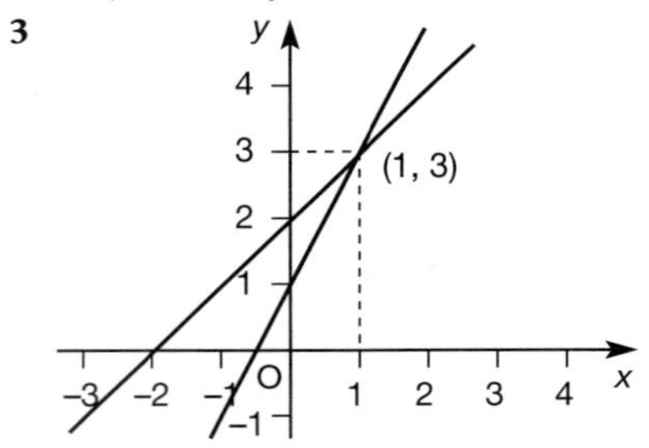

4

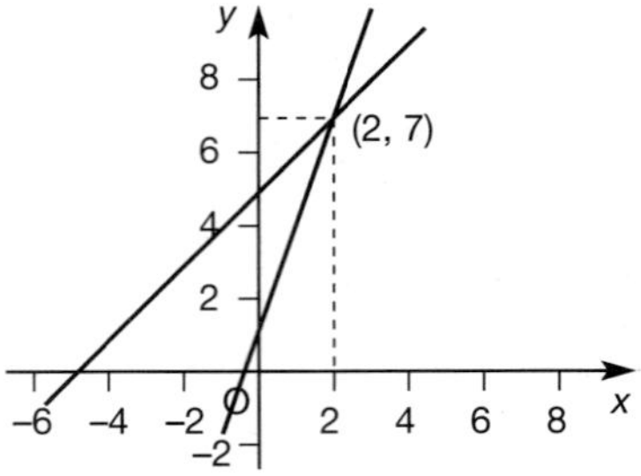

(2, 7)

5

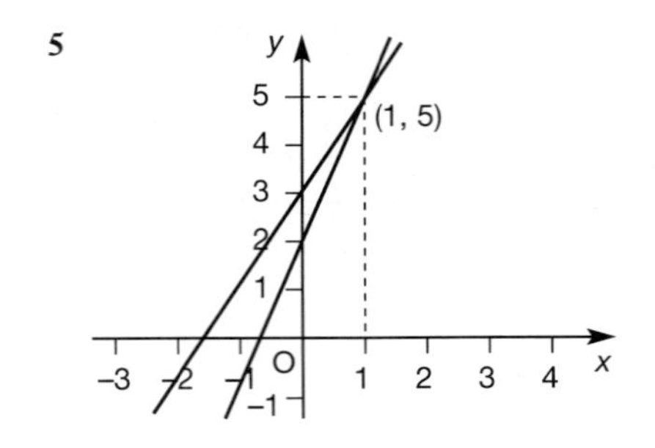

6

(3, 7)

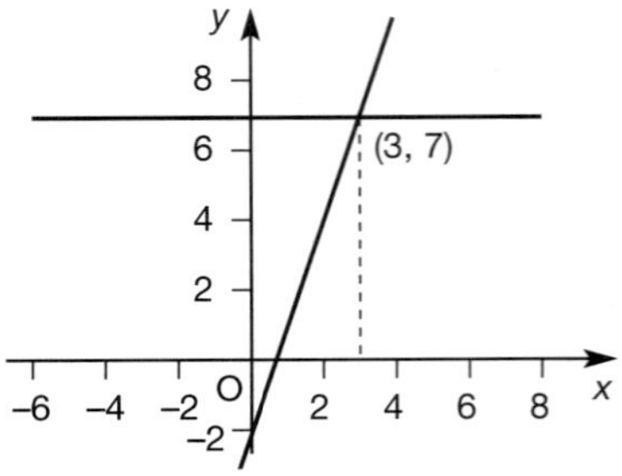

7

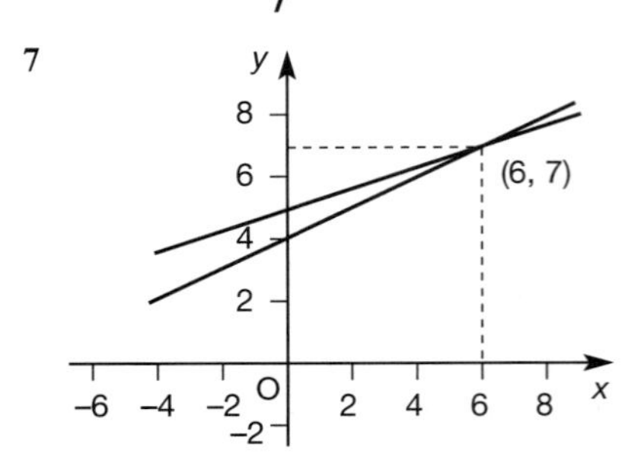

8

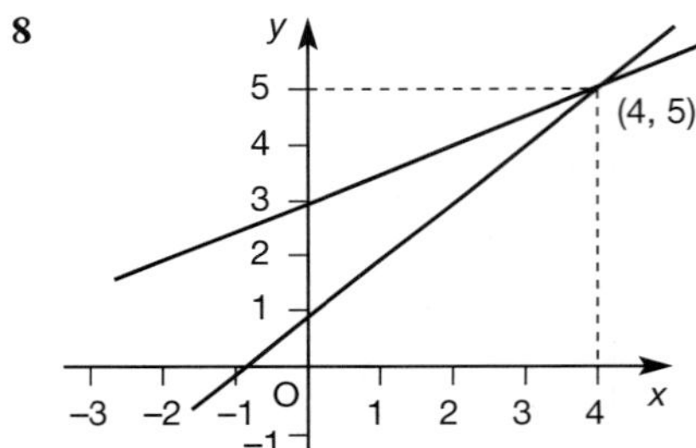

9

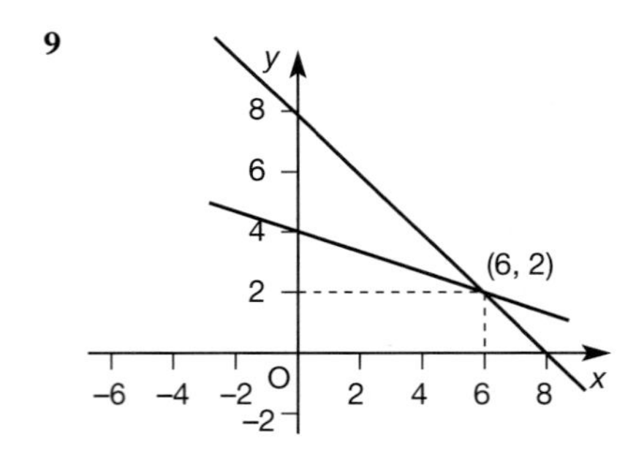

10

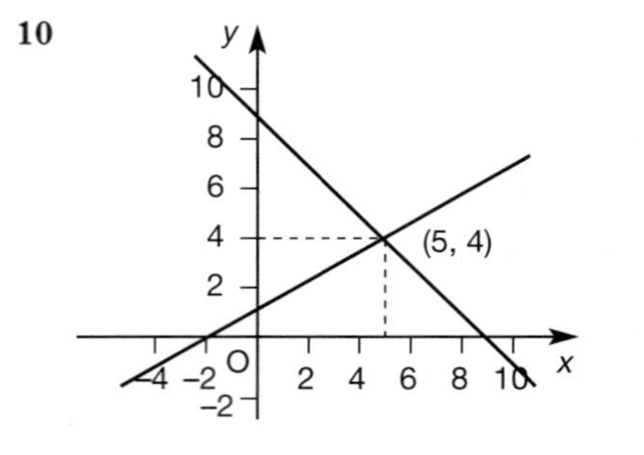

11

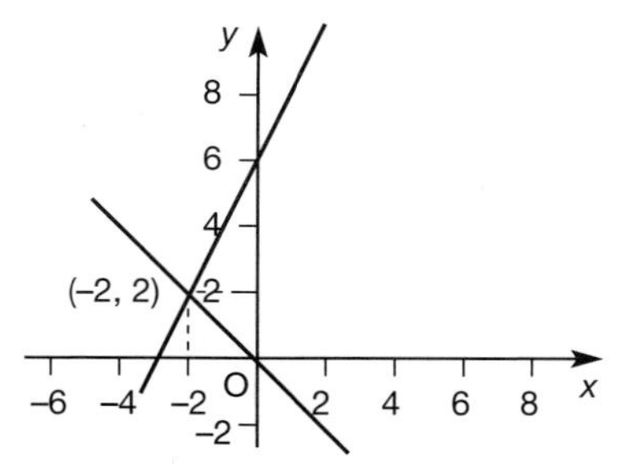

12

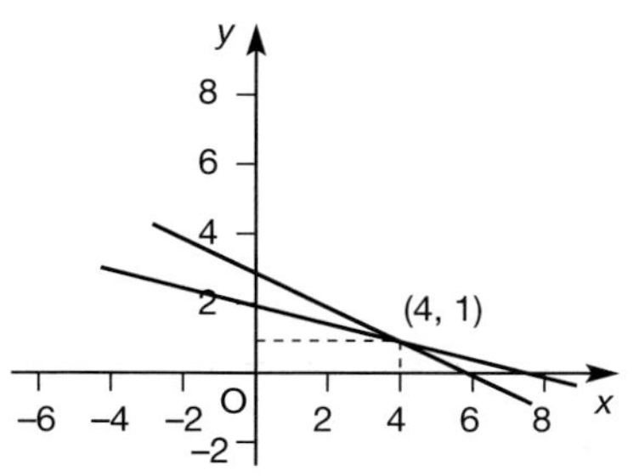

13

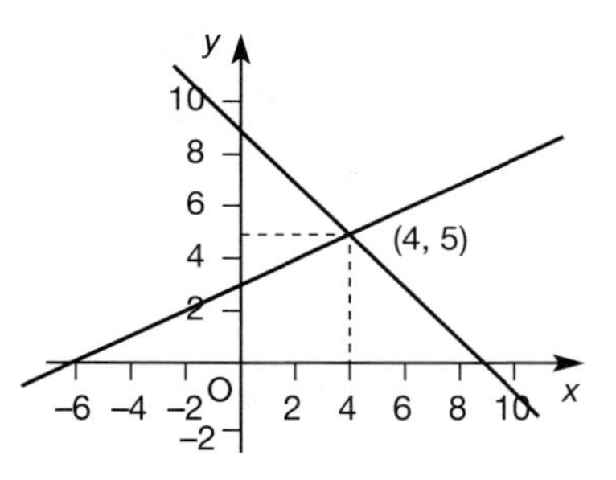

14

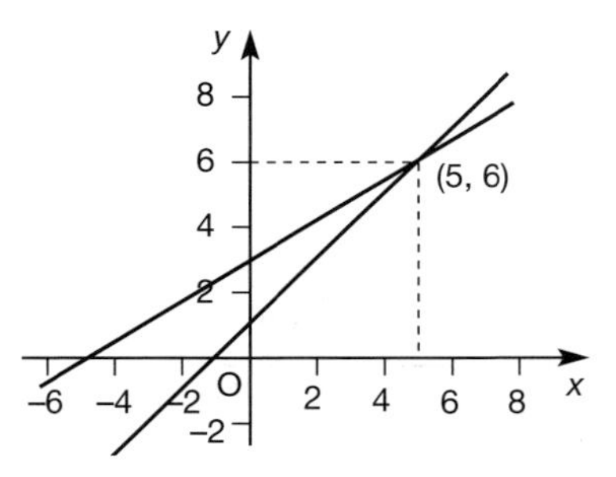

15

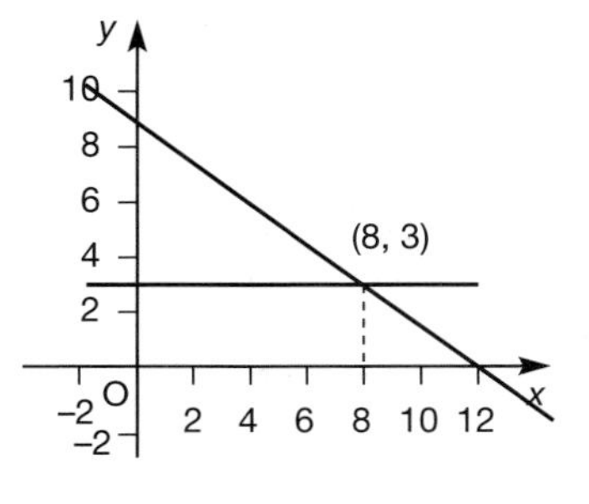

16

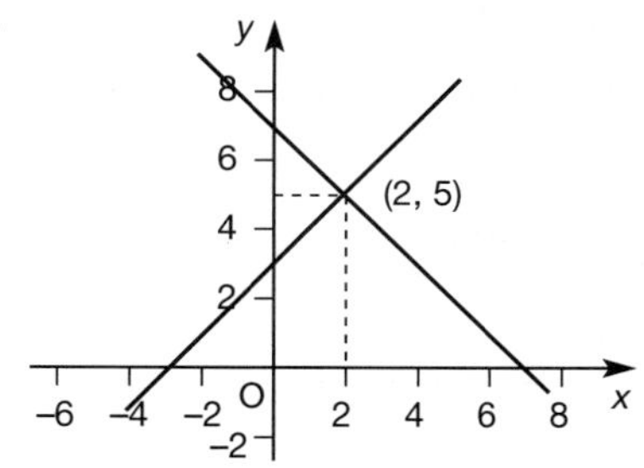

17

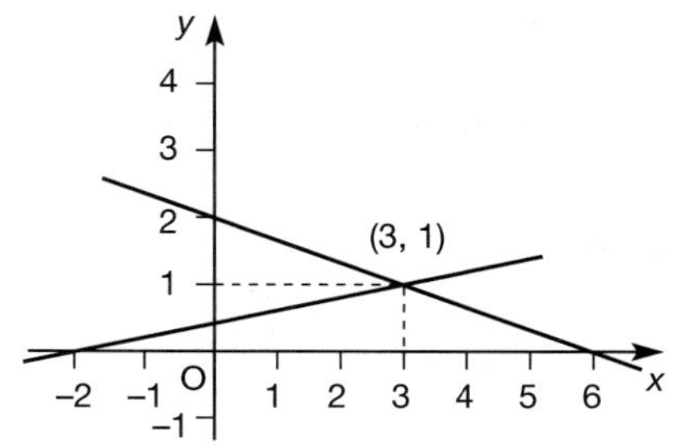

18

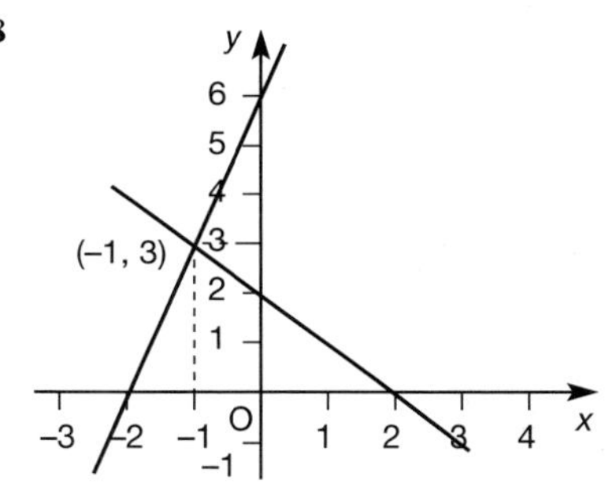

19

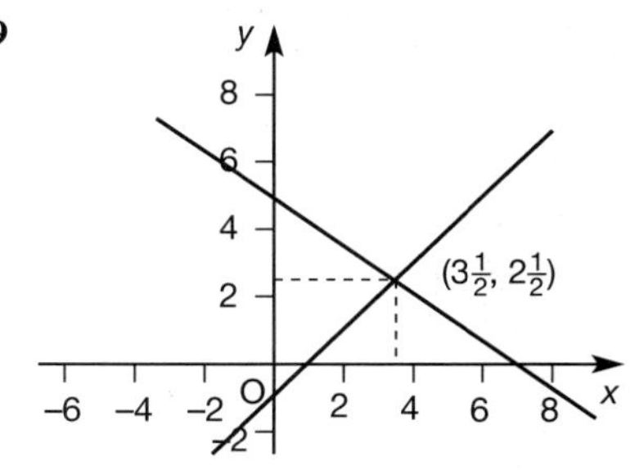

20

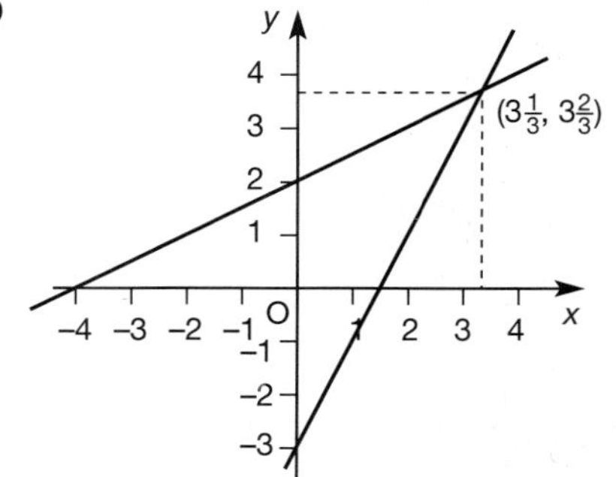

Exercise 55, page 117

1 (1, 9). **2** (1, 10). **3** (2, 9). **4** (2, 12).
5 (3, 27). **6** (4, 13). **7** (10, 1). **8** (6, 2).
9 (14, 5). **10** (6, −1). **11** (14, −4).
12 (−2, 4). **13** (−1, 1). **14** (−2, −6).
15 (−2, −7). **16** (3, 1). **17** (−2, −2).
18 (7, −3). **19** (3, 2). **20** (5, 2). **21** (4, 1).
22 (2, −1). **23** (3, −2). **24** (−2, −4).
25 (9, 23). **26** (10, 3). **27** (1, 3).
28 (7, −4). **29** (3, −2). **30** (7, 2).
31 50 weeks. **32** 10 minutes.

Exercise 56, page 121

1 23 and 17. **2** 30° and 60°. **3** 59° and 121°.
4 20 weeks. **5** 1782 mm^2. **6** £4·90.
7 adult £7, child £4. **8** £230.
9 fish supper £2·70, haggis supper £2·40.
10 22p per kg.

Exercise 57, page 122

1

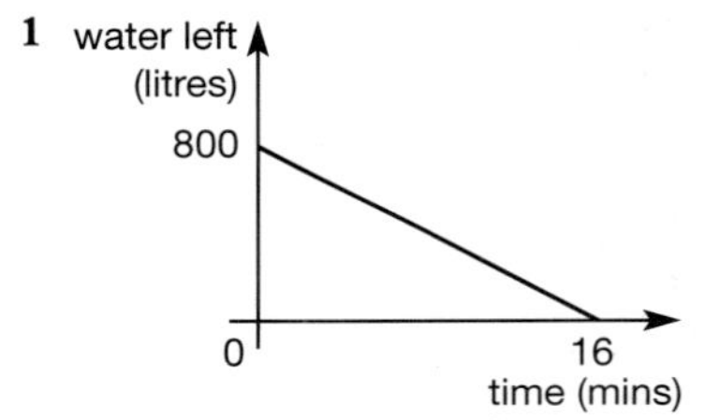

2 a)

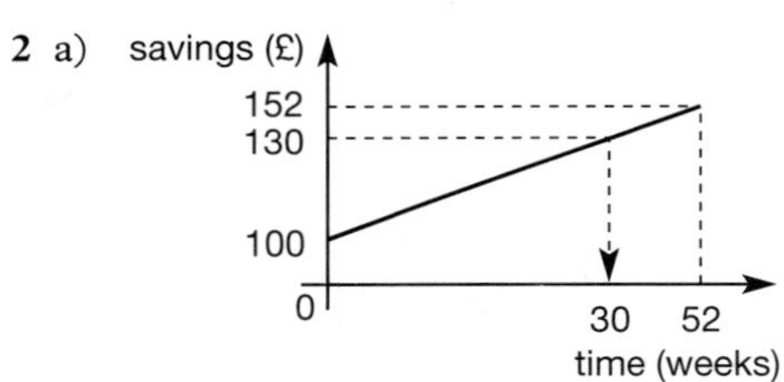

b) 30 weeks.

3 a)

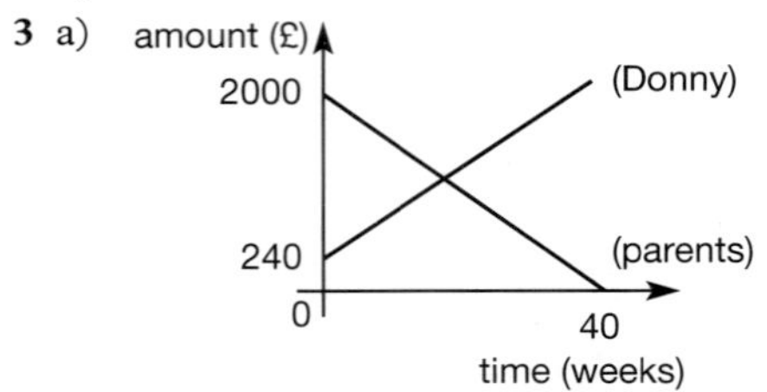

b) after 17 weeks (equal amounts at 16 weeks).

4 a)

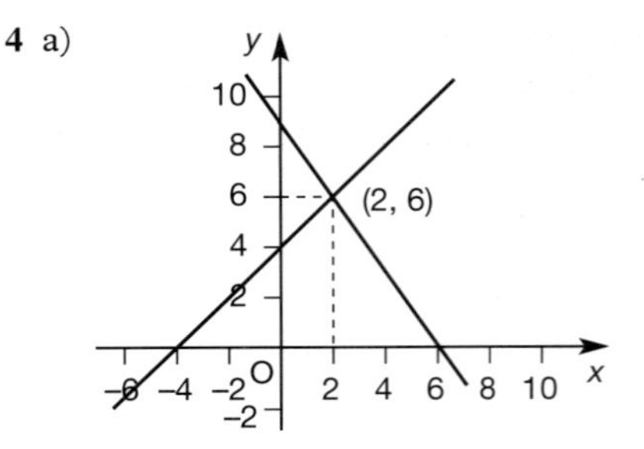

b)

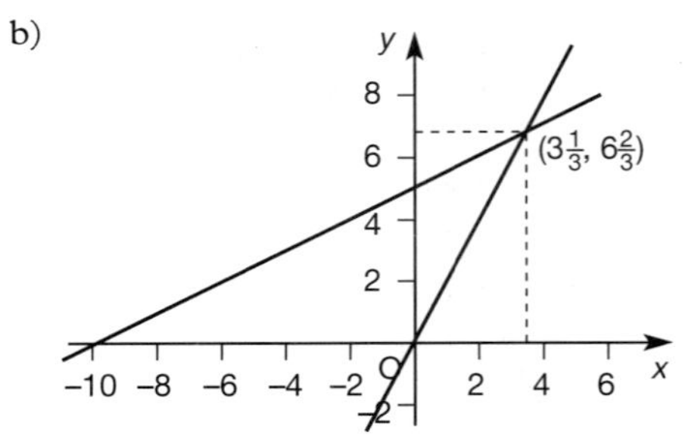

5 a) (−3, 4) b) (2, −4).

6 a)

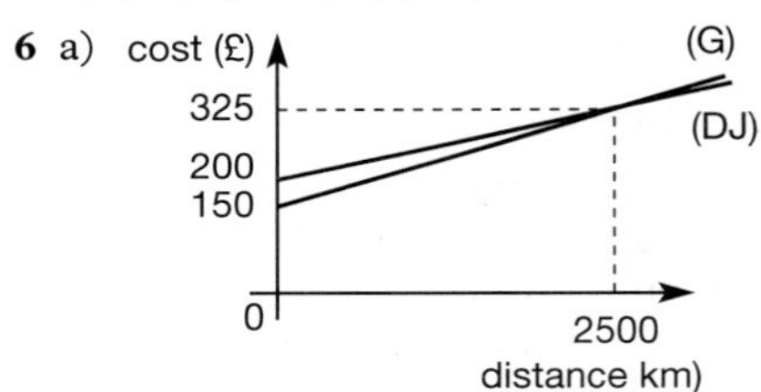

b) Gumshoe is cheaper for less than 2500 km.
Dirt Jeep is cheaper for more than 2500 km.
There is no difference for exactly 2500 km.

7 at 20 days. **8** 56 years old.

Exercise 58 (Test), page 123

1 a)

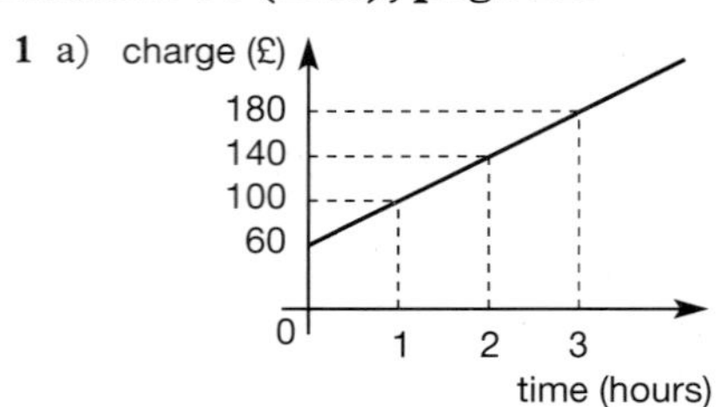

b) $2\frac{1}{2}$ hours.

2 a)

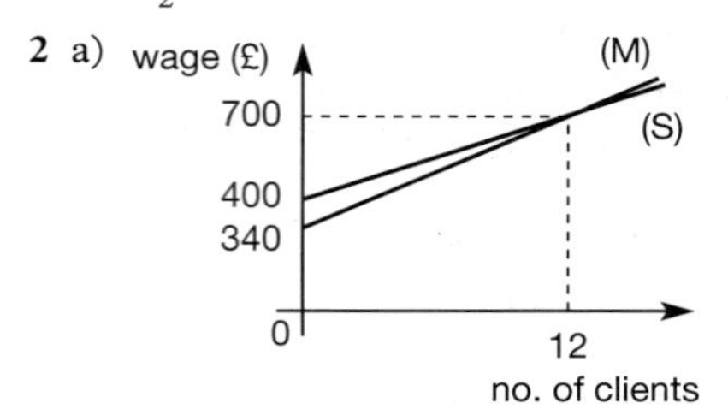

b) Sheila earns more for 0 to 11 clients.
(They both earn the same for 12 clients.)
Margaret earns more for 13 or more clients.

3 a)

b)

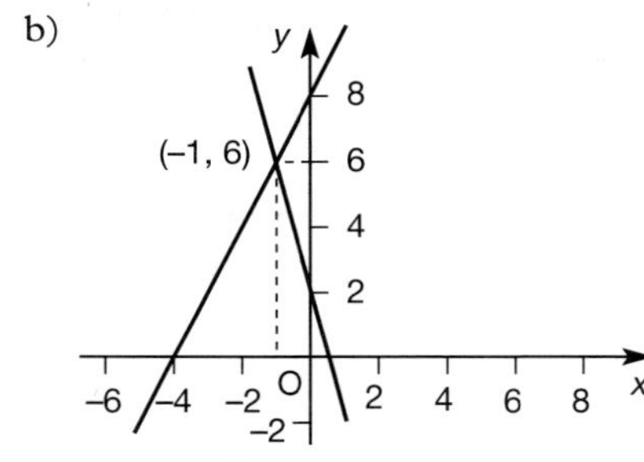

4 a) (3, −2) b) (4, 5).

5 Callwyde is cheaper for less than 500 units.
Allcall is cheaper for more than 500 units.
There is no difference for exactly 500 units.

6 290 mm. **7** 7 × 20p pieces and 8 × 50p pieces.

8 $\frac{2}{3}$.

Exercise 59, page 126

1 a) 27 b) 4 c) 6 d) 6 e) 2.

2 a) Friday b) Wednesday, Thursday
c) 35 d) $28\frac{1}{8}$%.

3 a) steak b) liver or kidney c) 18
d) 7 e) 5.

4 a) raspberry b) 16 c) vanilla d) 7
e) lemon f) 58.

5 a) 800 mm b) 1050 mm c) 9
d) 100 mm e) 200 mm.

6 a) (i) $2\frac{1}{2}$ km/l (ii) 3 km/l
b) 40 and 90 km/h
c) 25 and 100 km/h
d) (i) 70 km/h (ii) 20 km/h.

7 a) 1932 b) 1934 c) 44 d) 1933
e) 1933–4.

8 a) 87p b) 72p c) 80p d) March – April
e) January – February.

9 a) brown b) red c) 25
d) blonde and dark e) yes.

10 a) swimming b) tennis
c) athletics and hockey d) no e) 120°.

11 a) green and brown, black and yellow
b) red c) 30 d) 15°.

12 a) orange b) water
c) lemon, cherry, lime, ginger beer
d) 20 e) 60°.

13 a) 78 b) less c) no.

14 a) 1·81 m and 1·51 m b) 28% c) 36%.

15 a) 137, 103 b) more than c) 128.

16 a) 98 and 70 b) 24% c) 3.

17

0 0	0	0
9 7 3	1	2 7 9
9 5	2	4 7 8
5 4 1 0	3	3 5 8 9
0	4	5

Exercise 60, page 132

1

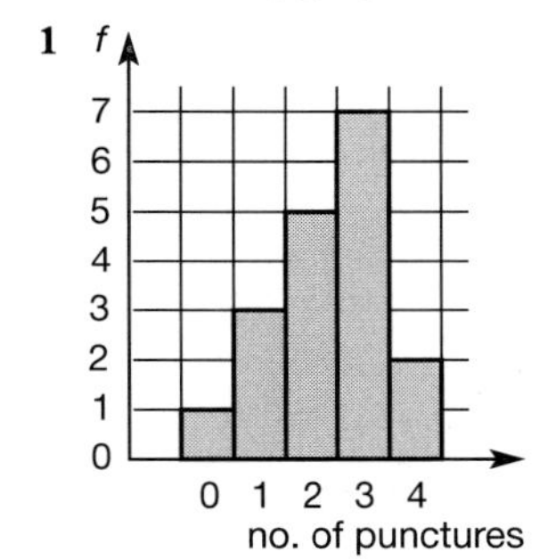

2 a)

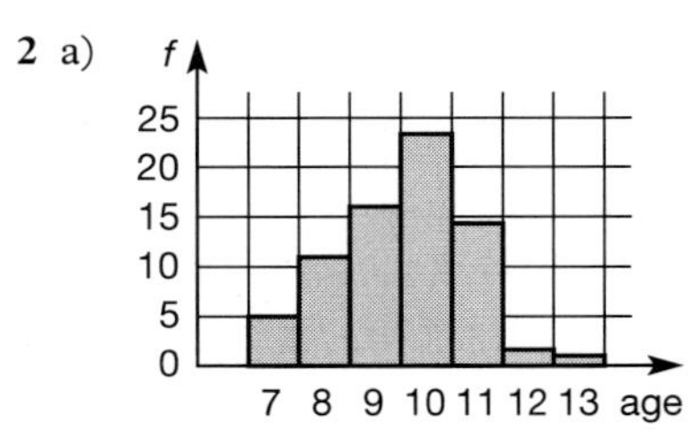

b) 10 years.

3

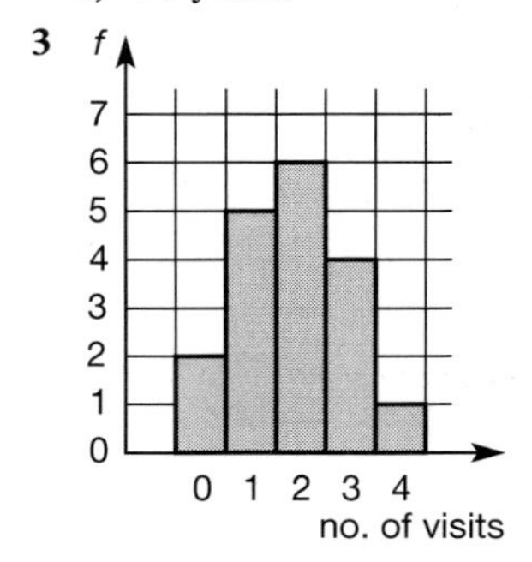

4 a)

(pupil no.)

7
6
5
4
3
2
1
0

0 1 2 3 4 5 6

(pens)

b) 7 pens.

5 a)

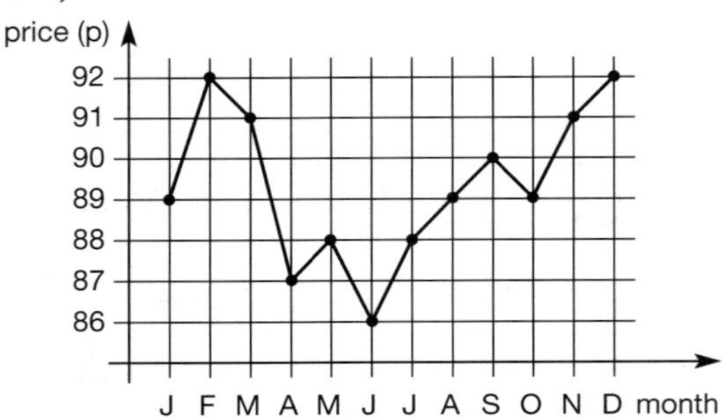

b) (i) June (ii) February and December.

6 a)

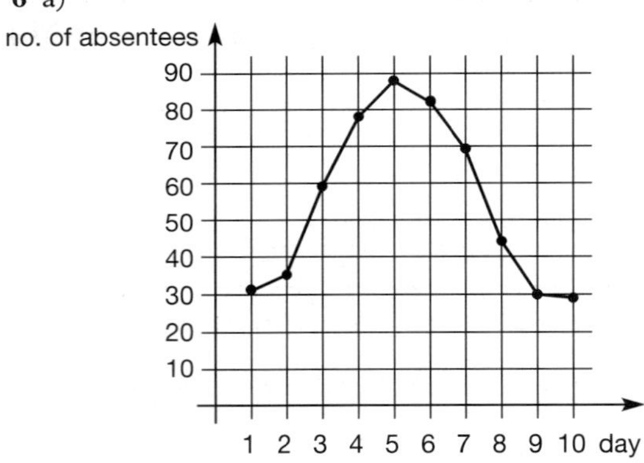

b) 40%

7 a) mass (kg)

6
5
4
3
2
1

0 2 4 6 8 10 12 14 16

age (weeks)

b) $12\frac{1}{2}$ to 13 weeks.

8

4	5 8 8 9 9
5	0 1 2 2 3 3 3 4 5 6 7 7 8 8
6	0 0 1 5 6 (4 | 5 denotes 45).

9

14	1 4 6 7 9 9
15	1 3 3 5 6 8 8 8 8
16	0 1 5 6 8 (14 | 1 denotes 141).

10

98	8 9
99	5 6 8
100	0 0 1 2 2 3 5 7 8 8 9
101	0 0 1 2 (101 | 2 denotes 1012).

Exercise 61, page 135

1 a) F has more income than E; F is more intelligent than E
b) G is older than H; H is taller than G
c) J is heavier than I; J takes bigger shoes than I
d) L is longer than K; L is more expensive than K
e) M and N cost the same; N is wider than M
f) B is faster than A; A is more economical than B
g) Q and R are the same size; Q is warmer than R
h) S has more rainfall than T; T has more sunshine than S
i) V is heavier than U; V is older than U
j) X is bigger than W; X costs more than W
k) Y is paid more than Z; Z works longer than Y
l) Q has a higher interest rate than P; Q is greater than P.

2

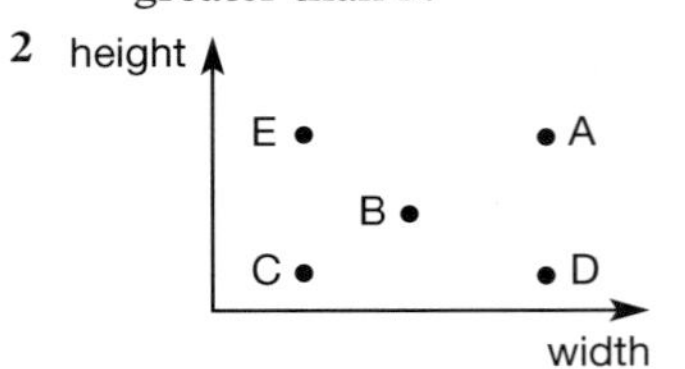

3

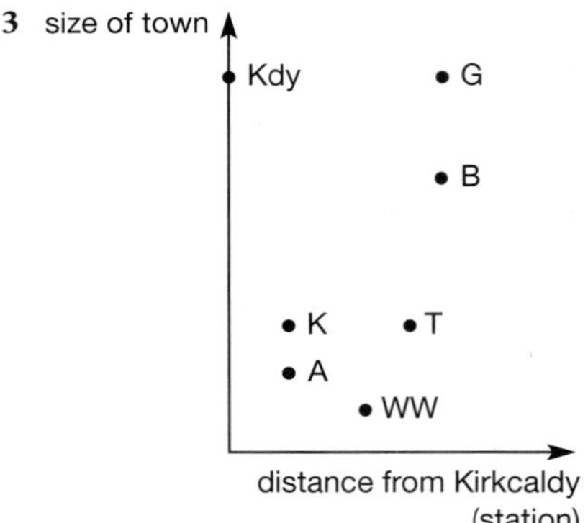

4 a)

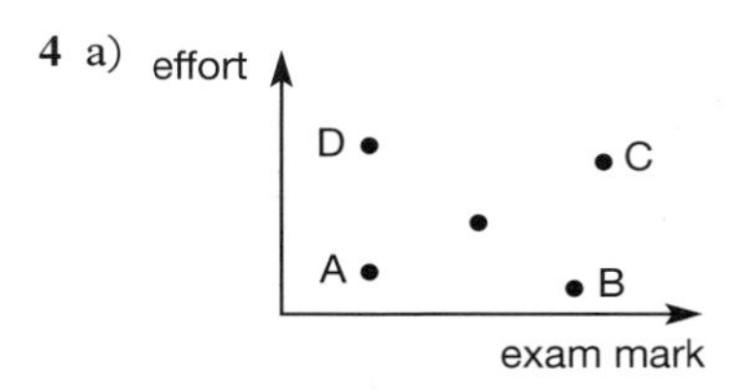

b) Ewan has performed reasonably well this term but would do even better by putting in a greater effort.

5 c) 800 mm.

6 c) 3

d) 15 days is extrapolation, 8 days is interpolation.

7 c) 75 kg.

Exercise 62, page 138

1 salmon 120°, gammon 90°, roast beef 150°.

2 walk 150°, bus 105°, car 60°, cycle 45°.

3 (0) 40°, (1) 100°, (2) 120°, (3) 80°, (4) 20°.

4 (0) 60°, (1) 75°, (2) 105°, (3) 45°, (4) 30°, (5) 30°, (6) 15°.

5 (7) 25°, (8) 55°, (9) 80°, (10) 115°, (11) 70°, (12) 10°, (13) 5°.

6 (red) 30°, (blue) 66°, (green) 102°, (yellow) 90°, (black) 48°, (white) 24°.

Exercise 63, page 140

1

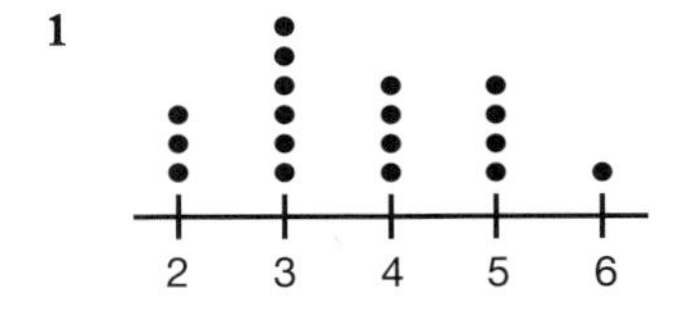

2

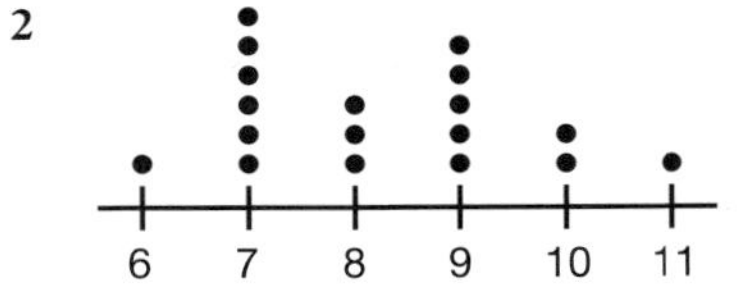

3

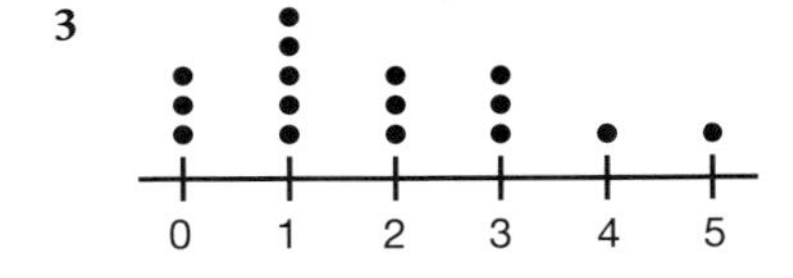

4

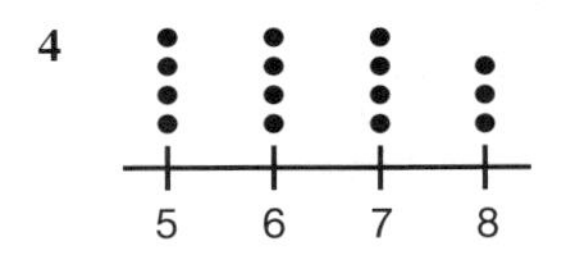

5

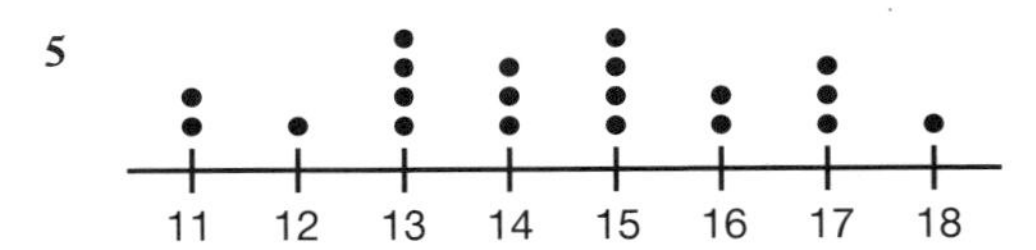

6

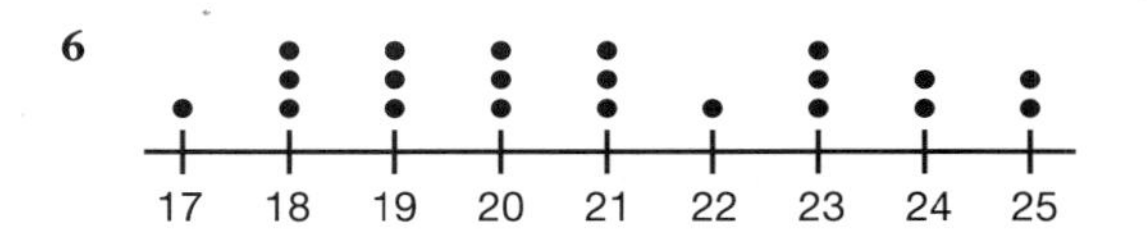

Exercise 64, page 142

1 a)

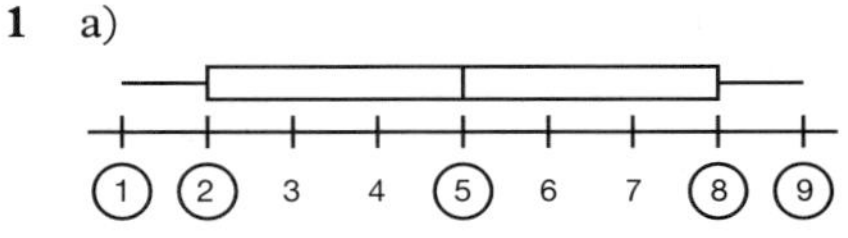

b)

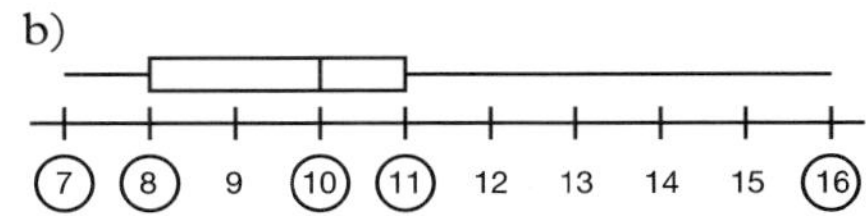

c)

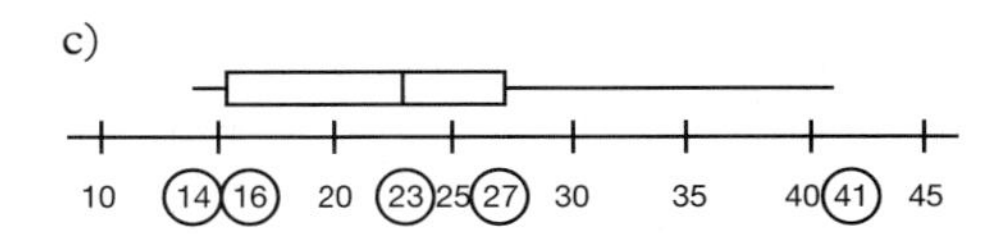

d)

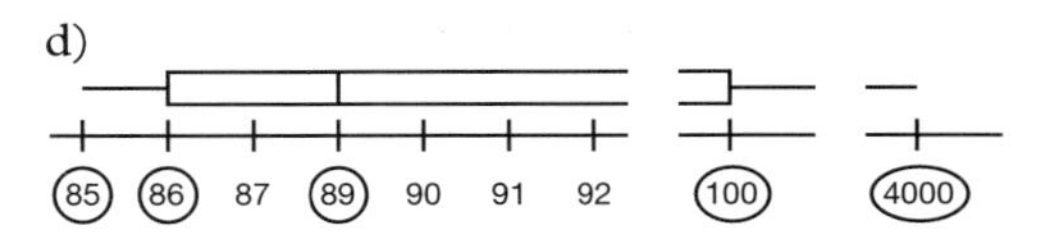

e)

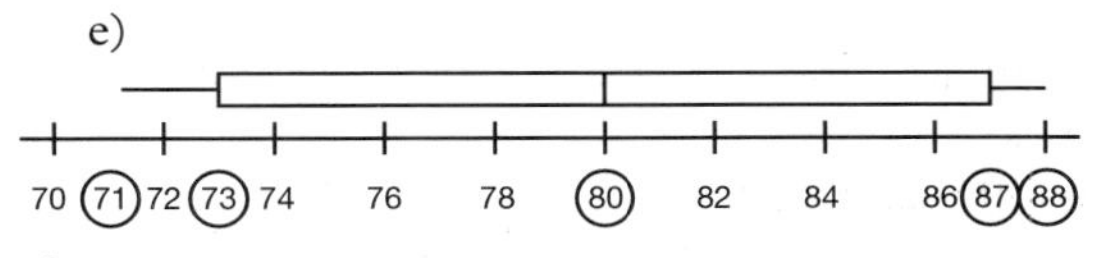

2

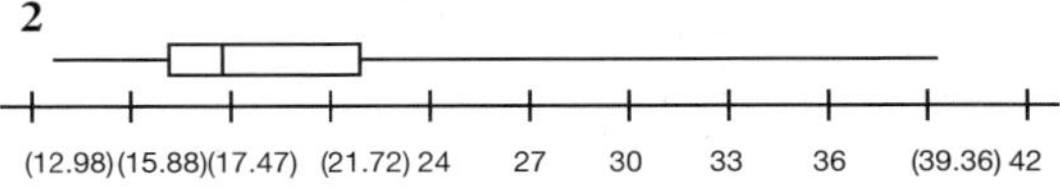

3 mode = 8 median = 8 mean = 8

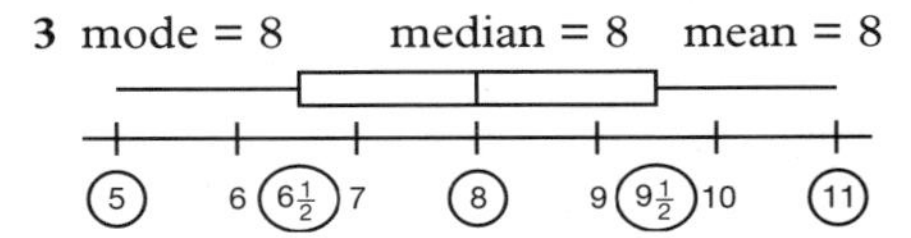

4 mode = 16 median = 15 mean = $14\frac{9}{11}$

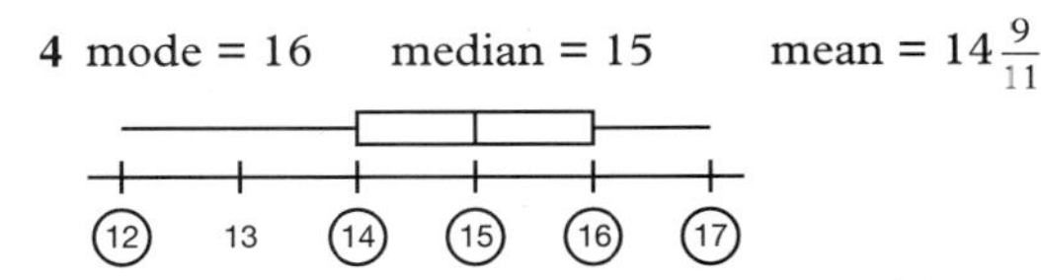

5 mode = 9 median = 8 mean = $7\frac{10}{11}$

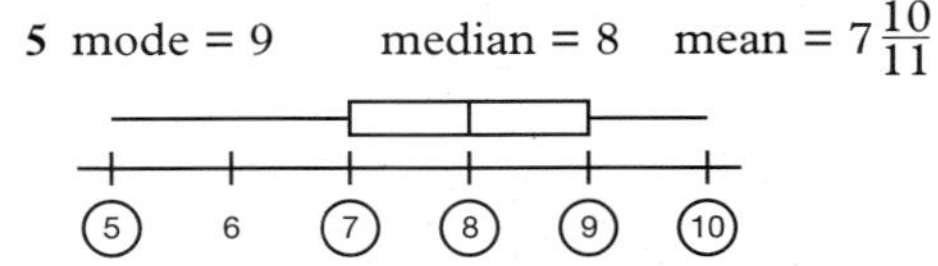

6 mode = 22 median = $19\frac{1}{2}$ mean = 19

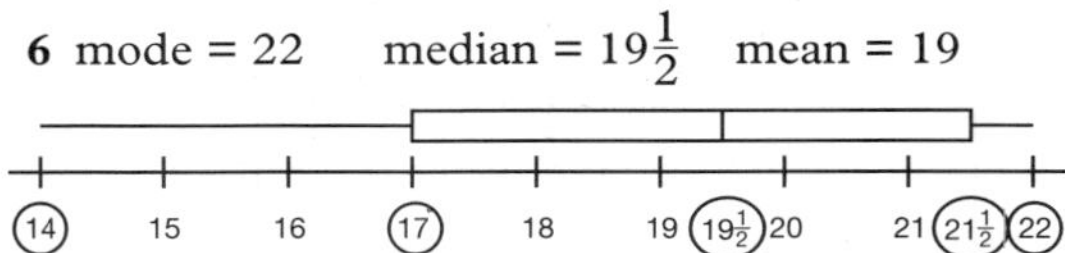

7 mode = 4 median = $3\frac{1}{2}$ mean = 3·2

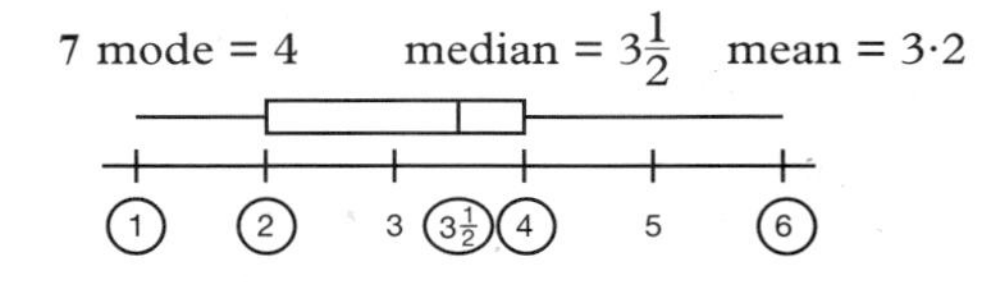

8

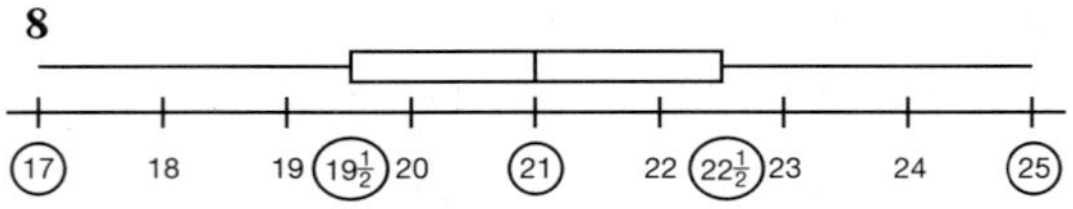

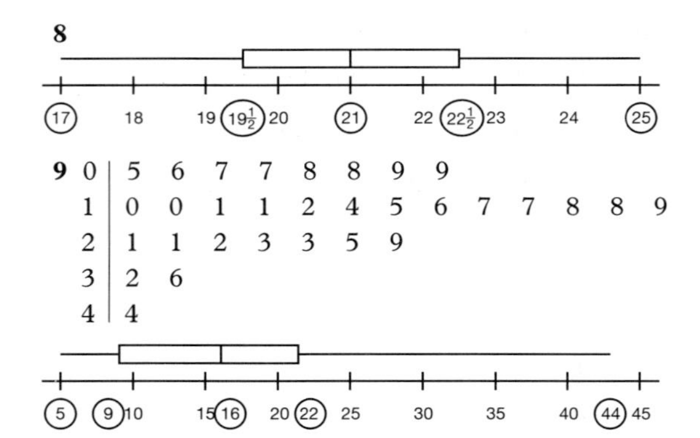

9

0	5 6 7 7 8 8 9 9
1	0 0 1 1 2 4 5 6 7 7 8 8 9
2	1 1 2 3 3 5 9
3	2 6
4	4

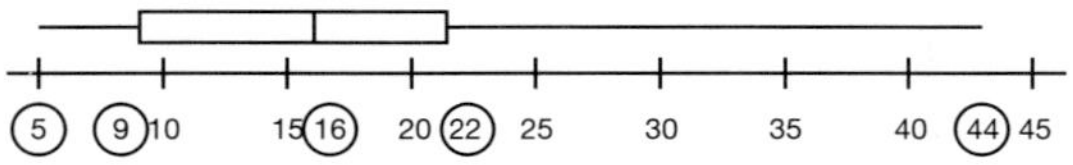

Exercise 65, page 144

1 a)

x	f	cum f
3	2	2
4	3	5
5	5	10
6	4	14
7	6	20
8	7	27
9	2	29
10	1	30

b)

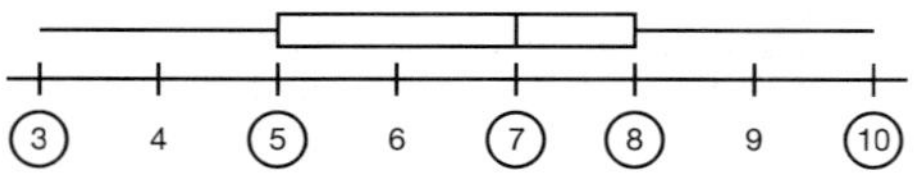

2 a)

x	f	cum f
4	2	2
5	2	4
6	3	7
7	5	12
8	7	19
9	6	25
10	5	30

b)

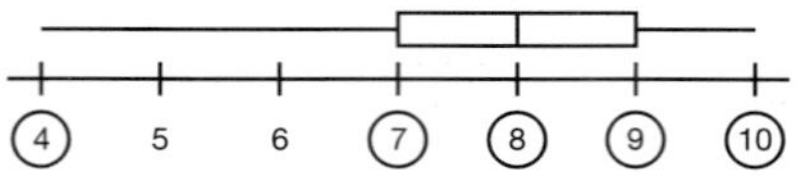

c) On the re–test, the median was 1 mark higher and 75% scored at least 7, whereas only 50% did so before.

3 Plot x on the x–axis and the cumulative frequency on the y–axis and join the points with a smooth curve.

4 a)

x	f	cum f
2	2	2
3	5	7
4	7	14
5	3	17
6	2	19
7	1	20
8	1	21

b)

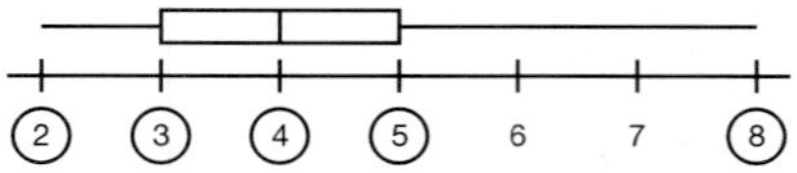

5 a)

x	f	cum f
0	11	11
1	10	21
2	8	29
3	6	35
4	3	38
5	1	39
6	0	39
7	1	40

b)

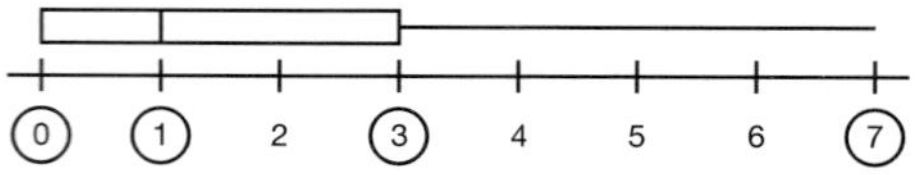

6

x	f	cum f
0	4	4
1	7	11
2	17	28
3	23	51
4	16	67
5	12	79
6	5	84

lower quartile = 2
median = 3
upper quartile = 4.

7

x	f	cum f
0	23	23
1	37	60
2	53	113
3	29	142
4	25	167
5	14	181
6	13	194
7	11	205

lower quartile = 1
median = 2
upper quartile = 4.

Exercise 66, page 147

1 a) $\frac{1}{2}$ b) 50.

2 a) $\frac{1}{6}$ b) 36.

3 a) $\frac{4}{5}$ b) 110.

4 a) $\frac{4}{9}$ b) $\frac{1}{3}$ c) $\frac{2}{9}$.

5 a) $\frac{3}{7}$ b) $\frac{1}{2}$.

6 a) $\frac{1}{3}$ b) $\frac{1}{4}$ c) $\frac{3}{8}$.

7 a) $\frac{1}{8}$ b) $\frac{1}{4}$ c) $\frac{3}{8}$ d) $\frac{3}{8}$ e) $\frac{7}{8}$.

8 (red blue)

1,1 2,1 3,1 4,1 5,1 6,1 a) $\frac{1}{6}$ g) 0

1,2 2,2 3,2 4,2 5,2 6,2 b) $\frac{1}{36}$ h) 1

1,3 2,3 3,3 4,3 5,3 6,3 c) $\frac{1}{6}$ i) $\frac{5}{6}$

1,4 2,4 3,4 4,4 5,4 6,4 d) $\frac{25}{36}$ j) $\frac{5}{12}$

1,5 2,5 3,5 4,5 5,5 6,5 e) $\frac{1}{2}$ k) $\frac{1}{4}$

1,6 2,6 3,6 4,6 5,6 6,6 f) $\frac{1}{12}$ l) $\frac{1}{9}$.

9 2345 3245 4235 5234 a) $\frac{1}{2}$

2354 3254 4253 5243 b) $\frac{1}{4}$

2435 3425 4352 5342 c) 0

2453 3452 4325 5324 d) 1.

2534 3542 4532 5423

2543 3524 4523 5432

10 a) 105 b) 0·86.

11 a) 0·45 b) 9.

12 a) 0·3 b) 21.

Exercise 67, page 149

1 a) (i) 82 (ii) 83 (iii) 86 (iv) 12 b) $\frac{1}{7}$.

2 a) (i) 45 (ii) 52 (iii) 59 (iv) 69 b) $\frac{3}{7}$.

3 a) (i) 16 (ii) 16 (iii) 16 (iv) 2 b) $\frac{1}{2}$.

4 a) (i) $24\frac{8}{11}$ (ii) 26 (iii) 26 (iv) 12
b) $\frac{3}{11}$.

5 a) (i) 530 (ii) 555 (iii) none (iv) 370
b) $\frac{1}{3}$.

6 a) (i) 3·5 (ii) 3·75 (iii) 4·2 (iv) 2·1
b) 0·3.

7 a) (i) 59·8 (ii) $60\frac{1}{2}$ (iii) 61 (iv) 19
b) 0·4.

8 a) (i) $34\frac{1}{2}$ (ii) $30\frac{1}{2}$ (iii) 33 (iv) 69 b) $\frac{1}{3}$.

Exercise 68, page 151

1 a) (i) 5·3 (ii) 5 (iii) 5 (iv) 6 b) $\frac{17}{45}$.

2 a) (i) 3 (ii) 3 (iii) 2 (iv) 5 b) $\frac{37}{180}$.

3 a) (i) 3·57 (ii) 3 (iii) 1 (iv) 21 b) $\frac{5}{7}$.

4 a) (i) 1·96 (ii) 2 (iii) 2 (iv) 11 b) $\frac{1}{25}$.

5 a) (i) 7·4 (ii) 7 (iii) 8 (iv) 13 b) $\frac{3}{20}$.

6 2·375. **7** 7·7. **8** 3.

9

x	0	1	2	3	4	5
f	5	11	9	7	3	1
cum f	5	16	25	32	35	36

b) mode = 1, range = 5

c)

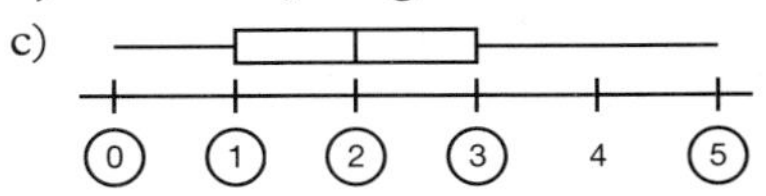

d) mean = 1·86 e) $\frac{5}{36}$.

10

x	0	1	2	3	4	7	8
f	3	5	11	10	8	2	1
cum f	3	8	19	29	37	39	40

b) mode 2, range = 8

c)

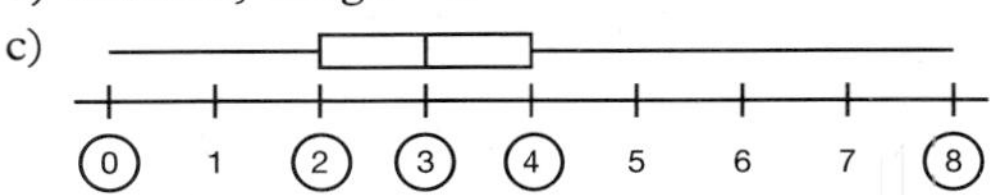

d) 2·775 e) $\frac{21}{40}$.

Exercise 69, page 154

1 a) $1\frac{1}{2}$ b) $\frac{7}{40}$. **2** a) 1 b) $\frac{1}{9}$. **3** $\frac{1}{2}$.
4 a) $\frac{1}{2}$ b) $\frac{1}{10}$. **5** a) 1 b) $\frac{8}{25}$. **6** 2.
7 a) $1\frac{1}{2}$ b) $\frac{3}{5}$.

Exercise 70, page 157

1 1·58. **2** 12·0. **3** 11·1. **4** 6·35.
5 1·76. **6** 7·76. **7** 2·38. **8** 8·35.
9 3·78.
10 a)

990 996 1003 1005 1009 1015

b) 25, 6·5 c) $\frac{5}{7}$.

11

(median)	(mean)	(standard deviation)
25	34	23·0
23	26	18·6.

Exercise 71, page 160

(Your answers may not agree exactly with those which follow.)

1 b) $y = 0{\cdot}025x + 0{\cdot}5$ c) 1·6 m.
2 b) $y = 50 - 2x$ c) 48
d) no, the pond would be ice
e) no, this value is not close enough to the given data set; you can't get '–2' ducks
f) extrapolation.
3 b) $y = 4x - 10$ c) 62
d) positive here, negative in question 2
e) the temperature at 3 p.m. might not reflect the weather all day; a visiting bus trip might boost the sales even in lower temperatures.
4 b) $y = 20x + 120$ c) (i) £280 (ii) £360.
5 b) $y = 13 - \frac{1}{2}x$
c) 8 km/h; reasonably confident as data all close together.
d) 13 km/h; less confident as 'no rucksack' gives greater freedom and causes less sweating, so he might be able to run even faster than this.
6 b) Teams which have higher goal differences have more points.
c) $y = 1{\cdot}4x - 70$
d) around 50 points, i.e. in the middle of the league.

Exercise 72, page 162

1 a)

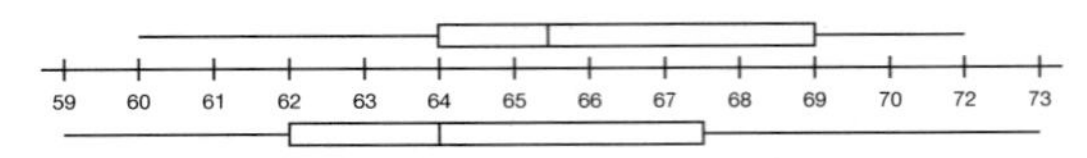

The teachers generally have higher pulse rates (75% of them 64 or over, but only 50% of the pupils as high as this) but the pupils are more widely spread.
b) teacher $\frac{3}{22}$ pupil $\frac{2}{7}$.
2 won 170°, drew 100°, lost 90°.
3 a) 9 6 3 9 3
b) 7 5 3 8 2·5
c) 12 7 $3\frac{1}{2}$ 10 $3\frac{1}{4}$.

4 a)

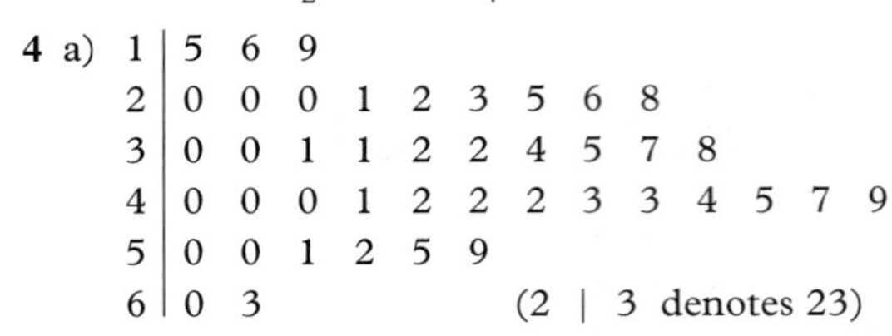

1	5 6 9
2	0 0 0 1 2 3 5 6 8
3	0 0 1 1 2 2 4 5 7 8
4	0 0 0 1 2 2 2 3 3 4 5 7 9
5	0 0 1 2 5 9
6	0 3

(2 | 3 denotes 23)

b) 15, 26, 38, 45, 63 c) 36·8.

5

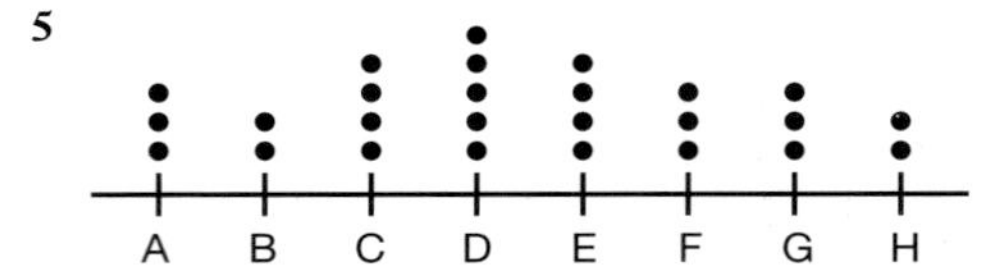

6 a) 7, 109, 105, 111, 3
b) 19, 27, 22, 30, 4
c) 16, 77, 73, 82, $4\frac{1}{2}$
d) 15, 97, $92\frac{1}{2}$, 99, $3\frac{1}{4}$
e) 64, 36, 30, 55, $12\frac{1}{2}$
f) 84, $56\frac{1}{2}$, 16, 85, $34\frac{1}{2}$.
7 a) (i) 27·5 (ii) 25 (iii) 25 (iv) 19
b) $\frac{2}{11}$.
8 mean = 353 S.D. = 7·87.
9 a) 46·4 (ii) 47 (iii) 47 (iv) 8 b) $\frac{1}{4}$.
10 a)

deviation	4	5	1	4	6	7	7	8	12	9	14	11
V	12	11	10	9	8	7	6	5	4	3	2	1

b) $y = 13 - x$ or similar c) £8
d) probably not; his theory is unlikely to be true. Correlation here is not good, especially for the larger and smaller prizes.

11 a) (i) $\frac{1}{16}$ (ii) $\frac{1}{8}$ (iii) $\frac{1}{2}$ b) 5

d)

x	1	2	3	4	5
f	8	4	2	1	1

mean = $1\frac{15}{16}$

e) This is not likely to make much profit for the charity as the ticket price of £2 is only marginally more than the mean score of £$1\frac{15}{16}$. It could even mean a loss, so raise the ticket price to £3 at least.

Exercise 73 (Test), page 164

1 a) $\frac{1}{52}$ b) $\frac{1}{13}$ c) $\frac{1}{2}$ d) $\frac{1}{4}$ e) $\frac{1}{26}$.

2 a) $\frac{1}{4}$ b) $\frac{12}{51}$ c) $\frac{38}{51}$.

3 a)

x	0	1	2	3	4	5	6	7	8
f	4	3	2	5	6	4	2	1	3

mean = 3·63.

4 a) 16 000 b) 20 000.

5 a)

Stem	Leaf
0	9 9
1	0 0 0 1 2 4 7 7 8 9
2	1 3 3 5 6 8 9 9
3	0 3 4 7 7 8 8 9 9
4	1 2 4 6 6 8 8 8
5	1 1 3 5 7 7 8
6	0 2 3 4 6 6 7

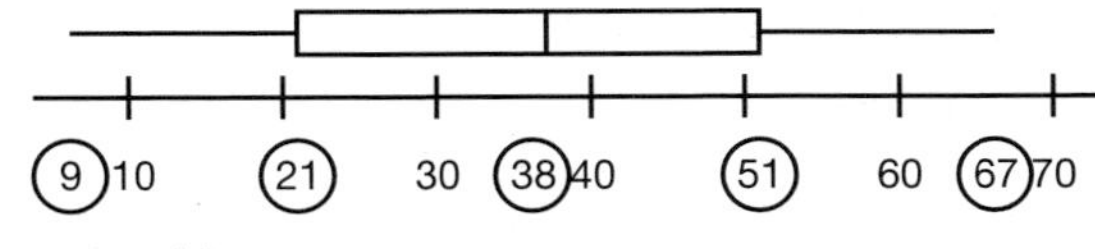

b) $\frac{12}{51}$.

6

x	0	1	2	3	4	5	6	7	8	9	10
f	1	2	2	4	5	5	7	6	3	2	1
cum f	1	3	5	9	14	19	26	32	35	37	38

a) 10 b) $1\frac{1}{2}$.

7 2·5, 3·29. **8** b) $y = 3x + 11$ c) 44.

Exercise 74, page 167

1 $\frac{12}{21} = \frac{20}{35}$, $\frac{15}{27} = \frac{25}{45}$, $\frac{7}{28} = \frac{11}{44}$.

2 $\frac{39}{54}$, $\frac{52}{72}$, $\frac{130}{180}$.

3 $\frac{4}{5} = \frac{8}{10} = \frac{12}{15} = \frac{24}{30} = \frac{36}{45} = \frac{72}{90} = \frac{216}{270}$.

4 a) $\frac{5}{6}$ b) $\frac{1}{6}$ c) $\frac{3}{7}$ d) $\frac{11}{13}$ e) $\frac{2}{13}$ f) $\frac{17}{20}$ g) $\frac{4}{9}$ h) $\frac{5}{7}$ i) $\frac{2}{3}$ j) $\frac{7}{9}$ k) $\frac{9}{14}$ l) $\frac{5}{11}$.

5 a) $\frac{2}{5}$ b) $\frac{4}{11}$ c) $\frac{2}{3}$ d) $\frac{5}{12}$ e) $\frac{4}{11}$ f) $\frac{6}{11}$ g) $\frac{5}{8}$ h) $\frac{20}{23}$ i) $\frac{2}{3}$ j) $\frac{9}{13}$ k) $\frac{4}{7}$ l) $\frac{6}{7}$.

6 a) $\frac{1}{y}$ b) $\frac{a}{b}$ c) q d) $\frac{1}{s}$ e) xy^2.

7 a) $2x + 3y$ b) $4p + 3q$ c) $\frac{1}{2u + 3v}$ d) $\frac{x}{x + 2y}$ e) $\frac{1}{2 + 4l}$.

8 a) 1 b) −1 c) −1 d) $-\frac{3}{4}$ e) $\frac{x}{3}$.

9 a) $\frac{1}{p + q}$ b) $\frac{m}{n}$ c) $\frac{x}{2}$ d) $\frac{3}{4}$ e) $\frac{x + 2}{y - 1}$.

10 a) $\frac{x - 2}{3}$ b) $\frac{x + 1}{4}$ c) $\frac{y + 3}{x}$ d) $\frac{z - 4}{4}$ e) $\frac{p + q}{2}$.

11 a) $\frac{x - 2}{6}$ b) $\frac{y - 2}{2}$ c) $\frac{p + 3}{3}$ d) $\frac{k - 1}{3}$ e) $\frac{z + 5}{5}$.

12 a) $\frac{4}{x + 4}$ b) $\frac{p + 3}{p + 1}$ c) $\frac{x + 3}{x + 1}$ d) $\frac{x}{x + 2}$ e) $\frac{3}{x(x + 1)}$.

Exercise 75, page 170

1 a) $\frac{3}{5}$ b) $\frac{1}{3}$ c) $\frac{1}{2}$ d) 1 e) $\frac{3}{4}$.

2 a) $\frac{3}{5}$ b) $\frac{3}{10}$ c) $\frac{2}{3}$ d) $\frac{3}{5}$ e) $\frac{1}{5}$.

3 a) $\frac{5}{8}$ b) $\frac{17}{30}$ c) $\frac{31}{35}$ d) $\frac{55}{63}$ e) $\frac{23}{24}$.

4 a) $\frac{3}{10}$ b) $\frac{5}{8}$ c) $\frac{7}{18}$ d) $\frac{7}{24}$ e) $\frac{4}{35}$.

5 a) $\frac{1}{6}$ b) $\frac{3}{4}$ c) $\frac{2}{3}$ d) $\frac{5}{6}$ e) $\frac{2}{15}$.

6 a) $\frac{4}{15}$ b) $\frac{2}{9}$ c) $\frac{1}{40}$ d) $\frac{2}{15}$ e) $\frac{9}{40}$.

7 a) $\frac{k + 1}{k}$ b) $\frac{2p + 3}{p}$ c) $\frac{pq + 1}{q}$ d) $\frac{7x}{3}$ e) $\frac{x + 2}{x + 1}$.

8 a) $\frac{2y - 1}{y}$ b) $\frac{pq - 2}{q}$ c) $\frac{3x^2 - 4}{x}$ d) $\frac{3xy - 4}{y}$ e) $\frac{z^3 - 1}{z}$.

9 a) $\frac{2x + 1}{x(x + 1)}$ b) $\frac{5y - 2}{y(y - 1)}$ c) $\frac{5z}{4}$ d) $\frac{2a}{a^2 - 1}$ e) $\frac{7b - 1}{b^2 - 1}$.

10 a) $\frac{1}{x(x + 1)}$ b) $\frac{4 - 3y}{y(y + 2)}$ c) $\frac{8 - z}{z(z - 2)}$ d) $\frac{2}{a^2 - 1}$ e) $\frac{b + 7}{b^2 - 1}$.

11 a) $\frac{7a + b}{a^2 - b^2}$ b) $\frac{a^2 - 2ab - b^2}{a^2 - b^2}$ c) $\frac{2p^2 - 1}{2p}$ d) $\frac{3q^2 + 3q + 1}{3q}$ e) $\frac{m^2 - n^2}{mn}$.

12 a) $\frac{x + y}{x - y}$ b) $\frac{1}{(x + 1)(x + 2)}$ c) $\frac{1}{(x - 3)(x - 4)}$ d) $\frac{3(y + 3)}{2}$ e) $\frac{1}{(x - 1)^2}$.

13 a) −1 b) $\frac{u - v}{u + v}$ c) $\frac{a + 2b}{2a - b}$ d) $\frac{1}{(x - 1)(x + 3)}$ e) $\frac{1}{(y - 1)^2}$.

14 a) $\frac{3x-1}{x^2-1}$ b) $\frac{5-x}{x^2-4}$ c) $\frac{4pq}{p^2-q^2}$
d) $\frac{2}{(k-1)(k^2-1)}$ e) $\frac{5x+2}{(x^2-4)(x-1)}$.

Exercise 76, page 175

1 $a = \frac{P}{3}$. **2** $R = \frac{V}{I}$. **3** $b = \frac{A}{l}$. **4** $f = \frac{P}{m}$.
5 $n = \frac{c}{P}$. **6** $V = \frac{RT}{P}$. **7** $m = \frac{E}{c^2}$.
8 $v = \frac{550H}{T}$. **9** $P = \frac{100I}{TR}$. **10** $b = \frac{2A}{h}$.
11 $m = \frac{2E}{v^2}$. **12** $p = g - c$. **13** $x = \frac{P-y}{2}$.
14 $l = \frac{P-2b}{2}$. **15** $u = \frac{2s-vt}{t}$.
16 $s = \frac{q-ar}{a}$. **17** $a = \frac{y-b}{x}$. **18** $h = \frac{2A}{a+b}$.
19 $h = \frac{A-2\pi r^2}{2\pi r}$. **20** $u = \frac{c-nv}{m}$.
21 $g = \frac{p-q}{2f}$. **22** $v = \frac{t-u}{u}$. **23** $q = \frac{3k-pt}{t}$.
24 $u = \frac{2s-ft^2}{2t}$. **25** $f = \frac{2(s-ut)}{t^2}$.
26 $f = \frac{v^2-u^2}{2s}$. **27** $r = \pm\sqrt{\frac{A}{\pi}}$.
28 $u = \pm\sqrt{v^2-2fs}$. **29** $g = \frac{4\pi^2 l}{T^2}$.
30 $a = \frac{L-b}{L}$. **31** $y = \pm\sqrt{r^2-x^2}$.
32 $p = \frac{mb+na}{m+n}$. **33** $t = \pm\sqrt{\frac{1-x}{1+x}}$.

Exercise 77, page 176

1 b), d), e), f), h), l). **2** a), b), d), e).
3 a) $2\sqrt{2}$ b) $2\sqrt{3}$ c) $2\sqrt{7}$ d) $3\sqrt{3}$ e) $3\sqrt{2}$
f) $5\sqrt{2}$ g) $5\sqrt{3}$ h) $2\sqrt{5}$ i) $2\sqrt{6}$ j) $3\sqrt{5}$
k) $4\sqrt{2}$ l) $3\sqrt{6}$ m) $7\sqrt{2}$ n) $10\sqrt{2}$
o) $6\sqrt{2}$ p) $7\sqrt{3}$ q) $11\sqrt{2}$ r) $15\sqrt{5}$
s) $20\sqrt{10}$ t) $6\sqrt{7}$ u) $6\sqrt{11}$ v) $100\sqrt{10}$
w) $28\sqrt{3}$ x) $45\sqrt{2}$.
4 a) $\sqrt{28}$ b) $\sqrt{45}$ c) $\sqrt{27}$ d) $\sqrt{8}$ e) $\sqrt{75}$
f) $\sqrt{32}$ g) $\sqrt{63}$ h) $\sqrt{500}$ i) $\sqrt{48}$ j) $\sqrt{98}$
k) $\sqrt{125}$ l) $\sqrt{99}$.
5 a) $8\sqrt{2}$ b) $5\sqrt{5}$ c) $12\sqrt{3}$ d) 0
e) $-3\sqrt{7}$ f) 0 g) $\sqrt{11}$ h) $5\sqrt{5}+4\sqrt{3}$
i) $15\sqrt{3}-6\sqrt{15}$ j) $6\sqrt{2}-5\sqrt{7}$
k) $\sqrt{3}+2\sqrt{5}$ l) $11+\sqrt{6}$.
6 a) $5\sqrt{2}+2\sqrt{5}$ b) $\sqrt{3}+2\sqrt{6}$ c) $\sqrt{5}+7\sqrt{3}$
d) $5\sqrt{6}-7\sqrt{5}$ e) $\sqrt{7}-8\sqrt{2}$ f) $11\sqrt{5}-7\sqrt{7}$
g) $\sqrt{3}-\sqrt{6}$ h) $\sqrt{2}+3\sqrt{3}$ i) $5\sqrt{6}+2\sqrt{7}$.
7 AC = $\sqrt{5}$ AD = $\sqrt{6}$. **8** a) $2\sqrt{2}$ b) $2\sqrt{3}$.
9 a) $\sqrt{13}, 2\sqrt{10}, 3\sqrt{5}$ b) 7.

Exercise 78, page 178

1 a) 7 b) 2 c) 5 d) 13 e) a f) $\sqrt{6}$
g) $2\sqrt{5}$ h) $\sqrt{6a}$ i) $3\sqrt{a}$ j) $\sqrt{5x}$ k) 4
l) 10 m) $3\sqrt{2}$ n) $9\sqrt{2}$ o) 4 p) 18
q) $15\sqrt{6}$ r) $8\sqrt{15}$ s) $4\sqrt{6}$ t) 270.
2 a) $3+\sqrt{3}$ b) $5-\sqrt{5}$ c) $\sqrt{2}-2$
d) $2\sqrt{5}+5$ e) $3\sqrt{2}+8$ f) 3 g) 1
h) -2 i) $3+2\sqrt{2}$ j) -4 k) $8+2\sqrt{15}$
l) $5-2\sqrt{6}$ m) $-4\sqrt{5}$ n) $12\sqrt{3}+8\sqrt{6}$
o) $12+2\sqrt{6}$.
3 a) 10 b) -2 c) 6.
4 a) $6\sqrt{3}$ b) 16 c) $4\sqrt{21}$ d) $82-10\sqrt{21}$.
5 a) 2 cm^2 b) $2\sqrt{3}$cm.
6 $k = 1$; OP = $\sqrt{3}$.
7 a) 2 b) 5 c) 3 d) $\sqrt{5}$ e) 2 f) $\frac{2\sqrt{3}}{3}$
g) $2\sqrt{2}$ h) 10 i) 1 j) $\frac{5}{2}$ k) $\frac{5}{6}$ l) $\frac{1}{2}$.
8 a) 2 b) 3 c) 5 d) $5\sqrt{2}$ e) $\frac{1}{4}$ f) $\frac{1}{8}$.
9 a) $\frac{\sqrt{2}}{2}$ b) $\frac{\sqrt{3}}{3}$ c) $\frac{\sqrt{5}}{5}$ d) $\sqrt{2}$ e) $2\sqrt{3}$
f) $2\sqrt{5}$ g) $\frac{3\sqrt{5}}{5}$ h) $10\sqrt{2}$ i) $\frac{3\sqrt{5}}{10}$
j) $\frac{2\sqrt{2}}{5}$ k) $\frac{3\sqrt{5}}{25}$ l) $\frac{\sqrt{2}}{4}$.
10 a) $\frac{\sqrt{5}}{10}$ b) $\frac{\sqrt{2}}{10}$ c) $\frac{5\sqrt{3}}{3}$ d) $\sqrt{2}$ e) $\frac{1}{2}$
f) $\frac{\sqrt{2}}{5}$.
11 a) $\frac{2\sqrt{3}}{3}$ b) $\frac{\sqrt{10}}{2}$ c) $\frac{3\sqrt{10}}{10}$ d) $\frac{\sqrt{6}}{2}$
e) $\frac{\sqrt{7}}{7}$ f) $\frac{\sqrt{15}}{5}$.
12 a) $4\sqrt{2}$ b) $\frac{6\sqrt{5}}{5}$ c) $\frac{\sqrt{3}}{2}$ d) $\frac{2\sqrt{3}}{3}$
e) $8+5\sqrt{2}$ f) $5\sqrt{3}$ g) $\frac{7\sqrt{2}}{2}$ h) $\frac{8\sqrt{3}}{15}$.
13 a) $\sqrt{2}-1$ b) $2(1+\sqrt{2})$ c) $\frac{3+\sqrt{3}}{2}$
d) $\frac{5(2+\sqrt{7})}{3}$ e) $\frac{2(4-\sqrt{3})}{13}$ f) $-\frac{3(5+\sqrt{11})}{14}$
g) $\frac{7(\sqrt{13}+3)}{4}$ h) $\frac{1}{7}(7-\sqrt{7})$ i) $\sqrt{5}-\sqrt{2}$
j) $4(\sqrt{7}+\sqrt{5})$ k) $\frac{1}{2}(\sqrt{10}+\sqrt{6})$
l) $\frac{7}{113}(11+2\sqrt{2})$.
14 a) $\frac{1}{2}(5-\sqrt{21})$ b) $2+\sqrt{3}$ c) $3-2\sqrt{2}$
d) $\frac{21+4\sqrt{5}}{19}$ e) $\frac{\sqrt{3}}{3}$ f) $\frac{2\sqrt{5}}{5}$ g) $-\sqrt{3}$.

Exercise 79, page 182

1 a) 3^3 b) 2^4 c) 5^3 d) 7^5 e) 10^3 f) a^4
g) x^3 h) a^3b^2 i) m^2n^3 j) $12a^2b$ k) a^2bc
l) $6p^3$ m) $2f^2g^2$ n) x^2y^3z o) $6m^2n^2$
2 a) $p \times p \times p \times p \times p$ b) $x \times x \times x$
c) $y \times y \times y \times y$ d) $m \times m \times n \times n \times n$
e) $3 \times m \times m \times m$ f) $2 \times x \times x \times y \times y \times y$

3 a) 3^5 b) 5^3 c) 2^6 d) 5^9 e) 7^{11} f) 10^9 g) a^5 h) a^4 i) p^5 j) p^9 k) m^8 l) y^7 m) $6y^8$ n) $12x^4$ o) $6a^9$ p) 10^6 q) x^{15} r) x^2y^5 s) $15a^7$ t) $12p^7$ u) $24x^3$ v) p^2q^6 w) a^7b^5 x) x^3y^5.

4 a) $\frac{3^3}{3^2} = \frac{3 \times 3 \times 3}{3 \times 3} = 3$ b) 2^2 c) 6 d) 5^3 e) x^3 f) y^2.

5 a) 2^4 b) 3^5 c) 10^2 d) 10^6 e) x^2 f) a g) p^4 h) q^3 i) x^3 j) x^5 k) a^5 l) p^5 m) $4m$ n) $3y^3$ o) $2z^5$ p) $2t^2$ q) $3p^2$ r) $4x^4$ s) $2ab$ t) $5pq^4$.

6 a) $\frac{1}{a^2}$ b) $\frac{1}{b}$ c) $\frac{1}{c^3}$ d) $\frac{1}{d^2}$ e) $\frac{2}{x^3}$ f) $\frac{3}{y}$ g) $\frac{z^2}{2}$ h) $\frac{1}{2k^2}$ i) $\frac{2}{p}$ j) $\frac{b}{a}$ k) $\frac{x}{y}$ l) $\frac{q^4}{p^2}$ m) $\frac{s}{r}$ n) $\frac{2}{3pq^2}$ o) $\frac{x}{2z^2}$ p) $\frac{3}{x}$.

7 a) $(2^2)^3 = (2^2)(2^2)(2^2)$
$= (2 \times 2) \times (2 \times 2) \times (2 \times 2)$
$= 2^6$
b) 3^6 c) 5^8 d) 2^{12} e) 5^6 f) 6^8.

8 a) x^6 b) y^8 c) z^{10} d) x^{12} e) p^{12} f) q^{15} g) r^{20} h) s^{12} i) t^{3n} j) u^{8m} k) v^{2x} l) w^{5p} m) x^{5a} n) y^{ab} o) m^{xy} p) n^{pq} q) w^{10a} r) x^{12s} s) y^{6ab} t) 2^{6xy}.

9 a) $(ab)^2 = (ab)(ab) = a \times a \times b \times b = a^2b^2$ b) p^3q^3 c) x^4y^4.

10 a) a^3b^3 b) p^4q^4 c) x^4y^2 d) p^6q^9 e) $16x^4$ f) $125x^3$ g) $32a^5$ h) $27p^6$ i) k^xm^x j) $c^{2p}d^p$ k) $f^{2x}g^{6x}$ l) $a^7b^7c^7$.

11 a) 3^7 b) 3^3 c) 2^3 d) 3^{10} e) 2^{12} f) 2^7 g) 10^{13} h) 10^3 i) 5^{18} j) a^{2m} k) a^{m+n} l) p^3 m) a^5 n) t^{10} o) u^8 p) v^3 q) w^3 r) x^6 s) 3^6 t) 8^2 u) 5^{11} v) $12x^5$ w) $\frac{3}{4}y$ x) 1 y) 10^2 z) 1 aa) p^3 bb) $\frac{6}{5}$ cc) $9a^3b$ dd) p^8q^7 ee) $\frac{2}{x^3}$ ff) $\frac{1}{2y^2z^4}$.

Exercise 80, page 186

1 a) $\frac{1}{2^8}$ b) $\frac{1}{10^3}$ c) $\frac{1}{2^4}$ d) $\frac{1}{10^6}$ e) $\frac{1}{10}$ f) $\frac{1}{3^2}$ g) $\frac{1}{8^4}$ h) $\frac{1}{3^4}$ i) $\frac{1}{a^3}$ j) $\frac{2}{b}$ k) $\frac{3}{k^3}$ l) $\frac{5}{t^2}$.

2 a) 5^{-1} b) 7^{-5} c) 9^{-3} d) 6^{-4} e) t^{-7} f) z^{-4}.

3 a) $\frac{1}{10^2}$ b) 10^{-3} c) 1 d) 10^{-1} e) 2^{-1} f) $\frac{1}{2^2}$ g) 2^{-4} h) $\frac{1}{2^3}$ i) 4^{-3} j) $\frac{1}{5^6}$ k) $\frac{1}{7^3}$ l) 5^{-1} m) 7^3 n) 10^{-5} o) 3^{-3} p) a^5 q) $\frac{1}{5^3}$ r) $\frac{1}{10^8}$ s) $\frac{2}{x^2}$ t) $\frac{1}{(2x)^2}$ u) $\frac{3}{z}$ v) $\frac{5}{y^4}$ w) $\frac{1}{(3x)^4}$ x) 7.

4 a) $\frac{1}{3}$ b) $\frac{1}{2^2}$ c) $\frac{1}{2^2}$ d) 3^2 e) $\frac{1}{5^2}$ f) $\frac{1}{6^6}$ g) $\frac{1}{2^3}$ h) $\frac{1}{5^2}$ i) t^2 j) $\frac{1}{u^3}$ k) $\frac{1}{v}$ l) $\frac{1}{w^2}$ m) x^{10} n) $\frac{1}{y^3}$ o) z^{10} p) 1 q) $2x^7$ r) $\frac{7}{y}$.

5 a) $6x$ b) $\frac{2b^5}{a^4}$ c) $\frac{24a^5}{b}$ d) $\frac{3}{2}ab^3$ e) $\frac{9x^2}{2y^5}$ f) $\frac{1}{72x^3y^2}$.

6 a) a^9 b) $2x^5$ c) $6y^7$ d) x^3 e) x^5 f) $\frac{2}{x^2}$ g) b^3 h) $\frac{1}{x^6}$ i) y^8 j) $2x^4$ k) $2y^5$ l) $\frac{x^2}{y^3}$ m) $\frac{1}{2x^5}$ n) 1 o) $\frac{1}{3^3}$ p) $\frac{y}{x^2}$.

7 a) $\frac{3}{4x^5}$ b) $\frac{3x}{2y^2}$ c) $\frac{c^2}{2}$ d) $3ab$ e) $\frac{4x}{y^2}$ f) a^5 g) $1 + x^{10}$ h) $1 + \frac{1}{x^4y^6}$ i) $\frac{1}{p^2}$.

Exercise 81, page 188

1 a) 27 b) 5 c) 16 d) $\frac{1}{4}$ e) $\frac{1}{8}$ f) 216 g) $\frac{1}{32}$ h) 27 i) $\frac{1}{25}$ j) $\frac{1}{100\,000}$.

2 a) $a^{\frac{7}{3}}$ b) b^4 c) c^2 d) d^2 e) $\frac{1}{\sqrt{y}}$ f) f^6 g) $6g$ h) $12h^{\frac{3}{2}}$ i) $i^{\frac{2}{3}}$ j) $j^{\frac{4}{5}}$ k) $k^{\frac{3}{4}}$ l) $l^{\frac{7}{4}}$ m) m n) n o) $2x$ p) $\frac{3}{p^{\frac{3}{2}}}$.

3 a) $a^{\frac{5}{6}}$ b) $b^{\frac{5}{6}}$ c) $c^{\frac{17}{12}}$ d) $d^{\frac{35}{24}}$ e) $y^{\frac{5}{12}}$ f) $f^{\frac{5}{6}}$ g) $g^{\frac{17}{6}}$ h) $h^{\frac{27}{14}}$ i) $\frac{6}{i}$ j) $12j^{\frac{17}{6}}$ k) $\frac{2}{k^{\frac{1}{6}}}$ l) $\frac{1}{3l}$.

4 a) 4 b) 9 c) $\frac{1}{16}$ d) 1 e) 36 f) 128 g) $\frac{4}{3}$ h) 1.

5 a) $y = 4$ b) $x = -3$ c) $z = \frac{5}{2}$ d) $k = -\frac{4}{3}$.

6 a) $x^2 + x$ b) $x^3 + 2x^2$ c) $x + 2x^2$ d) $6x^2 - 2x^3$ e) $1 - \frac{1}{y}$ f) $\frac{2}{y} + y$ g) $3z - \frac{6}{z^2}$ h) $10z^4 - 15$ i) $p^2 - p^{\frac{3}{2}}$ j) $2q - 3q^2$ k) $\frac{3}{r^{\frac{1}{3}}} + \frac{6}{r^{\frac{4}{3}}}$ l) $\frac{12}{s^{\frac{11}{4}}} - \frac{20}{s^{\frac{13}{12}}}$.

7 a) $\frac{4b^6}{a^4}$ b) $9ab^{\frac{2}{3}}$ c) $\frac{64b^2}{a^3}$ d) $\frac{125x^4}{y}$ e) $\frac{p}{3q^2}$ f) $\frac{r^{\frac{3}{2}}}{64s^3}$ g) $\frac{y^{\frac{2}{3}}}{x^{\frac{1}{5}}}$ h) $\frac{32m}{n^{10}}$.

8 $a = \frac{5}{6}$ $b = \frac{3}{2}$.

Exercise 82, page 189

1 a) $\frac{1}{2}$ b) $\frac{1}{3}$ c) $\frac{2}{3}$ d) $\frac{2}{5}$.

2 a) $\frac{x}{y^2}$ b) $4z^3$ c) $\frac{1}{(p^2+q^2)^2}$ d) $\frac{t}{2(t+2)}$.

3 a) $-\frac{4}{3}$ b) $\frac{3}{y-2}$ c) $\frac{x+4}{x-4}$ d) $\frac{x-2}{x-4}$.

4 a) $\frac{31}{20}$ b) $\frac{2t+3}{t}$ c) $\frac{x^2-x-1}{x^2-x}$

d) $\frac{2x^2+2x-1}{x^2+3x+2}$.

5 a) $\frac{7}{6}$ b) $2(x+1)$ c) $\frac{10}{9}$ d) $3x$.

6 a) $f=\frac{v-u}{t}$ b) $R=\frac{100I}{PT}$ c) $p=\frac{2a-qr}{r}$

d) $v=\pm\sqrt{\frac{2k}{m}}$.

7 a) $10\sqrt{2}$ b) $2\sqrt{2}-4\sqrt{3}$ c) $24\sqrt{15}$

d) $49+12\sqrt{5}$.

8 a) $\frac{2\sqrt{7}}{7}$ b) $\frac{\sqrt{15}}{5}$ c) $\frac{\sqrt{2}+\sqrt{6}}{2}$ d) $\frac{5+\sqrt{3}}{11}$.

9 a) x b) y c) $\frac{8x^3}{9y^2}$ d) $p^{\frac{1}{4}}$.

10 a) 10 b) $\frac{1}{3}$ c) 9 d) 5.

Exercise 83 (Test), page 190

1 a) $7\sqrt{3}$ b) $9\sqrt{2}-6\sqrt{3}$ c) $45\sqrt{35}$

d) $35-12\sqrt{6}$.

2 a) $\frac{3\sqrt{5}}{5}$ b) $\frac{\sqrt{6}}{3}$ c) $\frac{5\sqrt{3}-\sqrt{6}}{3}$ d) $\frac{7+\sqrt{5}}{4}$.

3 a) $p^{\frac{7}{3}}$ b) $q^{\frac{6}{5}}$ c) r^3 d) $6p^4q^8$.

4 a) 5 b) 64 c) $\frac{1}{9}$ d) 13.

5 a) $h=\frac{A}{2\pi r}$ b) $w=\frac{x-a}{t}$ c) $u=\frac{2a}{p-at}$

d) $R=\pm\sqrt{\frac{GME}{F}}$ e) $x=\pm\frac{\sqrt{w^2a^2-v^2}}{w}$.

6 a) (x^2+y^2) b) $\frac{3y}{x}$ c) $\frac{t}{2u}$ d) $\frac{5}{3}$.

7 a) $\frac{x}{3}$ b) $\frac{5}{y}$ c) $\frac{1-x}{2+x}$ d) $\frac{x-4}{2x+1}$.

8 a) $\frac{2x+5}{x}$ b) $\frac{x^2-1}{x^2}$ c) $\frac{4(y+1)}{y(y+2)}$

d) $\frac{2z^2+2z-1}{z^2+z-2}$.

9 a) 1 b) $\frac{x-3}{x-4}$ c) x d) $\frac{x^2-6x+5}{x^2}$.

Exercise 84, page 192

1 a) $y=2x^2$ b) $y=\frac{1}{4}x^2$ c) $y=-3x^2$

d) $y=-\frac{1}{2}x^2$.

2 a) 1 b) -1 c) 5 d) $-\frac{1}{3}$.

3 a) When $k>1$, $y=kx^2$ is $y=x^2$ stretched by a factor of k parallel to the y-axis.
When $0<k<1$, $y=kx^2$ is $y=x^2$ reduced by a factor of k parallel to the y-axis.
When $k<0$, say $k=-a$, then $y=-ax^2$ is $y=ax^2$ reflected in the x-axis.

b) When $k>1$, $y=kx$ is $y=x$ stretched by a factor of k parallel to the y-axis.
[The gradient of $y=kx$ is k. The gradient on the parabola is different at each point.]
When $0<k<1$, $y=kx$ is $y=x$ reduced by a factor of k parallel to the y-axis.
When $k<0$, say $k=-a$, then $y=-ax$ is $y=ax$ reflected in the x-axis.

4 a) $y=x^2+2$ b) $y=x^2-9$

c) $y=x^2-16$ d) $y=x^2+3$.

5 a) When $k>0$, $y=x^2+k$ is $y=x^2$ translated k units up the y-axis.
When $k<0$, say $k=-a$, $y=x^2-a$ is $y=x^2$ translated a units down the y-axis.

b) When $k>0$, $y=x+k$ is $y=x$ translated k units up the y-axis.
When $k<0$, say $k=-a$, then $y=x-a$ is $y=x$ translated a units down the y-axis.

6 a) $y=(x-2)^2$ b) $y=(x+3)^2$

c) $y=(x+1)^2$ d) $y=(x-3)^2$.

7 a) The graph of $y=(x-a)^2$ is that of $y=x^2$ translated a units to the right.
The graph of $y=(x+a)^2$ is that of $y=x^2$ translated a units to the left.

b) The graph of $y=x-a$ is that of $y=x$ translated a units to the right.
The graph of $y=x+a$ is that of $y=x$ translated a units to the left.

8 a) $y=(x+1)^2+1$ b) $y=(x-2)^2-1$

c) $y=(x-3)^2-7$ d) $y=(x+3)^2-2$.

9 a) $y=(x-1)^2-1$ b) $y=(x+2)^2-4$

c) $y=(x-3)^2-9$ d) $y=(x+1)^2-1$.

10 a) $y=4-(x-1)^2$ b) $y=9-(x+2)^2$

c) $y=-1-(x-2)^2$.

11 a) $y=2(x-3)^2$ b) $y=2x^2-8$

c) $y=3(x-2)^2+1$ d) $y=(x-2)^2+4$.

Exercise 85 page, 196

1 (0, 0) minimum $x = 0$
(0, 0) maximum $x = 0$
(–5, 0) minimum $x = -5$
(2, 0) minimum $x = 2$
(–4, 3) minimum $x = -4$
(3, 4) minimum $x = 3$
(–6, 3) maximum $x = -6$
(7, 8) maximum $x = 7$
(5, 6) minimum $x = 5$
(–5, 6) maximum $x = -5$
(0, 0) minimum $x = 0$.

2 a) (–3, –11) b) $x = -3$ c) (0, –2) d) $t = -7$
e) Q (–1, –7) f) $k = 5$ g) (–7, 5).

3 a) (1, 4) b) $x = 1$ c) (0, 3) d) (2, 3)
e) (3, 0).

4 a) (–2, 9) b) (4, 9) c) $y = 9 - (x - 4)^2$.

5 a) (1, –4) b) (–3, –4) c) $y = (x + 3)^2 - 4$.

Exercise 86, page 198

1 a)

16	9	4	1	0	1	4	9	16	25	36
12	9	6	3	0	–3	–6	–9	–12	–15	–18
–10	–10	–10	–10	–10	–10	–10	–10	–10	–10	–10
18	8	0	–6	–10	–12	–12	–10	–6	0	8

b) $x = -2, 5$ c) (1·5, –12·25).

2 a) 1, 3 (2, –1) b) –1, 3 (1, –4)
c) 2, –4 (–1, –9) d) –1, –3 (–2, –1)
e) 2, –3 $\left(-\frac{1}{2}, -\frac{25}{4}\right)$ f) ±5 (0, –25)
g) –2, 3 $\left(\frac{1}{2}, -\frac{25}{4}\right)$ h) 1, –4 $\left(-\frac{3}{2}, -\frac{25}{4}\right)$
i) –1, –2 $\left(-\frac{3}{2}, -\frac{1}{4}\right)$.

3 a) $\pm\frac{1}{2}$ b) $2\frac{1}{2}, -\frac{3}{4}$ c) $-\frac{3}{2}, \frac{4}{3}$ d) $-\frac{8}{7}, \frac{7}{9}$.

Exercise 87, page 199

1 a) 1, 3 b) 1, –2 c) 3, –4 d) 2, 3
e) 5, 7 f) –2, –3 g) 4, –5 h) 2, –2
i) 3, –3 j) 2, 2 k) 3, 3 l) 7, –4 m) 5, –5.
n) 7, 7 o) 3, –8.

2 a) ±2 b) 0, 4 c) ±3 d) 0, 9 e) 0, 6
f) 3, 3 g) ±6 h) 5, 5 i) –5, –5 j) –2, –8
k) 1, 2 l) 4, 5 m) 2, –3 n) 4, –3 o) 1, –5.

3 a) 0, –7 b) 2, –1 c) 2, 3 d) ±5 e) 3, 5
f) 7, –3 g) 0, –8 h) 2, 5 i) 4, –7 j) ±7.
k) 0, 0 l) none (it does not cross the x-axis).

4 a) 3, –2 b) 1, 2 c) 7, –2 d) 3, 4
e) 2, –3 f) 5, –7.

5 a) $-2, \frac{5}{2}$ b) $5, -\frac{3}{2}$ c) $-3, \frac{7}{2}$ d) $4, -\frac{1}{3}$
e) $3, \frac{5}{2}$ f) $-5, \frac{1}{5}$ g) $-2, \frac{4}{3}$ h) $-4, -\frac{5}{3}$
i) $2, \frac{3}{5}$ j) $1, -\frac{5}{6}$ k) $\frac{1}{2}, \frac{3}{2}$ l) $\frac{5}{2}, -\frac{1}{2}$
m) $\frac{1}{2}, -\frac{1}{3}$ n) $\frac{2}{5}, -\frac{4}{3}$ o) $\frac{3}{4}, -\frac{4}{3}$.

Exercise 88, page 201

1 a) 1, 1·5 b) –1, –1·333 c) 1·449, –3·449
d) 2·618, 0·382 e) 0·137, –3·637
f) 0·319, –1·569 g) 0·772, –7·772 h) 1, 6
i) no real solution j) 0·881, –0·681
k) 0·804, –1·554 l) no real solution
m) 5, –1 n) no real solution
o) 2·535, 0·131.

Exercise 89, page 203

1 a) $(x + 1)(x - 5)$ b) 5, –1
c) (5, 0) and (–1, 0) d) (2, –9) minimum.

2 a) C (1, 4) b) B (0, 3) c) A (–1, 0) D (3, 0).

3 a) Q (0, 8) b) P (–2, 8)
c) K (–5, –7) L (3, –7).

4 13 cm.

5 a) 0·363 inches b) 0·242 inches
c) proof: $0{\cdot}363 - 0{\cdot}242 = 0.121 = \frac{1}{2} \times 0{\cdot}242$.

6 a) 20 m b) 50 m.

7 a) (20, 0) b) 80 m c) 20 m.

8 a) (i) $y = (x - 1)^2$ (ii) $y = 2 - (x - 1)^2$
(iii) $y = \frac{1}{2}$
b) (i) 2·25 m (ii) $\sqrt{2}$ m.

Exercise 90, page 205

1 a) p: $y = 2x^2$ q: $y = 3x^2$
b) r: $y = \frac{1}{4}x^2$ s: $y = -x^2$
c) t: $y = x^2 + 1$ u: $y = x^2 + 2$
d) v: $(x - 1)^2$ w: $y = (x - 2)^2$
e) g: $y = (x - 1)^2 + 1$ h: $y = (x - 2)^2 + 4$
f) k: $y = 1 - (x - 1)^2$ m: $y = 4 - (x + 2)^2$.

2 (2, 3) minimum $x = 2$
(1, 4) maximum $x = 1$
(0, 0) maximum $x = 0$
(5, 0) minimum $x = 5$.

3 a) 1, –3 b) $\left(\frac{7}{2}, -\frac{3}{2}\right)$ c) $-1, \frac{1}{3}$ d) 1·3, 0·3.

4 a) 5, 6 b) 7, –3 c) 3, –5 d) $\frac{1}{2}, -\frac{2}{3}$.

5 a) 11, 13 b) 3·5, –1·667 c) 0·319, –1·569
d) 2·847, 0·820.

6 a)

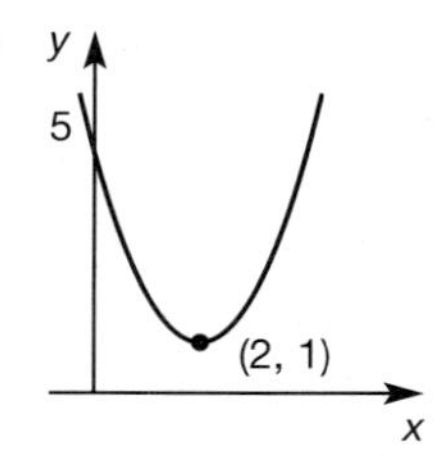

b)

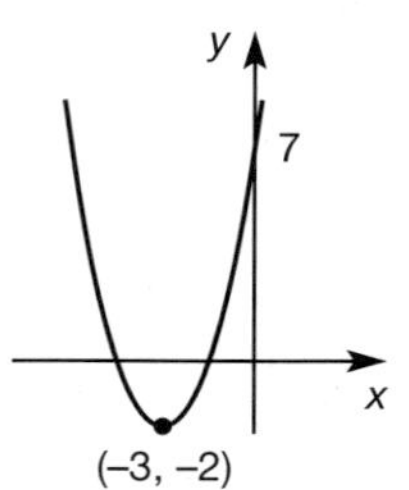

c)

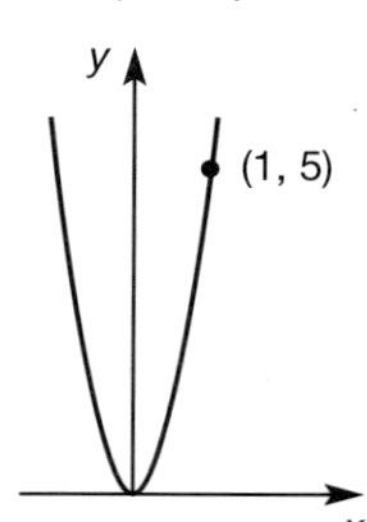

d)

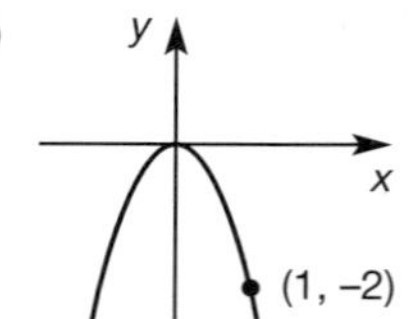

e)

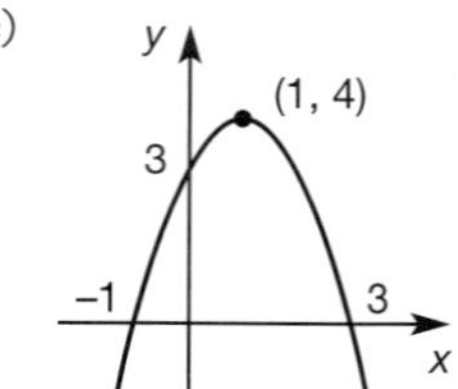

Exercise 91 (Test), page 206

1 a) $y = -3x^2$ b) $y = x^2 - 25$ c) $y = (x - 5)^2$
d) $y = 3 - (x - 4)^2$ e) $y = 4 - (x + 2)^2$
f) $y = (x - 4)^2 + 2$

2 a) (2, 5) minimum $x = 2$
b) (4, 7) maximum $x = 4$
c) (1, 9) maximum $x = 1$.

3 a) ±2·2 b) 0, –1·5 c) 1·4, 0·2.

4 a) 0, 12 b) $\frac{3}{2}, \frac{4}{3}$ c) $4, \frac{3}{2}$.

5 a) 3·41, 0·59 b) 4·79, 0·21 c) –0·63, –2·37.

6 a)

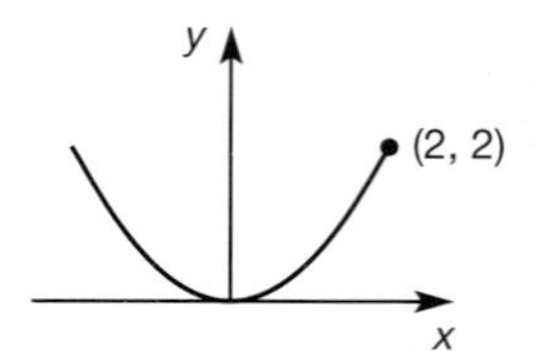

b)

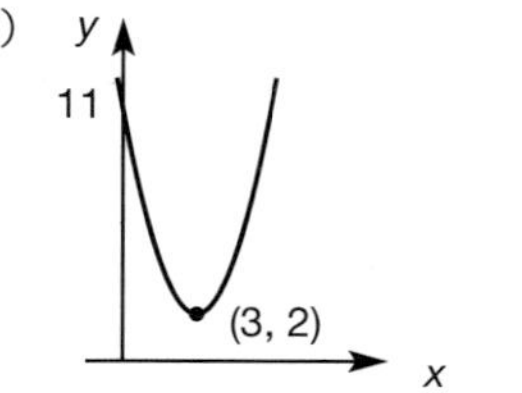

Exercise 92, page 208

1 a)

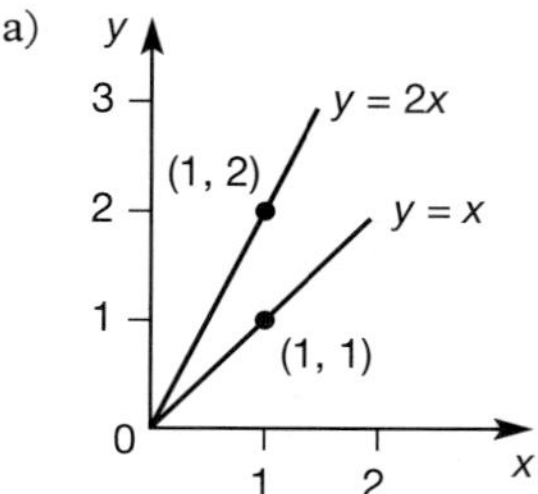

b)

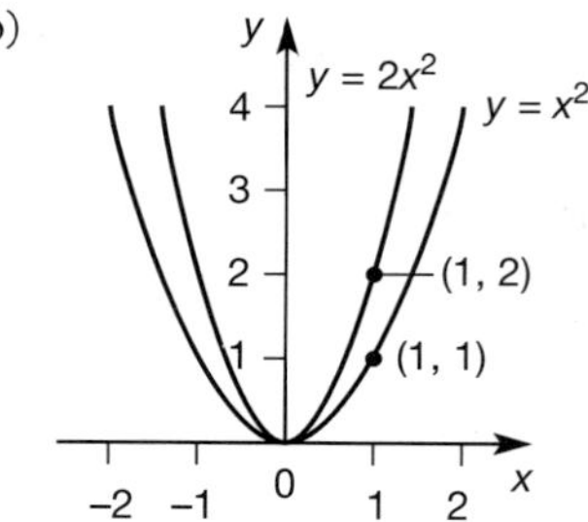

c)

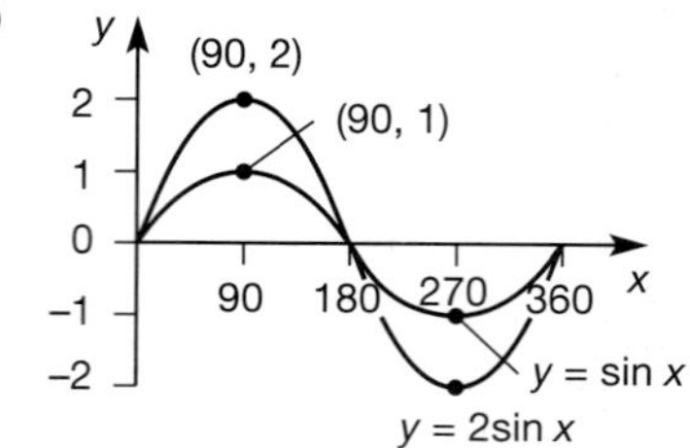

2

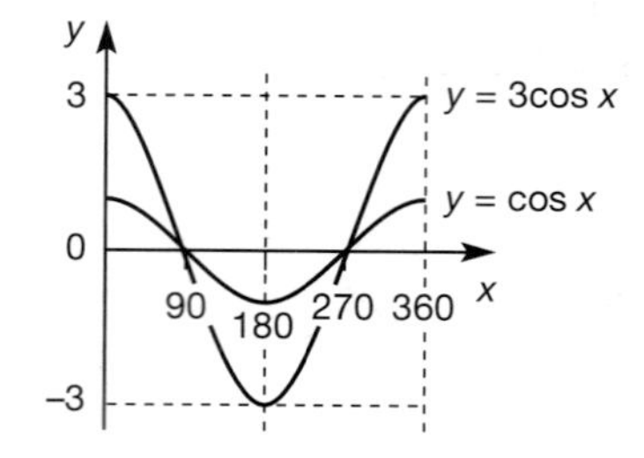

3

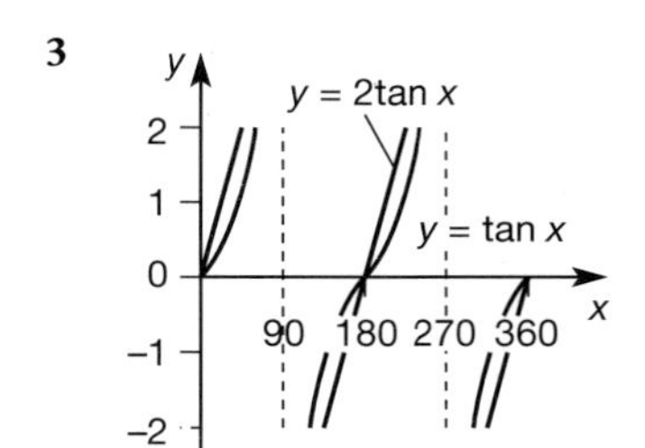

4

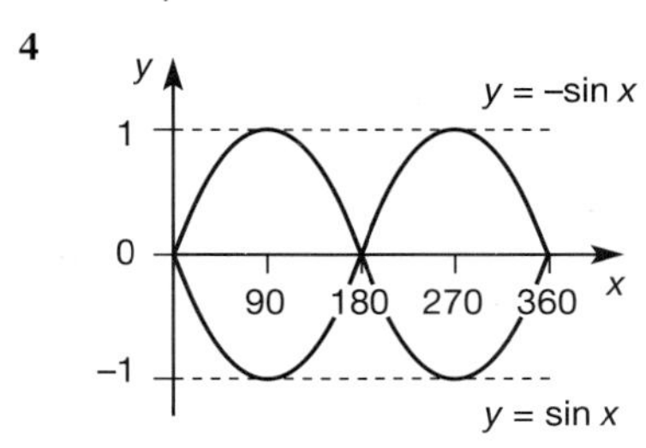

5

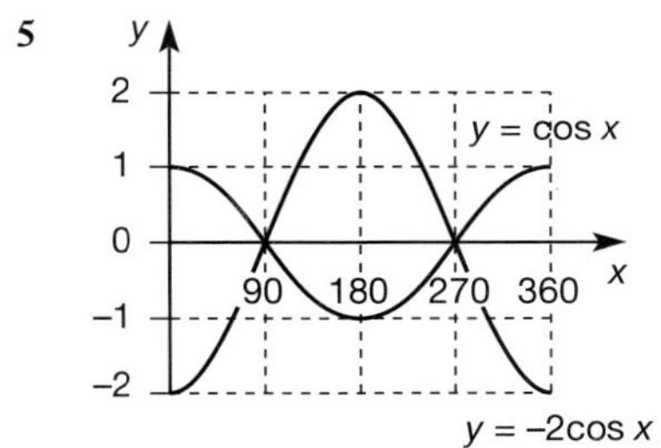

6 $a = 3$, $b = -4$.

Exercise 93, page 210

1 a)

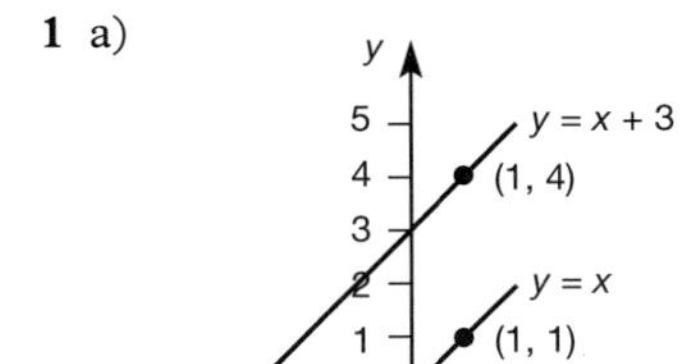

b)

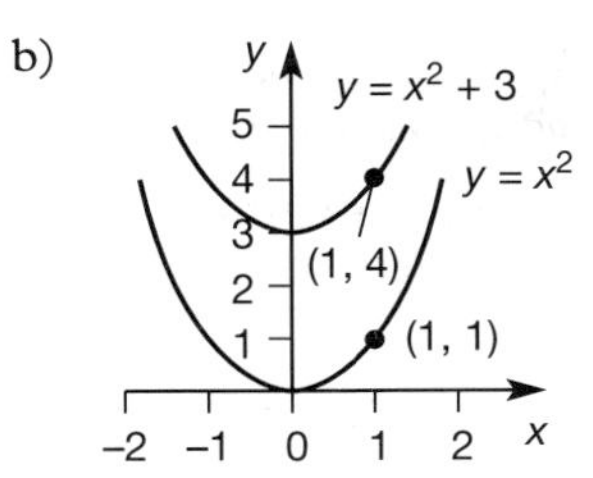

c)

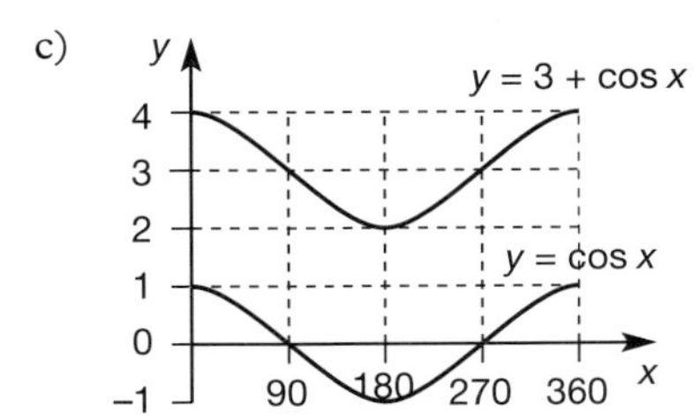

2

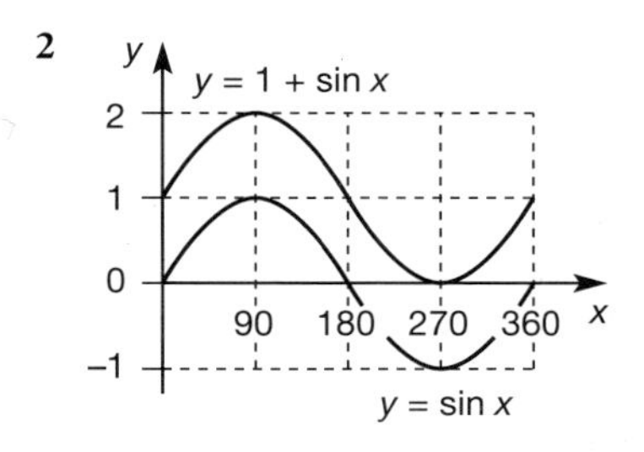

3

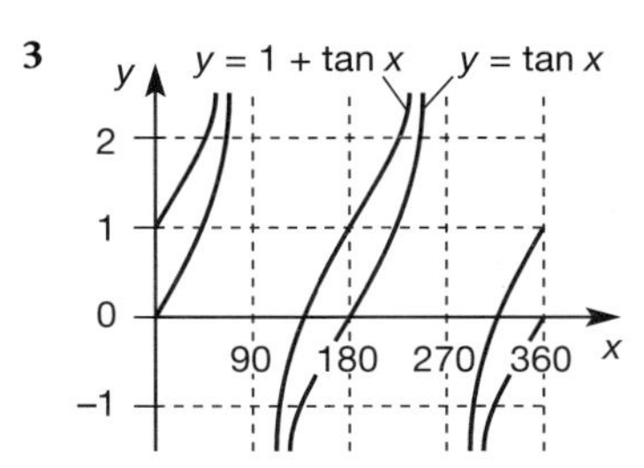

4

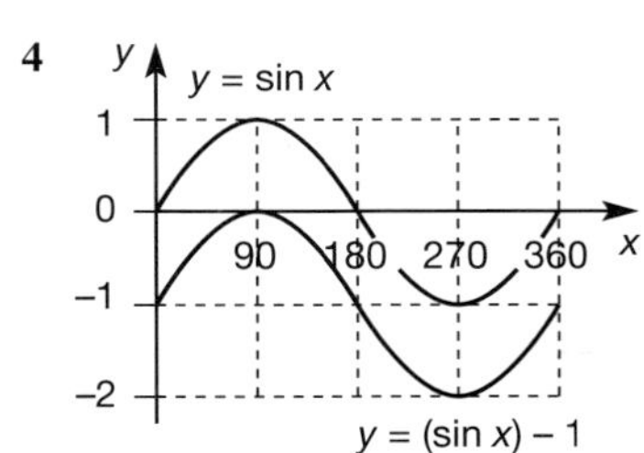

5

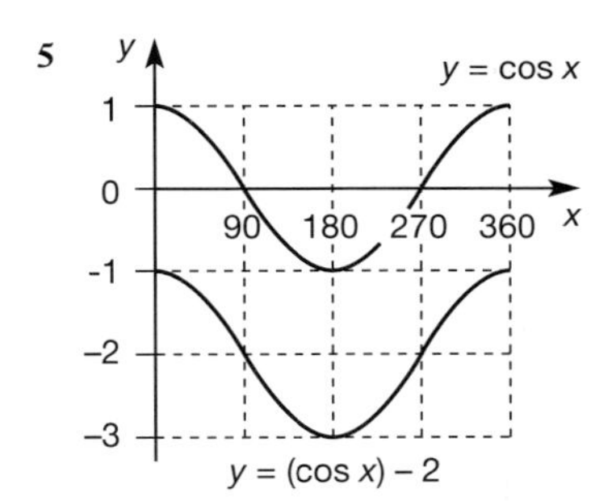

6 $a = 2$, $b = -1$.

Exercise 94, page 212

1 a) 360° b) 180° c) 72° d) 360° e) 90°
f) 180° g) 180° h) 60° i) 180°.

2 a)

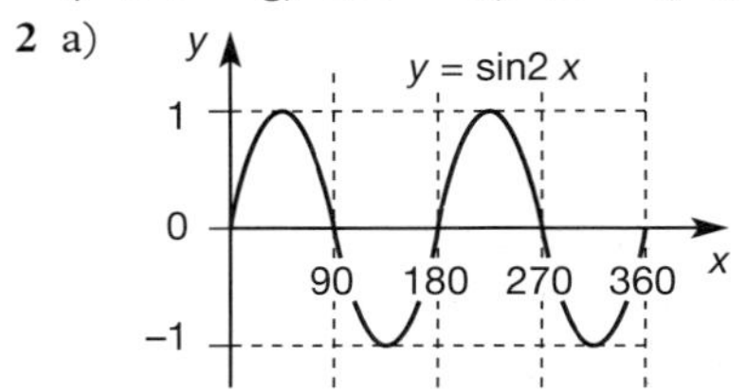

b)

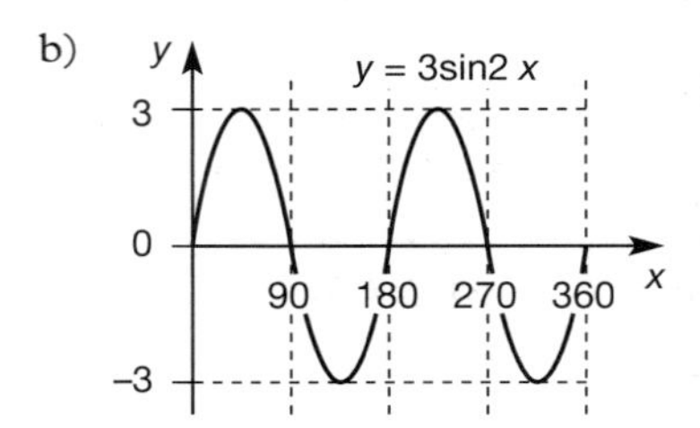

c)

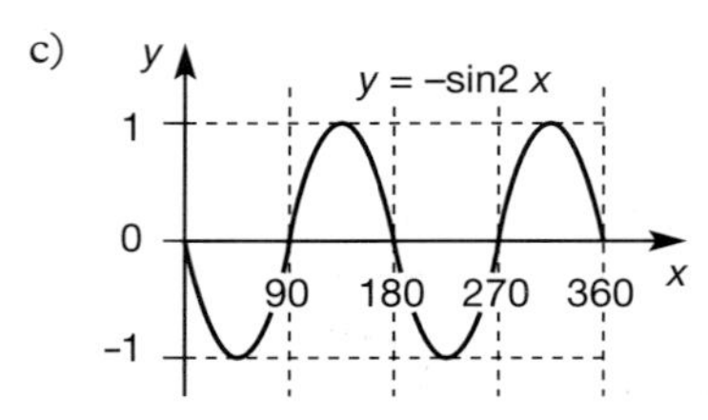

3 a)

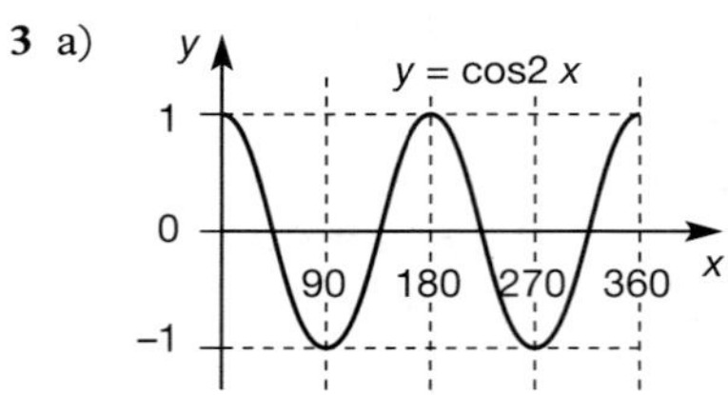

b)

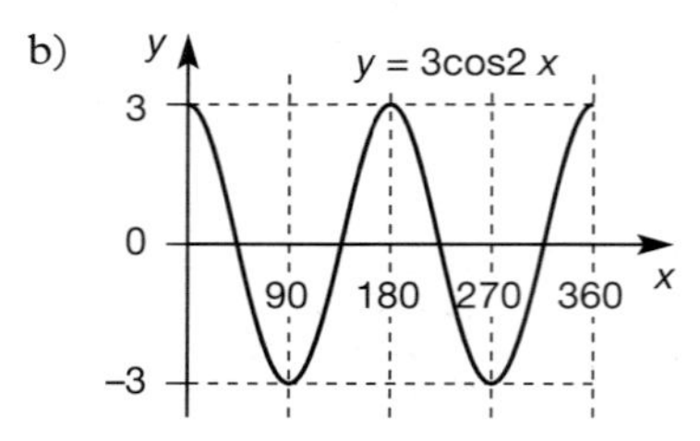

c)

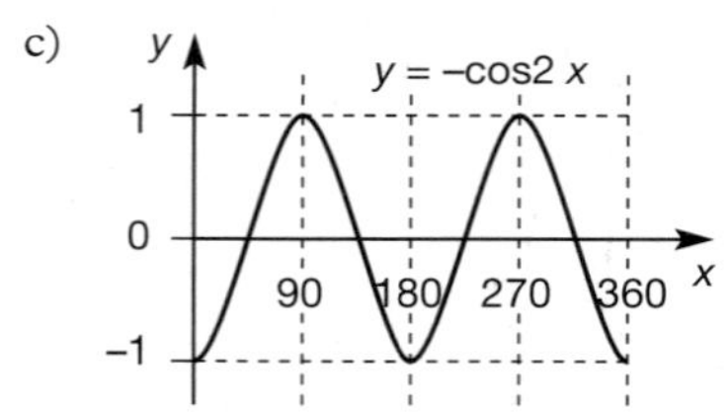

4 a)

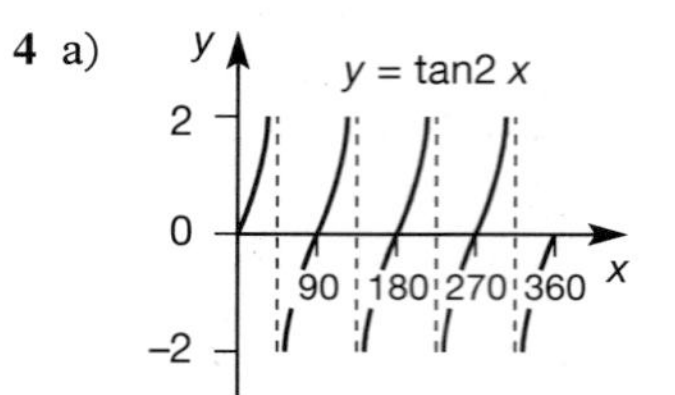

b)

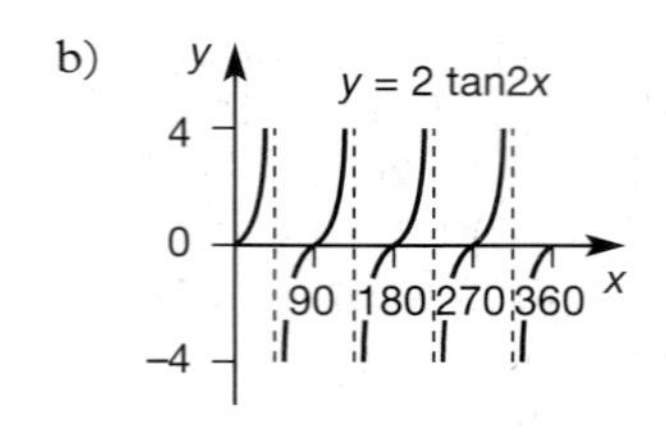

c)

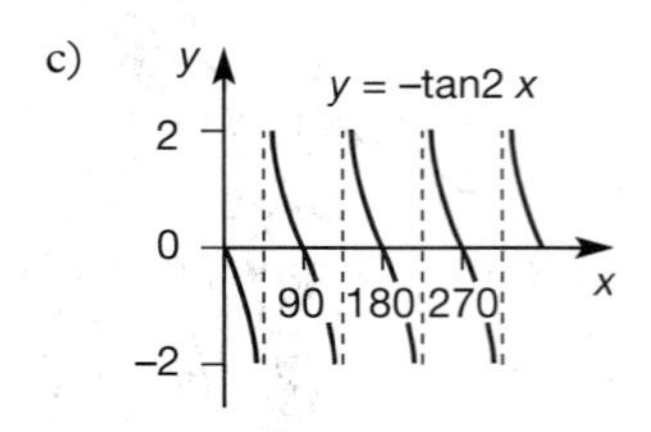

5 a)

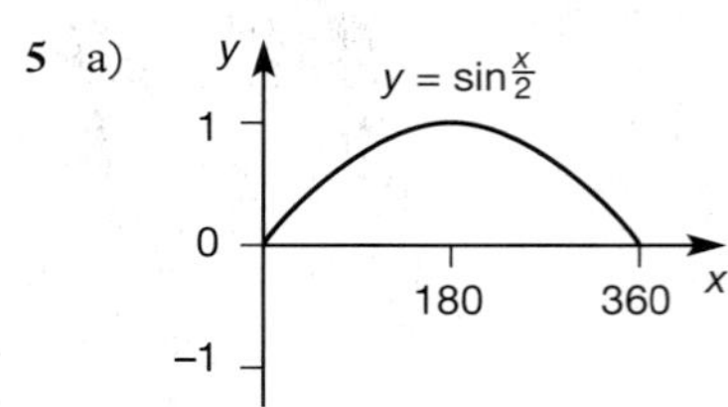

b)

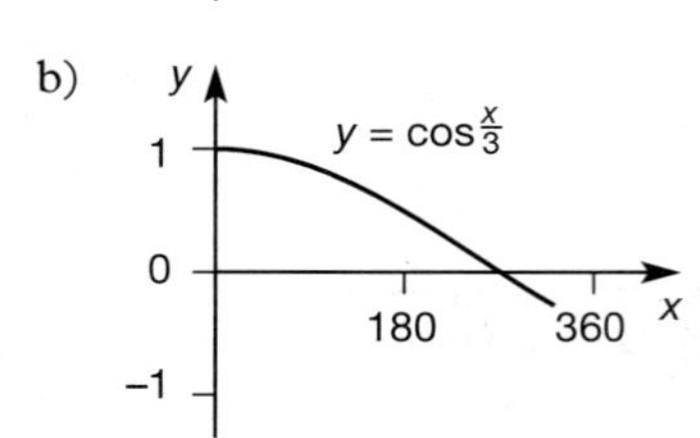

c)

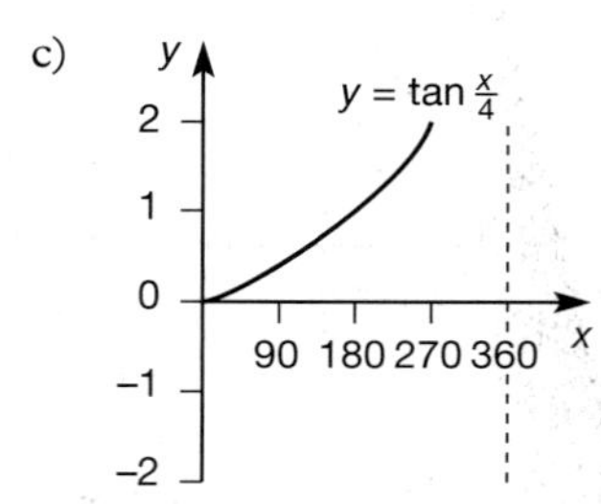

6 $a = 4, b = 5.$

7 $p = 3, q = 4; r = -2, s = 2; u = \frac{1}{2}; v = \frac{1}{2}, w = \frac{1}{3}.$

Exercise 95, page 214

1 a)

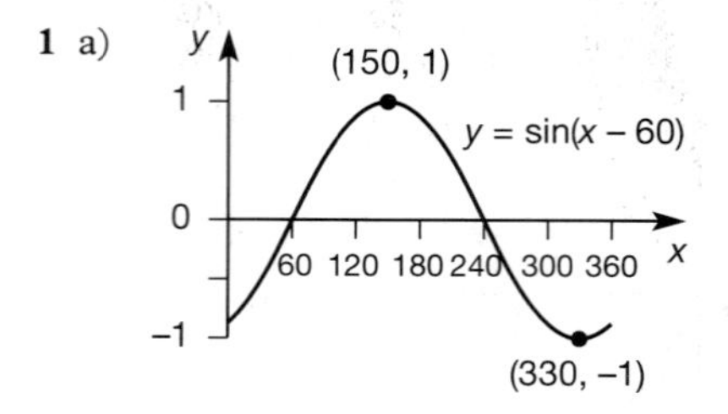

b)

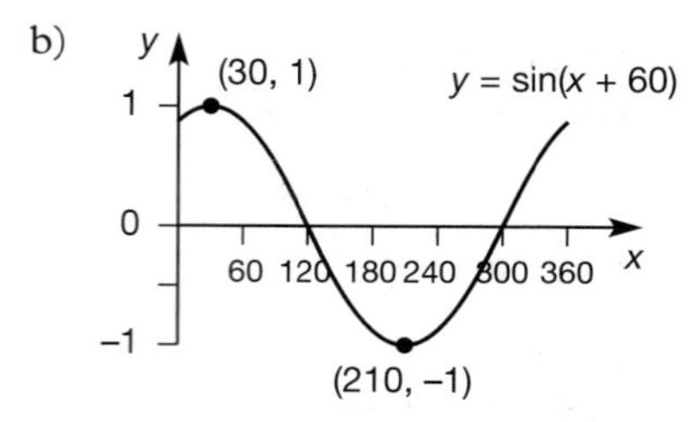

c)

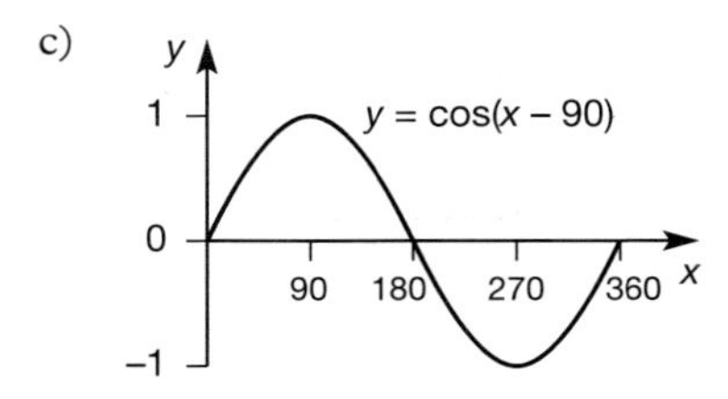

d)

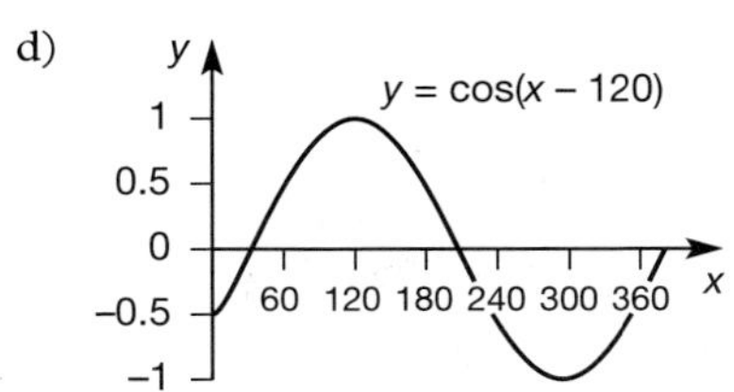

e)

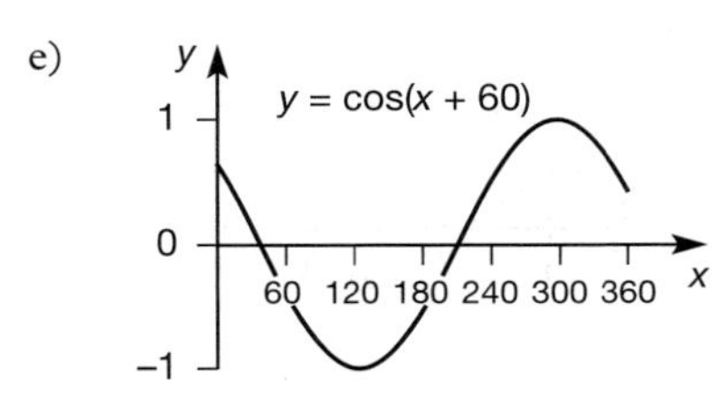

f)

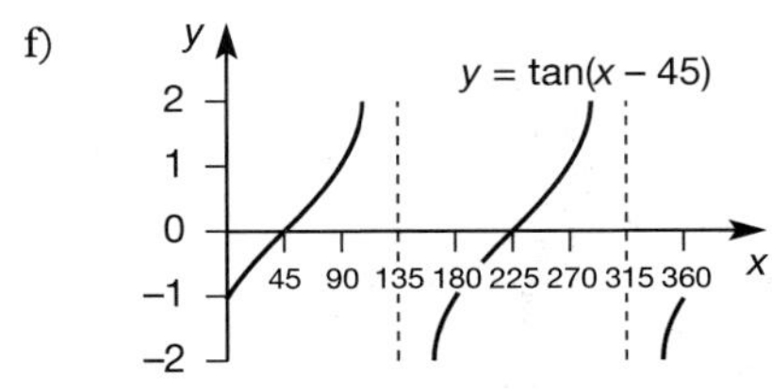

g)

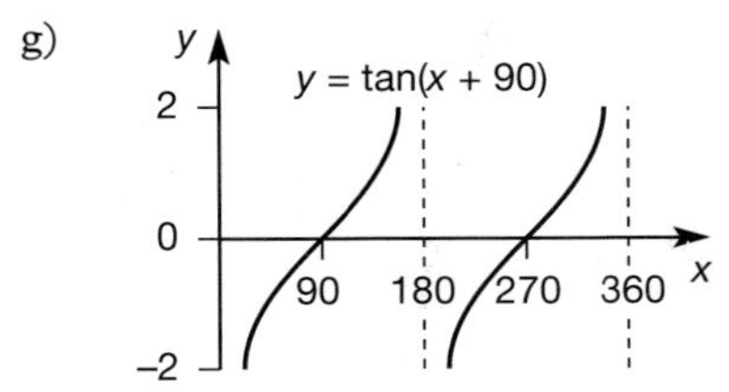

h)

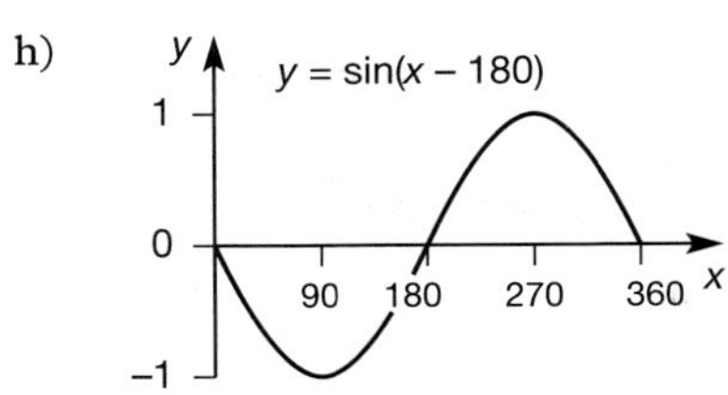

2 $p = -30$ (or 330) $q = 90$ (or -270) $r = -60$ (or 120) $s = 120$ (or -240).

3 a)

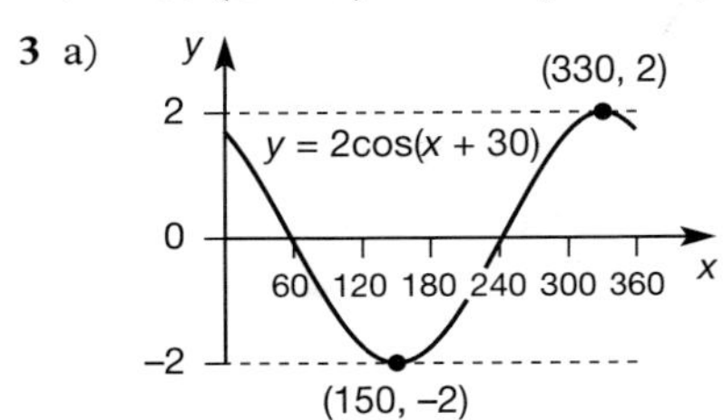

b)

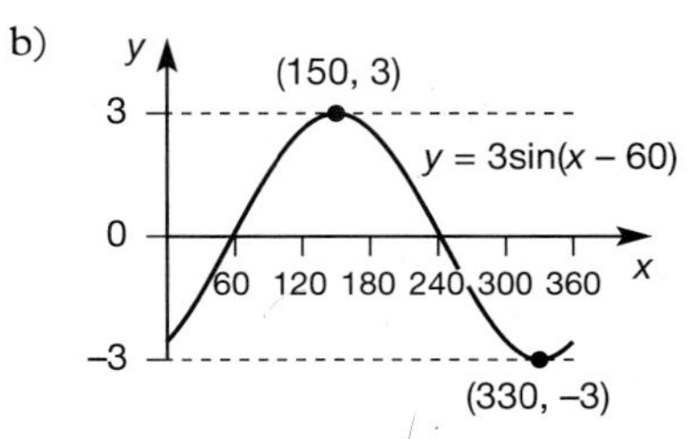

c)

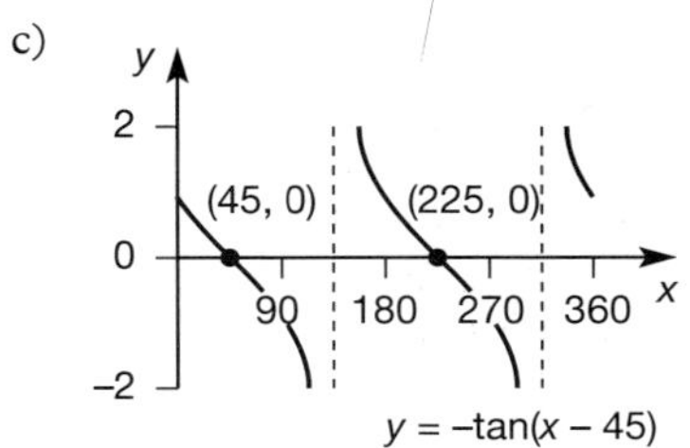

d)

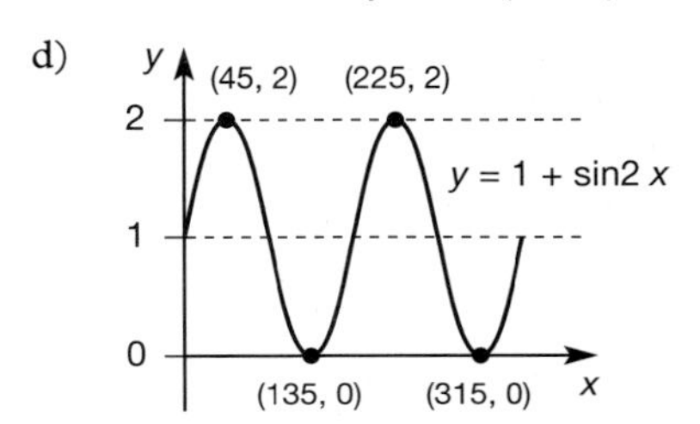

e)

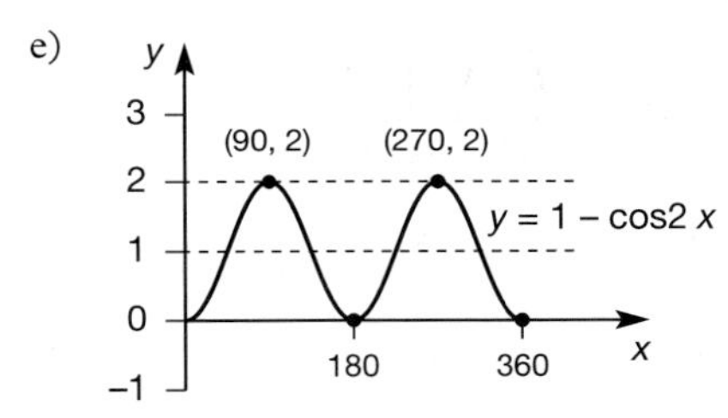

f)

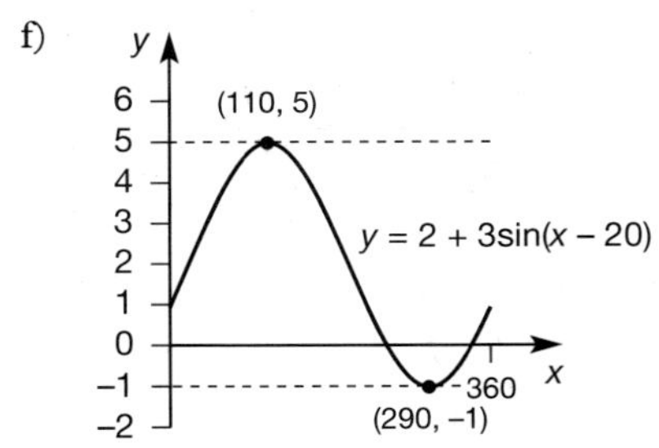

Exercise 96, page 216

1 a) 30, 150 b) 60, 300 c) 45, 225 d) 72, 108 e) 85, 275 f) 21, 201 g) 60, 300 h) 8, 172 i) 20, 200 j) 200, 340 k) 130, 230 l) 100, 280 m) 210, 330 n) 110, 250 o) 100, 280 p) 20, 160 q) 52, 308 r) 130, 310 s) 54, 234 t) 188, 352.

2 a) 45, 225 b) 135, 315 c) 60, 300
d) 120, 240 e) 60, 120 f) 120, 300
g) 225, 315 h) 26·6, 206·6 i) 75·5, 284·5
j) 30, 150 k) 19·5, 160·5 l) 150, 210
m) 210, 330 n) 135, 225 o) 104·5, 255·5
p) 46·8, 133·2 q) 5·8, 174·2 r) 72·4, 287·6
s) 70·0, 250·0 t) 200·2, 339·8
u) 104·7, 255·3 v) 123·7, 303·7
w) 225·6, 314·4 x) 38·9, 321·1.

3 a) 210, 330 b) 48·2, 311·8 c) 51·3, 231·3
d) 216·9, 323·1 e) 33·6, 326·4
f) 15·9, 195·9 g) 241·0, 299·0
h) 56·3, 303·7 i) 132·3, 312·3
j) 179·5, 359·5 k) 68·2, 111·8
l) 38·5, 321·5.

Exercise 97, page 217

(All answers are mathematical proofs.)

Exercise 98, page 218

1 a) 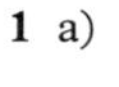

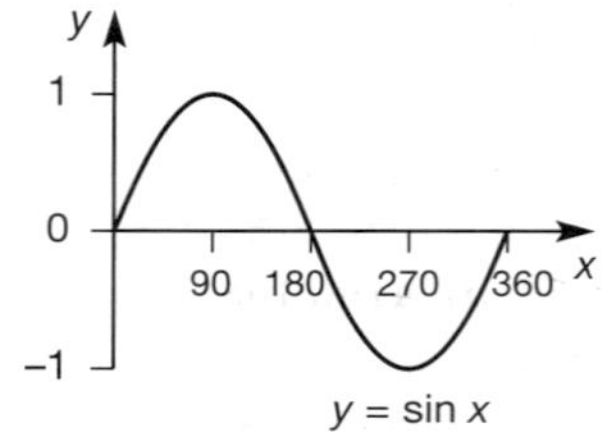

b)

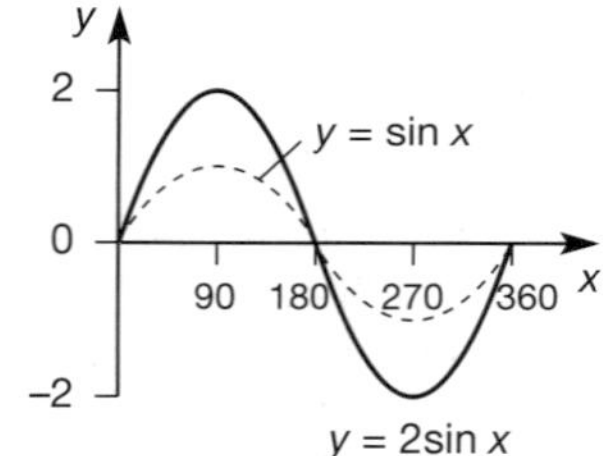

c)

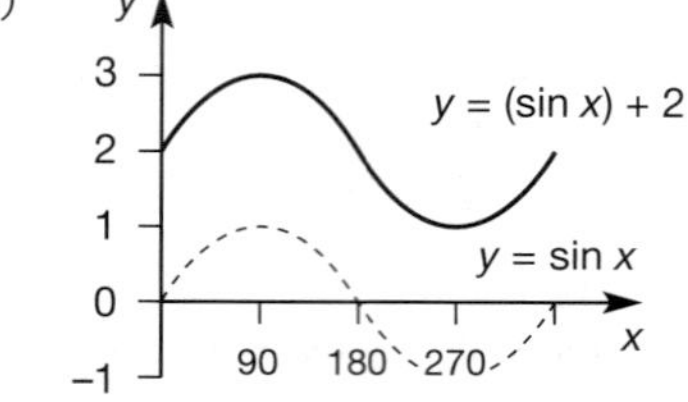

d)

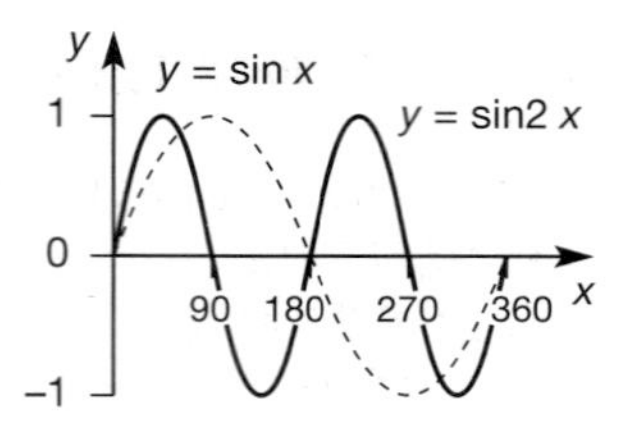

e)

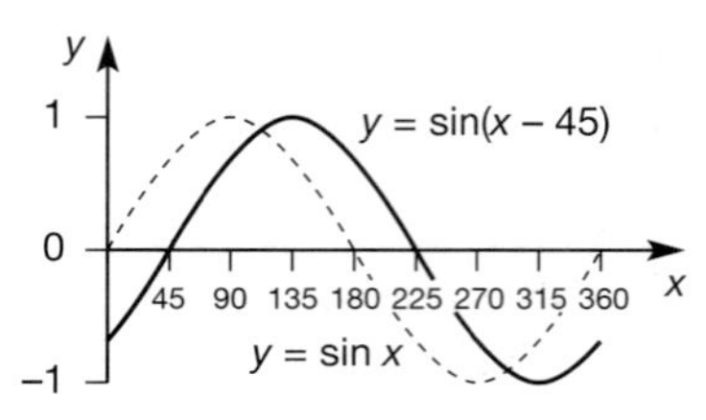

f)

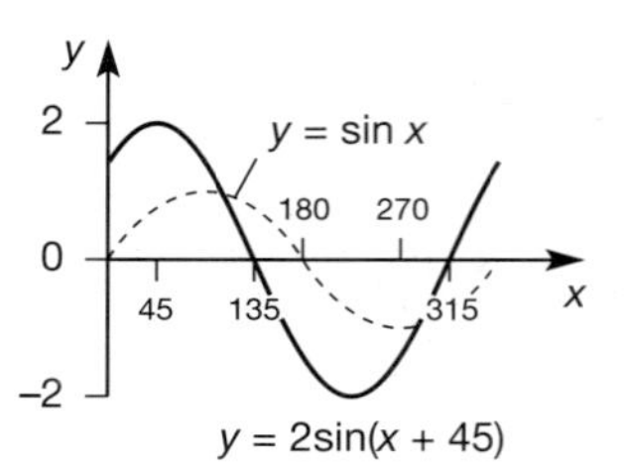

2 a)

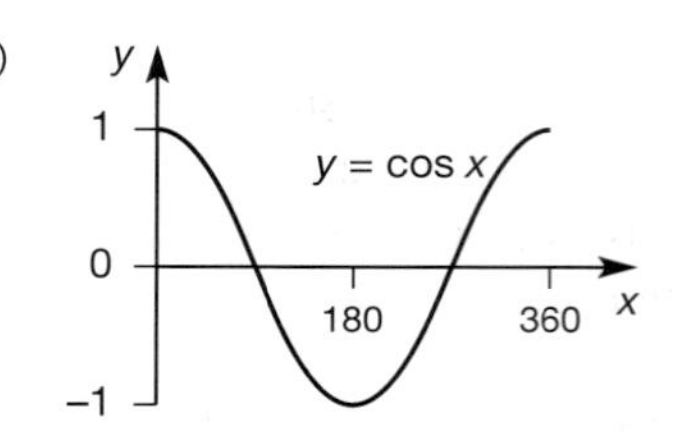

b)

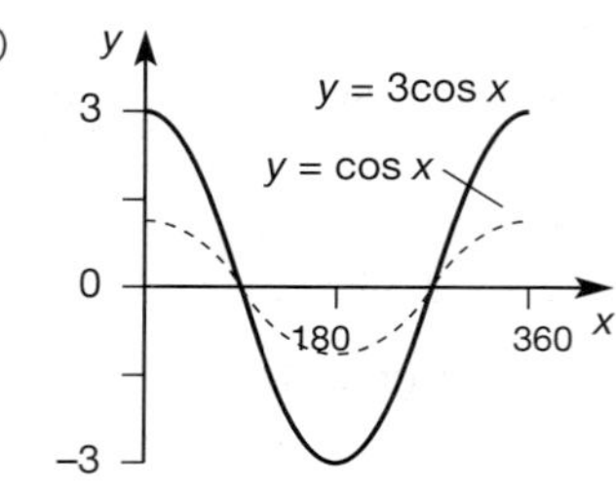

c)

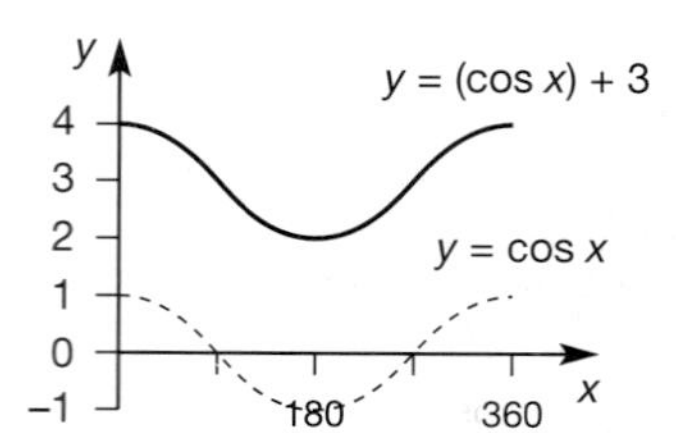

d)

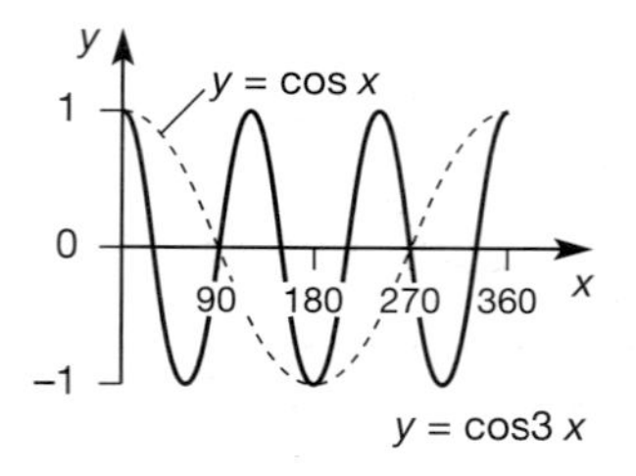

e)

y = cos(x + 45)
y = cos x

f)

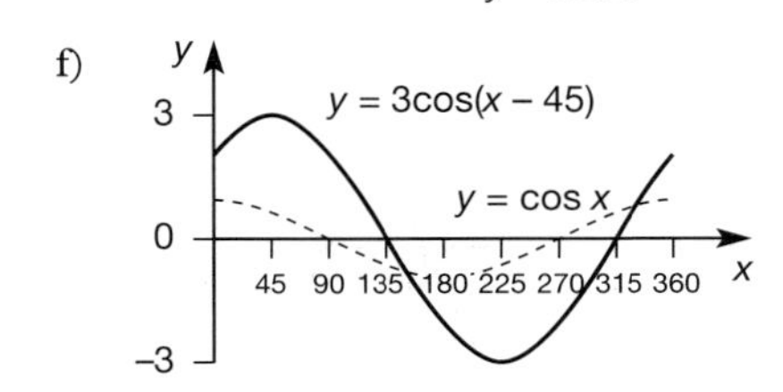

3 $a = 1,\ b = 1,\ c = 1 \quad p = 1,\ q = 1,\ r = 1.$
4 A = –20 B = 40.
5 $a = 2 \quad b = 25 \quad p = 3 \quad q = -50.$
6 $t = 2 \quad u = 3 \quad r = -1 \quad s = 2 \quad q = \frac{1}{2}.$
7 a) 180 b) 0, 180, 360 c) 90
d) 26·6, 206·6 e) 104·5, 255·5
f) 203·6, 336·4 g) 48·2, 311·8
h) 104·0, 284·0 i) 216·9, 323·1
j) no solutions k) 59·0, 223·0 l) 90, 270.
8 a) $4\sin^2 x - 1$ b) $2 - 5\cos^2 y$ c) $\frac{1}{\cos^2 A}$.
9 (proofs).
10 a) 12 m b) 80 s.
11 a) 5·6cm b) 2.25, 2.35 o'clock.

Exercise 99 (Test), page 220

1 a) 90° b) 120° c) 36° d) 1080°.
2 a)

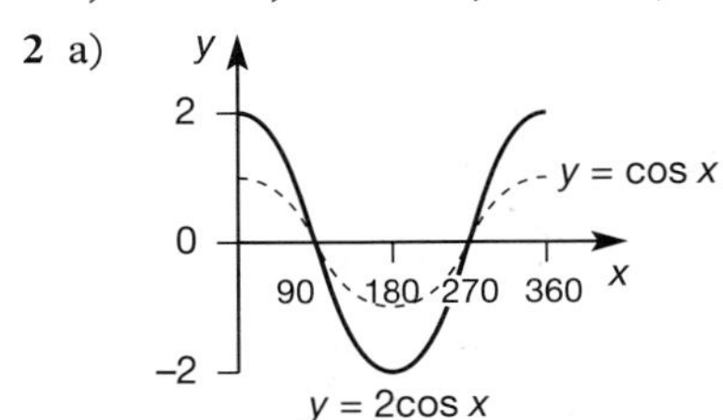

b)

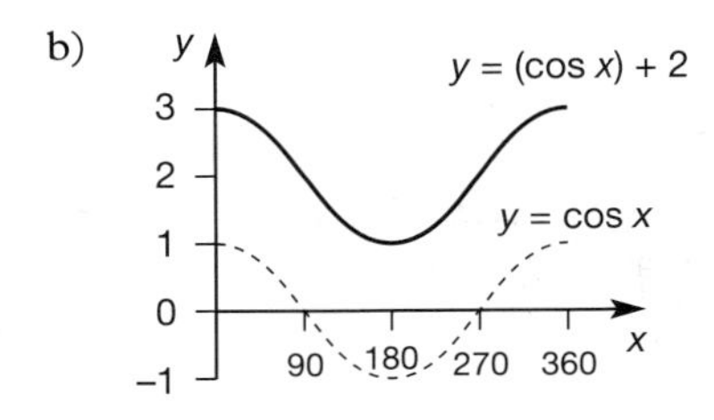

c)

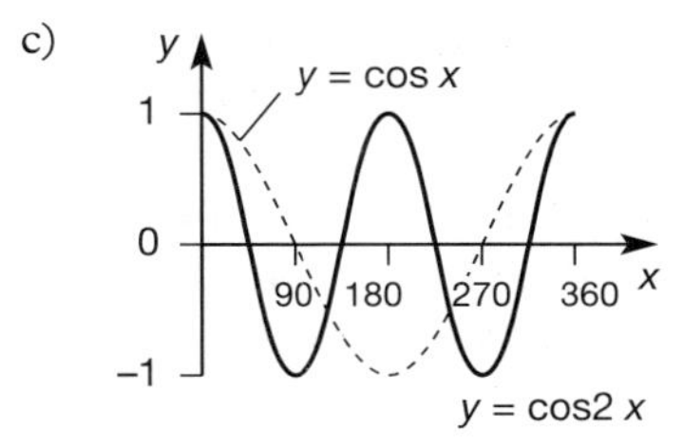

3 a)

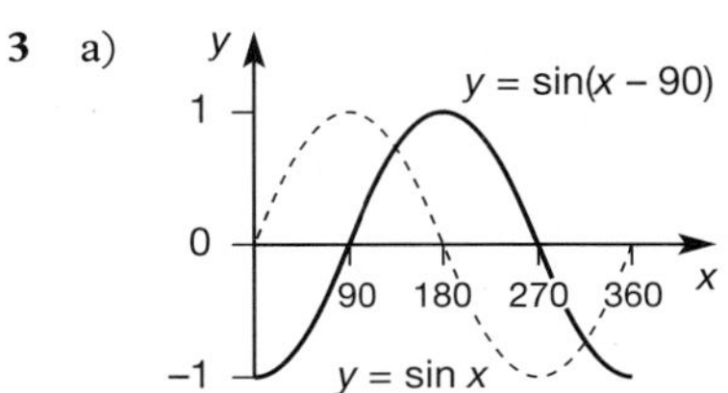

b)

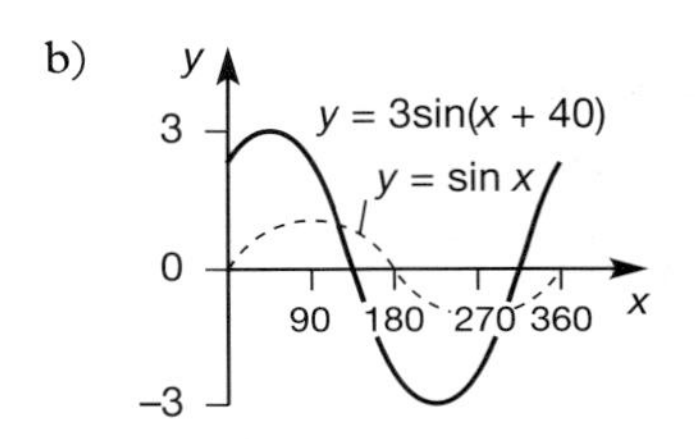

4 $a = -1, b = 1, p = \sqrt{2} \quad q = 3, s = 2 \quad t = 2 \quad u = 2.$
5 a) 48·6, 131·4 b) 103·3, 256·7
c) 125·5, 305·5 d) 41·8, 138·2
e) (no solutions).
6 a) $6 - \cos^2 x$ b) $7 + \sin^2 x$ c) $\sin^2 A \cos A$.
7 (proof).

Exercise 100, page 222

1 £1653·75. **2** 4%. **3** Holly by £929·61.
4 a) £91·50 b) £4087·20 c) £737·10.
5 Sam by £20. **6** £37 000.
7 £15 800. **8** £200 000 p.a. **9** 3%.
10 a) £112 b) £900 c) sales of £500, pay £100.
11 £17 368. **12** £427·50. **13** £416·52.
14 £492·13.
15 a) T £410, R £246, H £164 b) £10·25.
16 a) 11·25% b) £8·86.
17 A £60, B £52·50, C £37·50.

Exercise 101, page 226

1 A £231·06 B £261·60 C £327·13 D £483·90
E £680·63.
2 F £26·96 G £36·64 H £50·72 I £63·26
J £150·15.
3 a) K £1016·06 L £1099·81 M £1124·43
N £1236·03 O £1755·12
b) K £114·77 L £130·77 M £127·49
N £149·43 O £245·09.

4 a) 21·4% b) £1709·16.

5 A £484·18 B £43·25 C £81·78 D £125·03 E £359·15.

6 A £225 B £1250 C £94.05 D £167·81 E £988·14.

7 A £142·50 B £230·77 C £488·46 D £1988·27.

Exercise 102, page 229

1 a) £664·34 b) £198·04 c) £104·52. d) £458·39 e) £502·95.

2 a) £145·50 b) £3492 c) £492.

3 a) £142·02 b) £6816·96 c) £1816·96.

4 a) £122·16 b) £7329·60 c) £2329·60.

5 £2157·12.

6 £161·28. **7** £1091·84.

8 £2650·20. **9** £7·35. **10** £9·54.

11 £1·09. **12** £1·86. **13** £2·70.

14 A £41·50 B £1·92 C £257·49 D £7·72.

15 b) 15·5% APR. **16** b) 1·51% per month.

Exercise 103, page 232

1 a) £1560 b) £11 700 c) £17 513·60 d) £29 400.

2 a) £24 343·75 b) £351·19 c) £1982·75 d) £7·00.

3 Jimmy, by £20 p.a.

4 a) £16 432 b) £20 064 c) £18 600.

5 a) £18 500 b) £1541·67.

6 O £372 P £308·06 R £301·50.

7 £9·23.

8 a) Nil b) 44p c) £3·52 d) £33·99.

9 a) Nil b) £18·70 c) £89·10 d) £158·40.

10 a) £251·41 b) £286·43 c) £431·81 d) £472·51.

11 a) £171·27 b) £8220·96 c) £3220·96 d) £1133·76.

12 A £163·42 B £2·53 C £5·39 D £724·26 E £21·73.

Exercise 104 (Test), page 234

1 a) £100·12 b) £2402·88 c) £402·88.

2 £434·70. **3** £1286·50.

4 a) £200 b) £747·60 c) £3057·60 d) £8 207·60.

5 a) £25·74 b) £55·55.

6 a) £102·52 b) £242·11.

7 a) £19 020 b) £1212·60.

8 A £73·26 B £1·28 C £513·41 D £15·40.

Exercise 105, page 236

1 SH SR SP ST MH MR MP MT FH FR FP FT.

2 FGH FHG GHF GFH HFG HGF.

3 14 15 24 25 34 35.

4 PRTUW (14 miles) is the shortest.

5 PSTVW (57 mins) is quickest.

6 a) 1H 1T 2H 2T 3H 3T 4H 4T 5H 5T
b) $\frac{1}{5}$.

7 a) HH HT TH TT b) $\frac{1}{4}$
c) $P(1H \text{ and } 1T) = \frac{2}{4} = \frac{1}{2} > \frac{1}{4} = P(HH)$.

8 a) PQRST PQRTS PQSRT PQSTR PQTSR PQTRS
PRQST PRQTS PRSTQ PRSQT PRTSQ PRTQS; PRSTQ (56 km) is shortest
b) PQTSR or PRSTQ are 77 km.

Exercise 106, page 238

1 a) (i) EDGFC (ii) HIJK (iii) LMNPO
b)

vertex	C	D	E	F	G	H	I	J	K	L	M	N	O	P
degree	2	3	2	3	2	3	2	3	2	3	3	3	4	3

c) (i) D and F have odd degree ⇒ start at one and finish at the other, so getting back to the starting point means re-tracing some edges
(ii) DF
d) (i) DEFGDCF (ii) HIJKHJ
e) There are more than 2 (4 in fact) vertices with an odd degree.

2 a) and (ii) b) and (iii) c) and (i).

3 a)

vertex	A	B	C	D	E	P	Q	R	S	T	U	V	I	J	K	L	M	N	O
degree	2	4	2	3	3	2	4	2	2	3	3	2	2	3	2	3	2	3	3

b) (i) D and E (ii) T and U
c) (i) DCBDEABE (ii) TSRQPUQTVU
d) There are more than 2 (4 in fact) vertices with an odd degree.

4 a) 3, 3, 3, 3, 4
b)

5 a)

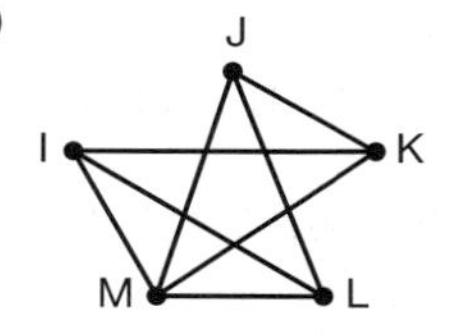

b) 8 c)

vertex	I	J	K	L	M
degree	3	3	3	3	4

d) sum of degrees = 16 = 2 × number of arcs.

6 a)

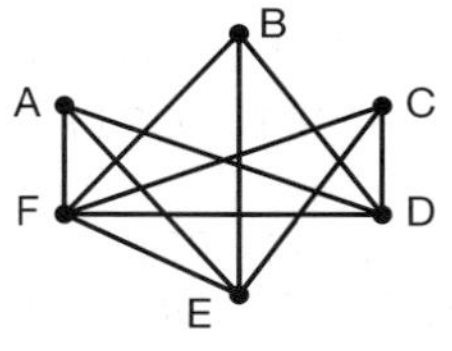

b)

vertex	A	B	C	D	E	F
degree	3	3	3	4	4	5

c) T = 22, E = 11 T = 2E.

7 a) C and D

b) (i) once (ii) 3 times (iii) twice

c) CBACDEFHBIGFDBGHD.

8 a)

Aberdeen					
67	Dundee				
125	58	Edinburgh			
147	80	46	Glasgow		
87	20	42	60	Perth	
119	52	32	28	32	Stirling

b) (i) Perth and Dundee

(ii) Aberdeen and Glasgow

c) (i) 4 miles

(ii) more dual carriageway

d)

A	D	E	P	S	G	I	J	K	L	M	N
1	3	4	3	3	2	1	3	3	3	2	4

e) I = A, J = D, K = P, L = S, M = G, N = E.

9 a)

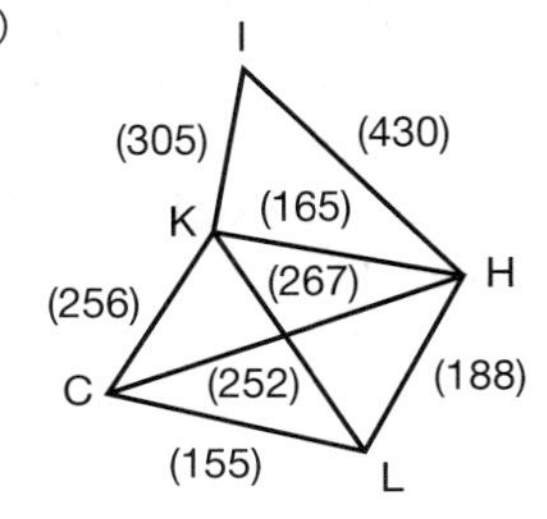

b) (i) 561 (ii) 572.

10

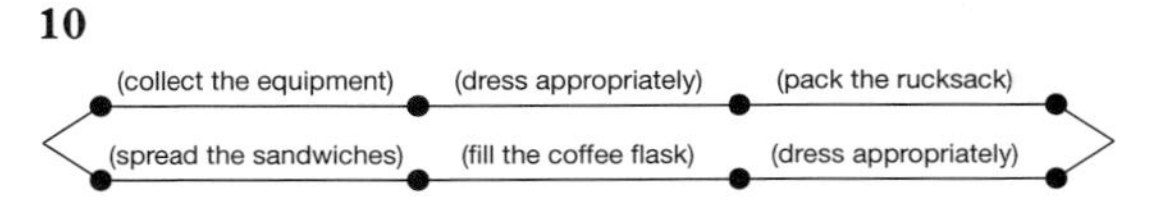

Exercise 107, page 243

1 a) 13 b) 285 c) 9·48.

2 a) –2 b) –4 c) 2 d) 86.

3 a) £1250 b) £1450 c) £1875.

4 a) £49·50 b) £162 c) £19 d) £58·50.

5 a) £2397·60 b) £200 c) £1517·60 d) £7105·60.

6 a) £1·25 b) £9·60 c) £5·08 d) £6.

7 a) 32p b) 40p c) 52p d) 58p e) 73p f) 54p.

Exercise 108, page 248

1 a) =AVERAGE(B4:B17) b) =SUM(B2:D2)

c) highlight cell E2 and fill down

d) highlight cell B18 and fill right

e) =E4/1·8 Format Menu Number FIXED; 0 decimal places

f) =E4/180 Format Menu Number PERCENT; 0 decimal places.

2 a) =B2*1·01 b) =C2–1250 c) filldown

d) cell D2: =C2–B2, fill down to D13; cell D14: =SUM(D2:D13).

3 a) Format Menu Number CURRENCY; 2 decimal places

b) D3: =C3 D4: =D3+C4

c) I3: =H3 I4: = I3+H4

d) (i) =D3–I3 (ii) =D11–I11

e) D14: £168·00 D15: £168·00; I14: £147·00 I15: £150·50 K14: £21·00 K15: £17·50.

4 a) (i) =C2/5 (ii) =C2–D2 (iii) =E2–B2 (iv) =F2/B2

b) Format Menu Number PERCENT; 0 decimal places

c) D4 £77 E4 £308 F4 £8 G4 3%
D5 £45 E5 £180 F5 £15 G5 9%

5 a) A2: =1 A3: =A2+1 and fill down from A3

b) (i) =A2–12 (ii) =A2*B2 (iii) =C2+35

c) =C9+35 d) 5, 7 e) $(x-5)(x-7)$.

6 a) =12*A2^2–53*A2+56 b) between 2 and 3

c) =12*D2^2–53*D2+56 d) 1·75, 2·67.

Exercise 109, page 253

1

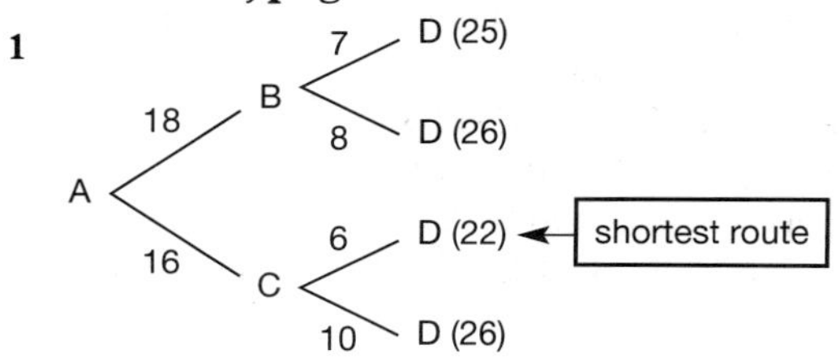

2 a) A B C D E F G H I J K L M N O
2 2 2 2 3 3 2 3 2 3 3 3 2 3 3
b) diagram (iii) because there are more than 2 vertices of odd degree
c) (i) EDCBFAEF (ii) JGHIJH.

3 a) £15·75 b) £35·72 c) £56·70.

4 a) (i) =B3–A3 (ii) =C3*0·0645
(iii) =12+D3 (iv) =0·05*E3
(v) =E3+F3
b) (i) A4 (ii) B3 (iii) A3
c) ← and ↓ in either order
d) (i) A5 and B5 would both contain 99 530, columns B and C would contain zeros column E would contain 12, column F: 0·60 and column G: 12·60.
(ii) A5 would become 99 530 and all the rest zeros, but correct entries for row 5 would be obtained on entering the new current reading in cell B5, and so on, a row at a time.

Exercise 110 (Test), page 254

1 a)

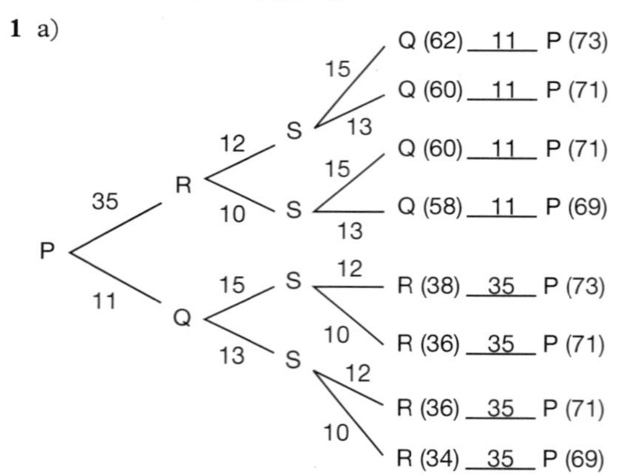

b) The 8 routes consist of 4 pairs of routes, the two routes in each pair being the reverse of each other.
c) PQSRSQP is only 68 km (11+13+10+10+13+11).

2 a) A B C D E F G H
2 2 6 2 4 2 4 2
b) sum of degrees = 24 ⇒ number of edges = 12
c) e.g. CABCDEFGHCEGC
d) e.g. ABCDEFGHCEGCAH.

3 a) £1480 b) £3633·75 c) £11 475.

4 a) (i) aligned left (ii) aligned right
b) (i) =AVERAGE(B2:B4)
(ii) =SUM(C2:C4)
c) Willie McNaught (took least throw-ins).

Exercise 111, page 257

1 a) 22 mm b) 114 mm c) 20·6 mm.
2 a) 9 mm b) 53 mm c) 11·4 mm.
3 a) 500 mm^2 b) 4687·5 mm^2 c) 305·5 mm^2.
4 a) (i) 12·6 mm (ii) 20·1 mm (iii) 40·8 mm
b) (i) 28·3 mm^2 (ii) 153·9 mm^2
(iii) 380·1 mm^2.
5 a) 1 mm^3 b) 8 mm^3 c) 140·6 mm^3.
6 a) $y = 2x + 1$ b) $y = -x + 2$ c) $y = 3x - 1$.
7 a) 4800 mm^3 b) 7770 mm^3 c) 7104 mm^3.
8 a) 420 mm^3 b) 250 mm^3 c) 191·1 mm^3.
9 a) £20 b) £300 c) £10.
10 a) 157 *l* b) 1060 *l* c) 3272 *l*.
11 a) 1440 mm^3 b) 24 400 mm^3
c) 4 450 000 mm^3.
12 a) 5·57 b) 5·46 c) 3·10.

Exercise 112, page 259

1 a) 8 b) 10 c) 22.
2 a) $\frac{1}{26}$ b) $\frac{1}{13}$ c) $\frac{4}{13}$.
3 a) 3 h 40 min b) 4 h 20 min c) 5 h 20 min.
4 a) 90 mg b) 145 mg c) 50 mg.
5 a) $y = 3x + 2$ b) $y = x - 5$ c) $y = -2x - 3$.
6 a) 8·66 mm^2 b) 17·23 mm^2 c) 22·75 mm^2.
7 a) 6 h 30 min b) 6 h 30 min c) 7 h 55 min.
8 a) 5 h b) 4 h 15 min c) 6 h 16 min.
9 a) $3\frac{1}{2}$°W b) 3°W c) 1°E.
10 a) 15·2% b) 14·5% c) 14·8%.

Exercise 113, page 262

1 a) 0·364 b) 0·754 c) 0·869.
2 a) 12 units b) 12 units c) 45 units.
3 a) 1 b) 0·866 c) 0.
4 a) 0·75 b) 0·5 c) 0·6.
5 a) 1 b) 0·37 c) 0·51.
6 a) 0·707 ms^{-1} b) 0·304 ms^{-1} c) 2·09 ms^{-1}.
7 a) 4·9 m b) 11·0 m c) 19·6 m.
8 a) 18·75 m; 7·5 m/s b) 13·33 m; 6·7 m/s
c) 36 m; 12 m/s.
9 a) 2 s b) 1·74 s c) 1·42 s.
10 a) (2, 1) b) $\left(\frac{3}{2}, \frac{5}{2}\right)$ c) $\left(-\frac{10}{3}, -\frac{2}{3}\right)$.
11 a) 5, 7 b) –3, 4 c) $\frac{3}{2}, \frac{5}{4}$.
12 a) 10·0 m; 2·1 m b) 40·2 m; 12·0 m
c) 19·9 m; 2·87 m.

Exercise 114, page 265

1 a) 5 m b) 15 m c) 32·5 m.
2 a) 10 m b) 5 m c) 12 m.
3 a) 20 m b) 16 m c) 30 m.
d) 0·4 m^2 e) 0·375 m^2 f) 0·625 m^2.
4 a) 0·8 m b) 1·3 m c) 750 mm.
5 a) 0·20 m b) 0·35 m c) 1·20 m.
6 a) 10 b) 7 c) 8 d) 4.
7 a) ±8 b) ±3 c) 6 d) 2.
8 a) 1·0 m b) 2·5 m c) 0·8 m.

Exercise 115, page 266

1 a) 314 mm^2 b) 2827 mm^2 c) 6648 mm^2.
2 a) 37 700 mm^2 b) 27 500 mm^2
c) 290 000 mm^2.
3 a) 4·5 km/h b) 6 km/h c) 5 km/h.
4 a) 8·9 b) 8 c) 13·3.
5 a) (i) 1·2 ohms (ii) 1·875 ohms
(iii) 4·3 ohms
b) (i) 0·97 ohms (ii) 1·48 ohms
(iii) 2·55 ohms.
6 a) 11 b) ±2 c) 11 d) ±8 e) 5 f) 7.

Exercise 116 (Test), page 267

1 a) 393 mm^2 b) 628 mm^2 c) 295 mm^2.
2 a) 800 m^3 b) 960 m^3 c) 360 m^3.
3 a) 10 000 mm^3 b) 11 250 mm^3
c) 15 000 mm^3.
4 a) (5, 6) b) $\left(\frac{11}{2}, 2\right)$ c) $\left(2, -\frac{7}{2}\right)$.
5 a) 30 mm b) 12 mm c) 25·5 mm.
6 a) 35 m^2 b) 3·9 m.

Exercise 117, page 272

[Answers for quartiles and modes need not necessarily agree exactly with what is given here.]

1 a) 16·45 b) 4; $11\frac{1}{2}$; $17\frac{1}{2}$; 22; 28 SIQR = $5\frac{1}{4}$
c) 18–20; 18.
2 a) 16·3 b) 9·3 c) 34·75 d) 25·96.
3 a) $12\frac{1}{2}$; 16; 19 b) 7; 10; 12 c) 27; 37; 42
d) 23; 26; 28.
4 a) 14 b) 10 c) 38 d) 28.
5 a) $16\frac{1}{6}$ b) 11; $16\frac{1}{2}$; $21\frac{1}{2}$; SIQR = $5\frac{1}{4}$
c) 15–19; 17 d) $\frac{11}{36}$.
6 a) 16–22; 21 b) 19·1 c) 14; $19\frac{1}{2}$; 25
d) 0·36.
7 a) 61 b) 58·3% c) 49; $58\frac{1}{2}$; 66
d) $26\frac{2}{3}$%.
8 a) 98·9 b) 88; 99; 110 c) 99
d) SIQR = 11 ≈ 0·7 × S.D.
9 a) 82·7; 83; 81 b) $69\frac{1}{2}$; 80; 83; 86; $93\frac{1}{2}$
c) 0·03.
10 a) 70 w.p.m. b) $64\frac{1}{2}$; $72\frac{1}{2}$; 77 $6\frac{1}{4}$
c) 75–79; 76 d) 0·4.